Curve and Surface Fitting

Innovations in Applied Mathematics

An international series devoted to the latest research in modern areas of mathematics, with significant applications in engineering, medicine, and the sciences.

SERIES EDITOR:
Larry L. Schumaker
Stevenson Professor of Mathematics
Vanderbilt University

PREVIOUSLY PUBLISHED TITLES INCLUDE

Curves and Surfaces with Applications in CAGD (1997)

Surface Fitting and Multiresolution Methods (1997)

Mathematical Methods for Curves and Surfaces II (1998)

Mathematical Models in Medical and Health Science (1998)

Approximation Theory IX Volume 1. Theoretical Aspects (1999)

Approximation Theory IX Volume 2. Computational Aspects (1999)

Curve and Surface Fitting

Saint-Malo 99

EDITED BY

Albert Cohen
Université Pierre et Marie Curie
Paris, France

Christophe Rabut
Institut National des Sciences Appliquées
Toulouse, France

Larry L. Schumaker
Department of Mathematics
Vanderbilt University
Nashville, Tennessee

VANDERBILT UNIVERSITY PRESS
Nashville

First Edition 2000

04 03 02 01 00   5 4 3 2 1

Library of Congress Cataloging-in-Publication Data

Curve and surface fitting : Saint-Malo 99 / edited by Albert Cohen, Christophe Rabut, Larry L. Schumaker.— 1st ed.
p. cm — (Innovations in applied mathematics)
Includes bibliographical references.
ISBN 0-8265-1357-3 (alk. paper)
1. Computer graphics. 2. Computer-aided design. 3. Curves, Algebraic—Data processing. 4. Surfaces, Algebraic—Data processing. I. Albert Cohen, 1965– II. Rabut, Christophe, 1951– III. Schumaker, Larry L., 1939– IV. Series.

T385 .C86 2000
620'.0042'0285—dc21   99-086093

Manufactured in the United States of America

CONTENTS

PREFACE

During the week of July 1-7, 1999, the Fourth International Conference on *Curves and Surfaces* was held in Saint-Malo (France). It was organized by the *Association Française d'Approximation*, (A.F.A.). The organizing committee consisted of L. Amodei (Toulouse), J.-L. Bauchat (Metz), A. Cohen (Paris), J.-C. Fiorot (Valenciennes), J. Gaches (Toulouse), G.-P. Bonneau (Grenoble), Y. Lafranche (Rennes), P.-J. Laurent (Grenoble), M.-L. Mazure (Grenoble), J.-L. Merrien (Rennes), C. Potier (Paris), C. Rabut (Toulouse), P. Sablonnière (Rennes), L.L. Schumaker (Nashville), C. Vercken (Paris).

The conference was attended by 275 mathematicians from 37 different countries, and the program included 10 invited one-hour lectures and 190 half-hour research talks or poster presentations. A number of research talks were presented in eight minisymposia organized by W. Dahmen, R. DeVore, D. Donoho, J. Hoschek, B. Lacolle, H. Pottmann, M. Sabin, and J. Stöckler.

The proceedings of this conference consists of this volume (containing 43 papers), and the companion volume *Curve and Surface Design: Saint-Malo 1999* (containing 45 papers).

We would like to thank the following institutions for their financial or technical support and their contribution to the success of this conference: Ministère de l'Education Nationale, de la Recherche et de la Technologie; European Office of Aerospace Research and Development (Air Force Office of Scientific Research, United States Air Force Research Laboratory); Institut National des Sciences Appliquées de Rennes; Institut d'Informatique et de Mathématiques Appliquées de Grenoble; Conseil Régional de Bretagne; Ministère de la Défense (contrat No 9960014, Direction des Systèmes de Forces et de la Prospective, Service de la Recherche et des Etudes Amont, Sous-direction Scientifique, Bureau de la Prospective Scientifique, Délégation Générale pour l'Armement); Université Pierre et Marie Curie (Paris); Laboratoire de Modélisation et Calcul de Grenoble; Institut National des Sciences Appliquées de Toulouse; Université Joseph Fourier (Grenoble); Vanderbilt University (Nashville); Ministère des Affaires Etrangères; Matra Datavision; Ecole Nationale Supérieure des Arts et Métiers de Metz; France Télécom; Ecole Nationale Supérieure des Télécommunications (Paris); Ecole Centrale de Nantes.

Finally, we would like to thank Gerda Schumaker for her assistance with the preparation of the proceedings.

Nashville, Tennessee April 5, 2000

CONTRIBUTORS

Numbers in parentheses indicate pages on which authors' contributions begin. Articles marked with * are in *Curve and Surface Design: Saint-Malo 1999*, and articles marked with ** are in *Curve and Surface Fitting: Saint-Malo 1999.*

GIAMPIETRO ALLASIA (**75), *Dipartimento di Matematica, Università di Torino, I-10123 Torino, Italy* [allasia@dm.unito.it]

IAIN J. ANDERSON (**1), *School of Computing and Mathematics, University of Huddersfield, Queensgate, Huddersfield, West Yorkshire, HD1 3DH, UK* [i.j.anderson@hud.ac.uk]

D. APPRATO (**9), *Université de Pau, Dpt de Mathématiques, IPRA, Av. de l'Université, 64000 Pau, France*

FRANCESC ARÀNDIGA (**19), *Departament de Matemàtica Aplicada, Universitat de València, 4610 Burjassot, València, Spain* [arandiga@uv.es]

DOMINIQUE ATTALI (*109), *Laboratoire LIS, ENSIEG, Domaine Universitaire, 38402 St-Martin-d'Hères, France* [Dominique.Attali@lis.inpg.fr]

SILVIA BACCHELLI (**27), *Dipartimento di Matematica Pura e Applicata, Università di Padova, Via Belzoni 7, Padova, Italy* [silvia@agamennone.csr.unibo.it]

EDUARDO ANTÔNIO BARROS DA SILVA (**153), *DEL/PEE/EE/COPPE, Universidade Federal do Rio de Janeiro, Brazil* [eduardo@lps.ufrj.br]

BRIAN A. BARSKY (**393), *Computer Science Division, University of California, Berkeley, CA 94720-1776, USA* [barsky@cs.berkeley.edu]

RICK BEATSON (**37,**47), *Department of Mathematics and Statistics, University of Canterbury, Private Bag 4800, Christchurch, New Zealand* [R.Beatson@math.canterbury.ac.nz]

MOHAMMED-NAJIB BENBOURHIM (**57), *Université Paul Sabatier (M.I.P), 118, Route de Narbonne, 31062 Toulouse Cedex 04, France* [bbourhim@cict.fr]

MICHEL BERCOVIER (*257), *The Hebrew University of Jerusalem, Givat Ram Campus, Institute of Computer Science, 991904 Jerusalem, Israel* [berco@cs.huji.ac.il]

STEFANO BERRONE (**65), *Dipartimento di Ingegneria Aeronautica e Spaziale, Politecnico di Torino, Corso Duca degli Abruzzi, 24, 10129 Torino, Italy* [sberrone@calvino.polito.it]

RENATA BESENGHI (**75), *Dipartimento di Matematica, Università di Torino, I-10123 Torino, Italy* [besenghi@dm.unito.it]

J. B. BETBEDER (**9), *Université de Pau, Dpt de Mathématiques, IPRA, Av. de l'Université, 64000 Pau, France*

GEORGES–PIERRE BONNEAU (**209,**237), *LMC–IMAG, Université Joseph Fourier, BP 53, 38041 Grenoble Cedex 9, France* [Georges-Pierre.Bonneau@imag.fr]

TINA BOSNER (*343), *Dept. of Mathematics, University of Zagreb, Bijenička 30, 10000 Zagreb, Croatia*

PIERRE BOULANGER (*381), *Institute for Information Technology, National Research Council of Canada, Montreal Road, Ottawa, Ontario, Canada K1A 0R6* [pierre.boulanger@iit.nrc.ca]

A. BOURAS (*407), *LIGIM, University Claude Bernard Lyon 1, 43 boulevard du 11 novembre 1918, 69622 Villeurbanne, France*

MIRA BOZZINI (**85), *Dipartimento di Matematica e Applicazioni, Università di Milano Bicocca, via Bicocca degli Arcimboldi 8, 20126 Milano, Italy* [bozzini@matapp.unimib.it]

PETER BROCKWELL (**343), *Department of Statistics, Colorado State University, Ft. Collins, CO 80523-1877 USA* [pjbrock@stat.colostate.edu]

J. M. BRUN (*407), *LIM/XAOLab, University Aix-Marseille 2, ESIL, Case 925, 13288 Marseille Cedex 9, France* [jmbrun@esil.univ-mrs.fr]

EMMANUEL CANDES (**95,**105), *Department of Statistics, Stanford University, Stanford, CA 94305-4065, USA* [emmanuel@stat.stanford.edu]

J. M. CARNICER (*1), *Departamento de Matemática Aplicada, Universidad de Zaragoza, 50009 Zaragoza, Spain* [carnicer@posta.unizar.es]

PAUL DE FAGET DE CASTELJAU (*9), *4 Avenue du Commerce, 78000 Versailles, France*

E. CHACKO (**37), *Department of Mathematics and Statistics, University of Canterbury, Private Bag 4800, Christchurch, New Zealand* [E.Chacko@math.canterbury.ac.nz]

PATRICK CHENIN (*183), *LMC-IMAG, Université Joseph Fourier, BP 53, 38.041 Grenoble Cedex 9, France* [pchenin@imag.fr]

J. B. CHERRIE (**47), *Department of Mathematics and Statistics, University of Canterbury, Private Bag 4800, Christchurch, New Zealand* [J.Cherrie@math.canterbury.ac.nz]

ENG-WEE CHIONH(*17), *School of Computing, National University of Singapore, Lower Kent Ridge Road, Singapore 119260* [chionhew@comp.nus.edu.sg]

H. Chiyokura (*389), *Faculty if Environmental Information, Keio University, 5322 Endo, Fujisawa, Kanagawa, 252-0816, Japan* [chiyo@sfc.keio.ac.jp]

Zbigniew Ciesielski (**121), *Instytut Matematyczny PAN, ul. Abrahama 18, 81-825 Sopot, Poland* [Z.Ciesielski@impan.gda.pl]

Blair Conrad (*27), *Computer Science Department, University of Waterloo, Waterloo, Ontario, Canada N2L 3G1* [beconrad@cgl.uwaterloo.ca]

C. Conti (**135), *Dipartimento di Energetica, Università di Firenze, Via C. Lombroso 6/17, I–50134 Firenze, Italy* [costanza@sirio.de.unifi.it]

Paolo Costantini (**143), *Dipartimento di Matematica " Roberto Magari", Via del Capitano 15, 53100 Siena, Italy* [costantini@unisi.it]

Mariantonia Cotronei (**27), *Dipartimento di Matematica, Università di Messina, Salita Sperone, 31, Messina, Italy* [marianto@dipmat.unime.it]

Marcos Craizer (**153), *Departamento de Matemática, Pontifícia Universidade Católica do Rio de Janeiro, R. Marquês de São Vicente, 225 - Gávea, Rio de Janeiro, RJ, 22453-900 Brazil* [craizer@mat.puc-rio.br]

Marc Daniel (*353), *XAOlab-ESIL/LIM, Campus de Luminy, case postale 925, 13288 Marseille Cedex 9, France* [Marc.Daniel@esil.univ-mrs.fr]

Oleg Davydov (**161,**171), *Justus-Liebig-Universität Giessen, Mathematisches Institut, 35392 Giessen, Germany* [oleg.davydov@math.uni-giessen.de]

Lubomir T. Dechevski (**181), *Département de mathématiques et statistique, Université de Montréal, C.P. 6128, Succursale A, Montréal (Québec) H3C 3J7, Canada* [dechevsk@dms.umontreal.ca]

Hafsa Deddi (**191), *Department of Mathematics and Computer Science, University of Lethbridge, 4401 University Drive, Lethbridge, Alberta, T1K 3M4, Canada* [deddi@cs.uleth.ca]

W. L. F Degen (*37), *University of Stuttgart, Mathematisches Institut B, Pfaffenwaldring 57, D-70569 Stuttgart, Germany* [degen@mathematik.uni-stuttgart.de]

Paul Dierckx (*45), *Celestijnenlaan 200A, 3001 Heverlee, Belgium* [paul.dierckx@cs.kuleuven.ac.be]

Fabrice Dodu (**201), *Institut National des Sciences Appliquées de Toulouse, U.M.R. 9974 C.N.R.S./I.N.S.A./M.I.P., Génie Mathématique* *et Modélisation, 135 Avenue de Rangueil, 31077 Toulouse Cedex 4, France* [dodu@gmm.insa-tlse.fr]

ROSA DONAT (**19), *Departament de Matemàtica Aplicada, Universitat de València, 4610 Burjassot. València, Spain* [donat@uv.es]

DAVID DONOHO (**105), *Department of Statistics, Stanford University, Stanford, CA 94305-4065, USA* [donoho@stat.stanford.edu]

EVA DYLLONG (*55), *Dept. of Computer Science, University of Duisburg, D-47048 Duisburg, Germany* [dyllong@informatik.uni-duisburg.de]

HAZEL EVERETT (**191), *LORIA - INRIA Lorraine, 615 rue du jardin botanique, B.P. 101, 54602 Villers-les-Nancy Cedex, France* [everett@loria.fr]

RIDA T. FAROUKI (*63), *Department of Mechanical and Aeronautical Engineering, University of California, Davis, CA 95616, USA* [farouki@ucdavis.edu]

LEONARD A. FERRARI (*307), *The Bradley Department of Electrical Computer Engineering, 340 Whittemore Hall, Virginia Tech, Blacksburg, VA 24061-0111, USA* [ferrari@vt.edu]

G. FIGUEROA (*91), *Universidad Central de Venezuela. Facultad de Ciencias, Escuela de Matemática, Laboratorio de Computación Gráfica y Geometría Aplicada, Apartado 47809, Los Chaguaramos, Caracas 1041-A, Venezuela* [giovanni@euler.ciens.ucv.ve]

JEAN-CHARLES FIOROT (*99), *Université de Valenciennes et du Hainaut-Cambrésis, Laboratoire MACS, B.P.311, 59304 Valenciennes Cedex, France* [fiorot@univ-valenciennes.fr]

DÉCIO ANGELO FONINI JR. (**153), *PEE/COPPE, Universidade Federal do Rio de Janeiro, Brazil* [fonini@lps.ufrj.br]

CÉDRIC GÉROT (*109), *Laboratoire LIS, ENSIEG, Domaine Universitaire, 38402 St-Martin-d'Hères, France* [Cedric.Gerot@lis.inpg.fr]

ALEXANDRE GERUSSI (**209), *LMC–IMAG, Université Joseph Fourier, BP 53, 38041 Grenoble Cedex 9, France* [Alexandre.Gerussi@imag.fr]

RONALD GOLDMAN (*17), *Department of Computer Science, Rice University, Houston, TX 77025, USA* [rng@cs.rice.edu]

TIM N. T. GOODMAN (**219), *Dept. of Mathematics, The University, Dundee DD1 4HN, U.K.* [tgoodman@mcs.dundee.ac.uk]

LAURA GORI (*119), *Dept. MeMoMat, Università La Sapienza, Via Antonio Scarpa 16 - 00161, Roma, Italy* [gori@dmmm.uniroma1.it]

A. ARDESHIR GOSHTASBY (**227), *Department of Computer Science and Engineering, Wright State University, Dayton, OH 45435, USA* [agoshtas@cs.wright.edu]

C. Gout (**9), *Université de Pau, Dpt de Mathématiques, IPRA, Av. de l'Université, 64000 Pau, France*
[chris_gout@alum.calberkeley.org]

Günther Greiner (*153), *Computer Graphics Group, University of Erlangen-Nürnberg, Am Weichselgarten 9, D-91058 Erlangen, Germany*
[greiner@informatik.uni-erlangen.de]

Gueorgui H. Gueorguiev (*127), *Faculty of Mathematics and Informatics, Shumen University, 9712 Shumen, Bulgaria*
[g.georgiev@shu-bg.net]

François Guibault (*213), *Département de génie électrique et de génie informatique, École Polytechnique de Montréal, C.P. 6079, Succ. Centre-Ville, Montréal (Québec), H3C 3A7, Canada*
[francois@cerca.umontreal.ca]

Stéphane Guillet (*135), *Laboratoire 3S, Domaine Universitaire, BP 53X, 38041 Grenoble Cedex 9, France*
[Stephane.Guillet@hmg.inpg.fr]

Stefanie Hahmann (**237), *Laboratoire de Modélisation et Calcul, CNRS-IMAG, B.P. 53, F-38041 Grenoble Cedex 9, France*
[hahmann@imag fr]

Stephen Hales (**247), *Mathematics & Computer Science Dept., University of Leicester, University Road, Leicester, LE1 7RH, U. K.*
[sjh16@mcs.le.ac.uk]

Hiroto Harada (*145), *Dept. of Mechanical Systems Engineering, Toyota Technological Institute, 2-12-1, Hisakata, Tempaku-ku, Nagoya 468-8511, Japan* [920043@nhkspg.co.jp]

Masatake Higashi (*145), *Dept. of Mechanical Systems Engineering, Toyota Technological Institute, 2-12-1, Hisakata, Tempaku-ku, Nagoya 468-8511, Japan* [higahsi@toyota-ti.ac.jp]

Kai Hormann (*153), *Computer Graphics Group, University of Erlangen-Nürnberg, Am Weichselgarten 9, D-91058 Erlangen, Germany*
[hormann@informatik.uni-erlangen.de]

Alfred Inselberg (*257), *School of Mathematical Sciences, Tel Aviv University, Tel Aviv 61396, Israel* [aiisreal@cs.tau.ac.il]

Kurt Jetter (**135), *Institut für Angew. Math. und Statistik, Universität Hohenheim, D–70593 Stuttgart, Germany*
[kjetter@uni-hohenheim.de]

Bert Jüttler (**385), *Darmstadt University of Technology, Dept. of Mathematics, Schloßgartenstr. 7, 64289 Darmstadt, Germany*
[juettler@mathematik.tu-darmstadt.de]

KĘSTUTIS KARČIAUSKAS (*163), *Dept. of Mathematics and Computer Science, Vilnius University, Naugarduko 24, 2600 Vilnius, Lithuania* [kestutis.karciauskas@maf.vu.lt]

KIYOTAKA KATO (*173), *Mitsubishi Electric Corporation, Industrial Electronics and Systems Lab, 8-1-1, Tsukaguchi-Honmachi, Amagasaki, Hyogo, Japan* [kato@sdl.melco.co.jp]

MOHAMMED KHACHAN (*183), *L.E.R.I (Laboratoire d'Etudes et de Recherches Informatiques), IUT-Léonard de Vinci, 51.059 Reims, France* [khachan@leri.univ-reims.fr]

RON KIMMEL (*193,*203), *Computer Science Department, Technion, Israel Institute of Technology, Haifa 32000, Israel* [ron@cs.technion.ac.il]

LEIF KOBBELT (*371), *Max-Planck-Institut für Informatik, 66123 Saarbrücken, Germany* [kobbelt@mpi-sb.mpg.de]

KOUICHI KONNO (*389), *RICOH corporation, 1-1-17 Koishikawa, Bunkyou, Tokyo, 112-0002, Japan* [konno@src.ricoh.co.jp]

V. V. KOVTUNETS (**255), *blvrd. Lesi Ukrainky 21a, ap.17, Kyiv, Ukraine 01133* [ktp@carrier.kiev.ua]

JERNEJ KOZAK (**263), *FMF and IMFM, University of Ljubljana, Jadranska 19, 1000 Ljubljana, Slovenia* [Jernej.Kozak@FMF.Uni-Lj.Si]

RIMVYDAS KRASAUSKAS (*163), *Dept. of Mathematics and Computer Science, Vilnius University, Naugarduko 24, 2600 Vilnius, Lithuania* [rimvydas.krasauskas@maf.vu.lt]

MITSURU KURODA (*145,**273), *Dept. of Mechanical Systems Engineering, Toyota Technological Institute, 2-12-1, Hisakata, Tempaku-ku, Nagoya 468-8511, Japan* [kuroda@toyota-ti.ac.jp]

BERNARD LACOLLE (*297), *LMC-IMAG, BP 53, 38041 Grenoble Cedex 9, France* [Bernard.Lacolle@imag.fr]

SYLVAIN LAZARD (**191), *LORIA - INRIA Lorraine, 615 rue du jardin botanique, B.P. 101, 54602 Villers-les-Nancy Cedex, France* [lazard@loria.fr]

DAMIANA LAZZARO (**27), *Dipartimento di Matematica, Università di Bologna, Piazza di Porta S. Donato, 5, Bologna, Italy* [lazzaro@csr.unibo.it]

MICHEL LÉGER (**281), *1 & 4 avenue de Bois-Préau, 92852 Rueil-Malmaison, France* [michel.leger@ifp.fr]

JEAN-CLAUDE LÉON (*135), *Laboratoire 3S, Domaine Universitaire, BP 53X, 38041 Grenoble Cedex 9, France* [Jean-Claude.Leon@hmg.inpg.fr]

JÉRÔME LÉPINE (*213), *Département de génie mécanique, École Polytechnique de Montréal, C.P. 6079, Succ. Centre-Ville, Montréal (Québec), H3C 3A7, Canada* [lepine@cerca.umontreal.ca]

JEREMY LEVESLEY (**247,**291), *Mathematics & Computer Science Dept., University of Leicester, University Road, Leicester, LE1 7RH, U. K.* [jl1@mcs.le.ac.uk]

ADI LEVIN (*221), *Tel-Aviv University, School of Mathematical Sciences, Tel-Aviv 69978, Israel* [adilev@inter.net.il]

KIA-FOCK LOE (**393), *Department of Computer Science, School of Computing, National University of Singapore, Lower Kent Ridge, Singapore 119260* [loekf@comp.nus.edu.sg]

H. LOPES (*229), *Pontifícia Universidade Católica do Rio de Janeiro, Departamento de Matemática, Rua Marquês de São Vicente, 225, Gávea, Rio de Janeiro, RJ, Brazil, CEP:22.453-900* [lopes@mat.puc-rio.br]

LIN-TIAN LUH (**301), *Dept. of Applied Mathematics, Providence University, 200 Chungchi Road, Shalu 43301, Taichung Hsien, Taiwan, Republic of China* [ltluh@pu.edu.tw]

WOLFRAM LUTHER (*55), *Dept. of Computer Science, University of Duisburg, D-47048 Duisburg, Germany* [luther@informatik.uni-duisburg.de]

DAVID LUTTERKORT (*239), *Dept. Computer Sciences, Purdue University, West Lafayette, IN, 47905-1398 USA* [lutter@cise.ufl.edu]

YVON MADAY (**309), *Laboratoire d'Analyse Numérique, Université Pierre et Marie Curie, 4, Place Jussieu, 75252, Paris Cedex 05, France* [maday@ann.jussieu.fr]

E. MAINAR (*1), *Departamento de Matemática Aplicada, Universidad de Zaragoza, 50009 Zaragoza, Spain* [esme@posta.unizar.es]

RAVI MALLADI (*203), *Lawrence Berkeley National Laboratory, University of California, Berkeley, CA 94720, USA* [malladi@math.lbl.gov]

PIERRE MALRAISON (*247), *Spatial Technology, Inc., 2425 55th Street, Suite 100, Boulder, CO 80301 USA* [pierre_malraison@spatial.com]

STEPHEN MANN (*27), *Computer Science Department, University of Waterloo, Waterloo, Ontario, Canada N2L 3G1* [smann@cgl.uwaterloo.ca]

CARLA MANNI (**143), *Dipartimento di Matematica, Via Carlo Alberto 10, 10123 Torino, Italy* [manni@dm.unito.it]

J. C. MASON (**1), *School of Computing and Mathematics, University of Huddersfield, Queensgate, Huddersfield, West Yorkshire, HD1 3DH, UK* [j.c.mason@hud.ac.uk]

TANYA MATSKEWICH (*257), *The Hebrew University of Jerusalem, Givat Ram Campus, Institute of Computer Science, 991904 Jerusalem, Israel* [fisa@cs.huji.ac.il]

CHRISTOPH MÄURER (*267), *TWT-GmbH, Finkenstr. 16, 70794 Filderstadt, Germany* [fa-twt.christoph.maeurer@daimlerchrysler.com]

KURT MEHLHORN (*277), *Max-Planck-Institut für Informatik, 66123 Saarbrücken, Germany* [mehlhorn@mpi-sb.mpg.de]

H. MICHAEL MÖLLER (**325,**333), *Fachbereich Mathematik, der Universität, 44221 Dortmund, Germany* [hmm@mathematik.uni-dortmund.de]

ANNICK MONTANVERT (*109), *Laboratoire LIS, ENSIEG, Domaine Universitaire, 38402 St-Martin-d'Hères, France* [Annick.Montanvert@lis.inpg.fr]

ROSSANA MORANDI (*287), *Dip. di Energetica "S. Stecco", Via Lombroso 6/17, 50134 Firenze, Italy* [morandi@de.unifi.it]

SHINJI MUKAI (**273), *Maebashi Institute of Technology, 460-1 Kamisanaru, Maebashi, Gunma 371-0816, Japan* [mukai@maebashi-it.ac.jp]

PHILIPPE NAVEAU (**343), *Geophysical Statistics Project, National Center for Atmospheric Research, Boulder, CO 80307 B.P. Box 3000 USA* [pnaveau@ucar.edu]

MANUELA NEAGU (*297), *LMC-IMAG, BP 53, 38041 Grenoble Cedex 9, France* [Manuela.Neagu@imag.fr]

L. NONATO (*229), *Pontifícia Universidade Católica do Rio de Janeiro, Departamento de Matemática, Rua Marquês de São Vicente, 225, Gávea, Rio de Janeiro, RJ, Brazil, CEP:22.453-900* [nonato@mat.puc-rio.br]

GÜNTHER NÜRNBERGER (**161), *Universität Mannheim, Fakultät für Mathematik und Informatik, A5, 68131 Mannheim, Germany* [nuern@euklid.math.uni-mannheim.de]

DOUG NYCHKA (**343), *Geophysical Statistics Project, National Center for Atmospheric Research, Boulder, CO 80307 B.P. Box 3000 USA* [nychka@ucar.edu]

M. PALUSZNY (*91), *Universidad Central de Venezuela. Facultad de Ciencias, Escuela de Matemática, Laboratorio de Computación Gráfica y Geometría Aplicada. Apartado 47809, Los Chaguaramos, Caracas 1041-A, Venezuela* [marco@euler.ciens.ucv.ve]

JAE H. PARK (*307), *The Bradley Department of Electrical Computer Engineering, 340 Whittemore Hall, Virginia Tech, Blacksburg, VA 24061-0111, USA* [park@vt.edu]

J. M. PEÑA (*315), *Departamento de Matemática Aplicada, Universidad de Zaragoza, 50006 Zaragoza, Spain* [jmpena@posta.unizar.es]

S. PESCO (*229), *Pontifícia Universidade Católica do Rio de Janeiro, Departamento de Matemática, Rua Marquês de São Vicente, 225, Gávea, Rio de Janeiro, RJ, Brazil, CEP:22.453-900* [pesco@mat.puc-rio.br]

MARTIN PETERNELL (**351), *Institut für Geometrie, TU Wien, Wiedner Hauptstrasse 8-10, A-1040 Wien, Austria* [peternell@geometrie.tuwien.ac.at]

JÖRG PETERS (*239), *Dept C.I.S.E., CSE Bldg, University of Florida, Gainesville, FL 32611-6120, USA* [jorg@cise.ufl.edu]

LAURA PEZZA (*119), *Dept. MeMoMat, Università La Sapienza, Via Antonio Scarpa 16 - 00161, Roma, Italy* [pezza@dmmm.uniroma1.it]

FRANCESCA PITOLLI (*119), *Dept. MeMoMat, Università La Sapienza, Via Antonio Scarpa 16 - 00161, Roma, Italy* [pitolli@dmmm.uniroma1.it]

MICHEL POCCHIOLA (*325), *Département d'Informatique, Ecole Normale Supérieure, 45, rue d'Ulm, 75230 Paris Cedex 05 - France* [Michel.Pocchiola@dmi.ens.fr]

HELMUT POTTMANN (*417,**351), *Institut für Geometrie, Wiedner Hauptstraße 8–10/113, A-1040 Wien, Austria* [pottmann@geometrie.tuwien.ac.at]

HARTMUT PRAUTZSCH (*335), *Institute for Operating- and Dialogsystems, Universität Karlsruhe, D-76128 Karlsruhe, Germany* [prau@ira.uka.de]

DAVID L. RAGOZIN (**47,**291), *University of Washington, Box 354350, Department of Mathematics, Seattle, WA 98195-4350 USA* [rag@math.washington.edu]

ALAIN RASSINEUX (*363), *Université de Technologie de Compiegne, U.R.A. CNRS 1505 LG2MS, BP 529, F-60205 Compiegne, Cedex, France* [alain.rassineux@utc.fr]

MLADEN ROGINA (*343), *Dept. of Mathematics, University of Zagreb, Bijenička 30, 10000 Zagreb, Croatia* [rogina@math.hr]

COLIN ROSS (**1), *School of Computing and Mathematics, University of Huddersfield, Queensgate, Huddersfield, West Yorkshire, HD1 3DH, UK* [c.ross@hud.ac.uk]

VINCENT ROSSIGNOL (*353), *IRIN (Institut de Recherche en Informatique de Nantes), 2, Rue de la Houssinière, BP 92208, 44322 Nantes Cedex 3, France* [Vincent.Rossignol@irin.univ-nantes.fr]

MILVIA ROSSINI (**85), *Dipartimento di Matematica e Applicazioni, Università di Milano Bicocca, via Bicocca degli Arcimboldi 8, 20126 Milano, Italy* [rossini@matapp.unimib.it]

THOMAS SAUER (**325,**333), *Mathematisches Institut, der Universität Erlangen–Nürnberg, Bismarckstr. 1 1/2, 91054 Erlangen, Germany* [sauer@mi.uni-erlangen.de]

J. M. SAVIGNAT (*363), *École des Mines, 35, rue Saint Honoré, 77 305 Fontainebleau Cedex, France* [savignat@cges.ensmp.fr]

DANIELA SCARAMELLI (*287), *Dip. di Matematica Pura ed Applicata, Via G. Belzoni 7, 35131 Padova, Italy* [daniela@sirio.de.unifi.it]

ROBERT SCHABACK (**359), *Institut für Numerische und Angewandte Mathematik, Universität Göttingen, Lotzestraße 16-18, D-37083 Göttingen, Germany* [schaback@math.uni-goettingen.de]

KARL SCHERER (**375), *Institut für Angewandte Mathematik, Universität Bonn, Bonn, 53115 Germany* [scherer@iam.uni-bonn.de]

LAURENT SCHIAVON (*99), *Université de Valenciennes et du Hainaut-Cambrésis, Laboratoire MACS, B.P.311, 59304 Valenciennes Cedex, France* [schiavon@univ-valenciennes.fr]

STEFAN SCHIRRA (*277), *Max-Planck-Institut für Informatik, 66123 Saarbrücken, Germany* [stschirr@mpi-sb.mpg.de]

ROBERT SCHNEIDER (*371), *Max-Planck-Institut für Informatik, 66123 Saarbrücken, Germany* [schneider@mpi-sb.mpg.de]

LARRY L. SCHUMAKER (**171), *Department of Mathematics, Nashville, TN 37240, USA* [s@mars.cas.vanderbilt.edu]

U. SCHWANECKE (**385), *Darmstadt University of Technology, Dept. of Mathematics, Schloßgartenstr. 7, 64289 Darmstadt, Germany* [schwanecke@mathematik.tu-darmstadt.de]

ALESSANDRA SESTINI (*287), *Dip. di Energetica "S. Stecco", Via Lombroso 6/17, 50134 Firenze, Italy* [sestini@de.unifi.it]

JAMES A. SETHIAN (*193), *Department of Mathematics, and Lawrence Berkeley National Laboratory, University of California, Berkeley, CA 94720, USA* [sethian@math.berkeley.edu]

CHANG SHU (*381), *Institute for Information Technology, National Research Council of Canada, Montreal Road, Ottawa, Ontario, Canada K1A 0R6* [chang.shu@iit.nrc.ca]

NIR A. SOCHEN (*203), *School of Mathematical Sciences, Tel-Aviv University, Tel-Aviv 69978, Israel* [sochen@math.tau.ac.il]

JUNJI SONE (*389), *Toshiba corporation, 1 Toshiba-tyou, Futyuu, Tokyo, 183-0043, Japan* [junji.sone@toshiba.co.jp]

OLIVIER STAB (*363), *Ecole des Mines - C.G.E.S, 35, rue saint Honore, 77 305 Fontainebleau Cedex, France* [stab@cges.ensmp.fr]

CHIEW-LAN TAI (**393), *Department of Computer Science, The Hong Kong University of Science & Technology, Clear Water Bay, Kowloon, Hong Kong* [taicl@cs.ust.hk]

RIADH TALEB (**237), *Laboratoire de Modélisation et Calcul, CNRS-IMAG, B.P. 53, F-38041 Grenoble Cedex 9, France* [talebr@imag.fr]

G. TAVARES (*229), *Pontifícia Universidade Católica do Rio de Janeiro, Departamento de Matemática, Rua Marquês de São Vicente, 225, Gávea, Rio de Janeiro, RJ, Brazil, CEP:22.453-900* [geovan@mat.puc-rio.br]

HOLGER THEISEL (**403), *University of Rostock, Computer Science Department, PostBox 999, 18051 Rostock, Germany* [theisel@informatik.uni-rostock.de]

F. TOVAR (*91), *Universidad Central de Venezuela. Facultad de Ciencias, Escuela de Matemática, Laboratorio de Computación Gráfica y Geometría Aplicada. Apartado 47809, Los Chaguaramos, Caracas 1041-A, Venezuela* [ftovar@euler.ciens.ucv.ve]

JEAN-YVES TRÉPANIER (*213), *Département de génie mécanique, École Polytechnique de Montréal, C.P. 6079, Succ. Centre-Ville, Montréal (Québec), H3C 3A7, Canada* [jyves@cerca.umontreal.ca]

KENJI UEDA (*399), *Ricoh Company, Ltd., 1-1-17, Koishikawa, Bunkyo-ku, Tokyo 112-0002, Japan* [ueda@src.ricoh.co.jp]

GEORG UMLAUF (*335), *Dept. C.I.S.E., CSE Bldg, University of Florida, Gainesville, FL 32611-6120, USA* [umlauf@cise.ufl.edu]

KARSTEN URBAN (**65), *Institut für Geometrie und Praktische Mathematik, RWTH Aachen, Templergraben 55, 52056 Aachen, Germany* [urban@igpm.rwth-aachen.de]

MARIE-GABRIELLE VALLET (*213), *CERCA, 5160 boul. Decarie, #400, Montréal (Québec), H3X 2H9, Canada* [vallet@cerca.umontreal.ca]

GERT VEGTER (*325), *Department of Mathematics and Computing Science, University of Groningen, P. O. Box 800, 9700 AV Groningen, The Netherlands* [gert@cs.rug.nl]

S. VEIRA-TESTE (**9), *TOPCAD SA, BP 521, 31674 Labege, France*

PIERRE VILLON (*363), *Université de Technologie de Compiegne, U.R.A. CNRS 1505 LG2MS, BP 529, F-60205 Compiegne, Cedex, France* [pierre.villon@utc.fr]

G. Wahu (*407), *LIGIM, University Claude Bernard Lyon 1, 43 boulevard du 11 novembre 1918, 69622 Villeurbanne, France* [gwahu@ligim.univ-lyon1.fr]

Johannes Wallner (*417), *Institut für Geometrie, Wiedner Hauptstraße 8–10/113, A-1040 Wien, Austria* [wallner@geometrie.tuwien.ac.at]

Joe Warren (**411), *Rice University, Department of Computer Science, P. O. Box 1892, Houston, TX 77251-1892, USA* [jwarren@rice.edu]

Henrik Weimer (**411), *Rice University, Department of Computer Science, P.O. Box 1892, Houston, TX 77251-1892, USA* [henrik@rice.edu]

Holger Wendland (**359), *Institut für Numerische und Angewandte Mathematik, Universität Göttingen, Lotzestraße 16-18, D-37083 Göttingen, Germany* [wendland@math.uni-goettingen.de]

Wolfgang L. Wendland (**181), *Mathematisches Institut A, Universität Stuttgart, Pfaffenwaldring 57, 70569 Stuttgart, Germany* [wendland@mathematik.uni-stuttgart.de]

Joris Windmolders (*45), *Celestijnenlaan 200A, 3001 Heverlee, Belgium* [joris.windmolders@cs.kuleuven.ac.be]

Hans J. Wolters (*433), *Hewlett Packard Laboratories, Palo Alto, CA, USA* [wolters@hpl.hp.com]

Emil Žagar (**263), *FRI and IMFM, University of Ljubljana, Tržaška 25, 1000 Ljubljana, Slovenia* [Emil@Gollum.Fri.Uni-Lj.Si]

Frank Zeilfelder (**161), *Universität Mannheim, Fakultät für Mathematik und Informatik, A5, 68131 Mannheim, Germany* [zeilfeld@fourier.math.uni-mannheim.de]

Ming Zhang (*17), *Department of Computer Science, Rice University, Houston, TX 77025, USA* [mzhang@cs.rice.edu]

Extending Lawson's Algorithm to Include the Huber M-Estimator

Iain J. Anderson, John C. Mason,
and Colin Ross

Abstract. When fitting a curve to experimental data, there is no guarantee that the data obtained are as accurate as might be expected. The effect of outside influences may cause the data set to contain outliers. These outliers can have a significant effect on any curve which is fitted to such data. The ℓ_∞-norm, which is particularly appropriate for fitting data with uniformly distributed errors, is extremely sensitive to such outliers, since it minimises the maximum error from the data to the curve. Therefore, a technique which approximates a data set using the ℓ_∞-norm, without being adversely affected by outliers, would be a useful addition to the array of tools available. We present numerical examples to illustrate the use of such a technique and also some practical applications to justify its use.

§1. Introduction

It is widely accepted that the ℓ_∞-norm is the most appropriate measure of the error when approximating data which are very accurate or have errors sampled from a uniform distribution. Unfortunately, because the ℓ_∞ norm is extremely sensitive to outliers, it is not suitable for use in fitting experimental data containing such points. Nevertheless, it may be the case that the ℓ_∞-norm is the most appropriate error measure for the non-outlying data, and so we present an algorithm for finding an ℓ_∞ fit to the non-outliers of a data set.

The algorithm itself is based on a combination of the Huber M-estimator [6] and Lawson's algorithm [7]. There is considerable literature on both techniques as separate subjects, and we mention here only a selection. Lawson's algorithm was first analysed by Lawson [7] in 1961, and was later studied by Rice and Usow [11], Cline [2] and Ellacott [4]. Similarly, the Huber M-estimator was developed by Huber [6] in 1964 and has received considerable attention in the form of algorithms for its solution as well as analyses of its behaviour. Papers by Clark and Osborne [1], Ekblom [3], Madsen and Nielsen

Curve and Surface Fitting: Saint-Malo 1999
Albert Cohen, Christophe Rabut, and Larry L. Schumaker (eds.), pp. 1–8.

ISBN 0-8265-1357-3.

[9], Michelot and Bougeard [10] and Li [8] all look at the Huber M-estimator either in its own right or as one of a class of robust estimators.

In this paper, we consider the problem of fitting a function of the linear form $f(x) = \sum_{j=1}^{n} c_j \phi_j(x)$ to a set of data $\{(x_i, y_i)\}_{i=1}^{m}$, where the $\{\phi_j\}$ are a set of basis functions. To this end, we minimise the residuals $r_i = y_i - f(x_i)$. What our algorithm achieves in practice is to obtain an ℓ_∞ fit for those r_i such that $|r_i|$ is less than the Huber parameter γ, say, and effectively to ignore the remaining data.

The circumstances that require such an algorithm occur in practice, particularly in metrology where extremely accurate readings can be obtained (by, for example, a CD reader) but are subject to the occasional outlier (due, for example, to optical effects). These outliers usually only appear in groups of one or two, so they are isolated, which leads to an easier problem than if they appeared in larger groups. Another metrological situation where this algorithm can be applied is in the measurement of a cylinder in an automotive engine where there is approximately 95% very accurate data, and 5% outliers. Naturally, these problems might require a slightly different fitting technique, but this algorithm is a useful starting point from which more general fitting procedures may be developed in future work.

§2. Background

In this section, we discuss some aspects of both Huber estimation and Lawson's algorithm. In the next section we describe how to combine the two techniques to create a new algorithm which satisfies our requirements.

The Huber M-estimator

The Huber M-estimator is based on the Huber function

$$\rho(t) = \begin{cases} t^2/2, & \text{if } |t| \le 1, \\ |t| - 1/2, & \text{if } |t| > 1, \end{cases} \tag{1}$$

introduced by Huber [6] in 1964, and is defined in the following straightforward way:

$$E = \sum_{i=1}^{m} \rho(r_i/\gamma), \tag{2}$$

where r_i is the residual in the ith datum, and γ is the Huber threshold defining the distinction between "accurate" and "inaccurate" data.

There are several algorithms to solve the problem of minimising (2) with respect to $\mathbf{c}$, several of which are described by Li [8]. However in this paper, we limit ourselves to the Newton method. This involves solving [8]

$$\frac{1}{\gamma^2} A^T D A \mathbf{p} = \frac{1}{\gamma} A^T \mathbf{v}$$

at each iteration, where A is the design matrix with entries $A_{ij} = \phi_j(x_i)$, D is a diagonal matrix with entries $D_{ii} = 1$ if $|r_i| \le \gamma$ and $D_{ii} = 0$ if $|r_i| > \gamma$, and $\mathbf{v}$ has entries $v_i = \rho'(r_i/\gamma)$. Solving this system gives an update vector $\mathbf{p}$ which should provide a better estimate $\mathbf{c}+\mathbf{p}$ of the solution parameters $\mathbf{c}^*$. In order to ensure convergence, we also incorporate a line search which involves finding a scalar α which is the solution to the equation

$$(A\mathbf{p})^T \rho' \left(\frac{\mathbf{r} + \alpha A\mathbf{p}}{\gamma} \right) = 0.$$

Having found α, we then obtain a new estimate of $\mathbf{c}^*$ by setting $\mathbf{c} := \mathbf{c} + \alpha \mathbf{p}$. We repeat this procedure, updating D and $\mathbf{v}$ as necessary, until we have obtained $\mathbf{c}^*$ to sufficient accuracy.

Weighting

We choose to generalise (2) by introducing weights to obtain a weighted Huber M-estimator of the form $F = \sum_{i=1}^{m} w_i \rho(r_i/\gamma)$, where γ is the Huber threshold, w_i is the weight associated with the ith datum, and r_i is the residual associated with the ith datum. It may be necessary to introduce weights in this way in order to deal with non-identically distributed errors in the data, in which case the weights may be chosen to be the reciprocals of the standard deviations of the underlying probability distributions.

Many algorithms exist to find unweighted Huber fits, and in general, adapting them to find a weighted Huber fit is a straightforward task. As an example, we show how to adapt a Newton-like method.

Weighted Huber algorithm.

1) Calculate $v_i = w_i \rho'(r_i/\gamma)$.
2) If $\frac{1}{\gamma} A^T D_w A\mathbf{p} = -A^T \mathbf{v}$ is consistent, define $\mathbf{p} := -\frac{1}{\gamma}(A^T D_{\mathbf{w}} A)^+ A^T \mathbf{v}$,

   Otherwise, define $\mathbf{p} := -\frac{1}{\gamma} P^{-1} A^T \mathbf{v}$, where P is a positive definite matrix.
3) Find a steplength $\alpha > 0$ such that $(A\mathbf{p})^T D_{\mathbf{w}} \rho'((\mathbf{r} + \alpha A\mathbf{p})/\gamma) = 0$.
4) Set $\mathbf{c} := \mathbf{c} + \alpha \mathbf{p}$.

Here, A is the $m \times n$ matrix representing the underlying linear model, $D_{\mathbf{w}}$ is a diagonal matrix with entries

$$(D_{\mathbf{w}})_{ii} = \begin{cases} w_i, & \text{if } |r_i/\gamma| \le 1, \\ 0, & \text{if } |r_i/\gamma| > 1. \end{cases}$$

P is usually the identity matrix, I and Y^+ denotes the pseudo-inverse of a matrix Y, defined so that Y^+ is that matrix X of the same dimensions as Y^T such that $YXY = Y$, $XYX = X$ and YX and XY are symmetric.

We note here that there are many other algorithms for finding a Huber fit, and that most, if not all, can be adapted just as easily.

Lawson's algorithm

This algorithm, analysed by Lawson [7] in 1961, enables an ℓ_∞ fit to be obtained by repeated weighted ℓ_2 fits. The algorithm itself is very straightforward, and involves updating the weights at each iteration according to

$$w_i^{(l+1)} := \frac{w_i G(r_i^{(l)})}{\sum_{k=1}^{m} w_k G(r_k^{(l)})}, \tag{3}$$

where $G(t) = |t|$. The denominator is a normalisation term to ensure that the weights sum to unity. The numerator has the effect of weighting data with large residuals more heavily, with the result that, in the limit, only those data with a maximal residual will have any weight attached to them.

Lawson's algorithm finds the points of extreme oscillation and weights these accordingly to obtain the best ℓ_∞ approximation. The other weights are not important, and in fact converge to zero.

Initial values for the weights are usually chosen to be $w_i^{(1)} = 1/m$, as this treats all the data equally and satisfies the condition that the sum of the weights must be unity. Proofs of convergence require that the $\{\phi_i(x)\}$ form a Chebyshev set, but experimental results (see, for example, [4]) suggest that the algorithm is more generally applicable.

A summary of Lawson's algorithm.

1) Set all weights equal (with the sum of weights equal to unity).
2) Perform a weighted least-squares fit.
3) Calculate the residuals from the weighted least-squares fit.
4) Update the weights according to Lawson's formula (3).
5) Return to Step 2 until convergence is obtained.

§3. The Algorithm

We are concerned with the solution of the problem

$$\min_{\mathbf{c}} \max_{\{r_i : |r_i| \le \gamma\}} |r_i|,$$

where r_i is the residual for the datum (x_i, y_i), and γ is the Huber threshold value. In order to solve this problem, we reformulate it as

$$\min_{\mathbf{c}} \sum_{i=1}^{m} w_i \rho(r_i/\gamma),$$

where ρ is defined as in equation (1), and we adopt an iterative procedure to find $\mathbf{c}$ by performing successive weighted Huber fits. The weights are updated after each iteration in a manner similar to Lawson's original algorithm. While Lawson's algorithm is concerned with finding a minimax fit via a sequence of weighted least-squares fits, this new algorithm finds a minimax fit to the non-outlying data via a sequence of weighted Huber fits.

Unfortunately, Lawson's rule for updating the weights cannot be used in this new algorithm since the rule would weight the outliers too heavily. As a result, the outliers would be fitted more accurately at the next iteration. The essential point of Lawson's update is to weight those datum points which correspond to the maximal errors of the minimax fit. To maintain this general trend, we need an update in which the function G in (3), rather than being monotonic, instead increases to a peak and then decays, with the peak corresponding to the residual with the largest magnitude which does not exceed γ. The latter is termed the "γ-maximal residual" and denoted by γ_{MR}.

The function we choose in place of $|t|$ is a negative exponential of the form

$$G^{(l)}(t) = \begin{cases} |t|, & \text{if } |t| \leq \gamma_{MR}, \\ \gamma_{MR} e^{-\frac{1}{\gamma_{MR}}(|t|-\gamma_{MR})}, & \text{if } |t| > \gamma_{MR}, \end{cases}$$

and we update the weights at each iteration according to (3). (Note that $G^{(l)}$ changes with the iteration l.)

For $|t| > \gamma_{MR}$, the γ_{MR} factor in $G^{(l)}(t)$ is needed to ensure continuity at $|r_i^{(k)}| = \gamma_{MR}$ and the $-\gamma_{MR}^{-1}$ term in the exponential is used so that the left and right derivatives of $G_i(t)$ are continuous at γ_{MR}. The reason for this second condition is to ensure that points with residuals just over γ_{MR} and those with residuals just less than γ_{MR} are treated similarly.

§4. Convergence

We have obtained favourable results with this algorithm, provided that certain conditions are met. Firstly, the form of the approximating model needs to be appropriate. For example, trying to approximate a set of data corresponding to a quadratic by a straight line will probably lead to problems, as it is likely that a considerable number of the data will be treated as outliers. Secondly, γ needs to be chosen carefully. If γ is chosen to be too small, then there may be many solutions and it may not be possible to predict to which solution the algorithm will converge — if it converges at all.

We therefore conclude that in order to use this algorithm effectively, we first need to have some details of the problem we are to tackle. If we are unsure as to what sort of model to fit to the data, then γ should be chosen to be larger than we might initially require. If we are unsure what value of γ to choose, then some sort of γ-reduction procedure may be effective for finding an appropriate value. An initial value of γ may be chosen by use of the formula $\gamma = 1.9906 \times \mathbf{median}(|r_i - \mathbf{median}(r_i)|)$ (see, for example, Ross et al, [12]).

The effect of using a Lawson-like update with a non-monotonic factor is to increase the weights at the extrema of the minimax approximation and reduce all other weights, including those of the outliers. In practice, the algorithm produces a minimax approximation to a subset of the data with the aim that this subset should be the non-outlying data. Unfortunately, we have been unable thus far to prove convergence for this algorithm. However, it should be noted that the convergence rate would be expected to be similar to that of Lawson's original algorithm as they essentially do the same task.

§5. Acceleration Schemes

Although the algorithm as it stands is acceptable for small problems, it nevertheless takes a considerable length of time to achieve relaxed convergence conditions. This is no surprise as one of the shortcomings of Lawson's algorithm is its slowness to converge. More specifically, the convergence of Lawson's algorithm is linear with a ratio of τ^* [11], where

$$\tau^* = \max \left[\tau = \frac{|\mathbf{r}^*|}{\max |\mathbf{r}^*|} : \tau < 1 \right],$$

with $\mathbf{r}^*$ defined to be the vector of residuals from the optimal ℓ_∞ fit. In many situations, this ratio can be very close to one, leading to rather slow convergence.

One technique to increase the rate of convergence is to use the fact that, upon convergence, the weights corresponding to non-extremal residuals are zero. Specifically, after a set number of normal iterations to allow the weights to settle a little, we may set $w_i = 0$ if $|r_i| \le \sigma^2/\|r_i\|_\infty$, where $\sigma = \left[\sum_{i=1}^m w_i^{(k)} (r_i^{(k)})^2\right]^{1/2}$. This latter technique is the one presented by Rice and Usow [10], although Ellacott [4] found that it could cause the algorithm to fail.

Of course in the case of this new algorithm, these schemes cannot be applied directly. We need to compensate for those data which are being treated as outliers, thus this scheme is not valid. If it were possible to find some analogue of σ for this new algorithm, then it may be possible to use that analogue in an acceleration scheme.

§6. Numerical Results

We have tested this algorithm extensively and now present some numerical results to illustrate it. In Figure 1, we show a synthesised data set consisting of 95 points lying close to the polynomial $f(x) = 2x^2 - 3x + 1$ with 5 outliers. Figure 1 also shows the best fitting quadratic polynomial to the data obtained by a least-squares fit, by an ℓ_∞ fit and by the new algorithm presented in this paper. The noise in the data is taken from a uniform distribution on $[-0.1, 0.1]$ and we thus choose $\gamma = 0.1$. Table 1 shows the results from the various fits performed. It is clear that both the ℓ_2 and ℓ_∞ fits are unsuitable and are affected by the outliers. However, the new algorithm succeeds in identifying the outliers and successfully ignoring them. Comparing the results from the new algorithm with those from performing least-squares and minimax fits to the data without outliers, we see that they are much more in agreement. In fact, as we would expect, the new algorithm has generated an almost identical fit to the ℓ_∞ fit on the accurate data.

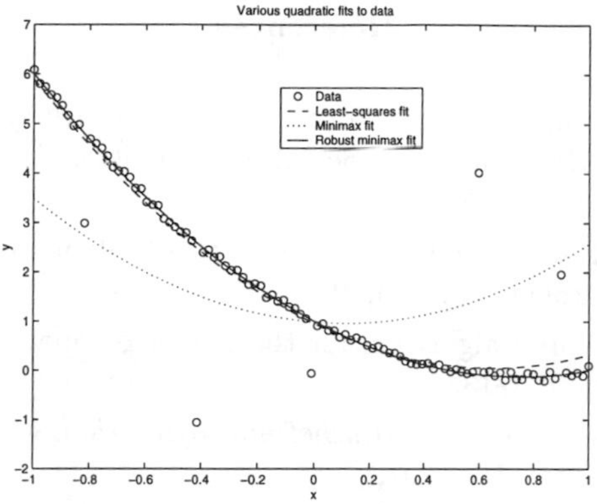

Fig. 1. Various quadratic polynomial fits to a set of data with outliers.

We also note that, while the new algorithm seems to be significantly faster in this example, this is not the case in general. In fact with stricter convergence criteria, Lawson's original algorithm applied to the non-outlying data converges in fewer iterations than the new algorithm. The reason that the minimax fit to the data containing outliers takes fewer iterations is due to the result in Section 5 involving τ^*, which, because of the outliers, is actually quite small ($\tau^* = 0.9497$) compared to $\tau^* = 0.9894$ for the case of the new algorithm.

	ℓ_2	ℓ_∞	New	ℓ_2 (NO)	ℓ_∞ (NO)
c_0	+0.9499	+0.9839	+0.9958	+1.0059	+0.9971
c_1	−2.7868	−0.4571	−2.9951	−3.0007	−3.0001
c_2	+2.1568	+2.0492	+2.0057	+2.0050	+2.0035
Iterations	1	46	39	1	140

Tab. 1. Numerical results: fitting a quadratic (NO : No outliers).

The convergence criterion was the same for both Lawson-like algorithms, namely that the magnitude of the four largest γ-maximal residuals should agree with a relative error of less than 10^{-2}. In addition, no acceleration schemes were used since we needed to obtain a measure of how fast the algorithms were in their unaccelerated form.

§7. Conclusions

We have presented an algorithm for fitting a linear form to data containing uniform noise, contaminated by outliers. Future work will concentrate on three main areas. Firstly, acceleration of the convergence of the algorithm. Secondly, extension to non-linear forms. Thirdly, extension to general ℓ_p norms rather than solely to the ℓ_∞-norm.

References

1. Clark, D. I., and M. R. Osborne, Finite algorithms for Huber's M - estimator, SIAM Journal on Scientific and Statistical Computing **7**(1) (1986), 72–85.
2. Cline, A. K., Rate of convergence of Lawson's algorithm, Mathematics of Computation **26**(117) (1972), 167–176.
3. Ekblom, H., A new algorithm for the Huber estimator in linear models, BIT **28** (1988), 123–132.
4. Ellacott, S. W., *Linear Chebyschef approximation*, M.Sc. thesis, University of Manchester, UK, 1972.
5. Hermey, D., and G. A. Watson, Fitting data with errors in all variables using the Huber M-estimator, SIAM Journal on Scientific Computing **20**(4) (1999), 1276–1298.
6. Huber, P. J., Robust estimation of a location parameter, Annals of Mathematical Statistics **35** (1964), 73–101.
7. Lawson, C. L., *Contributions to the theory of linear least maximum approximation*, Ph.D. thesis, University of California, Los Angeles, CA, USA, 1961.
8. Li, W., Numerical algorithms for the Huber M-estimator problem. In *Approximation Theory VIII*, C. K. Chui and L. L. Schumaker (eds.), World Scientific, New York, NY, USA, 1995, 325–334.
9. Madsen, K., and H. B. Neilsen, Finite algorithms for robust linear regression, BIT **30** (1990), 682–699.
10. Michelot, M. L., and M. L. Bougeard, Duality results and proximal solutions of the Huber M-estimator problem, Applied Mathematics and Optimization **30** (1994), 203–221.
11. Rice, J. R., and K. H. Usow, The Lawson algorithm and extensions, Mathematics of Computation **22** (1968), 118–127.
12. Ross, C., I. J. Anderson, J. C. Mason, and D. A. Turner, Approximating coordinate data that has outliers, in *Advanced Mathematical and Computational Tools in Metrology IV*, P. Ciarlini, A. B. Forbes, F. Pavese, and D. Richter (eds.), World Scientific, Singapore, 2000, 210–219.

I. J. Anderson, J. C. Mason, and C. Ross
School of Computing and Mathematics
University of Huddersfield
Queensgate, Huddersfield, West Yorkshire
HD1 3DH, UK

i.j.anderson@hud.ac.uk
j.c.mason@hud.ac.uk
c.ross@hud.ac.uk

A Segmentation Method under Geometric Constraints after Pre-processing

D. Apprato, J. B. Betbeder, C. Gout, and S. Vieira-Teste

Abstract. For a geophysical image with homogeneous grey levels, we propose a method of segmentation that could be subdivided into two parts: the first one concerns a pre-processing of the image which provides an enhancement of some features present on the image. The originality of the method consists in using a scale transformation applied to the pixel values of the image. The second part presents a segmentation method using deformable surfaces. The originality of this segmentation method is that it considers the active contour model as a set of articulated curves, which corresponds to the interfaces between different layers and faults. Moreover, the a priori knowledge of well data allows us to make some geometric constraints on the model. The solution is obtained by minimization of a nonlinear functional under constraints in a suitable convex set. Solving the minimization problem consists in particular in a k-order Taylor formula applied to linearize the nonlinear term.

§1. Segmentation Pre-processing

Image segmentation is one of the most important steps leading to the analysis of processed image data. Its main goal is to divide an image into parts that have strong correlation with objects or areas of the real world contained in the image. The image is divided into separate regions that are homogeneous with respect to a chosen property such as brightness, color, reflectivity, context, etc. However, in certain cases, the grey levels of an image could be homogeneous and make the segmentation more difficult to realize. This is particularly true in the case of geophysical and medical images (cf. [14,15]). In the first part of this work, we propose a method to solve this problem using families of scale transformations. The use of scale transformations is common in imaging. The aim of this pre-processing is an improvement of the image function data that suppresses unwilling distortions, or enhances some image features important for further processing. It provides improvement of the contrasts, and it represents a tool to pre-process images used in most computer algorithms today.

Curve and Surface Fitting: Saint-Malo 1999
Albert Cohen, Christophe Rabut, and Larry L. Schumaker (eds.), pp. 9–18.

ISBN 0-8265-1357-3.

According to Sonka, Haclav and Boyle [15], the pre-processing of images may be classified into four categories (pixel brightness transformations, geometric transformations, pre-processing methods that use a local neighborhood of the processed pixel, and image restoration that requires knowledge about the entire image) according to the size of the pixel neighborhood that is used for the calculation of a new pixel brightness. The transformation of the brightness and of the contrast of an image allows us to focus on phenomena that are hard to see in the plain image.

For a given image, we are going to consider the pixel values as a topographic map: the brightness value of each pixel is the height of the (hyper-) surface at that point. For a data set of pixels $(x_i, y_i, z_i, A(x_i, y_i, z_i))_i$, we apply the following functions:

- ζ_d: $A(x_i, y_i, z_i)_i \subset [0, 255] \longrightarrow [0, 255]$,
- $T^d(\varphi_d \circ (\varsigma_d o A)) \in H^m(\Omega, \mathbb{R})$,
- $\psi_d(T^d(\varphi_d \circ (\varsigma_d o A)))$ converges to $\zeta \circ A$ when d converges to 0,

where A is an attribute function introduced in Section 2.1, ς_d (resp. φ_d and ψ_d) are scale transformations converging to ς (resp. φ and ψ), and T^d is a D^m*spline* operator (see Arcangéli [2]).

The scale transformation ς_d converges to a usual brightness transformations ς (see Apprato and Gout [1]): for instance, ζ could be a scale transformation which enhances the image contrast between brightness values p_1 and p_2.

Let us consider the subdivision $\{u_1, u_2, ..., u_i, ..., u_{p(d)}\}_{i=1,...,p(d)}$ of the interval $[0, 255]$ satisfying $\zeta(A(x_i, y_i, z_i)) = u_i$, $p(d)$ being the number of different pixel values of the image (≤ 255 for a grey scale image). The function ζ_d is defined, for any $x \in [A(x_i, y_i, z_i) = w_i, A(x_{i+1}, y_{i+1}, z_{i+1}) = w_{i+1}]$, and for an integer $1 \leq i \leq p(d) - 1$, by

$$\begin{aligned}\zeta_d(x) = {} & u_i q^0_{0m}\left[(x - w_i) / (w_{i+1} - w_i)\right] + u_{i+1} q^1_{0m}\left[(x - w_i) / (w_{i+1} - w_i)\right] \\ & + \alpha_1(w_i)(w_{i+1} - w_i)\, q^0_{1m}\left[(x - w_i) / (w_{i+1} - w_i)\right] \\ & + \alpha_1(w_{i+1})(w_{i+1} - w_i)\, q^1_{1m}\left[(x - w_i) / (w_{i+1} - w_i)\right],\end{aligned}$$

where the q^l_{jm}, for $l = 0, 1$, and $j = 1, ..., m$, are the Hermite finite element basis functions, and where $\alpha_1(w_i) = (u_{i+1} - u_i) / (w_{i+1} - w_i)$ and $\alpha_1(w_{p(d)}) = (w_{p(d)} - w_{p(d)-1}) / (u_{p(d)} - u_{p(d)-1})$. Then, Gout [9] showed that for any $d \in D$, for an integer i, $1 \leq i \leq p(d) - 1$, $\zeta_d(w_i) = u_i$ and $\zeta_d \in C^m([0, 255])$.

Likewise, in order to recover a finer image, it is useful to apply the "large variations" algorithm introduced in [9]. In fact, after having applied the function ζ_d to improve the contrast of the image (and thus increasing the variations of the corresponding pixel values), it is very useful to use a method that takes into account these rapidly varying data. Let us note that even without using the scale transformations ζ_d, an image often has large variations (this occurs for example when a dark zone is near a brighter one). That is why we propose to use the "Large variations" algorithm. This algorithm uses two-scale transformations, namely φ_d for the pre-processing, and ψ_d for the post-processing.

The first one, φ_d, is used to suppress the oscillations of the data. The pre-processing function φ_d is such that the data do not present large variations, and therefore a usual spline operator T^d (e.g. [2]) can subsequently be applied without generating significant oscillations. The second scale transformation ψ_d is then applied to the approximated values to map them back and obtain the initial approximated pixel values. It is important to underline that the proposed scale transformations do not create parasitic oscillations. Moreover, this method is applied without any particular knowledge of the location of the large variations in the dataset.

So, for pre-processing, we propose two algorithms: in the first one, we just apply a scale transformations ζ_d as a brightness transformation for contrast enhancement, in the second one, we also apply the "large variations" algorithm in order to obtain a finer represention of the image which represents the main advantage of this approach.

The reader is referred to [1,8,10] for a complete study of this method, including its convergence and numerical results. Let us note that this method is also efficient for noise removal as shown in [1].

§2. Segmentation Method

We use deformable models (external forces, evolution term, see Kass, Witkin and Terzopoulos [16,17]) and classical approximation techniques such as spline theory (see de Boor [3], Laurent [12], Schumaker [13]) and the finite element method [4].

We propose an analytic approach which uses deformable models instead of a geometrical one as done for instance in Sethian [14]. We recall that the principle of the deformable model method lies in attracting the representation towards the structure using forces:

- Internal forces describing properties of elasticity and rigidity of the representation, connected to its derivatives (e.g., the energy of thin plates);
- External forces coming from potentials which characterize the elements of the structure with respect to the attributes data.

Geometrical constraints are associated with well interpolation conditions (case of geophysical images with well data). Deformable models provide a way of interactively acting on the representation by adding a dynamic term in the minimization problem (see for instance Cohen and Cohen [5], Cohen, Cohen and Ayache [6], and Cohen, Bardinet and Ayache [7]), that permits upgrading the models to the solution of the minimization problem introduced.

In this section, we first give the geophysical data and then the minimization problem is studied. The nonlinear problem and its discretization are given in the subsequent sections.

2.1. The data on the structure

Two types of data are available: attribute data and well data. For each attribute A, the attribute data are $(x_i, y_i, z_i, A(x_i, y_i, z_i))$, where (x_i, y_i, z_i) are the coordinates of the barycentre of a voxel, and $A(x_i, y_i, z_i)$ is the attribute value A in this voxel. The well data are depth data: $(x_j, y_j, z_j)_{j=1,\dots,N} = a_j$ where N is the number of interpolation points. This model allows a conceptual representation of the structure by identification of its various elements, and permits topological connections between those elements. This model induces the parameterization of the structure. Each element of the structure (layer, fault, etc.) is identified with a connection of four points with a label Σ. Furthermore, each quadruplet is connected by two points (which can be thought as a common side of a "quadrilateral" represented by the four points) with another quadruplet. Practically, the a priori model can be constructed by introducing a 3D block and a regular grid of this block. The aim is to find a space of admissible representations consistent with the a priori model and the criteria connected with the data. Therefore, it is necessary to choose a space of functions characterized by a domain of definition connected with the a priori model and regularity conditions connected to the data. The idea is to transform the a priori model into a normalized model called the model of reference (denoted by M'). For example, we can choose $M' \subset \overline{\Omega} = [0,1] \times [0,1] \times [0,1]$. The model M' is then the image by transformations of the set of vertical and horizontal closed sides of the a priori model as done in [18]. Let γ be the union of the common edges of any two sides of M', we define by M the interior of $M' \setminus \gamma$. All the functional spaces needed in this work are given in Vieira-Testé [18].

2.2. Minimization criterion

2.2.1. Internal forces: The criterion associated with the internal forces is a classical one. Modelling this criterion bring us to the following energy functional: for any $v \in V = H^2(M, \mathbb{R}^3) \cap C^0(M', \mathbb{R}^3)$,

$$E_1(v) = [v]_{1,M}^2 + [v]_{2,M}^2 ,$$

where

$$[v]_{1,M} = \left(\sum_{\Sigma \subset M} \varepsilon_1(\Sigma) \int_M \left[\left\langle \frac{\partial v}{\partial s} \right\rangle_3^2 + \left\langle \frac{\partial v}{\partial r} \right\rangle_3^2 \right] dsdr \right)^{1/2}$$

and

$$[v]_{2,M} = \left(\sum_{\Sigma \subset M} \varepsilon_2(\Sigma) \int_M \left[\left\langle \frac{\partial^2 v}{\partial s^2} \right\rangle_3^2 + \left\langle \frac{\partial^2 v}{\partial r \partial s} \right\rangle_3^2 + \left\langle \frac{\partial^2 v}{\partial r^2} \right\rangle_3^2 \right] dsdr \right)^{1/2}$$

with $\varepsilon_i(\Sigma) \geq 0, \quad \forall i = 1,2, \forall \Sigma \in M$. The term $[v]_{1,M}$ corresponds to an approximation of the elastic deformation of the model while the term $[v]_{2,M}$ corresponds to an approximation of the rigid deformation of the model (cf. Cohen, Cohen and Ayache [6]).

2.2.2. External forces: External forces are issued from potentials connected with attributes. We introduce the following energy, for any $v \in V$,

$$E_2(v) = \sum_{\Sigma \subset M} \int_{\Sigma} P_{\Sigma}\left(v_{/\Sigma}(s,r)\right) dsdr,$$

where P_{Σ} is the potential associated with the element parameterized by Σ.

The modelling we propose consists in minimizing the previous energies E_1 and E_2 as we will see in subsection 2.2.4. In the case of the velocity attribute, we use the following potential to define the layers:

$$P(x,y,z) = -\alpha \left\| \overrightarrow{\nabla A}(x,y,z) \right\|^2, \qquad \alpha \geq 0$$

where A is the attribute "velocity of propagation of the seismic wave".

2.2.3. Interpolation data: If we suppose some parameterization $(s_j, r_j) \in M$ of each interpolating point $a_j = (x_j, y_j, z_j)$ is known, then we require that $v \in V$ satisfies $v(s_j, r_j) = a_j$ for any $j = 1, ..., N$.

2.2.4. Minimization criterion: Using the notation and definitions introduced above, we consider the functional E defined on V by

$$E(v) = [v]_{1,M}^2 + [v]_{2,M}^2 + \sum_{\Sigma \subset M} \int_{\Sigma} P_{\Sigma}\left(v_{/\Sigma}(s,r)\right) dsdr$$

for any $v \in V$. We consider the set K associated with the interpolation constraints, and defined by

$$K = \{v \in V, \quad \forall j = 1, ..., N, \quad v(s_j, r_j) = a_j\}.$$

This set is convex and closed in V. We also introduce the following linear mapping (continuous on V with the norm $\|\cdot\|_{2,M}$)

$$\rho_0 : v \in V \mapsto \rho_0 v = \left(v(s_j, r_j)\right)_{j=1,\dots,N} \in \left(\mathbb{R}^3\right)^N.$$

We consider the following minimization problem: find $\sigma \in K$ satisfying

$$\forall v \in K, \quad E(\sigma) \leq E(v).$$

We note that this problem is nonlinear on the convex set K with respect to σ. There are two techniques to treat this problem. The first one consists in linearizing the nonlinear term (linked to the potentials) in the functional E. The second one consists in using the deformable models technique as done in the following subsection: we suppose that the solution is a function of time, which leads to a new evolution problem that will be discretized both in time and space.

2.3. The nonlinear problem

In this subsection, we give the nonlinear minimization problem and its discretization. Let us recall that the deformable models technique consists in assuming that σ depends on time, and so consists in adding a dynamic term to the functional $E(\sigma)$

$$\frac{1}{2}\frac{\partial}{\partial t}\int_M \varepsilon(M)\,\sigma^2(t,s,r)\,dsdr,$$

where $(\varepsilon(M))_{/\Sigma} = \varepsilon(\Sigma) > 0$. This term allows the control at each time of the deformation of the surfaces.

2.3.1. Evolution problem: Let $T > 0$. We note

$$W(0,T,V) = \left\{ w \in L^2(]0,T],V),\ \frac{\partial w}{\partial t} \in L^2(]0,T],V') \right\}.$$

We then consider the following evolution problem defined on $[0,T]$. For any $t \in]0,T]$ and any $\omega \in W(0,T,V)$, find $\sigma \in W(0,T,V), \sigma(t) \in K$, satisfying $(\mathbf{P_t})$:

$$E(\sigma)+\frac{1}{2}\frac{\partial}{\partial t}\int_M \varepsilon(M)\,\sigma^2(t,s,r)\,dsdr \le E(\omega)+\frac{1}{2}\frac{\partial}{\partial t}\int_M \varepsilon(M)\,\omega^2(t,s,r)\,dsdr,$$

with

$$\sigma(0) = \sigma_0 \in L^2\left(M, \mathbb{R}^3\right).$$

We are currently studying existence and uniqueness of $(\mathbf{P_t})$ using a Lipschitz approximation of the sign function.

Likewise, for any $t \in]0,T]$, we consider the term

$$L_{\sigma(t)}(v) = -\sum_{\Sigma\subset M}\int_\Sigma P_\Sigma\left(v_{/\Sigma}(s,r)\right)dsdr.$$

The variational formulation of the problem $(\mathbf{P_t})$ with Kuhn and Tucker's relation is, taking as test function v on the stationary space V (necessary condition without uniqueness), for any $t \in]0,T]$ and any $v \in V$, find $(\sigma,\lambda) \in W(0,T,V) \times C^0\left([0,T], (\mathbb{R}^3)^N\right), \sigma(t) \in K$, satisfying $(\widetilde{\mathbf{P}})$:

$$\int_M \varepsilon(M)\frac{\partial \sigma(t,s,r)}{\partial t}v(s,r)dsdr + a(\sigma(t),v) + \langle \lambda(t), \rho_0 v\rangle_{N,3} = L_{\sigma(t)}(v)$$

under conditions

$$\sigma(0) = \sigma_0 \in L^2\left(\mathbb{R}^3\right)$$

and

$$\lambda(0) = \lambda_0 \in \left(\mathbb{R}^3\right)^N,$$

where

$$\frac{1}{2}a(u,v) = [v]_{1,M}^2 + [v]_{2,M}^2 .$$

2.3.2. Discretization in time: In this subsection, we discretize $(\widetilde{\mathbf{P}})$ both in time and space. The originality of this discretization consists in using a k-order Taylor development which allows us to take into account many more voxels and so to improve the convergence of the method (see Vieira-Teste [18] for more details). We cut the interval $]0,T]$ into sub intervals with length $\triangle t$. Consider

$$t_m = m \triangle t, \quad m = 1, ..., D_T.$$

We use the following approximation of the time derivative:

$$\frac{\partial \sigma}{\partial t}(t_m) \simeq \frac{\sigma(t_m) - \sigma(t_{m-1})}{\Delta t}.$$

Assuming that $\sigma^m = \sigma(t_m)$ and $\lambda^m = \lambda(t_m)$, we approximate the variational problem as follows: For any $m = 1, ..., D_T$ and any $v \in V$, find $(\sigma^m, \lambda^m) \in V \times (\mathbb{R}^3)^N, \sigma^m \in K$, satisfying $(\mathbf{P_m})$:

$$\int_M \varepsilon(M)\sigma^m v ds dr + \Delta t\left[a(\sigma^m, v) + \langle \lambda^m, \rho_0 v\rangle_{N,3}\right]$$

$$= \int_M \varepsilon(M)\sigma^{m-1} v ds dr + \Delta t L_{\sigma^m}(v)$$

with $\sigma^0 = \sigma_0 \in L^2(M, \mathbb{R}^3)$ and $\lambda^0 = \lambda_0 \in (\mathbb{R}^3)^N$.

The previous problem is implicit and nonlinear with respect to the solution σ^m. We propose to replace $L_{\sigma^m}(v) = L_{\sigma,v}(t_m)$ by a Taylor series expansion of order $k \geq 0$ about the time t_m. We suppose that σ is in $C^k([0,T], L^2(M, \mathbb{R}^3))$. We have

$$L_{\sigma^m}(v) = L_{\sigma,v}(t_m)$$

and $L_{\sigma,v}(t_m) \simeq L_{\sigma,v}(t_{m-1}) + \Delta t D L_{\sigma,v}(t_{m-1}) + \frac{(\Delta t)^2}{2} D^2 L_{\sigma,v}(t_{m-1}) + \cdots + \frac{(\Delta t)^k}{k!} D^k L_{\sigma,v}(t_{m-1})$. We note that the problem $(\mathbf{P_m})$ is linear and explicit with respect to σ^m. The following result is based on the Lax-Milgram Lemma.

Theorem. *The problem* $(\mathbf{P_m})$ *has a unique solution* (σ^m, λ^m).

2.3.3. Discretization in space: Let H be a nonempty bounded subset in $\mathbb{R}_+^*$ for which 0 is an accumulation point. For any $h \in H$, we solve the minimization problem $(\mathbf{P_m})$ in the finite element space $(V_h)^3 \subset V$. The generic finite element are the Hermite finite element of class C^1 for snakes and the Bogner-Fox-Schmit finite element rectangle of class C^1 (see [4]) for deformable surfaces. To have $(V_h)^3 \subset V$, it is necessary to have a C^0 connection on γ. To do that, it is sufficient to divide some degrees of liberty connected to derivatives as done in [18].

We denote by $\left(\alpha_1^m, ..., \alpha_{M_h}^m\right)$ the coordinates of σ^m in the basis of V_h and by $(\lambda_1^m, ..., \lambda_N^m)$ the coordinates of λ^m $\quad (M_h = dim(V_h))$. If σ^m is a solution of the discretized problem (P_m) in $(V_h)^3$, we can write σ^m in the basis of $(V_h)^3$: $\forall m = 1, ..., D_T, \quad \forall q = 1,2,3,$

$$(\sigma^m)^q = \sum_{j=1}^{M_h} \left(\alpha_j^m\right)^q \varphi_j$$

with $\left(\alpha_j^m\right)^q \in \mathbb{R}$ and where the $(\varphi_j)_{j=1,..,M_h,}$ are the basis functions of V_h.

In the following, we miss out q and h. Taking $v = \varphi_l$ in $(\widetilde{\mathbf{P}})$, we have to solve (for $q = 1,2,3$, in the linear problem $(\widetilde{\mathbf{P}})$) a system of $(M_h + N)$ equations with $(M_h + N)$ unknowns. We easily show that this system has a unique solution, and that the matrix $R = [C, B,^t B, 0]$ (first line : C, B; second line : ${}^tB, 0$) of the system is symmetrical and sparse with

$$C_{j,l} = [\varphi_j, \varphi_l]_{0,M} + \Delta ta\,(\varphi_j, \varphi_l)\,, \qquad i,j,l = 1,...,M_h$$

$$B_{j,i} = \Delta t \cdot \varphi_j\,(s_i, r_i)\,, \qquad j = 1,...,M_h, \quad i = 1,...,N,$$

where for any $u, v \in (V_h)^2$,

$$[u,v]_{0,M} = \int_M \varepsilon\,(M) \cdot u(s,r) \cdot v(s,r) dsdr,$$

and where $T = \left(\alpha_1^m, ..., \alpha_{M_h}^m, \lambda_1, ..., \lambda_N\right)$ is the unknown vector. We obtain a linear system $RT = L$, where the lines of L are

$$\int_M \varepsilon\,(M)\,\sigma^{m-1}\varphi_1 dsdr + \Delta t L_{\sigma^m}\,(\varphi_1)\,,$$

$$\vdots$$

$$\int_M \varepsilon\,(M)\,\sigma^{m-1}\varphi_{M_h} dsdr + \Delta t L_{\sigma^m}\,(\varphi_{M_h}\cdot)\,,$$

$$\Delta ta_1,$$

$$\vdots$$

$$\Delta ta_N.$$

This method has been implemented in fortran, C and C++. Numerical examples on real data are given in [11,18].

Acknowledgments. The authors are very grateful to the Région Aquitaine and ELF which have supported this work.

References

1. Apprato, D., and C. Gout, Noise removal in Medical Imaging, preprint, PAM 1000, University of California at Berkeley, 1998.
2. Arcangéli, R., Some applications of discrete D^m-splines, in *Mathematical Methods in Computer Aided Geometric Design*, T. Lyche and L. L. Schumaker (eds.), Academic Press, New York, 1989, 35–44.
3. de Boor, C., *A Practical Guide to Splines*, Springer Verlag, Berlin - Heidelberg, 1978.
4. Ciarlet, P. G., *The finite element method for elliptic problems*, North Holland, Amsterdam, 1979.
5. Cohen, I. and L. D. Cohen, A finite element method applied to new active contour models and 3D reconstruction from cross sections, Technical report 1245, INRIA, 1990.
6. Cohen, I., L. D. Cohen, and N. Ayache, Using deformable surfaces to segment 3D images and infer differential strutures, CVGIP : Image Understading, vol.56, n°2, September, 1992, 242–263.
7. Cohen, L. D., E. Bardinet, and N. Ayache, Surface reconstruction using active contour models, INRIA report 1824, 1992.
8. Gout, C., Segmentation in Medical Imaging from images with homogeneous level of colors, preprint, UPRES A 5033 98/05, Université de Pau, 1998.
9. Gout, C., Approximation using scale transformations, in *Approximation Theory IX, Vol. 1*, Charles K. Chui and Larry L. Schumaker (eds.), Vanderbilt Univ. Press, 1998, 151–158.
10. Gout, C., and A. Guessab, A new family of extended Gauss quadratures with an interior constraint, Journal of Computational and Applied Mathematics, to appear.
11. Gout, C. and S. Vieira-Teste, 3D surfaces modeling using deformable models techniques, in preparation.
12. Laurent, P. J., *Approximation et optimisation*, Hermann, Paris, 1972.
13. Schumaker, L. L., Fitting surfaces to scattered data, in *Approximation Theory II*, G. G. Lorentz, C. K. Chui, and L. L. Schumaker (eds.), Academic Press, New York, 1976, 203–269.
14. Sethian, J. A., *Level Set Methods: Evolving interfaces in Geometry, Fluid Mechanics, Computer vision, and Material Science*, Cambridge University Press, 1996.
15. Sonka, M., V. Haclav, and R. Boyle, *Image, Processing, Analysis and Machine Vision*, Chapman & Hall, 1993.

16. Terzopoulos, D., A. Witkin, and M. Kass, Constraints on deformable models: Recovering 3D shape and nonrigid motions, Artificial Intelligence **36** (1988), 91–123.

17. Terzopoulos, D., A. Witkin and M. Kass, Symmetryseeking models for 3D object reconstruction, in Proceedings, First International Conference on Computer Vision, London, England, IEEE, Piscataway, NJ, 1987, 269–276.

18. Vieira-Teste, S., Représentations de structures géologiques à l'aide de modèles déformables sous contraintes géométriques, Thèse de Doctorat, Université de Pau, 1997.

D.A., J.B.B., C.G.
Université de Pau - Dpt de Mathématiques
IPRA - Av.de l'Université - 64000 Pau, France
`chris_gout@alum.calberkeley.org`

S.V.T.
TOPCAD SA - BP 521 - 31674 Labege, France

Building Adaptive Multiresolution Schemes within Harten's Framework

Francesc Aràndiga and Rosa Donat

Abstract. We consider the cell-average framework described by A. Harten in [5], and build the prediction operator using two nonlinear interpolation techniques. We test the resulting nonlinear, adaptive, multiresolution scheme, and compare it with a linear scheme of the same accuracy. The nonlinear prediction processes we develop can also be used in the context of iterative refinement. Numerical tests show that this is also a viable alternative for piecewise smooth data.

§1. Introduction

The goal of a multi-scale decomposition of a discrete set of data is a "re-arrangement" of its information content in such a way that the new discrete representation, exactly equivalent to the old one, is more "manageable" in some respects. Some of the best known applications of multi-scale decompositions derive from their compression capabilities: a multiresolution representation of a function, i.e., of a discrete set which represents the function in some sense, can be highly compressed with minimal loss of information content. Precisely because of this potential, multi-scale techniques have an emergent role in numerical analysis, where the multi-scale idea has been used successfully over the years, from multigrid techniques to hierarchical bases in finite element spaces or subdivision schemes in Computer-Aided Design (CAD).

In the late 80's and early 90's, ideas from all these fields, together with a wide experience in the numerical solution of Hyperbolic Conservation Laws (HCL) lead Ami Harten to develop a *General Framework* for multiresolution representation of discrete data. The building blocks of a multiresolution scheme *à la Harten* are two operators which connect discrete and continuous data: The *discretization operator* obtains discrete information from a given signal (belonging to a particular function space) at a given resolution level; the *reconstruction operator* produces an approximation to that signal (in the same function space) from its discretized values.

Curve and Surface Fitting: Saint-Malo 1999
Albert Cohen, Christophe Rabut, and Larry L. Schumaker (eds.), pp. 19–26.

ISBN 0-8265-1357-3.

Harten's point of view is that the way in which the discrete data is generated, i.e., the discretization process, determines its *nature* and should provide an adequate setting for a multiresolution analysis. Once the setting is specified, the choice of an appropriate reconstruction operator provides the key step to the construction of a multiresolution scheme.

The reconstruction process lies at the very heart of a multiresolution scheme built *à la Harten*, and adaptivity can be introduced in the multiresolution scheme at this level. A nonlinear, adaptive reconstruction technique which fits the approximation to the local nature of the data will lead to a nonlinear adaptive multiresolution algorithm with improved compression capabilities.

The aim of this study is to examine a particular class of nonlinear adaptive multiresolution schemes, those using the Essentially Non Oscillatory (ENO) interpolatory techniques of [6] in the reconstruction step. Numerical experiments [1,2] show that ENO-MR schemes have larger compression rates than linear ones when the original signal or image is composed of smooth parts joined together by singularities. ENO techniques can be used to construct very accurate interpolants, which in turn lead to multiresolution schemes with high compression capabilities. When the original signal is geometric, nonlinear schemes can be used as loss-less compression techniques, and we show some application of this in the last section of this paper.

The nonlinear prediction process can be used also in the context of subdivision refinement. This amounts to setting to zero all scale coefficients and using the prediction operator to proceed by dyadic refinement. Preliminary tests show that these nonlinear subdivision schemes lead to non-oscillatory limiting functions when applied to piecewise smooth data with jumps, and open up an interesting alternative for iterative refinement of piecewise smooth data.

§2. Cell Average Multiresolution Analysis

When dealing with discrete data coming from a piecewise smooth function, the simplest discretization process, that of considering the *point-values* of the function, might not be well defined, especially at jump discontinuities. On the other hand, the discretization by cell-averages procedure acts naturally on the space of integrable functions, and it provides a more adequate setting to deal with piecewise smooth signals. Because of this, we shall carry out our numerical study within the cell-average framework.

Images are considered here as two-dimensional signals, and we use the usual tensor-product approach to design our two-dimensional algorithms. Thus, we only describe the essential features of the one-dimensional setting for the sake of completeness. The interested reader can find the missing details in this section in [2] or [5].

Let us consider a set of nested dyadic grids in [0,1]:

$$X^k = \{x_i^k\}_{i=0}^{N_k}, \quad x_i^k = ih_k, \quad h_k = 2^{-k}/N_0, \quad N_k = 2^k N_0, \quad k = L, \ldots, 0,$$

where N_0 is some integer. The discretization by cell average operator is defined as follows:

$$\mathcal{D}_k : L^1[0,1] \longrightarrow V^k, \quad \bar{f}_i^k = (\mathcal{D}_k f)_i = \frac{1}{h_k}\int_{x_{i-1}^k}^{x_i^k} f(x)dx, \quad 1 \leq i \leq N_k, \quad (1)$$

where $L^1[0,1]$ is the space of absolutely integrable functions in $[0,1]$ and V^k is the space of sequences with N_k components.

Due to the relation

$$\bar{f}_i^{k-1} = \frac{1}{h_{k-1}}\int_{x_{i-1}^{k-1}}^{x_i^{k-1}} f(x)dx = \frac{1}{2h_k}\int_{x_{2i-2}^k}^{x_{2i}^k} f(x)dx = \frac{1}{2}(\bar{f}_{2i-1}^k + \bar{f}_{2i}^k),$$

it is easy to see that $\{\bar{f}_i^k\}_{i=1}^{N_k}$, $k = L-1, \ldots, 0$, can be evaluated directly from $\{\bar{f}_i^L\}_{i=1}^{N_L}$ without using explicitly (1) (i.e., without knowledge of the original function $f(x)$).

To define an appropriate reconstruction operator for this setting (in fact, a whole family of them), we consider the sequence $\{F_i^k\}$ on the k-th grid defined from the cell values $\{\bar{f}_i^k\}$ as follows:

$$F_i^k = h_k \sum_{s=1}^{i} \bar{f}_s^k = \int_0^{x_i^k} f(x)dx = F(x_i^k) \quad \Rightarrow \quad \bar{f}_i^k = \frac{F_i^k - F_{i-1}^k}{h_k}. \quad (2)$$

The function $F(x)$ ($\in \mathcal{C}[0,1]$) is, in fact, a primitive of the original function $f(x)$, and the sequence $\{F_i^k\}$ represents a point value discretization of $F(x)$ on the k-th grid (with $F_0^k = 0$). Notice that (2) establishes a one-to-one correspondence between $\{\bar{f}_i^k\}_{i=0}^{N_k}$ and $\{F_i^k\}_{i=0}^{N_k}$.

Let us denote by $\mathcal{I}(x; F^{k-1})$ an interpolatory reconstruction of the set $\{F^{k-1}\}$ on the grid X^{k-1}, i.e., $\mathcal{I}(x_i^{k-1}; F^{k-1}) = F_i^{k-1}$. Then, we obtain an approximation , $\tilde{f}_i^k$, to $\bar{f}_i^k$ using (2) as follows:

$$\tilde{f}_i^k = (\mathcal{I}(x_i^k, F^{k-1}) - \mathcal{I}(x_{i-1}^k, F^{k-1}))/h_k. \quad (3)$$

Since $F_{2i}^k = F(x_{2i}^k) = F(x_i^{k-1}) = F_i^{k-1}$, we obtain

$$\tilde{f}_{2i-1}^k = \frac{1}{h_k}(\mathcal{I}(x_{2i-1}^k, F^{k-1}) - F_{i-1}^{k-1}) \quad \textit{and} \quad \tilde{f}_{2i}^k = \frac{1}{h_k}(F_i^{k-1} - \mathcal{I}(x_{2i-1}^k, F^{k-1})). \quad (4)$$

Let us define the prediction errors as $e_i^k := \bar{f}_i^k - \tilde{f}_i^k$. Using (2) and (4), we easily obtain

$$\bar{f}_{2i-1}^k - \tilde{f}_{2i-1}^k = -(\bar{f}_{2i}^k - \tilde{f}_{2i}^k).$$

Thus, we can simply store only the prediction errors with odd indexes; these are the scale coefficients, $d_i^k = e_{2i-1}^k$, of the multiresolution transform.

The multiscale decomposition of the original data $\bar{f}^L$ is described by the encoding algorithm:

$$\begin{cases} Do & k = L, \dots, 1 \\ & \bar{f}_i^{k-1} = \frac{1}{2}(\bar{f}_{2i-1}^k + \bar{f}_{2i}^k) & 1 \le i \le N_{k-1} \\ & d_i^k = \bar{f}_{2i-1}^k - (\mathcal{I}(x_{2i-1}^k; F^{k-1}) - F_{i-1}^{k-1})/h_k & 1 \le i \le N_{k-1}. \end{cases} \quad (5)$$

We recover the original data with the decoding algorithm:

$$\begin{cases} Do & k = 1, \dots, L \\ & \bar{f}_{2i-1}^k = (\mathcal{I}(x_{2i-1}^k; F^{k-1}) - F_{i-1}^{k-1})/h_k + d_i^k & 1 \le i \le N_{k-1} \\ & \bar{f}_{2i}^k = 2\bar{f}_i^{k-1} - \bar{f}_{2i-1}^k & 1 \le i \le N_{k-1}. \end{cases} \quad (6)$$

In our study we consider only local interpolation techniques with Lagrangian polynomials, i.e.,

$$\mathcal{I}(x; F^k) = q_i(x; F^k) \qquad x \in [x_{i-1}^k, x_i^k],$$

where $q_i(x; F^k)$ is a polynomial of degree r satisfying $q_i(x_{i-1}^k; F^k) = F_{i-1}^k$ and $q_i(x_i^k; F^k) = F_i^k$.

When the stencil of points used to construct $q_i(x)$ is symmetric around the ith interval (i.e., $r = 2s-1$, $\mathcal{S} = \{x_{i-s}^k, \cdots, x_{i+s-1}^k\}$), we obtain a centered interpolation technique. Centered interpolation techniques are very often used in approximation theory because they minimize the interpolation error, thus leading to very accurate reconstructions of smooth signals. It turns out that the multiresolution schemes obtained from (5) and (6) with Lagrangian piecewise polynomial centered interpolation techniques are equivalent to the Biorthogonal Wavelet (BOW) schemes of [4] (with the box function as the scaling function).

The compression properties of BOW schemes have been widely analyzed in the literature, but from an approximation theory standpoint, it is very easy to study the behavior of the coefficients in terms of the smoothness of the underlying signal and the accuracy of the interpolation technique. Notice that the scale coefficients d_i^k are related to interpolation errors at the odd points of the k-th grid. In fact,

$$d_i^k = (F_{2i-1}^k - \mathcal{I}(x_{2i-1}^k; F^{k-1}))/h_k.$$

Thus, if $f(x)$ is sufficiently smooth at $[x_{i-1}^{k-1}, x_i^{k-1}]$, we have $d_i^k = O(h_{k-1}^r)$. However, the presence of an isolated singularity $x_d \in [x_{i-1}^{k-1}, x_i^{k-1}]$ induces a loss of accuracy in the polynomial pieces whose stencils cross the singularity. The accuracy loss is related to the strength of the singularity as follows [2]: if $[f^{(p)}]_{x_d} = f^{(p)}(x_d+) - f^{(p)}(x_d+) = O(1)$ $(p \le r)$, and f is smooth everywhere else, we have

$$d_l^k = \begin{cases} O([f^{(p)}])h_{k-1}^p, & l = i-s, \dots, i+s-1, \\ O(h_{k-1}^r), & \text{otherwise.} \end{cases} \quad (7)$$

Thus, centered interpolation techniques lead to relatively large regions of poor accuracy around singularities, and therefore to large detail coefficients at those locations where the accuracy loss takes place. The consequence is a loss in efficiency for the multiresolution-based compression scheme.

It seems reasonable to improve the efficiency of the multiresolution-based compression scheme by improving the accuracy of the interpolatory technique used in the reconstruction step. Notice that when the convex hull of the stencil used to construct a polynomial interpolant is contained within a region of smoothness of the underlying signal, the interpolation error (and the corresponding detail coefficient) becomes small. Thus, it is clear that the key point is the construction of polynomial pieces that *avoid the singularity.*

In the literature related to the numerical solution of conservation laws, where discontinuities can spontaneously develop, we find an interpolation procedure with all the features we need: Essentially Non Oscillatory (ENO) interpolatory techniques [6] (it is not surprising that Harten was one of the developers of these techniques).

ENO interpolatory techniques lead to piecewise polynomial interpolants that are fully accurate except in those intervals that contain singularities. The essential feature of ENO interpolatory techniques is a stencil selection procedure that attempts to choose each stencil $\mathcal{S}_i$ within the same region of smoothness of $F(x)$. The stencil selection process uses the divided differences of the discrete set to be interpolated as smoothness indicators: Large divided differences indicate a possible loss of smoothness. The selection process is such that it tends to *look away* from large gradients, when this is feasible.

ENO interpolatory techniques are nonlinear, because the stencil used to construct each polynomial piece depends on the function being interpolated. When the singularities are sufficiently well separated (this means that there are at least $r+1$ points in each smoothness region), ENO techniques lead to stencils such that (assuming the singularity is located at the ith cell) $\mathcal{S}_{i-1} \cap \mathcal{S}_{i+1} = \emptyset$. Hence, the detail coefficients satisfy

$$d_l^k = \begin{cases} O([f^p])h_{k-1}^p, & l = i, \\ O(h_{k-1}^r), & \text{otherwise.} \end{cases} \tag{8}$$

Thus ENO interpolants have a nearly optimal high accuracy region, which should in turn improve the efficiency of the corresponding multiresolution-based compression algorithms.

The case of a corner of f (i.e., a jump in f') is especially interesting because it is possible to construct an even better (in terms of local accuracy) interpolant: the ENO-SR interpolant.

The Subcell Resolution (SR) technique (also due originally to Harten [3]) allows us to obtain an approximation to the *location* of an isolated corner in a continuous function up to the order of the truncation error. The approximated value is then used to modify locally (in the interval where the discontinuity lies) the definition of the piecewise polynomial interpolant in such a way that the interpolation error is small except for an $O(h^{r+1})$ band around the corner.

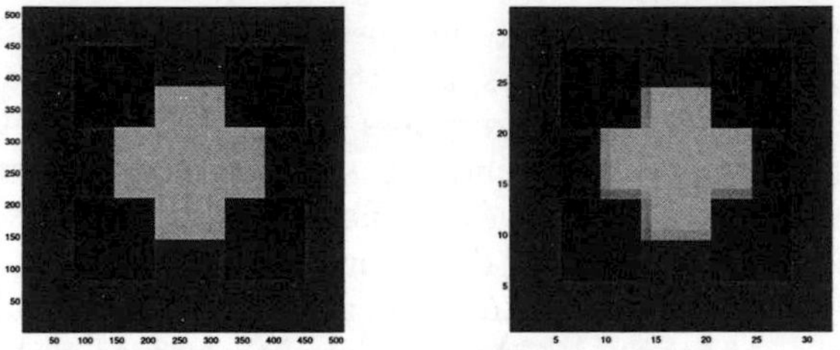

Fig. 1. Left original, right coarse version.

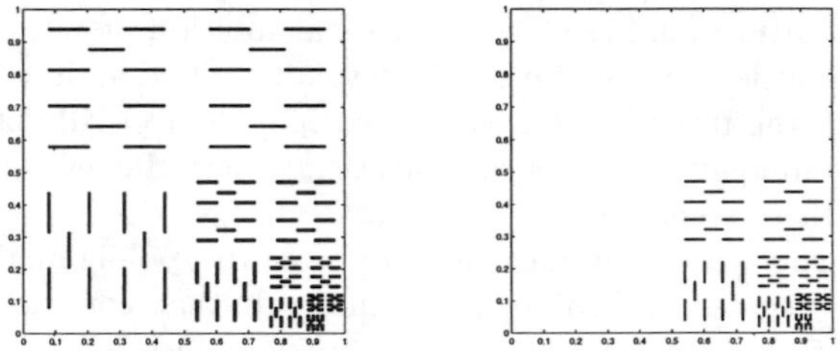

Fig. 2. Left linear, right ENO.

Recall that in the cell-average setting, the interpolation process is applied to the primitive function. Since jumps in $f(x)$ become corners in $F(x)$, using an ENO-SR interpolant in the reconstruction step lead to detail coefficients satisfying

$$d_l^k = O(h_{k-1}^r), \quad \text{except when} \quad x_{2l-1}^k \approx x_d. \tag{9}$$

The SR technique is, thus, appropriate to increase the efficiency of multiresolution-based compression algorithm for piecewise continuous functions with jumps.

§3. Numerical Experiments

Let us consider a purely geometric image as shown in Figure 1 (left), and apply to it the tensor-product version of algorithm (5). We consider piecewise polynomial interpolants of degree 4, thus the accuracy of the reconstruction in the cell-average framework is 3. In Figure 2 we display the location of non-zero scale coefficients in the multiresolution representation of the signal. When using the ENO-SR technique, and because all discontinuities are "aligned with the (tensor-product) grid", all scale coefficients are zero. This is a direct consequence of the fact that the ENO-SR reconstruction commits no error at the odd points in each one of the resolution levels considered ($L = 4$ in this example). In the case of the ENO scheme, the scale coefficients at the highest resolution level are all zero. This is a consequence of the nature of the data (the point-values of the signal at the highest resolution level), which locates all discontinuities at the cell end-points. The ENO technique is perfectly

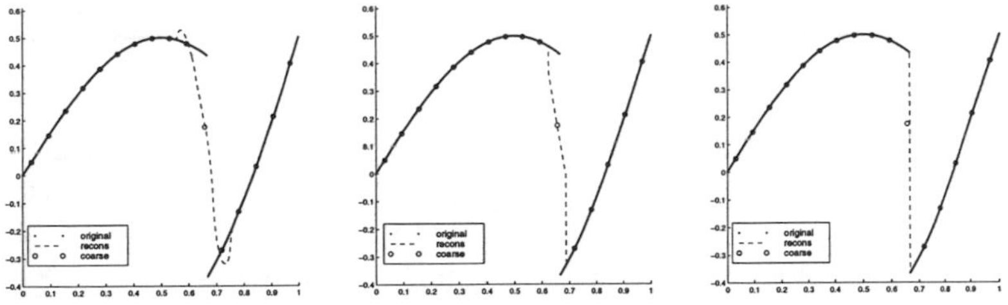

Fig. 3. 1d Linear, ENO, ENO-SR.

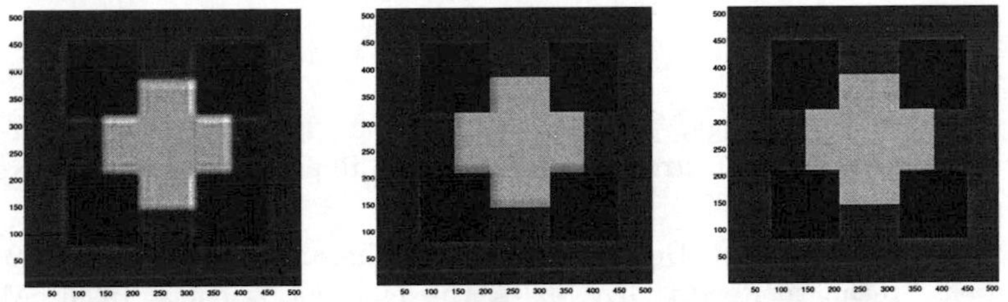

Fig. 4. Reconstruction from the coarse: linear, ENO, ENO-SR.

accurate when the discontinuity is located at a grid point. The technique is fully accurate at all the lower resolution levels except at the interval where the discontinuity is located. Thus, there is only one scale coefficient, located at the point which is closest to the singularity. This should be compared with the 3 scale coefficients per singularity obtained with the linear scheme (the lines showing the location of non-zero scale coefficients are thicker for the linear scheme). For the sake of comparison, the number of non-zero scale coefficients in each case is: Linear 8554, ENO 1688, ENO-SR 0.

We turn next to the nonlinear subdivision scheme obtained by considering the ENO and ENO-SR in the prediction process. In Figure 3 we show a univariate process. Starting from the the cell-averages of a piecewise smooth function at a very coarse level (16 points), we proceed by dyadic refinement until we obtain 1024 data. The numerical results clearly indicate that no overshoots or undershoots are obtained with the non-linear techniques. Again, the excellent properties of the ENO-SR technique in terms of approximation lead to the best results.

In Figure 4 we show a simple multivariate test, the reconstruction of the geometric figure considered before from a very coarse representation (right in Figure 1). The Gibbs-like oscillations typical of linear schemes in the presence of discontinuities lead to the blurring of the edges observed in Figure 4 (left). There is no blurring in the reconstructed image obtained with the nonlinear techniques. Again, the ENO-SR scheme leads, in this simple case, to the exact original image. One dimensional cuts of the reconstructed figures are displayed in Figure 5.

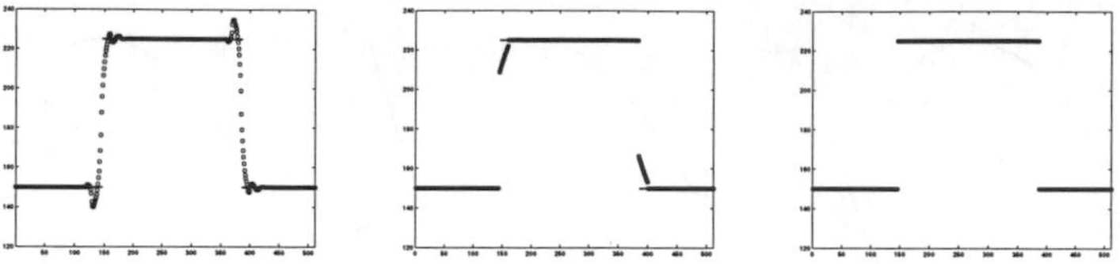

Fig. 5. Horizontal cuts linear, ENO, ENO-SR.

Acknowledgments. Supported in part by DGICYT PB97-1402 and in part by TMR project ERBFMRX-CT98-0184.

References

1. Amat, S., F. Aràndiga, A. Cohen, and R. Donat, Tensor product multiresolution analysis with error control, Technical Report GrAN-99-4, University of València, 1999.
2. Aràndiga, F., and R. Donat, A class of nonlinear multiscale decompositions, Technical Report GrAN-99-1, University of València, 1999. (Also submitted to Numerical Algorithms).
3. Harten, A., ENO schemes with subcell resolution, J. Comput. Phys. **83** (1989), 148–184.
4. Cohen, A., I. Daubechies, and J. C. Feauveau, Biorthogonal bases of compactly supported wavelets, Comm. Pure Applied Math. **45** (1992), 485–560.
5. Harten, A., Multiresolution representation of data II: General framework, SIAM J. Numer. Anal. **33** (1996), 1205–1256.
6. Harten, A., B. Engquist, S. Osher, and S. Chakravarthy, Uniformly high order accurate essentially non-oscillatory schemes III, J. Comput. Phys. **7** (1987), 231–303.

F. Aràndiga and R. Donat
Departament de Matemàtica Aplicada.
Universitat de València.
4610 Burjassot. València (Spain)
`donat@uv.es`
`arandiga@uv.es`

A Recursive Approach to the Construction of k-Balanced Biorthogonal Multifilters

Silvia Bacchelli, Mariantonia Cotronei,
and Damiana Lazzaro

Abstract. In this paper we discuss some numerical aspects of a particular construction of balanced biorthogonal multifilters by means of the lifting scheme. This construction allows, by simply solving linear equations, to obtain multifilters which do not need prefiltering, and for which the discrete versions of polynomial preservation/annihilation properties, are respectively, satisfied by their low and high-pass branches. In particular, we conduct experiments on how a parameter which appears in our recursive definition of balancing can be chosen to suitably influence the spectral behaviour of the multifilter low-pass branch, making it more effective in image compression problems.

§1. Introduction

Multiwavelets are a new addition to the classical scalar wavelet theory, and have been extensively studied in the last six years [5,9,11,17]. The main motivation for multiwavelets is that, unlike the scalar wavelet case, they can simultaneously possess desirable properties which are found to be useful for image compression applications, such as orthogonality and symmetry, short support, linear phase, a high approximation order, a high number of vanishing moments, etc. This combination would not be possible in any real-valued scalar wavelet. In fact, all real-valued scalar wavelets, with only one scaling function and one mother wavelet, can never possess all the above properties at the same time. This flexibility of vector-valued wavelet functions is due to the fact that multiwavelets satisfy conditions in which matrix rather than scalar coefficients are involved.

However, multiwavelets lack some attributes that scalar wavelets possess, and this becomes apparent when one implements the discrete multiwavelet transform. In particular, in the scalar case, a scaling low-pass filter with an approximation order k refers to the ability of the low-pass filter to reproduce

Curve and Surface Fitting: Saint-Malo 1999
Albert Cohen, Christophe Rabut, and Larry L. Schumaker (eds.), pp. 27–36.

ISBN 0-8265-1357-3.

discrete-time polynomials up to a degree $k-1$, while the corresponding wavelet high-pass filter annihilates discrete-time polynomials up to the same degree. This property, which is very important in many applications, does not hold in the multiwavelet case. In fact, the approximation power property does not assure the preservation and annihilation of discrete-time polynomials by the low-pass and the high-pass branch of a multiwavelet-based filter bank, respectively.

Moreover, because the approximation order for multiwavelets is not accompanied by the additional properties mentioned above, in applications using multiwavelets, a preprocessing or prefiltering step is necessary to obtain an efficient signal or image compression. A detailed investigation of prefiltering methods can be found in the literature [12,21,7].

Recently, to overcome these problems, Lebrun and Vetterli, and Selesnick [15,18] introduced the concept of balanced multiwavelets. They constructed orthogonal multiwavelet bases whose multifilter coefficients satisfy the discrete version of the approximation and zero-moments properties, and, at the same time, avoid the use of prefilters in implementing the discrete multiwavelet transform. This is a great advantage because the preprocessing step is a crucial point in multiwavelet-based algorithms. In fact, this initialization can sometimes destroy the very properties a multiwavelet basis is designed to have. Nevertheless, the above authors' construction of orthogonal balanced multifilters implies the resolution of non-linear equations that are solved by the Gröbner basis method.

Following the previous authors' idea, in order to avoid the difficulties due to the above-mentioned non-linearity, in [2] we have given a simple algebraic construction of k-balanced biorthogonal multifilters making use of the well-known tool called the lifting scheme. As shown in [19], the lifting scheme provides a simple method for constructing new biorthogonal filters with requested properties, starting from an assigned set of biorthogonal analysis-synthesis filters. In [2] we have extended the lifting scheme to the multifilter case, and in so doing, we have exploited the additional degrees of freedom left in the multifilter construction after satisfying the perfect reconstruction condition in order to easily construct finite k-balanced multifilters. Our results have been stated using the algebraic framework of banded block recursive matrices, exploiting this flexible mathematical tool to translate both the k-balancing conditions and other desirable properties in terms of simple linear conditions on the multifilter coefficients.

In this paper, we discuss some numerical aspects of the procedure for the construction of biorthogonal balanced multifilters given in [2], and analyze in particular the effect of the choice of the shift constant p which appears in our definition of k-balancing on the compression capabilities of this kind of filters. In fact, shift constant p plays an important role, and it can be used as a further degree of freedom.

Starting from Lazy multifilters, k-balanced multifilters of order 2 and 4 are constructed, and their effectiveness in image compression is tested on the Lena image. Using numerical experiments, we observe that the p parameter influences the shape of the Fourier transform of the scalar filters associated with the low-pass matrix coefficients, and we determine the value of p in such a way that the spectral behaviour of the newly constructed low-pass filters is as close as possible to the optimal shape. With this selection of p, we obtain the best compression results.

We remark that the aim of this paper is essentially to show the flexibility of our tool in building multifilters which do not need prefiltering and which are easily found by solving simple linear equations.

§2. Balanced Biorthogonal Multifilters

Let $\{\boldsymbol{\mathcal{H}} = \sum_i \boldsymbol{H}_i t^i, \boldsymbol{\mathcal{W}} = \sum_i \boldsymbol{W}_i t^i\}$ and $\{\widetilde{\boldsymbol{\mathcal{H}}} = \sum_i \widetilde{\boldsymbol{H}}_i t^i, \widetilde{\boldsymbol{\mathcal{W}}} = \sum_i \widetilde{\boldsymbol{W}}_i t^i\}$ be two pairs of block Laurent polynomials associated, respectively, with the analysis and the synthesis phase of a FIR multifilter bank, where $\{\boldsymbol{H}_i\}$, $\{\boldsymbol{W}_i\}$, $\{\widetilde{\boldsymbol{H}}_i\}$, $\{\widetilde{\boldsymbol{W}}_i\}$ are finite sequences of $r \times r$ matrices. In the following section we will refer to $\{\boldsymbol{\mathcal{H}}, \boldsymbol{\mathcal{W}}\}$, $\{\widetilde{\boldsymbol{\mathcal{H}}}, \widetilde{\boldsymbol{\mathcal{W}}}\}$ as analysis multifilters and synthesis multifilters, respectively, where we can think of them either as the sequences of matrix coefficients or as their associated block Laurent polynomials.

Let $R(t^2, \boldsymbol{\mathcal{H}}), R(t^2, \boldsymbol{\mathcal{W}}), R(t^2, \widetilde{\boldsymbol{\mathcal{H}}}), R(t^2, \widetilde{\boldsymbol{\mathcal{W}}})$ be the block banded Hurwitz matrices whose generating functions are $\boldsymbol{\mathcal{H}}, \boldsymbol{\mathcal{W}}, \widetilde{\boldsymbol{\mathcal{H}}}, \widetilde{\boldsymbol{\mathcal{W}}}$, respectively. With these matrices, we can give an algebraic description of the action of the analysis-synthesis system on a block Laurent polynomial σ given as input, in the following way:

Analysis:

$$[\boldsymbol{\sigma}^{(0)}] = R(t^2, \boldsymbol{\mathcal{H}})[\sigma]$$
$$[\boldsymbol{\sigma}^{(1)}] = R(t^2, \boldsymbol{\mathcal{W}})[\sigma]$$

Synthesis:

$$[\hat{\sigma}] = R(t^2, \widetilde{\boldsymbol{\mathcal{H}}})^T[\sigma^{(0)}] + R(t^2, \widetilde{\boldsymbol{\mathcal{W}}})^T[\sigma^{(1)}],$$

where $\boldsymbol{\sigma}^{(0)}$ and $\boldsymbol{\sigma}^{(1)}$ represent the output of the analysis phase, while $\hat{\boldsymbol{\sigma}}$ represents the output of the synthesis phase and therefore of the whole FIR system.

Given any pair of multifilters $\{\boldsymbol{\mathcal{H}}, \boldsymbol{\mathcal{W}}\}$, define the 2-decimated matrix

$$\Delta_{(\boldsymbol{\mathcal{H}}, \boldsymbol{\mathcal{W}})} = \begin{bmatrix} \boldsymbol{\mathcal{H}}_0 & \boldsymbol{\mathcal{H}}_1 \\ \boldsymbol{\mathcal{W}}_0 & \boldsymbol{\mathcal{W}}_1 \end{bmatrix},$$

whose elements are the 2-decimated block series related to $\boldsymbol{\mathcal{H}}, \boldsymbol{\mathcal{W}}$, that is

$$\boldsymbol{\mathcal{H}}_k = \sum_i \boldsymbol{H}_{2i+k} t^i, \ \boldsymbol{\mathcal{W}}_k = \sum_i \boldsymbol{W}_{2i+k} t^i, \quad k = 0, 1.$$

Definition 1. *We say that the pairs* $\{\boldsymbol{\mathcal{H}}, \boldsymbol{\mathcal{W}}\}$, $\{\widetilde{\boldsymbol{\mathcal{H}}}, \widetilde{\boldsymbol{\mathcal{W}}}\}$ *are* biorthogonal multifilters *or duals to each other or, equivalently, that they satisfy the* Perfect Reconstruction (PR) property *if*

$$\Delta^{T*}_{(\widetilde{\boldsymbol{\mathcal{H}}}, \widetilde{\boldsymbol{\mathcal{W}}})} \times \Delta_{(\boldsymbol{\mathcal{H}}, \boldsymbol{\mathcal{W}})} = \Delta_{(\boldsymbol{\mathcal{H}}, \boldsymbol{\mathcal{W}})} \times \Delta^{T*}_{(\widetilde{\boldsymbol{\mathcal{H}}}, \widetilde{\boldsymbol{\mathcal{W}}})} = I.$$

In this case, if furthermore $\mathcal{H}$ and $\widetilde{\mathcal{H}}$ admit a convergent subdivision scheme, then it is possible to define corresponding multiscaling functions and multiwavelets from the well-known matrix two-scale relations:

$$\Phi(x) = \sqrt{2}\, R(t^2, \mathcal{H})\, \Phi(2x), \quad \widetilde{\Phi}(x) = \sqrt{2}\, R(t^2, \widetilde{\mathcal{H}})\, \widetilde{\Phi}(2x),$$

$$\Psi(x) = \sqrt{2}\, R(t^2, \mathcal{W})\, \Phi(2x), \quad \widetilde{\Psi}(x) = \sqrt{2}\, R(t^2, \widetilde{\mathcal{W}})\, \widetilde{\Phi}(2x),$$

where $\Phi(x), \Psi(x), \widetilde{\Phi}(x), \widetilde{\Psi}(x)$ represent the vector containing the translates of the r-vectors $\boldsymbol{\phi} = [\phi_0, \ldots, \phi_{r-1}]^T$, $\boldsymbol{\psi} = [\psi_0, \ldots, \psi_{r-1}]^T$, $\tilde{\boldsymbol{\phi}} = [\tilde{\phi}_0, \ldots, \tilde{\phi}_{r-1}]^T$, $\tilde{\boldsymbol{\psi}} = [\tilde{\psi}_0, \ldots, \tilde{\psi}_{r-1}]^T$.

We now extend the concept of balancing order (introduced in [15]) to biorthogonal multifilters. We require that the multifilters associated with the analysis system must satisfy the discrete versions of both the polynomial preserving and zero moment properties.

Definition 2. *A pair of multifilters $\{\mathcal{H}, \mathcal{W}\}$ related to the analysis phase of a FIR system is said to be* balanced of order k *(or k-balanced), if there exists at least one real number p such that the following relations hold:*

$$\begin{aligned} R(t^2, \mathcal{H}) \times [\pi_n] &= \sqrt{2}\, 2^n [(\pi + p)_n], \\ R(t^2, \mathcal{W}) \times [\pi_n] &= 0, \end{aligned} \qquad n = 0, \ldots, k-1, \tag{1}$$

where $[\pi_n]$ and $[(\pi + p)_n]$ are bi-infinite column vectors which can also be seen as r-block vectors associated with the formal block series

$$\boldsymbol{\pi}^n = \sum_i \begin{bmatrix} (ri)^n \\ (ri+1)^n \\ \vdots \\ (ri+r-1)^n \end{bmatrix} t^i, \quad (\boldsymbol{\pi} + p)^n = \sum_i \begin{bmatrix} (ri+p)^n \\ (ri+1+p)^n \\ \vdots \\ (ri+r-1+p)^n \end{bmatrix} t^i.$$

In [2] an equivalent condition to (1) has been given which turns out to be more useful in practice:

Theorem 3. *A pair of FIR multifilters $\{\mathcal{H}, \mathcal{W}\}$ is balanced of order k if and only if*

$$\sum_{j=l_1}^{l_2} \boldsymbol{H}_j \begin{bmatrix} (rj)^n \\ (rj+1)^n \\ \vdots \\ (rj+r-1)^n \end{bmatrix} = \sqrt{2}\, 2^n \begin{bmatrix} p^n \\ (p+1)^n \\ \vdots \\ (p+r-1)^n \end{bmatrix}, \tag{2}$$

$$\sum_{j=m_1}^{m_2} \boldsymbol{W}_j \begin{bmatrix} (rj)^n \\ (rj+1)^n \\ \vdots \\ (rj+r-1)^n \end{bmatrix} = 0, \tag{3}$$

for $n = 0, \ldots, k-1$.

§3. Construction with the Lifting Scheme

The lifting scheme (introduced by Sweldens [19]) is a flexible tool for the construction of biorthogonal bases. In [2] an extension to the multifilter setting has been given. In short, given a set $\{\mathcal{H}, \widetilde{\mathcal{H}}, \mathcal{W}, \widetilde{\mathcal{W}}\}$ of biorthogonal multifilters, then the new multifilters

$$\begin{cases} \widetilde{\mathcal{H}}^{new} = \widetilde{\mathcal{H}} + (\boldsymbol{S} \circ t^2)\widetilde{\mathcal{W}} \\ \mathcal{W}^{new} = \mathcal{W} - (\boldsymbol{S}^{*T} \circ t^2)\mathcal{H}, \end{cases} \tag{4}$$

where $\boldsymbol{S}$ is any block Laurent polynomial, gives rise to a new set $\{\mathcal{H}, \widetilde{\mathcal{H}}^{new}, \mathcal{W}^{new}, \widetilde{\mathcal{W}}\}$ of biorthogonal multifilters.

Analogously, by simply changing the roles of the previous multifilters,

$$\begin{cases} \mathcal{H}^{new} = \mathcal{H} + (\widetilde{\boldsymbol{S}} \circ t^2)\mathcal{W} \\ \widetilde{\mathcal{W}}^{new} = \widetilde{\mathcal{W}} - (\widetilde{\boldsymbol{S}}^{*T} \circ t^2)\widetilde{\mathcal{H}} \end{cases} \tag{5}$$

gives rise to a new set $\{\mathcal{H}^{new}, \widetilde{\mathcal{H}}, \mathcal{W}, \widetilde{\mathcal{W}}^{new}\}$ of biorthogonal multifilters. We call (4) and (5) respectively the lifting scheme and the dual lifting scheme.

In [2] some useful conditions are given which allow the new multifilters to inherit symmetry/antisymmetry properties from the starting multifilters.

We can take advantage of the previous scheme to construct new balanced biorthogonal multifilters. In fact, unlike the orthogonal case where the balancing and the orthogonal conditions give rise to non-linear equations, which in [15,18], for example, are solved with a Gröbner basis approach, our balancing conditions (2) and (3) applied to the lifted (or dual lifted) multifilters give rise to linear conditions. The main steps of our approach are:

1) Construct the new low-pass multifilter coefficients, using the dual lifting scheme;
2) Apply the balancing condition (2), and solve the linear equations to find the coefficients of the unknown dual lifting matrix polynomial;
3) Construct the new high-pass multifilter coefficients, with the lifting scheme;
4) Apply the balancing condition (3), and solve the linear equations to find the coefficients of the unknown lifting polynomial;
5) Construct the corresponding dual low and high-pass multifilters.

It is important to note that in applying the balancing condition (2), a value must be assigned to the shift parameter p. In our experiments, it turns out that p influences on the effectiveness of the multifilters in their applications.

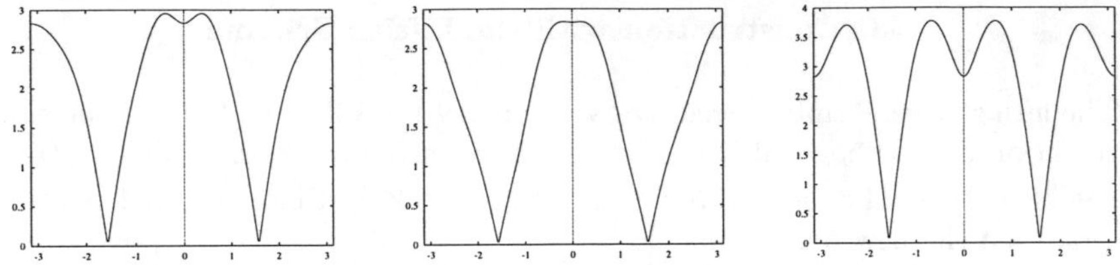

Fig. 1. $|\hat{h}^0| + |\hat{h}^1|$ associated to the Lazy 2-balanced low-pass multifilter with a varying p: from left to right, $p = -1, 0.25, 1$.

§4. Examples

In the following example, we start from Lazy multifilters to obtain balanced biorthogonal multifilters of order 2 and 4, which, furthermore, are of the type symmetric/antisymmetric (see [2]).

We restrict to the case $r = 2$, and define Lazy multifilters as follows:

$$2\mathcal{H} = \widetilde{\mathcal{H}} = \sqrt{2}\, I, \quad \mathcal{W} = 2\widetilde{\mathcal{W}} = \sqrt{2}\, I\, t.$$

In order to show the influence of p on the performance of the new multifilters, we introduce the notation

$$h^m_{2k+n} = [\boldsymbol{H}_k]_{m,n}, \quad k \in \mathbb{Z}, \; m, n = 0, 1,$$

which give the two low-pass scalar filters $h^0 = \{h^0_k\}_{k\in\mathbb{Z}}, h^1 = \{h^1_k\}_{k\in\mathbb{Z}}$ obtained by reorganizing the set of 2×2 low-pass matrix multifilter $\{\boldsymbol{H}_k\}_{k\in\mathbb{Z}}$, as a multichannel scalar filter bank.

As shown in the following figures, the shift constant p influences the shape of the Fourier transforms of $h^{0,new}, h^{1,new}$, making them more or less suitable for application problems.

In Figure 1, we show 3 graphs of the sum $|\hat{h}^{0,new}| + |\hat{h}^{1,new}|$, with p varying in $\{-1, 1/4, 1\}$. It can be seen that the choice $p = 1/4$ gives visually a better low-pass behaviour. In this case the new coefficients (except for a factor $\sqrt{2}$) are

$$\mathcal{H}^{new} = \begin{bmatrix} \frac{1}{4} & \frac{-1}{8} \\ \frac{-3}{8} & \frac{1}{4} \end{bmatrix} t^{-1} + \begin{bmatrix} \frac{1}{2} & 0 \\ 0 & \frac{1}{2} \end{bmatrix} + \begin{bmatrix} \frac{1}{4} & \frac{1}{8} \\ \frac{3}{8} & \frac{1}{4} \end{bmatrix} t,$$

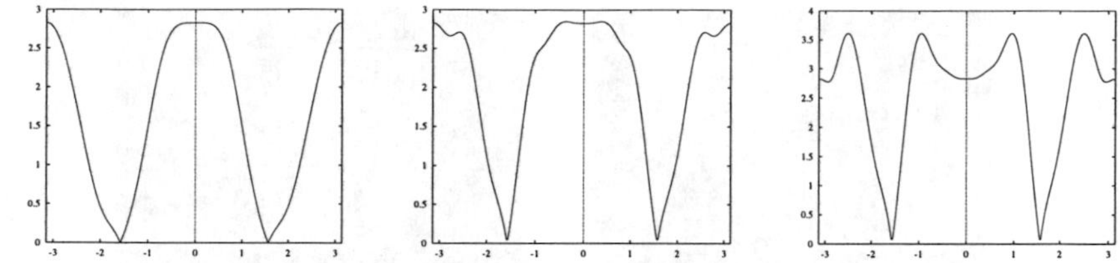

Fig. 2. $|\hat{h}^0| + |\hat{h}^1|$ associated to the Lazy 4-balanced low-pass multifilter with a varying p: from left to right, $p = 0, 0.4, 0.8$.

$$\mathcal{W}^{new} = \begin{bmatrix} -\frac{5}{64} & \frac{1}{32} \\ \frac{3}{32} & -\frac{5}{64} \end{bmatrix} t^{-1} + \begin{bmatrix} -\frac{1}{4} & -\frac{1}{16} \\ -\frac{3}{16} & -\frac{1}{4} \end{bmatrix} + \begin{bmatrix} \frac{21}{32} & 0 \\ 0 & \frac{21}{32} \end{bmatrix} t +$$

$$\begin{bmatrix} -\frac{1}{4} & \frac{1}{16} \\ \frac{3}{16} & -\frac{1}{4} \end{bmatrix} t^2 + \begin{bmatrix} -\frac{5}{64} & -\frac{1}{32} \\ -\frac{3}{32} & -\frac{5}{64} \end{bmatrix} t^3,$$

$$\widetilde{\mathcal{H}}^{new} = \begin{bmatrix} -\frac{5}{64} & -\frac{3}{32} \\ -\frac{1}{32} & -\frac{5}{64} \end{bmatrix} t^{-2} + \begin{bmatrix} \frac{1}{4} & -\frac{3}{16} \\ -\frac{1}{16} & \frac{1}{4} \end{bmatrix} t^{-1} + \begin{bmatrix} \frac{21}{32} & 0 \\ 0 & \frac{21}{32} \end{bmatrix} +$$

$$\begin{bmatrix} \frac{1}{4} & \frac{3}{16} \\ \frac{1}{16} & \frac{1}{4} \end{bmatrix} t + \begin{bmatrix} -\frac{5}{64} & \frac{3}{32} \\ \frac{1}{32} & -\frac{5}{64} \end{bmatrix} t^2,$$

$$\widetilde{\mathcal{W}}^{new} = \begin{bmatrix} -\frac{1}{4} & -\frac{3}{8} \\ -\frac{1}{8} & -\frac{1}{4} \end{bmatrix} + \begin{bmatrix} \frac{1}{2} & 0 \\ 0 & \frac{1}{2} \end{bmatrix} t + \begin{bmatrix} -\frac{1}{4} & \frac{3}{8} \\ \frac{1}{8} & -\frac{1}{4} \end{bmatrix} t^2.$$

A second example (Fig. 2) shows the behaviour of the scalar low-pass filters associated to the Lazy 4-balanced multifilters, with different choices for the parameter p. In this case, the choice $p = 0.4$ provides the best behaviour of the low-pass filters.

We have experimented with the above multifilters in an image compression example (on the Lena image), by making use of a multiwavelet-based embedded coding [6]. Results obtained with the best choices of 2 and 4-balanced Lazy multifilters are compared, at same compression ratio 1:16, with those produced by Chui-Lian (CL) [5] and Geronimo-Hardin-Massopust (GHM) [9] multifilters. CL and GHM multifilter both have approximation order 2, but need prefiltering. For comparison purposes, this prefiltering step has been

Fig. 3. Reconstructions of Lena compressed with different bases. From the top left corner: 2-balanced Lazy multifilter with $p = 0.25$; 4-balanced Lazy multifilter with $p = 0.4$; CL without prefiltering; GHM without prefiltering.

C.R.	2-bal. Lazy	4-bal. Lazy	CL without pref.	GHM without pref.
16	26.61	28.12	11.79	21.72

Tab. 1. PSNR values (in dB) with different multifilters.

omitted, in order to show how a prefiltering is absolutely necessary when dealing with non-balanced multifilters.

These results are shown in Table 1. Figure 3 shows the reconstruction of Lena compressed with the above-mentioned bases. It can be seen from the table and the figure that 2-balanced Lazy multifilters behave better than CL and GHM. Better results are of course achieved by the 4-balanced multifilters.

In the above experiments, we have not taken into account the orthonormal balanced multiwavelets of Lebrun-Vetterli [16] (which definitely give the best results, due to their good spectral properties), since our aim was not to construct the best possible filters, but to show the flexibility of our tool in building multifilters which do not need prefiltering and which are easily found by solving simple linear equations. One can obtain more effective filters with this procedure by extending the length of the lifting polynomials, and by using one of the many well-known good strategies for filter construction.

Acknowledgments. This work was supported by COFIN 97 project: "Analisi Numerica: Metodi e Software Matematico" and CNR project 98.01017.CT01.

References

1. Bacchelli, S., Block Toeplitz and Hurwitz matrices: a recursive approach, Advances in Appl. Math. **23** (1999), 199–210.
2. Bacchelli, S., M. Cotronei, and D. Lazzaro, An algebraic construction of k-balanced multiwavelets via the lifting scheme, submitted to Numer. Algorithms.
3. Bacchelli, S., and D. Lazzaro, Some practical applications of block recursive matrices, International Journal of Computers and Mathematics with Applications, to appear.
4. Barnabei, M., and L. Montefusco, Recursive properties of Toeplitz and Hurwitz matrices, Linear Algebra Appl. **274** (1998), 367–388.
5. Chui, C. K., and J. Lian, A study of orthonormal multi-wavelets, Appl. Numer. Math. **20** (1996), 273–298.
6. Cotronei, M., D. Lazzaro, L. B. Montefusco, and L. Puccio, Image compression through embedded multiwavelet transform coding, IEEE Trans. on Image Process, to appear.
7. Cotronei, M., L. B. Montefusco, and L. Puccio, Multiwavelet analysis and signal processing, IEEE Trans. on Circuits and Systems II: Analog and Digital Signal Processing **45** (1998), 970–987.
8. Daubechies, I., and W. Sweldens, Factoring wavelet transform into lifting steps, Tech. Report, Bell Labs, 1996.
9. Geronimo, J. S., D. P. Hardin, and P. R. Massopust, Fractal functions and wavelet expansions based on several scaling functions, J. Approx. Theory **78(3)** (1994), 373–401.
10. Goh, S. S., Q. Jiang, and T. Xia, Construction of biorthogonal multiwavelets using the lifting scheme, preprint.
11. Goodman, T. N. T., and S. L. Lee, Wavelets of multiplicity r, Trans. Amer. Math. Soc.. **342(1)** (1994), 307–324.
12. Hardin, D. P., and D. W. Roach, Multiwavelet prefilters I: orthogonal prefilters preserving approximation order $p \leq 2$, IEEE Trans. on Circuits and Systems II: Analog and Digital Signal Processing **45** (1998), 1106–1112.
13. Jiang, Q., On the construction of biorthogonal multiwavelet bases, preprint.
14. Lazzaro, D., Biorthogonal M-band filter construction using the lifting scheme, Numer. Algorithms, to appear.
15. Lebrun, J., and M. Vetterli J. Lebrun and M. Vetterli, Balanced multiwavelets: theory and design, IEEE Trans. on Signal Process. **46** (1998).

16. Lebrun, J., and M. Vetterli, High order balanced multiwavelets: theory, factorization and design, preprint.
17. Plonka, G., and V. Strela, Construction of multiscaling functions with approximation and symmetry, SIAM J. Math. Anal. (1995).
18. Selesnick, I., Multiwavelet bases with extra approximation properties, IEEE Trans. on Signal Process. **46** (1998), 2898–2908.
19. Sweldens, W., The lifting scheme: a custom-design construction of bi-orthogonal wavelets, Appl. Comput. Harmonic Anal. **3** (1996), 186–200.
20. Vaidyanathan, P. P., *Multirate Systems and Filter Banks*, Prentice-Hall, 1995.
21. Xia, X. G., J. S. Geronimo, D. P. Hardin, and B.W. Suter, Design of prefilters for discrete multiwavelet transforms, IEEE Trans. on Signal Process. 44, (1996), 25–35.

Silvia Bacchelli
Dipartimento di Matematica Pura e Applicata
Università di Padova
Via Belzoni, 7
Padova, Italy
`silvia@agamennone.csr.unibo.it`

Mariantonia Cotronei
Dipartimento di Matematica
Università di Messina
Salita Sperone, 31
Messina, Italy
`marianto@dipmat.unime.it`

Damiana Lazzaro
Dipartimento di Matematica
Università di Bologna
Piazza di Porta S. Donato, 5
Bologna, Italy
`lazzaro@csr.unibo.it`

Fast Evaluation of Radial Basis Functions: A Multivariate Momentary Evaluation Scheme

R. K. Beatson and E. Chacko

Abstract. This paper presents a scheme for fast evaluation of splines, or "radial" basis functions, of the form $s(\cdot) = p(\cdot) + \sum_{i=1}^{N} \lambda_i \Phi(\cdot - z_i)$. Here p is a low degree polynomial and $\Phi : \mathbb{R}^n \to \mathbb{R}$ is a function that need not be radial. This multivariate momentary evaluation scheme is a generalization of the fast multipole method in which calculations with far field expansions are replaced by calculations involving moments of the data. The primary advantage of this new algorithm is that it is highly adaptive to changes in Φ.

§1. Introduction

This paper presents a scheme for fast evaluation of splines, or "radial" basis functions, of the form

$$s(\cdot) = p(\cdot) + \sum_{i=1}^{N} \lambda_i \Phi(\cdot - z_i). \tag{1}$$

Here p is a low degree polynomial, and $\Phi : \mathbb{R}^n \to \mathbb{R}$ is a function that need not be radial. This multivariate momentary evaluation scheme is a generalization of the fast multipole method in which calculations with far field expansions are replaced by calculations involving moments of the data. The primary advantage of this new algorithm is that it is highly adaptive to changes in Φ. More precisely, changing to a new basic function Φ only requires coding a one or two line function for the (slow) evaluation of Φ. In contrast, adapting a conventional fast multipole code to a different Φ requires considerable analysis of appropriate expansions and transformation theorems, followed by writing a very specific code. The current algorithm reduces the incremental cost of a single extra evaluation from $\mathcal{O}(N)$ to $\mathcal{O}(1)$ operations, and the cost of a matrix-vector product (that is, evaluation at all centres) from $\mathcal{O}(N^2)$ to

Curve and Surface Fitting: Saint-Malo 1999
Albert Cohen, Christophe Rabut, and Larry L. Schumaker (eds.), pp. 37–46.

ISBN 0-8265-1357-3.

$\mathcal{O}(N \log N)$ operations. The algorithm can be viewed as a spline fitter in that the approximation it produces is a piecewise polynomial supplemented by appropriate direct evaluations. The method described is a multivariate generalisation of the method of [4].

In outline the setup phase of method is as follows. Firstly, space is divided in a hierarchical manner. For example, in a 2-D setting an initial square could be divided into a quadtree. Then centres are associated with the panels they lie in. Next, proceeding up the tree level by level, the moments of the coefficients (the λ_i's) about panel centres are calculated. Next working down the tree for each level, a number of approximations to Φ are formed. Then for each panel within a level, polynomial approximations to that part of s due to far away centres are formed by combining moments and the approximations to Φ. The evaluation phase first identifies the childless panel to which the evaluation point belongs. Then it approximates the far field by evaluating the polynomial associated with that panel, and adds to that approximation the directly calculated near field part of $s(x)$. For reasons of space, we will not detail suitable methods for subdividing space, or the process of evaluation. These matters are well understood in the context of the fast multipole method, see for example [2].

The paper is organized as follows. The necessary mathematics for forming polynomial approximations to s from moments and approximations to Φ, and for translating moments is given in Secton 3 below. Section 4 contains symmetry results that can substantially reduce the amount of work required to form approximations to Φ at each level. Section 5 contains numerical results obtained with a preliminary implementation of the method.

§2. Notation

We will need the following notation. A multi-index $\alpha = (\alpha_1, \ldots, \alpha_n)$ is an n-tuple of nonegative integers. If x is an element of $\mathbb{R}^n$, we will write its components as $x_1, x_2, \ldots, x_n$. We will also need sequences of points in $\mathbb{R}^n$. In an effort to make the meaning of all subscripted symbols transparent, we will write all such sequences of vectors as $\{z_m\}$, and z will never be used unsubscripted to denote a single point in $\mathbb{R}^n$.

If a and b are elements of $\mathbb{R}^n$, then we will say a is less than or equal to b, and write $a \le b$, if $a_i \le b_i$ for all $1 \le i \le n$. We also define for $x \in \mathbb{R}^n$ and α, β multi-indices $|\alpha| = \alpha_1 + \alpha_2 + \cdots + \alpha_n$, $\alpha! = \alpha_1!\alpha_2! \cdots \alpha_n!$, $x^\alpha = x_1^{\alpha_1} x_2^{\alpha_2} \cdots x_n^{\alpha_n}$, and for $0 \le \beta \le \alpha$ take $\binom{\alpha}{\beta} = \frac{\alpha!}{(\alpha-\beta)!\beta!}$. The (multivariate) Binomial Theorem then assumes the form

$$(x+t)^\alpha = \sum_{0 \le \beta \le \alpha} \binom{\alpha}{\beta} x^{\alpha-\beta} t^\beta,$$

for all multi-indices $0 \le \alpha \in \mathcal{Z}^n$ and points $x, t \in \mathbb{R}^n$. We further define the normalized monomial

$$V_\alpha(x) = x^\alpha / \alpha! \,, \tag{2}$$

and the α-th normalized moment about t of the data $\{(z_m, \kappa_m)\}_{m=1}^M$

$$\sigma_{t,\alpha} = \sum_{m=1}^{M} \kappa_m V_\alpha(t - z_m). \tag{3}$$

Also define $[a,b]$ to be the n dimensional box $\{x \in \mathbb{R}^n : a \leq x \leq b\}$, and e to be the n-vector $[1,1,\ldots,1]^T$.

§3. Moment Expansions

Lemma 1 below shows that we can form a polynomial approximation s_1 to a spline s of the form (1) by combining moments of the weights λ with the coefficients of a polynomial approximation q to Φ on a "double width" panel.

Lemma 1. Approximation via moments - correlations. *Let $c,\ d \in \mathbb{R}^n$ with $c,\ d > 0$. Let $t \in \mathbb{R}^n$ with $|t_i| > c_i + d_i$ for $1 \leq i \leq n$. Let $\epsilon > 0$ and Φ be a function in $C[t-(c+d),\ t+(c+d)]$. Let*

$$q(x) = \sum_{\{\alpha:0\leq\alpha\leq ke\}} a_\alpha V_\alpha(x-t)$$

be a polynomial of coordinate degree k such that

$$\|\Phi - q\|_{L^\infty[t-(c+d),t+(c+d)]} \leq \epsilon\ .$$

Given centres $z_1, z_2, \ldots z_M$ with $z_m \in [-d,d] \subset \mathbb{R}^n$ for $1 \leq m \leq M$, and weights $\kappa_1, \kappa_2, \ldots, \kappa_m \in \mathbb{R}$, let the corresponding "radial" basis function

$$s(x) = \sum_{m=1}^{M} \kappa_m \Phi(x - z_m), \tag{4}$$

be approximated by

$$s_1(x) = \sum_{m=1}^{M} \kappa_m q(x - z_m)\ . \tag{5}$$

Then

$$\|s - s_1\|_{L^\infty[t-c,t+c]} \leq \epsilon \|\kappa\|_1\ . \tag{6}$$

Moreover,

$$s_1(x) = \sum_{\{\beta:0\leq\beta\leq ke\}} b_\beta V_\beta(x-t), \tag{7}$$

where

$$b_\beta = \sum_{\{\alpha:\beta\leq\alpha\leq ke\}} a_\alpha \sigma_{0,\alpha-\beta}. \tag{8}$$

Remark: Often in our applications of this lemma, Φ will be truly radial, i.e. of the form $\Phi(\cdot) = \phi(\|\cdot\|_2)$, for some function ϕ of one variable.

Proof: If $-c \le x - t \le c$, then for $1 \le m \le M$ we have

$$-(c+d) \le (x - z_m) - t \le c + d.$$

This shows all but the expression for s_1 in terms of the moments of the data. To see the latter, write

$$\begin{aligned}
s_1(x) &= \sum_{m=1}^{M} \kappa_m q(x - z_m) \\
&= \sum_{m=1}^{M} \kappa_m \sum_{0\le\alpha\le ke} a_\alpha V_\alpha(x - z_m - t) \\
&= \sum_{m=1}^{M} \kappa_m \left\{ \sum_{0\le\alpha\le ke} \frac{a_\alpha}{\alpha!} \left((x-t) - z_m\right)^\alpha \right\} \\
&= \sum_{m=1}^{M} \kappa_m \left\{ \sum_{0\le\alpha\le ke} \frac{a_\alpha}{\alpha!} \sum_{0\le\beta\le\alpha} \binom{\alpha}{\beta} (x-t)^\beta (-z_m)^{\alpha-\beta} \right\} \\
&= \sum_{m=1}^{M} \kappa_m \left\{ \sum_{0\le\alpha\le ke} a_\alpha \sum_{0\le\beta\le\alpha} \frac{(x-t)^\beta}{\beta!} \frac{(-z_m)^{\alpha-\beta}}{(\alpha-\beta)!} \right\} \\
&= \sum_{0\le\alpha\le ke} a_\alpha \left\{ \sum_{0\le\beta\le\alpha} V_\beta(x-t)\sigma_{0,\alpha-\beta} \right\} \\
&= \sum_{0\le\beta\le ke} \left\{ \sum_{\beta\le\alpha\le ke} a_\alpha \sigma_{0,\alpha-\beta} \right\} V_\beta(x-t). \qquad \square
\end{aligned}$$

An efficient way to form the approximation s_1 is to use real FFTs to compute the correlation of equation (8). Nominally, such a process involves three FFTs per correlation. However, things may be arranged so that the transforms of moments of panels, and those of the required approximations to Φ, are computed once and used many times. Also sequences of coefficients can be summed in the transform domain rather than the function domain. This lowers the average number of FFTs per correlation dramatically. Thus, in 2 dimensions the work per correlation is reduced to approximately $\mathcal{O}(k^2) + bk^2 \log k$ operations, where b is small.

A proof similar to that of Lemma 1 shows the following total degree version of the approximation via moments lemma.

Lemma 2. *Let $c, d \in \mathbb{R}^n$ with $c, d > 0$. Let $t \in \mathbb{R}^n$ with $|t_i| > c_i + b_i$, $1 \leq i \leq n$. Let $\epsilon > 0$ and Φ be a function in $C[t-(c+d),\ t+(c+d)]$. Let*

$$q(x) = \sum_{\{\alpha:0\leq\alpha \text{ and } |\alpha|\leq k\}} a_\alpha V_\alpha(x-t),$$

be a polynomial of total degree k such that

$$\|\Phi - q\|_{L^\infty[t-(c+d),t+(c+d)]} \leq \epsilon \ .$$

Given centres $z_1, z_2, \ldots, z_M$ with $z_m \in [-d, d]$ for $1 \leq m \leq M$ and weights $\kappa_1, \kappa_2, \ldots, \kappa_M \in \mathbb{R}$, let the corresponding "radial" basis function

$$s(x) = \sum_{m=1}^{M} \kappa_m \Phi(x - z_m),$$

be approximated by

$$s_1(x) = \sum_{m=1}^{M} \kappa_m q(x - z_m) \ .$$

Then

$$\|s - s_1\|_{L^\infty[t-c,t+c]} \leq \epsilon \|\kappa\|_1 \ .$$

Moreover,

$$s_1(x) = \sum_{\{\beta:0\leq\beta,\ |\beta|\leq k\}} b_\beta V_\beta(x-t),$$

where

$$b_\beta = \sum_{\{\alpha:\beta\leq\alpha,|\alpha|\leq k\}} a_\alpha \sigma_{0,\alpha-\beta} \ .$$

The next lemma shows that shifted moments can be expressed as a convolution of moments about a given point. This result will be used in generating the moments corresponding to a larger panel of centres indirectly from the moments corresponding to subpanels. The indirect process will be more efficient than direct formation when the number of centres is large because the operation count for the indirect shift depends on the order of the moments, not on the number of centres.

Lemma 3. Indirect shifting of moments - convolutions. *Let $z_1, \ldots, z_M$ be given points in $\mathbb{R}^n$ and $\kappa_1, \ldots, \kappa_M$ be corresponding weights. Let $\sigma_{v,\alpha}$ be the α-th normalized moment of the data defined in equation (3). Then for all $v, u \in \mathbb{R}^n$ and multi-integers α*

$$\sigma_{v,\alpha} = \sum_{0\leq\beta\leq\alpha} V_\beta(v-u)\sigma_{u,\alpha-\beta} \ . \tag{9}$$

Proof:

$$\begin{aligned}\sigma_{v,\alpha} &= \frac{1}{\alpha!}\sum_{m=1}^{M}\kappa_m(v-z_m) = \frac{1}{\alpha!}\sum_{m=1}^{M}\kappa_m(v-u+u-z_m)^\alpha \\ &= \frac{1}{\alpha!}\sum_{m=1}^{M}\kappa_m\sum_{0\le\beta\le\alpha}\binom{\alpha}{\beta}(v-u)^\beta(u-z_m)^{\alpha-\beta} \\ &= \sum_{m=1}^{M}\kappa_m\sum_{0\le\beta\le\alpha}V_\beta(v-u)V_{\alpha-\beta}(u-z_m) \\ &= \sum_{0\le\beta\le\alpha}V_\beta(v-u)\sigma_{u,\alpha-\beta}. \quad \square\end{aligned}$$

In using Lemma 3 to translate moments we can reduce the operation count by using either FFT convolution or a tensor product approach.

We discuss firstly the tensor product approach. The tensor product approach may be viewed as making a shift from u to v not in a single step, but as a series of shifts in the coordinate directions. For simplicity we will discuss only the 2-dimensional case.

Consider the formula (9). If $v-u=(x,0)$, then we see immediately that $V_\beta(v-u)$ is nonzero only when the second component of β, β_2, is zero. Thus for v having the same second component as u,

$$\sigma_{v,\alpha} = \sum_{0\le\beta_1\le\alpha_1}V_{\beta_1}(v_1-u_1)\sigma_{u,\alpha-(\beta_1,0)}\ . \tag{10}$$

Considering $\sigma_{u,\gamma}$ as an array indexed by γ and the calculation of moments of degree not exceeding k, equation (10) above implies that each row of $\sigma_{v,\alpha}$ may be calculated in $\mathcal{O}(k^2)$ flops. Thus if v and u have the same second component all moments of degree not exceeding k can be shifted in $\mathcal{O}(k^3)$ flops. Similar remarks apply to $\{\sigma_{w,\alpha}\}$ and $\{\sigma_{v,\alpha}\}$ when w and v differ only in their second components. Thus, in the 2-dimensional case this tensor product strategy reduces the flop count for a single shift of all moments of degree not exceeding k from $\mathcal{O}(k^4)$ to $\mathcal{O}(k^3)$.

An alternative is to use FFT convolution to compute the transformation of the moment shifting lemma, Lemma 3. The corresponding operation count is $\mathcal{O}(k^2\log k)$ in the 2-dimensional case.

§4. Symmetry and Approximations to Φ

In this section we will show how symmetry considerations can greatly reduce the number of approximations to Φ that need to be computed. In the 2-dimensional situation, with a quad tree subdivision of space, and without clumping, there are 40 different geometries of source and target for each level. The method requires approximations to Φ on all the corresponding double rectangles. However, for most choices of Φ the number of approximations that need to be computed from scratch is reduced to 7 by the symmetry relations of Lemma 4 below. Related symmetry considerations for the fast multipole method are discussed in Wang and LeSar [7].

Lemma 4. Symmetries and approximations of "radial" functions. *Let $\mathcal{J}_1, \ldots, \mathcal{J}_n$, be subsets of $\mathbb{R}$ and $1 \le p \le \infty$, and let Φ and q be functions in $L^p(\mathcal{J}_1 \times \cdots \times \mathcal{J}_n)$.*

- *Suppose Φ is an even function of the k-th component of x, x_k. Define a function $\tilde{q}$ by*

$$\tilde{q}(x_1, \ldots, x_k, \ldots, x_n) = q(x_1, \ldots, -x_k, \ldots, x_n), \tag{11}$$

for all $(x_1, \ldots, x_n) \in \mathcal{J}_1 \times \ldots \times -\mathcal{J}_k \times \ldots \times \mathcal{J}_n$. Then

$$\|\Phi - \tilde{q}\|_{L^p(\mathcal{J}_1 \times \cdots \times -\mathcal{J}_k \times \cdots \times \mathcal{J}_n)} = \|\Phi - q\|_{L^p(\mathcal{J}_1 \times \cdots \times \mathcal{J}_k \times \cdots \times \mathcal{J}_n)}. \tag{12}$$

- *Suppose $\Phi(y)$ is unchanged by permutation of the components $y_1, \ldots, y_n$, of y. Let π be any permutation of the integers $\{1, 2, \ldots, n\}$, let π^{-1} be its inverse, and define $\tilde{q}$ by*

$$\tilde{q}(y_1, \ldots, y_n) = q(y_{\pi^{-1}(1)}, \ldots, y_{\pi^{-1}(n)}), \tag{13}$$

for all $y \in \mathcal{J}_{\pi(1)} \times \cdots \times \mathcal{J}_{\pi(n)}$. Then

$$\|\Phi - \tilde{q}\|_{L^p(\mathcal{J}_{\pi(1)} \times \cdots \times \mathcal{J}_{\pi(n)})} = \|\Phi - q\|_{L^p(\mathcal{J}_1 \times \cdots \times \mathcal{J}_n)}. \tag{14}$$

Proof: Let Φ, q and $\tilde{q}$ be as in the statement of the first part of the lemma. Let $x = (x_1, \cdots, x_k, \cdots x_n)$ be a point in $\mathcal{J}_1 \times \ldots \times -\mathcal{J}_k \times \ldots \times \mathcal{J}_n$. Using the evenness of Φ in the k-th component of x,

$$\begin{aligned}
|\Phi(x_1, \ldots, x_k, \ldots, x_n) &- \tilde{q}(x_1, \ldots, x_k, \ldots, x_n)| \\
&= |\Phi(x_1, \ldots, -x_k, \ldots, x_n) - q(x_1, \ldots, -x_k, \ldots, x_n)| \\
&= |\Phi(x_1, \ldots, u, \ldots, x_n) - q(x_1, \ldots, u, \ldots, x_n)| \ ,
\end{aligned}$$

where $(x_1, \cdots, u, \cdots, x_n)$ is a point in $\mathcal{J}_1 \times \cdots \times \mathcal{J}_n$. The first part of the lemma follows by using this equality in the appropriate p^{th} power integrals and essential supremum.

We turn now to the second part of the lemma. Let Φ, q, and $\tilde{q}$ be as in the statement of that part of the lemma. Let $y = (y_1, \ldots, y_n)$ be a point in $\mathcal{J}_{\pi(1)} \times \cdots \times \mathcal{J}_{\pi(n)}$. Using that $\Phi(y)$ is unchanged by permutations of the components of y, and defining $x = (x_1, \ldots, x_n) = (y_{\pi^{-1}(1)}, \ldots, y_{\pi^{-1}(n)})$,

$$\begin{aligned}
|\Phi(y_1, \ldots, y_n) - \tilde{q}(y_1, \ldots, y_n)| &= \left|\Phi(y_1, \ldots, y_n) - q(y_{\pi^{-1}(1)}, \ldots, y_{\pi^{-1}(n)})\right| \\
&= |\Phi(x_1, \ldots, x_n) - q(x_1, \ldots, x_n)| \,,
\end{aligned}$$

where $(x_1, \ldots, x_n)$ is a point in $\mathcal{J}_1 \times \cdots \times \mathcal{J}_n$. The result follows by using this equality in the appropriate p^{th} power integrals and essential supremum. □

Remark: Suppose $\Phi(\cdot) = \phi(\|\cdot\|_p)$ for some function of one variable ϕ and some p-norm for $\mathbb{R}^n$, $1 \le p \le \infty$. Then Φ is even in all the components of x,

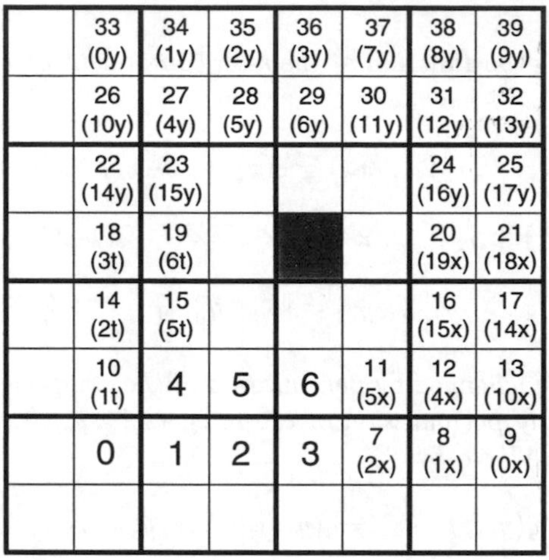

Fig. 1. Symmetry in the 2D setting.

and is also unchanged by permutations of the components of x. Hence, this lemma applies to all such *generalized radial* functions and allows us to use symmetry to obtain approximations to Φ on new regions from those on old.

Figure 1 shows the geometry of source panels to target panels in $\mathbb{R}^2$ when we use a quad tree subdivision of $\mathbb{R}^2$ without clumping. The solid black square is the target, or evaluation panel, and the possible source panels are numbered 0 through 39. Actually, the sources in the left-most column and bottom-most row would not be used for the illustrated position of the evaluation panel within its parent. However, they would be used for different positions of the target. If the source is $[-d, d]$ and the target is $[t-c, t+c]$, then Lemma 1 requires an approximation q to Φ on the "double width" rectangle $[t-(c+d), t+(c+d)]$. If Φ has all the symmetries of Lemma 4, then at each level only the seven approximations corresponding to source panels $0, \ldots, 6$ need be calculated directly. The 33 other approximations are easily obtained by symmetry. The relevant symmetries to use on a previously calculated approximation are indicated in parentheses in panels 7 to 39. For example, the notation $(2x)$ in source panel 7 indicates that the approximation for source panel 7 is obtained from that for source panel 2 by symmetry in x. In the function domain this corresponds to negating the coefficients of odd powers of x. In the Fourier domain it corresponds to a block rearrangement of columns or rows, depending on which correspond to x. Similarly, the notation $(1t)$ in source panel 10 indicates that the approximation for source panel 10 can be obtained from that for source panel 1 by symmetry in x and y. This corresponds to a transpose operation on the coefficients in both the function and Fourier domains.

§5. Numerical Results

Some numerical results from a primitive implementation of the algorithm are given in Tables 1–4. In this implementation the core tensor product polynomial approximations q to Φ, employed as in Lemma 1, are formed by interpolation at shifted and scaled Chebyshev nodes followed by economisation. The moments, and the coefficients of the approximations to Φ, typically have a wide dynamic range when the polynomial degree is 15 or more. Consequently, some extra device is needed in order to make the algorithm stable, especially when the FFT is used. In the current code the method used is a scaling of moments and polynomial coefficients analogous to that suggested by Greengard and Rokhlin [6] for the 2–D fast multipole method. The device may be viewed as scaling every panel at every level to be $[-2,2]^2$.

In the calculations reported, the centres are approximately uniformly distributed on $[0,1]^2$. If the number of centres is N, then in Table 1 the number of evaluation points is the smallest perfect square bigger than N, and in Table 2 the number of evaluation points is the smallest perfect square bigger than $10N$. In both cases $\Phi([x,y]) = \sqrt{c^2+x^2+y^2}$, where $c = 1/\sqrt{N}$, and all the coefficients κ_m of the spline (4) are 1. The piecewise tensor product bivariate polynomials used are of coordinate degree 7.

Tab. 1. Moment based method versus direct evaluation.

# of centres	Ord. Alg. time	FFT Alg. time	Direct time	Ratio	Abs. error	Rel. error
4000	0.41	0.27	3.57	13.16	3.99E-04	1.30E-07
8000	0.84	0.58	14.90	25.51	2.42E-03	3.95E-07
16000	1.78	1.23	59.98	48.82	1.71E-03	1.40E-07
32000	3.64	2.65	237.47	89.75	9.30E-03	3.79E-07

Tab. 2. Moment based method versus direct evaluation.

# of centres	Ord. Alg. time	FFT Alg. time	Direct time	Ratio	Abs. error	Rel. error
4000	1.37	0.97	35.14	36.28	5.49E-04	1.79E-07
8000	2.88	2.00	148.98	74.49	4.39E-03	7.16E-07
16000	5.87	4.17	593.40	142.23	2.07E-03	1.69E-07
32000	12.51	8.98	2356.92	262.33	1.82E-02	7.43E-07

The timings in the tables are in seconds on an Intel pentium based machine. Timings are given for direct evaluation and for the algorithm both with and without the speed benefits of FFT convolutions and correlations. Tables 3 and 4 repeat the runs with polynomials of coordinate degree 15.

Tab. 3. Moment based method versus direct evaluation.

# of centres	Ord. Alg. time	FFT Alg. time	Direct time	Ratio	Abs. error	Rel. error
4000	1.19	0.81	3.57	4.42	3.47E-08	1.13E-11
8000	2.97	1.70	14.90	8.74	1.74E-07	2.84E-11
16000	5.93	3.55	59.98	16.88	1.35E-07	1.10E-11
32000	13.19	7.74	237.54	30.69	6.17E-07	2.52E-11

Tab. 4. Moment based method versus direct evaluation.

# of centres	Ord. Alg. time	FFT Alg. time	Direct time	Ratio	Abs. error	Rel. error
4000	4.20	2.19	35.14	16.06	2.49E-08	8.11E-12
8000	9.32	4.51	148.37	32.89	1.09E-07	1.77E-11
16000	19.16	9.32	593.40	63.65	8.48E-08	6.91E-12
32000	39.70	19.52	2356.92	120.76	3.86E-07	1.57E-11

References

1. Beatson, R. K. and G. N. Newsam, Fast evaluation of radial basis functions I, Comput. Math. Applic., **24** (1992), 7–19.
2. Beatson, R. K. and L. L. Greengard, A short course on fast multipole methods, in *Wavelets, Multilevel Methods and Elliptic PDEs*, M. Ainsworth and J. Levesley and W. A. Light and M. Marletta (eds), Oxford University Press, 1997, 1–37.
3. Beatson, R. K. and W. A. Light, Fast evaluation of radial basis functions: methods for two-dimensional polyharmonic splines, IMA J. Numer. Anal., **17** (1997), 343–372.
4. Beatson, R. K. and G. N. Newsam, Fast evaluation of radial basis functions: Moment based methods, SIAM J. Sci. Comput., **19** (1998), 1428–1449.
5. Greengard, L. L. and V. Rokhlin, A fast algorithm for particle simulations, J. Comput. Phys., **73** (1987), 325–348.
6. Greengard, L. L. and V. Rokhlin, On the efficient implementation of the fast multipole algorithm, Research report 602, Yale University Department of Computer Science, 1988.
7. Wang, H. Y. and R. LeSar, An efficient fast multipole algorithm based on expansion in solid harmonics, J. Chem. Phys., **104** (1995), 4173–4179.

Department of Mathematics and Statistics,
University of Canterbury,
Private Bag 4800, Christchurch, New Zealand
{R.Beatson,E.Chacko}@math.canterbury.ac.nz

Polyharmonic Splines in $\mathbb{R}^d$: Tools for Fast Evaluation

R. K. Beatson, J. B. Cherrie, and David L. Ragozin

§1. Introduction

As is now well known, hierarchical and fast multipole-like methods can greatly reduce the storage and operation counts for fitting and evaluating radial basis functions. In particular, for spline functions of the form

$$s(x) = p(x) + \sum_{i=1}^{N} \lambda_i \phi(|x - x_i|), \tag{1}$$

p a low degree polynomial, the cost of a single extra evaluation can be reduced from $O(N)$ to $O(1)$ operations, and the cost of a matrix-vector product (that is, evaluation at all centers) can be decreased from $O(N^2)$ to $O(N)$.

This paper outlines some of the mathematics required to implement methods of these types for polyharmonic splines in $\mathbb{R}^d$, d even, that is for splines s corresponding to ϕ chosen from the list

$$\phi_\ell(r) = \begin{cases} r^{2(\ell+1-d/2)}, & \ell = 0, \dots, d/2 - 2, \\ r^{2(\ell+1-d/2)} \log(r), & \ell = d/2 - 1, \dots . \end{cases} \tag{2}$$

We carry out most of our work in the general $\mathbb{R}^d$ setting and then specialize to $d = 4$. We refer the reader to our more detailed work [1] which contains all the details of this special case. We are currently working on developing all the details for the general $\mathbb{R}^d$ case.

A key technique in our development is the exploitation of the rotation group invariance of radial basis functions. This means that we exploit the fact that any kernel $k(x, y) = \phi(|x - y|)$ will be rotation invariant in the sense that

$$k(gx, gy) = k(x, y), \qquad \text{for all orthogonal } g \in O(d). \tag{3}$$

Invariance leads to many crucial simplifications and efficiencies in developing and manipulating the polyharmonic expansions which lie at the heart of the

Curve and Surface Fitting: Saint-Malo 1999
Albert Cohen, Christophe Rabut, and Larry L. Schumaker (eds.), pp. 47–56.

ISBN 0-8265-1357-3.

hierarchical and fast multipole expansions. Related development of general spherical harmonic expansions based on these techniques can be found in [6, 7,8].

We will not detail the basic framework of hierarchical and fast multipole methods within which this mathematics sits. However, we do recall that an essential component of the method is the grouping of approximations to summands like (1) into subsums, which are approximations to the influence of that part of (1) associated with centers in a single panel or cluster. The key steps to obtain them requires:

- Finding explicit Taylor/Laurent *expansions*

$$\text{For each } x, \quad \phi(|x - x_<|) = \sum_{m \geq m_0} \widetilde{p_m}(x, x_<), \quad |x_<| < |x|, \tag{4}$$

  with $\widetilde{p_m}$ homogeneous polynomials of degree m in $x_<$.
- Finding an efficient *separation* of the $x, x_<$ influence in $\widetilde{p_m}$, *i.e.*, expanding

$$\widetilde{p_m}(x, x_<) = \sum_i f_i^m(x) g_i^m(x_<), \tag{5}$$

  for some good choice of (basis) functions $\{f_i^m(x)\}$ and $\{g_i^m(x)\}$.

These expansion and separation results provide the approximations to subsums which are the far and near field expansions. Other essential components are the tools to manipulate these expansions, namely error estimates, uniqueness theorems, and translation formulae. In this paper we concentrate on the algebraic tools and give some extensive general results on (4) (Theorem 6) and (5) ((20) in Section 3) and for $\mathbb{R}^4$ we give the appropriate far and near field expansions for the ϕ_ℓ (Theorems 7 and 8), and a brief indication of the dual basis leading to (20). Analogous results for polyharmonic splines in $\mathbb{R}^2$ appear in [3]. The reader unfamiliar with the framework of the fast multipole method may wish to refer to the original paper of Greengard and Rokhlin [4], or to the introductory short course [2].

§2. Polyharmonic Functions and Homogeneous Polynomials

First we record some detailed facts relating to the Laplacian Δ, and its actions on special homogeneous functions and the logarithm of the distance. In particular, we show why our basic functions ϕ_ℓ in (2) are polyharmonic or more specifically $(\ell + 1)$-harmonic in the sense that $\Delta^{\ell+1}\phi_\ell = 0$.

Lemma 1. *Let $|\cdot|$ be the 2-norm on $\mathbb{R}^d$, d even.*

i) *If $f : \mathbb{R}^d \setminus \{0\} \to \mathbb{R}$ is a non–trivial harmonic function that is homogeneous of integral degree m, then*

$$\Delta(|\cdot|^{2\ell} f) = 2\ell\,(d + 2\ell + 2m - 2)\,|\cdot|^{2(\ell-1)} f. \tag{6}$$

Hence $|\cdot|^{2\ell} f$ *is polyharmonic of exact order*

$$\begin{cases} \ell+1, & \text{for } \ell \geq 0, m > -\frac{d}{2} \text{ or } m < 1-\ell-\frac{d}{2}, \\ \ell+m+\frac{d}{2}, & \text{for } \ell < 0, m \geq 1-\ell-\frac{d}{2}. \end{cases}$$

In particular $|\cdot|^{2\ell}$ *is* $\begin{cases} (\ell+1)\text{-harmonic} & \text{for } \ell \geq 0, \\ (\ell+d/2)\text{-harmonic} & \text{for } -d/2 < \ell < 0. \end{cases}$

ii) $|\cdot|^{2\ell} \log|\cdot|$ *is* $(\ell+d/2)$*-harmonic for* $\ell \geq 0$. *More generally,*

$$\Delta|\cdot|^{2\ell}\log|\cdot| = |\cdot|^{2(\ell-1)}(2\ell(2\ell+d-2)\log|\cdot| + (4\ell+d-2)). \quad (7)$$

Proof: For the first part of (i), just apply the product rule for the Laplacian,

$$\Delta(fg) = (\Delta f)g + 2(\nabla f)\cdot(\nabla g) + f(\Delta g),$$

and the Euler relation for a function f that is homogeneous of degree m,

$$x \cdot (\nabla f)(x) = mf(x).$$

Then observe how many applications of Δ are required to reduce one of the multipliers to 0. Specializing to the case $f = 1$ yields the last result of (i).

The first part of (ii) follows from (7) in combination with (i) and its proof. (7) follows from (i), the product rule for the Laplacian, and the computation of $\nabla \log|\cdot|$ and $\Delta \log|\cdot|$. □

From the detailed eigenvalue-like information on the Laplacian map in (6), we can get a decomposition theorem for Π_n, the homogeneous polynomials of degree n, in terms of the spherical harmonics of degree n:

$$\mathbb{H}_n = \{p \in \Pi_n : \ \Delta p = 0\} = \ker(\Delta) \cap \Pi_n.$$

This is useful in understanding the structure of the homogeneous polynomial terms in any Taylor/Laurent type series expansions for the $\phi(|x - x_i|)$. In view of Lemma 1 the decomposition splits Π_n into its harmonic, biharmonic, triharmonic, etc., parts.

Lemma 2. $\Pi_n = \bigoplus_{\ell=0}^{\lfloor n/2 \rfloor} |\cdot|^{2\ell}\mathbb{H}_{n-2\ell}$. *In particular,* $\mathbb{H}_n \cap |\cdot|^2\Pi_{n-2} = \{0\}$.

Proof: Note that for $n = 0, 1$, $\Pi_n = \mathbb{H}_n$, so the base for an inductive proof is true. Assume the decomposition for $n-2$, some $n \geq 2$, so for each $p \in \Pi_n$,

$$\Delta p = \sum_{\ell=0}^{\lfloor (n-2)/2 \rfloor} |\cdot|^{2\ell} h_{n-2-2\ell}, \text{ some } h_i \in \mathbb{H}_{n-2-2\ell}.$$

Then by (6), if

$$h_n = p - \sum_{\ell=0}^{\lfloor (n-2)/2 \rfloor} \frac{1}{2}(\ell+1)^{-1}(d+2n-2(\ell+2))^{-1}|\cdot|^{2(\ell+1)} h_{n-2(\ell+1)},$$

then $\Delta h_n = 0$. So $h_n \in \mathbb{H}_n$ and the decomposition of Π_n is proved by induction. □

Some additional consequences of (6) come when we study what happens for negative ℓ. Here we have noted that $|\cdot|^{2-2m-d}f$ is harmonic whenever f is m-homogeneous and harmonic. But bringing the factor of $(|\cdot|^{-2})^m$ inside the m-homogeneous function f shows $|\cdot|^{2-d}f(\cdot/|\cdot|^2)$ is harmonic for any homogeneous harmonic function f. In fact this construction is independent of the homogeneity order of f. In general the Kelvin transform, defined by $Kf(x) = |x|^{2-d}f(x/|x|^2)$, which arises from inversion in the sphere followed by multiplication by $|x|^{2-d}$, maps harmonic functions to harmonic functions. On Π_n $Kf = |\cdot|^{2-d-2n}f$. Associated with the Kelvin transform are the spaces of negative degree $2-d-m$

$$\bigoplus_{\ell=0}^{\infty} |\cdot|^{2\ell} K\mathbb{H}_{m+2\ell} = \bigoplus_{\ell=0}^{\infty} |\cdot|^{2-d-2m-2\ell}\mathbb{H}_{m+2\ell}, \tag{8}$$

which are useful for analysing Laurent (far field) series. An application of Lemma 1 shows that Equation (8) displays the space under consideration split into its harmonic, biharmonic, triharmonic, etc., parts.

§3. Rotation Invariance and Simplified Taylor Expansions

The decompositions of polynomial spaces in the previous section already simplify the Taylor/Laurent type expansions (4) we need to determine. To make further progress, we want to exploit the *rotation invariance* of $\phi(|x-x_i|)$. When we come to combine subsums in (1), we will want to fix x and concentrate on rotations (orthogonal matrices) which fix x. When we are given a pole $p \in \mathbb{R}^d$, we let $G_p = \{g : g \in O(d),\ gp = p\}$ denote the rotations about the ray through p. So the function $f_\phi^x(x_<) = \phi(|x-x_<|)$ satisfies $f_\phi^x(gx_<) = f_\phi^x(x_<)$, for all $g \in G_x$. We refer to any function f which is unchanged by rotations in G_p as a p-zonal function. In particular we have the p-zonal harmonics

$$\mathbb{H}_n^p = \{h \in \mathbb{H}_n : h(gy) = h(y) \text{ for all } g \in G_p\}, \tag{9}$$

and the p-zonal homogeneous polynomials Π_n^p. Now the Taylor/Laurent expansion of f_ϕ^x as in (4), will have $\widetilde{p_m}(x, gx_<) = \widetilde{p_m}(gx, gx_<) = \widetilde{p_m}(x, x_<)$, for $g \in G_x$ since the homogeneous terms must remain unchanged under rotations (see Theorem 6). Thus these terms will be x-zonal polynomials as a function of $x_<$. What is the general stucture of Π_n^x and $\mathbb{H}_n^x$?

Theorem 3. *Fix a pole $x \in \mathbb{R}^d \backslash \{0\}$. Let $\chi_0^x(\cdot) = 1$, $\chi_1^x(\cdot) = 2\langle x, \cdot\rangle$. Then there exist a unique set of constants a_m, $m > 1$, such that the inductively defined sequence of homogeneous polynomials,*

$$\chi_{m+1}^x = \chi_1^x\chi_m^x - a_{m+1}|x|^2|\cdot|^2\chi_{m-1}^x, \qquad m \geq 1, \tag{10}$$

consists of harmonic functions. Moreover,

i) *χ_m^x is an m-homogeneous x-zonal harmonic function, which is rotation invariant in the sense that $\chi_m^{gx}(g\cdot) = \chi_m^x(\cdot)$ for all $g \in O(d)$.*

ii) *The constants* a_{m+1} *are independent of* x. *Hence*

$$\chi_m^x(x_<) = \chi_m^{x_<}(x), \qquad 0 \neq x, x_< \in \mathbb{R}^d, \tag{11}$$

and the $\chi_m^x(x_<)$ *are also homogeneous and harmonic as functions of* x.

iii) *If* $x \neq 0$, $\{|\cdot|^{2\ell}\chi_{m-2\ell}^x(\cdot),\ \ell = 0, \ldots, \lfloor m/2 \rfloor\}$, *form a basis for* Π_m^x. *In particular,* χ_j^x *is the unique (up to a scalar multiple) element of* $\mathbb{H}_j^x$.

iv) *For* m, n *nonnegative integers, the kernels*

$$\begin{aligned} k_{m,\ell,n}(x, x_<) = &|x|^{2\ell+m-\kappa}|x_<|^{2\ell+n-\kappa}\chi_{\kappa-2\ell}^x(x_<), \quad \kappa = \min(m,n), \\ &m \equiv n \bmod 2, \quad \ell = 0, \ldots, \lfloor \kappa/2 \rfloor, \end{aligned} \tag{12}$$

form a basis for the space of all rotation invariant polynomial kernels, $p_{m,n}(x, x_<)$ *which are homogeneous of degree* m *in* x, *and degree* n *in* $x_<$.

Proof: The proof of the existence of a_m and (i), (ii), and (iii) is by induction. For $m = 0$ (i) and (iii) are trivially true. Let $0 \neq h \in \mathbb{H}_1^x$. Then h has the form $h(\cdot) = c\langle p, \cdot\rangle$. Since h is x-zonal $c\langle p, \cdot\rangle = c\langle p, g\cdot\rangle = c\langle g^{-1}p, \cdot\rangle$, for all $g \in G_p^x$. This implies p has the same direction as x. Hence h is a multiple of χ_1^x and (i) and (iii) follow for $m = 1$.

Now induction shows that (10) defines m-homogeneous x-zonal polynomials which are m-homogeneous in x, for any choice of a_{m+1}. Also they are rotation invariant. To complete the inductive step for (i) with a fixed x we need only show that there is a unique a_{m+1} that makes $\chi_{m+1}^x(\cdot)$ harmonic.

From the homogeneity in x, we may assume $|x| = 1$. Since $\chi_1^x\chi_m^x$ is a homogenous polynomial of degree $m+1$, Lemma 2 asserts that there exist unique homogenous harmonic polynomials $q_{m+1-2\ell}$ such that $\chi_1^x\chi_m^x = \sum_{\ell=0}^{\lfloor (m+1)/2 \rfloor} |\cdot|^{2\ell} q_{m+1-2\ell}$. Since $\nabla\chi_1^x = 2x^T$, the product rule for the Laplacian and the inductive assumption that χ_m^x is harmonic show that

$$\Delta(\chi_1^x\chi_m^x) = 4x^T \cdot \nabla\chi_m^x = 4\,\partial_x\chi_m^x, \quad |x| = 1, \tag{13}$$

where ∂_x denotes the directional derivative in the (fixed) direction x. Since $\Delta\partial_x\chi_m^x = \partial_x\Delta\chi_m^x = 0$, it follows that $\chi_1^x\chi_m^x$ is bi-harmonic and

$$\Delta(\chi_1^x\chi_m^x) = \Delta(|\cdot|^2 q_{m-1}) = 2(d + 2(m-1))q_{m-1}.$$

Since Δ maps x-zonal functions to x-zonal functions it follows that $q_{m-1} \in \mathbb{H}_{m-1}^x$ and therefore by part (iii) of the inductive hypothesis q_{m-1} is a multiple of χ_{m-1}^x. Thus the existence and uniqueness of a_{m+1} making χ_{m+1}^x harmonic is proved.

We now turn to the inductive step in the proof of (ii). Using the rotation invariance part of the inductive hypothesis, $\chi_{m+1}^x(g^{-1}\cdot)$ is

$$\begin{aligned} \chi_1^x(g^{-1}\cdot)\chi_m^x(g^{-1}\cdot) - a_{m+1}|x|^2|g^{-1}\cdot|^2\chi_{m-1}^x(g^{-1}\cdot) = \\ \chi_1^{gx}\chi_m^{gx} - a_{m+1}|gx|^2|\cdot|^2\chi_{m-1}^{gx}. \end{aligned} \tag{14}$$

Since rotations and Δ commute, the left-hand side of the above is harmonic. Thus the right-hand side of (14) equals χ^{gx}_{m+1} and a_{m+1} is independent of g. Using homogeneity it is also independent of $|x|$. Hence a_{m+1} is independent of x. The symmetry in $x, x_<$ of (10) then implies (11), and hence the homogeneity and harmonicity of $\chi^x_{m+1}(x_<)$ as a function of x.

We now turn to the inductive step in the proof of (iii). Since $\Pi^x_m = \Pi^{x/|x|}_m$ and $\chi^{ax}_\kappa = a^\kappa \chi^x_\kappa$ we may assume $|x| = 1$. It suffices to show that $\dim \Pi^x_{m+1} \leq \lfloor m/2 \rfloor + 1$ since, by Lemma 1, $\{|\cdot|^{2\ell}\chi^x_{m+1-2\ell},\ \ell = 0, ..., \lfloor m/2 \rfloor\}$ is an independent set in Π^x_{m+1}. Since we can rotate e_1 to x by some orthogonal map, which will isomorphically map Π^x_{m+1} to $\Pi^{e_1}_{m+1}$, we prove our dimensionality statement for $x = e_1$. To analyze the value $f(y)$ of any e_1-zonal function f, we choose orthogonal $g_\pm \in G_{e_1}$ which transform y into the coordinate plane spanned by the first two basis vectors. The two possible values for the transformed y are

$$g_\pm y = \langle y, e_1 \rangle e_1 \pm \sqrt{|y|^2 - \langle y, e_1 \rangle^2}\, e_2.$$

Then $f(y) = f(g_\pm y) = f(y_1, \pm\sqrt{|y|^2 - y_1^2}, 0, \ldots, 0)$. In particular, f must be even in its second variable. If $f \in \Pi^{e_1}_{m+1}$,

$$f(y) = \sum_{\ell=0}^{\lfloor (m+1)/2 \rfloor} c_\ell y_1^{m+1-2\ell} \left(\sqrt{|y|^2 - y_1^2} \right)^{2\ell}, \qquad \text{for some } c_\ell.$$

Hence the functions $y_1^{m+1-2\ell}(|y|^2 - y_1^2)^\ell$ span $\Pi^{e_1}_{m+1}$.

For (iv) we just note that $p_{m,n}(x, \cdot) = |x|^m p_{m,n}(x/|x|, \cdot) \in \Pi^{x/|x|}_n$ by the homogeneity assumptions. Thus the basis facts from (iii) imply there are functions $b_{n,\ell}(x/|x|)$ with

$$p_{m,n}(x, \cdot) = \sum_{\ell=0}^{\lfloor n/2 \rfloor} b_{n,\ell}(x/|x|)|x|^m |\cdot|^{2\ell} \chi^{x/|x|}_{n-2\ell}.$$

The rotation invariance of $p_{m,n}$ and of the terms $|x|^m |\cdot|^{n-2\ell} \chi^{x/|x|}_{2\ell}$ implies $b_{n,\ell}(gx/|gx|) = b_{n,\ell}(x/|x|)$ for all rotations g. Rotating $x/|x|$ to e_1 shows $b_{n,\ell}(x/|x|) = b_{n,\ell}(e_1)$, *i.e.*, the $b_{n,\ell}$ are constants. Moreover, the homogeneity in x of χ^x_j shows

$$p_{m,n}(x, \cdot) = \sum_{\ell=0}^{\lfloor n/2 \rfloor} b_{n,\ell} |x|^{2\ell-(n-m)} |\cdot|^{2\ell} \chi^x_{n-2\ell}. \tag{15}$$

Since the left hand side is a polynomial of degree m in x, and the $|\cdot|^{2\ell}\chi^x_{n-2\ell}$ are independent, each $|x|^{2\ell-(n-m)}\chi^x_{n-2\ell}$ associated with a nonzero coefficient must be a polynomial of degree m in x. Hence, $n - m = 2j$ must be even. Also, applying the second part of Lemma 2, $2(\ell - j) = 2\ell - (n-m) \geq 0$. If $m \geq n$ the proof of (iv) is done. If $m < n$, then reindexing the sum in terms of $(\ell - j)$ yields (12). □

The following result is known [5], but is included for the sake of completeness.

Theorem 4. *Define a rotation invariant inner product (pairing) for functions on the unit ball $B \subset \mathbb{R}^d$ by*

$$[f,h] = c_V \int_{\{|y|\leq 1\}} f(y)h(y)dy, \quad c_V^{-1} = \text{vol}\{|y| \leq 1\}. \tag{16}$$

i) *If f, h are homogeneous of degrees m,n, respectively with $fh \in L^1(B)$, then $[f,h] = 0$ if and only if $\int_{\{|y|=1\}} f(y)h(y)dA = 0$, i.e., the integral of fh over the sphere S^{d-1} is zero.*

ii) *If $m \neq n$, then $|\cdot|^{2i}\mathbb{H}_n$ and $|\cdot|^{2j}\mathbb{H}_m$ are orthogonal with respect to this inner product(pairing), provided $m+n+2(i+j) > -d$.*

Proof: For (i) just introduce polar coordinates $(r,y) \in [0,1] \times S^{d-1}$. Then, by homogeneity and scaling properties of the area of $\{x : \ |x| = r\}$,

$$[f,h] = c_V \int_{r=0}^{1} r^{m+n+d-1} \int_{S^{d-1}} f(y)h(y)dA\,dr, \qquad m+n+d > 0,$$

and the result follows. By (i) it suffices to prove (ii) when $i = j = 0$, since integrals of a product $|\cdot|^{2(i+j)}fh$ on S^{d-1} do not depend on i,j. Let $f \in \mathbb{H}_m$ and $h \in \mathbb{H}_n$. Then by Green's Theorem and the Euler relation for homogeneous functions

$$\begin{aligned}0 &= \int_{\{|y|\leq 1\}} \Big(f(y)\Delta h(y) - h(y)\Delta f(y)\Big)dy \\ &= \int_{\{|y|=1\}} \Big(f(y)\nabla h(y) - h(y)\nabla f(y)\Big)\cdot \mathbf{n}dA = \int_{\{|y|=1\}} (n-m)f(y)h(y)dA.\end{aligned}$$

Thus $\int_{\{|y|=1\}} f(y)h(y)dA = 0$, and the analogous relation holds for integration over the ball by part (i). □

An application of the above gives

Lemma 5. *The constants a_m, $m \geq 2$, in the 3-term recurrence (10) defining the x-zonal harmonics χ_m^x of Theorem 3 are positive.*

Proof: By Theorem 4 and (10),

$$\begin{aligned}0 = [\chi_{m+1}^x, \chi_{m-1}^x] &= [\chi_1^x\chi_m^x, \chi_{m-1}^x] - a_{m+1}[|\cdot|^2\chi_{m-1}^x, \chi_{m-1}^x] \\ &= [\chi_m^x, \chi_1^x\chi_{m-1}^x] - a_{m+1}[|\cdot|^2\chi_{m-1}^x, \chi_{m-1}^x] \\ &= [\chi_m^x, \chi_m^x + a_m|\cdot|^2\chi_{m-2}^x] - a_{m+1}[|\cdot|^2\chi_{m-1}^x, \chi_{m-1}^x] \\ &= [\chi_m^x, \chi_m^x] - a_{m+1}[|\cdot|^2\chi_{m-1}^x, \chi_{m-1}^x].\end{aligned}$$

Hence, $a_{m+1} = [\chi_m^x, \chi_m^x]/[|\cdot|^2\chi_{m-1}^x, \chi_{m-1}^x] > 0$. □

Now part (iii) of Theorem 3 leads quite directly to the structure of near and far field expansions of general *rotation invariant* kernels $\psi(x, x_<)$. The heuristic that a far field expansion of $\psi(x, x_<)$ with respect to x can be found from a Taylor expansion with respect to $x_<$, has been known to us for some while. Theorem 6 below gives a proof that the underlying idea is correct inimportant special cases. In fact, we have the following result for such ψ which are jointly homogeneous ($\psi(ax, ax_<) = a^{2n}\psi(x, x_<)$) for some even integral power and are analytic about $x_< = 0$, such as

$$\psi_n(x, x_<) = |x - x_<|^{2n}(\log(|x - x_<|^2) - \log(|x|^2)). \tag{17}$$

Theorem 6. *Let $\psi(x, x_<)$, $x, x_< \in \mathbb{R}^d$ be rotation invariant, jointly homogeneous of degree $2n$ and analytic in $x, x_<$, for $|x_<| < |x|$. Then there exist constants $c^n_{m,\ell}$ such that the Taylor expansion of ψ about $x_< = 0$ has the form*

$$\psi(x, x_<) = \sum_{m=0}^{\infty} \sum_{\ell=0}^{\lfloor m/2 \rfloor} c^n_{m,\ell}|x|^{2(n+\ell-m)}|x_<|^{2\ell}\chi^x_{m-2\ell}(x_<) \tag{18}$$

$$= \sum_{m=0}^{\infty} \sum_{\ell=0}^{\lfloor m/2 \rfloor} c^n_{m,\ell}|x|^{2(n+\ell-m)}|x_<|^{2\ell}\chi^{x_<}_{m-2\ell}(x). \tag{19}$$

When ψ is $(k+1)$-harmonic in $x_<$, the upper limit on ℓ in (18) or (19) is $\min\{k, \lfloor m/2 \rfloor\}$. *If $\psi(x, 0) = 0$ then the lower limit on m in (18) or (19) is* 1.

Proof: The terms $\widetilde{p_m}(x, x_<)$ in (4), the Taylor series of $\psi(x, x_<)$ with respect to $x_<$, are degree m homogeneous polynomials in $x_<$. When any Taylor series is grouped by homogeneity with respect to $x_<$, each group is uniquely determined. Since only the term $\widetilde{p_m}(gx, gx_<)$ in the series for $\psi(gx, gx_<)$ has homogeneity m in $x_<$, the rotation invariance implies that $\widetilde{p_m}$ is also rotation invariant. Similarly the joint homogeneity of ψ yields $\widetilde{p_m}(ax, ax_<) = a^{2n}\widetilde{p_m}(x, x_<)$. Since for any $x, x_<$ there is a rotation g (or reflection if $d = 2$) which interchanges the rays through x and $x_<$, *i.e.*, $g(x/|x|) = (x_</|x_<|)$ and $g(x_</|x_<|) = (x/|x|)$, it follows that

$$|x|^{2(m-n)}\widetilde{p_m}(x, x_<) = |x|^{2(m-n)}\widetilde{p_m}(|x|x_</|x_<|, |x_<|x/|x|)$$
$$= \frac{|x|^{2m}}{|x_<|^{2n}}\widetilde{p_m}\left(x_<, \left(\frac{|x_<|}{|x|}\right)^2 x\right) = |x_<|^{2(m-n)}\widetilde{p_m}(x_<, x).$$

Since the final right side in this string of equalities is an m-homogeneous polynomial in x, we see that the terms in (4) have the form $|x|^{2(n-m)}p_m(x, x_<)$ with p_m a rotation invariant m-homogeneous polynomial in each of $x, x_<$. Hence (18) follows by Theorem 3.(iii). □

The *separation* properties in (5) can now be achieved by further use of rotation invariance. Each of the subspaces $|\cdot|^{2\ell}\mathbb{H}_j$, $j + 2\ell = n$, which occur in the decomposition of Π_n is rotation invariant. Hence it has a (unique) rotation invariant reproducing kernel

$$k(x, y) = |x|^{2\ell}|y|^{2\ell} \sum_{i=0}^{\dim \mathbf{H}_j} f_i(x)\tilde{f}_i(y), \tag{20}$$

where $\{|\cdot|^{2\ell} f_i\}$ and $\{|\cdot|^{2\ell}\tilde{f}_i\}$ are any bases for this subspace which are bi-orthogonally dual with respect to some rotation invariant inner product, *e.g.* the inner product (16) (see [8].) But by (12) in Theorem 3.(iii), since $k(x,y)$ is (exactly) $(\ell+1)$-harmonic as a function of y, and is homogeneous of degree $2\ell+j$ in both x, y,

$$k(x,y) = c_{j,\ell}|x|^{2\ell}|y|^{2\ell}\chi_j^x(y), \tag{21}$$

for some $c_{j,\ell} > 0$. Equating (20) and (21) provides separation of the influence of x, $x_<$ in the expression for $\chi_j^x(x_<)$. A consequence is separation of x, $x_<$ in the far and near field expansions given by Theorem 6, thus allowing the combination of the expansions for several centers $x_< = x_i, i = 1,\ldots,N$, into one expansion about 0.

§4. Expansions in $\mathbb{R}^4$

In this section we use the results of Section 3 in the $\mathbb{R}^4$ case to outline the explicit *expansion* formulae for the ϕ_ℓ in (2) for $d = 4$. We start with the far field expansion of the potential function $|x - x_<|^{-2}$.

Theorem 7. *For $x, x_< \in \mathbb{R}^4$ with $|x_<| < |x|$,*

$$|x - x_<|^{-2} = \sum_{m=0}^{\infty} |x|^{-2(m+1)} c_m \chi_m^{x_<}(x), \quad c_m = 1. \tag{22}$$

Proof: Since $|x - x_<|^{-2}$ is harmonic in $\mathbb{R}^4$, an expansion of this form holds for some constants c_m by Theorem 6. Using $\chi_m^{x_<}(x) = \chi_m^x(x_<)$, multiplication by $|x - x_<|^2 = |x|^2 + |x_<|^2 - \chi_1^x(x_<)$ yields

$$1 = \sum_{m=0}^{\infty} (c_m - c_{m-1})|x|^{-2m}\chi_m^x(x_<)$$
$$+ (c_{m-2} - c_{m-1}a_m)|x|^{-2(m-1)}|x_<|^2\chi_{m-2}^x(x_<),$$

when the recurrence (10) is used and the geometrically convergent expansion is rearranged to group terms of common homogeneity in $x_<$. Then equating coefficients using (iii) of Theorem 3 shows $c_0 = 1$, $c_m = c_{m-1}$ and $c_{m-2} = c_{m-1}a_m$. These must be consistent so $a_m = c_m = 1$ for all m. □

We now outline the expansion of $\psi_n(x, x_<)$ from (17). This gives us the bulk of the far field expansion for ϕ_{n+2}.

Theorem 8.

$$\psi_n(x, x_<) = \sum_{m=1}^{\infty} \sum_{\ell=0}^{\min\{n+1,\lfloor m/2\rfloor\}} c_{m,\ell}^n |x|^{2(\ell+n-m)}|x_<|^{2\ell}\chi_{m-2\ell}^{x_<}(x), \tag{23}$$

where the non-zero coefficients $c_{m,\ell}^n$ are given by the formulae $c_{m,\ell}^0 = -\frac{(-1)^\ell}{m}$, and the recurrence $c_{m,\ell}^{n+1} = c_{m,\ell}^n - c_{m-1,\ell}^n - c_{m-1,\ell-1}^n + c_{m-2,\ell-1}^n$.

Proof: The form of all the expansions follow from (18), since $\psi_n(x,0) = 0$. The explicit determination of the $c_{m,\ell}^0$, the $n = 0$ case, is done in Lemma 4.4

of [1]. The recurrence for the $c^n_{m,\ell}$ follows as in Theorem 7 from (10) with $a_m = 1$ upon multiplication of the ψ_n case by $|x - x_<|^2$. The details are in Lemma 4.6 of [1]. □

The explicit construction of bases for $\mathbb{H}_j$ (and dual bases) which are needed for the separation results can also be significantly simplified by use of the rotation invariance perspective. A detailed development in $\mathbb{R}^4$ is in our previously cited work.

Acknowledgments. The work of R.K. Beatson and J.B. Cherrie was partially supported by PGSF subcontract DRF601. D. Ragozin's work was supported by NSF grant DMS-9972004.

References

1. Beatson, R. K., J. B. Cherrie, and D. L. Ragozin, Fast evaluation of radial basis functions: methods for four-dimensional polyhamonic splines, preprint, 1999.
2. Beatson, R. K. and L. L. Greengard, A short course on fast multipole methods, in *Wavelets, Multilevel Methods and Elliptic PDEs,* M. Ainsworth, J. Levesley, W.A. Light, M. Marletta (eds), Oxford University Press, New York, 1997, 1–37.
3. Beatson, R. K. and W. A. Light, Fast evaluation of radial basis functions: methods for two-dimensional polyharmonic splines, IMA J. Numer. Anal. **17** (1997), 343–372.
4. Greengard, L. L. and V. Rokhlin, A fast algorithm for particle simulations, J. Comput. Phys. **73** (1987), 325–348.
5. Müller, C., *Spherical Harmonics,* Lecture Notes in Mathematics **17**, Springer Verlag, New York, 1966.
6. Ragozin, David L., Uniform convergence of spherical harmonic expansions, Math. Ann. **195** (1972), 87–94.
7. Ragozin, D. L. and J. Levesley, Zonal kernels, approximations, and positive definiteness on spheres and compact homogeneous spaces, in *Surface Fitting and Multiresolution Methods,* A. LeMéhauté, C. Rabut, and L. L. Schumaker (eds), Vanderbilt University Press, Nashville TN, 1997, 143–150.
8. Ragozin, D. L. and J. Levesley, The density of translates of zonal functions on compact homogeneous spaces, J. Approx. Theory **103** (2000), 252–268.

R. K. Beatson and J. B. Cherrie
Dept. of Mathematics and Statistics,
Private Bag 4800,
Christchurch, New Zealand
R.Beatson@math.canterbury.ac.nz
J.Cherrie@math.canterbury.ac.nz

David L. Ragozin
University of Washington, Box 354350
Department of Mathematics,
Seattle, WA 98195-4350 USA
rag@math.washington.edu

Constructive Approximation by (V,f)-Reproducing Kernels

Mohammed-Najib Benbourhim

Abstract. In this paper we propose a constructive method to build reproducing kernels. We define the notion of (V, f)-reproducing kernel, and prove that every reproducing kernel is a (V, f)-reproducing kernel. We study the minimal approximation by these (V, f)-reproducing kernels for different choices of V and f. Examples to which our results apply include curve and surface fitting.

§1. (V,f)-Reproducing Kernels

For any set (respectively locally compact set) Ω, we denote by $\mathbb{R}^\Omega$ (respectively $\mathcal{C}^m(\Omega)$) the space of real-valued functions (respectively m-times continuously differentiable functions) defined on Ω equipped with the topology of pointwise convergence (respectively uniform convergence on the compact subsets of Ω). Let us recall some definitions.

Definition 1.1. *A real-valued function H defined on $\Omega \times \Omega$ is a* reproducing kernel *on $\Omega \times \Omega$ if*

1) *H is symmetric: $H(t,s) = H(s,t)$ for all t, $s \in \Omega$,*
2) *H is of positive type:*

$$\sum_{k,l=1}^{k,l=N} \lambda_k \lambda_l H(t_k, t_l) \geq 0,$$

for any finite point set $\{t_k\}_{k=1}^N$ of Ω and real numbers $\{\lambda_k\}_{k=1}^N$.

Definition 1.2. *A vector subspace $\mathcal{H}$ of $\mathbb{R}^\Omega$ is said to be a* hilbertian subspace *of $\mathbb{R}^\Omega$ (respectively $\mathcal{C}^m(\Omega)$) if*

1) *$\mathcal{H}$ is a Hilbert space,*
2) *The natural injection from $\mathcal{H}$ into $\mathbb{R}^\Omega$ (respectively $\mathcal{C}^m(\Omega)$) is continuous.*

We review some important results on reproducing kernels which are studied in [4].

Curve and Surface Fitting: Saint-Malo 1999
Albert Cohen, Christophe Rabut, and Larry L. Schumaker (eds.), pp. 57–64.

ISBN 0-8265-1357-3.

Theorem 1.1.

1) *A Hilbert space $\mathcal{H}$ (respectively a real-valued function H defined on $\Omega\times\Omega$) is a hilbertian subspace of $\mathbb{R}^\Omega$ (respectively a reproducing kernel on $\Omega\times\Omega$) if and only if there exists one and only one reproducing kernel H on $\Omega\times\Omega$ (respectively hilbertian subspace $\mathcal{H}$ of $\mathbb{R}^\Omega$) such that*

$$u(t) = \langle u \mid H(\cdot, t)\rangle_{\mathcal{H}}, \qquad \forall t \in \Omega, \quad \forall u \in \mathcal{H}.$$

$\mathcal{H}$ is called the hilbertian subspace *associated with H.*

2) *For any hilbertian basis $(f_i)_{i\in I}$ of $\mathcal{H}$: $H(t,s) = \sum_{i\in I} f_i(t)f_i(s)$.*
3) *If H is separately m-times continuously differentiable, then $\mathcal{H}$ is a hilbertian subspace of $\mathcal{C}^m(\Omega)$.*
4) *The vector space $\mathcal{H}_0$ =span$\{(H(\cdot,t))_{t\in\Omega}\}$ is dense in $\mathcal{H}$.*

Let $(V, \langle\cdot \mid \cdot\rangle_V)$ be a Hilbert space, Ω be a set and f be a function from Ω into V.

Definition 1.3. *For all $f : \Omega \longrightarrow V$, we define a (V, f)-*reproducing kernel *H_f by*

$$H_f(t,s) = \langle f(t) \mid f(s)\rangle_V, \qquad \forall (t,s) \in \Omega\times\Omega. \tag{1.1}$$

We have the following result:

Theorem 1.2. *H_f defined by (1.1) is a reproducing kernel on $\Omega\times\Omega$ and its associated hilbertian subspace $\mathcal{H}_f$ of $\mathbb{R}^\Omega$ is*

$$\mathcal{H}_f = \Big\{w \in \mathbb{R}^\Omega \mid \exists u \in V: \ w(t) = \langle u \mid f(t)\rangle_V, \ \forall t \in \Omega\Big\}.$$

Proof: One can easily verify that H_f is a reproducing kernel.
Let $\tilde{H}_f : V \longrightarrow \mathbb{R}^\Omega$ be defined by $(\tilde{H}_f u)(t) = \langle u \mid f(t)\rangle_V$. The mapping $\tilde{H}_f$ is linear, and the inequality

$$\mid (\tilde{H}_f u)(t) \mid \le \mid u \mid_V \mid f(t) \mid_V = \mid u \mid_V \ H_f(t,t)^{\frac{1}{2}},$$

for all $t \in \Omega$ and for all $u \in V$ implies that it is continuous. Let $\mathcal{M}$ be the closure in V of the vector space span$\{(f(t))_{t\in\Omega}\}$, and $P_{\mathcal{M}}$ the orthogonal projector on $\mathcal{M}$. We define on $\mathcal{H}_f = \tilde{H}_f(V)$ the bilinear form

$$\langle \tilde{H}_f u \mid \tilde{H}_f v\rangle_{\mathcal{H}_f} = \langle P_{\mathcal{M}} u \mid P_{\mathcal{M}} v\rangle_V.$$

It is easy to see that this form is a scalar product on $\mathcal{H}_f$. Then the linear mapping $\tilde{H}_f : \mathcal{M} \longrightarrow \mathcal{H}_f$ is an isometry, and consequently $(\mathcal{H}_f, \langle\cdot \mid \cdot\rangle_{\mathcal{H}_f})$ is a Hilbert space. For all $t \in \Omega$, the function

$$H_f(t,\cdot) : s \in \Omega \longrightarrow H_f(t,s) = \langle f(t) \mid f(s) \rangle_V,$$

is an element of $\mathcal{H}_f$, and satisfies the reproducing formula

$$(\tilde{H}_f u)(t) = \langle \tilde{H}_f u \mid H_f(t,\cdot) \rangle_{\mathcal{H}_f}, \qquad \forall u \in V.$$

Consequently (see Theorem 1.1), $\mathcal{H}_f$ is a hilbertian subspace of $\mathbb{R}^\Omega$ and admits H_f as reproducing kernel. □

Theorem 1.3. *Let ω be a set and c a mapping from ω to Ω. Then*

$$(Hc)_f(y,z) = H_f(c(y),c(z)) = \langle f(c(y)) \mid f(c(z)) \rangle_V$$

is a reproducing kernel on $\omega \times \omega$.

Proof: For all $y \in \omega$, $f(c(y))$ is in V. The function $(Hc)_f$ is symmetric and is of positive type:

$$\sum_{k,l=1}^{k,l=N} \lambda_k \lambda_l (Hc)_f(y_k, y_l) = \langle \sum_{k=1}^{k=N} \lambda_k f(c(y_k)) \mid \sum_{k=1}^{k=N} \lambda_k f(c(y_k)) \rangle_V \ \geq 0. \ \square$$

Example 1.1. Let $V = L^2(a,b)$, Ω a subset of $\mathbb{R}$ and $f(t)(x) = \exp(cxt)$ where c is a real constant. Then

$$H_f(t,s) = \begin{cases} (\exp(cb(t+s)) - \exp(ca(t+s)))/(c(t+s)), & \text{if } (t+s) \neq 0, \\ b-a, & \text{otherwise.} \end{cases}$$

Example 1.2. Let $V = L^2(\mathbb{R}^+)$, and suppose Ω is a subset of $\mathbb{R}^n$.

For all functions $c : \Omega \longrightarrow (0,+\infty)$, we have

(i) If $f(t)(x) = \frac{2}{\pi^{\frac{1}{4}}} \exp^{-c(t)|x|^2}$, then $H_f(t,s) = \dfrac{1}{\sqrt{c(t)+c(s)}}$.

(ii) If $f(t)(x) = \exp^{-c(t)x}$, then $H_f(t,s) = \dfrac{1}{c(t)+c(s)}$, and in particular if

$c(t) = \dfrac{P(t)}{Q(t)}$ (with $P(t)$ and $Q(t)$ polynomials), we obtain the rational reproducing kernel

$$H_f(t,s) = \frac{Q(t)Q(s)}{P(t)Q(s) + P(s)Q(t)}.$$

§2. (V,f)-Reproducing Kernels of Convolution Type

We consider the case where

1) $V = L^2(\mathbb{R}^n)$ and $\Omega = \mathbb{R}^n$.

2) $f(t)(x) = f(t-x)$, with f in the familar Sobolev space $H^m(\mathbb{R}^n)$.

Then $H_f(t,s) = \displaystyle\int_{\mathbb{R}^n} f(t-x) f(s-x) dx.$

Theorem 2.1. *We have the following properties:*

1) $H_f(t,s) = h_f(t-s)$ *with* $h_f(\xi) = (f * \check{f})(\xi) = \mathcal{F}(\mid \mathcal{F}f \mid^2)(\xi)$, *where* $\check{f}(x) = f(-x)$ *and* $\mathcal{F}f$ *is the Fourier transform of* f.
2) $h_f \in \mathcal{C}_0^m(\mathbb{R}^n) = \left\{u \in \mathcal{C}^m(\mathbb{R}^n) \mid \lim_{|t|\to\infty} (D^\alpha u)(t) = 0, \quad 0 \leq \mid \alpha \mid \leq m\right\}$.
3) *The associated hilbertian subspace of* H_f *is*

$$\mathcal{H}_f = f * L^2(\mathbb{R}^n) \quad \hookrightarrow \mathcal{C}_0^m(\mathbb{R}^n) \text{ (continuous embedding)}.$$

4) *In particular, if* $\mid \mathcal{F}f \mid > 0$, *then*

$$\mathcal{H}_f = \left\{w \in \mathcal{S}' \mid \mathcal{F}w \in L^1_{loc}(\mathbb{R}^n), \quad \frac{\mathcal{F}w}{\mathcal{F}f} \in L^2(\mathbb{R}^n)\right\},$$

equipped with the scalar product

$$\langle w_1 \mid w_2\rangle_{\mathcal{H}_f} = \int_{\mathbb{R}^n} \frac{\mathcal{F}w_1(\xi)\,\overline{\mathcal{F}w_2(\xi)}}{\mid \mathcal{F}f(\xi) \mid^2} d\xi.$$

5) If f *is radial, then* h_f *is radial:* $H_f(t,s) = h_f(\mid t-s \mid)$.
6) For all distinct points $\{t_k\}_{k=1}^N$ *in* $\mathbb{R}^n$, *the matrix* $H_N = \big(H_f(t_k,t_l)\big)_{1\leq k,l\leq N}$ *is invertible (strictly positive definite).*

Proof:
1) We have

$$H_f(t,s) = (2\pi)^{-\frac{n}{2}} \int_{\mathbb{R}^n} e^{-i\langle t-s|\xi\rangle} \mid \mathcal{F}f(\xi)|^2 d\xi = \mathcal{F}(\mid \mathcal{F}f \mid^2)(t-s).$$

2) $f \in H^m(\mathbb{R}^n) \Rightarrow D^\alpha h_f = (D^\alpha f) * \check{f} \in \mathcal{C}_0^0(\mathbb{R}^n)$ for $0 \leq \mid \alpha \mid \leq m$, (see [2]).
3) is a consequence of Theorem 1.2 and the property given in 1).
4) Since $\mathcal{H}_f \hookrightarrow \mathcal{C}_0^m(\mathbb{R}^n) \hookrightarrow \mathcal{S}'$, we have the equivalences:

$$\begin{aligned}\{w \in \mathcal{H}_f\} &\Leftrightarrow \{\exists u \in L^2(\mathbb{R}^n) : \mathcal{F}w = \mathcal{F}u\mathcal{F}f\} \\ &\Leftrightarrow \{\mathcal{F}w \in L^1_{loc}(\mathbb{R}^n), \ \frac{\mathcal{F}w}{\mathcal{F}f} \in L^2(\mathbb{R}^n)\}.\end{aligned}$$

From Theorem 1.2, we have

$$w(t) = \langle u \mid f(t-\cdot)\rangle_{L^2(\mathbb{R}^n)} = \langle P_{\mathcal{M}}u \mid f(t-\cdot)\rangle_{L^2(\mathbb{R}^n)} = (P_{\mathcal{M}}u * f)(t).$$

Then $\mathcal{F}w = \mathcal{F}P_{\mathcal{M}}\, u\mathcal{F}f$, and

$$\begin{aligned}\langle w_1 \mid w_2\rangle_{\mathcal{H}_f} &= \int_{\mathbb{R}^n} P_{\mathcal{M}}u_1(x) P_{\mathcal{M}}u_2(x) dx \\ &= \int_{\mathbb{R}^n} \mathcal{F}P_{\mathcal{M}}u_1(\xi)\, \overline{\mathcal{F}P_{\mathcal{M}}u_2(\xi)} d\xi = \int_{\mathbb{R}^n} \frac{\mathcal{F}w_1(\xi)\,\overline{\mathcal{F}w_2(\xi)}}{\mid \mathcal{F}f(\xi) \mid^2} d\xi.\end{aligned}$$

5) For any orthogonal matrix A,

$$h_f(At) = \int_{\mathbb{R}^n} f(x)f(At-x)dx = \int_{\mathbb{R}^n} f(Ax)f(A(t-x))dx$$
$$= \int_{\mathbb{R}^n} f(x)f(t-x)dx = h_f(t),$$

since $f(Ax) = f(x)$ and $\mid detA \mid = 1$.

6) We suppose that $f \not\equiv 0$ in $L^2(\mathbb{R}^n)$. Since the matrix $H_N = \Big(\int_{\mathbb{R}^n} f(t_k - x)f(t_l - x)dx\Big)_{1\le k,l\le N}$ is a Gram matrix, it is invertible if and only if the system $\{f(t_k - \cdot)\}_{k=1}^{N}$ is lineary independent in $L^2(\mathbb{R}^n)$. If $\sum_{k=1}^{k=N} c_k f(t_k - x) = 0$ in $L^2(\mathbb{R}^n)$, for $c_k \in \mathbb{R}$, $1 \le k \le N$, then by the Fourier transform we get $\big(\sum_{k=1}^{k=N} c_k e^{-i\langle t_k|\xi\rangle}\big)\mathcal{F}f(\xi) = 0$ in $L^2(\mathbb{R}^n)$.

The Lebesgue measure of the set $\mathcal{N} = \{\xi \in \mathbb{R}^n \mid \sum_{k=1}^{k=N} c_k e^{-i\langle t_k|\xi\rangle} = 0\}$ is equal to zero. Then $\mathcal{F}f$ vanishes outside $\mathcal{N}$, i.e: $\mathcal{F}f \equiv 0$ in $L^2(\mathbb{R}^n)$ and by the inverse Fourier transform, $f \equiv 0$ in $L^2(\mathbb{R}^n)$, which complete the proof. □

Example 2.1. Let $u(x) = (1- \mid x \mid)_+$ and $\mathcal{F}u = v$. We have $v(x) = \dfrac{sin^2(\frac{x}{2})}{x^2}$.

(i) Taking $f = \mathcal{F}(\mid u \mid^{\frac{1}{2}})$, $H_f(t,s) = (1- \mid t-s \mid)_+$.

(ii) Taking $f = \mathcal{F}(\mid v \mid^{\frac{1}{2}})$, $H_f(t,s) = \dfrac{sin^2(\frac{t-s}{2})}{(t-s)^2}$.

Example 2.2. (Bessel reproducing kernels) For $n \in \mathbb{N}$, $\alpha \in \mathbb{R}$ and $\alpha > n$; consider $G_\alpha \in L^2(\mathbb{R}^n)$ defined by $\mathcal{F}(G_\alpha)(x) = (1+ \mid x \mid^2)^{-\frac{\alpha}{2}}$.

(i) Taking $f = \mathcal{F}(\mid G_\alpha \mid^{\frac{1}{2}})$, $H_f(t,s) = (1+ \mid t-s \mid^2)^{-\frac{\alpha}{2}}$.

(ii) Taking $f = \mathcal{F}(\mid \mathcal{F}G_{n+1} \mid^{\frac{1}{2}})$, $H_f(t,s) = \frac{\pi^{\frac{1-n}{2}}}{2^n\Gamma(\frac{n+1}{2})} \exp(-\mid t-s \mid)$ and $\mathcal{H}_f = H^{\frac{n+1}{2}}(\mathbb{R}^n)$ (Sobolev space).

Example 2.3. (ν-B-spline reproducing kernels) Let

1) $Y_l(x) = \frac{1}{l!}x_+^l$.
2) $\nu \in \mathcal{E}'$ (distributions with compact support) such that $\nu(p) = 0$, for all polynomial p in $\mathcal{P}_l(\mathbb{R})$.
3) $f = \nu * Y_l$.

For such functions f, we give the following theorem without proof.

Theorem 2.2. *We have:*

1) $f \in L^2(\mathbb{R})$.
2) *For all* u *in* $V^{l+1}(\mathbb{R}) = \{v \in L^2_{loc}(\mathbb{R}) / \ v^{(l+1)} \in L^2(\mathbb{R})\}$ *(Beppo-Levi space) we have* $\int_{\mathbb{R}} u^{(l+1)}(x) f(t-x)dx = (\nu * u)(t)$.
3) $H_f(t,s) = (-1)^l(\check{\nu} * \nu * Y_{2l+1})(t-s)$ *and* $\mathcal{H}_f = \nu * V^{l+1}(\mathbb{R})$.

In the particular case of divided differences, ν is defined as the mth-iterated convolution $\nu = \frac{(\delta_a - \delta_b)^{*m}}{(b-a)}$, and $\check{\nu} = \frac{(\delta_{-a} - \delta_{-b})^{*m}}{(b-a)}$.

§3. Data Fitting by (V,f)-Reproducing Kernels

Let $\{t_k\}_{k=1}^N$ a set of distinct points in Ω, and define a linear operator A_N from $\mathcal{H}_f$ into $\mathbb{R}^N$ by $A_N(u) = \big(u(t_k)\big)_{1 \le k \le N}$.

Definition 3.1. *For all* $z_N \in \mathbb{R}^N$ *and* $\epsilon \in [0,1)$ *we define a spline to be any solution of the following minimal approximation problem:*

$$\Big(P_\epsilon(z_N)\Big): \qquad \inf_{u \in \mathcal{C}_\epsilon} \Big((1-\epsilon)\langle u \mid u\rangle_{\mathcal{H}_f} + \epsilon \|A_N u - z_N\|^2_{\mathbb{R}^N}\Big),$$

where

$$C_\epsilon = \begin{cases} A_N^{-1}\{z_N\}, & \text{if } \epsilon = 0 \text{ (Interpolation)}, \\ \mathcal{H}_f, & \text{if } \epsilon \in]0,1[\text{ (Smoothing)}. \end{cases}$$

The following theorem gives the spline in the case $\epsilon \neq 0$.

Theorem 3.1. *For all* $(\epsilon, z_N) \in]0,1[\times \mathbb{R}^N$, *the problem* $P_\epsilon(z_N)$ *(Smoothing) admits a unique solution*

$$\sigma^\epsilon(t) = \sum_{k=1}^{k=N} \lambda_k^\epsilon H_f(t, t_k),$$

where the coefficients ${}^t\Lambda^\epsilon = (\lambda_1^\epsilon, \ldots, \lambda_N^\epsilon) \in \mathbb{R}^N$ *are the solution of the system*

$$(H_N + \frac{\epsilon}{1-\epsilon} I_N)\Lambda_N^\epsilon = z_N,$$

with $H_N = (H_f(t_k, t_l))_{1 \le k,l \le N}$ *and* I_N *is the identity matrix.*

Proof: 1) From the continuous embedding: $\mathcal{H}_f \hookrightarrow \mathbb{R}^\Omega$ (see Theorem 1.2), we deduce that A_N is continuous. 2) $A_N(\mathcal{H}_f)$ is closed as a vector subspace of $\mathbb{R}^N$. Then from the general spline theory (see [1,3]) we get the theorem. □

Theorem 3.2. *The following two properties are equivalent:*
1) For all $z_N \in \mathbb{R}^N$, the problem $P_0(z_N)$ admits a unique solution.
2) The system $\{f(t_k)\}_{k=1}^N$ is linearly independent in V.

Proof: For all $z_N \in \mathbb{R}^N$, the problem $P_0(z_N)$ admits a unique solution if and only if the matrix $H_N = \big(H_f(t_k, t_l)\big)_{1\le k,l\le N}$ is invertible (see [1,3]). Since the matrix $H_N = \big(\langle f(t_k) \mid f(t_l)\rangle_V\big)_{1\le k,l\le N}$ is a Gram matrix, it is invertible if and only if the system $\{f(t_k)\}_{k=1}^N$ is lineary independent in V. □

Furthermore, for the particular case $V = L^2(\mathbb{R}^n)$ and $f \in H^m(\mathbb{R}^n)$, Theorem 3.2 and the property (6) of Theorem 2.1 imply the following theorem.

Theorem 3.3. *For all $f \in H^m(\mathbb{R}^n)$ and $z_N \in \mathbb{R}^N$, the problem $P_0(z_N)$ (Interpolation) admits a unique solution*

$$\sigma^0(t) = \sum_{k=1}^{k=N} \lambda_k^0 H_f(t, t_k),$$

where the coefficients ${}^t\Lambda^0 = (\lambda_1^0, \dots, \lambda_N^0) \in \mathbb{R}^N$ are the solution of the system

$$H_N \Lambda_N^0 = z_N,$$

with $H_N = (H_f(t_k, t_l))_{1\le k,l\le N}$ and I_N is the identity matrix.

§4. Data Fitting Preserving Polynomials

Let $\mathcal{P}_d(\mathbb{R}^n)$ the vector space of polynomials of degree at most d. We suppose:

(H1) For all $p \in \mathcal{P}_d(\mathbb{R}^n)$ the subset $\{t_k\}_{k=1}^N$ of $\mathbb{R}^n$ is such that

$$\{p(t_k) = 0, \quad 1 \le k \le N\} \Longleftrightarrow p \equiv 0.$$

(H2) $\mathcal{H}_f \cap \mathcal{P}_d(\mathbb{R}^n) = \{0\}$.

We remark that in the case $V = L^2(\mathbb{R}^n)$ and $f \in H^m(\mathbb{R}^n)$, the hypothesis (H2) is satisfied because $\mathcal{H}_f \subset \mathcal{C}_0^m(\mathbb{R}^n)$ (see Theorem 2.1(2)), and

$$\mathcal{C}_0^m(\mathbb{R}^n) \cap \mathcal{P}_d(\mathbb{R}^n) = \{0\}.$$

Let $\mathcal{H}_f^d$ be the Hilbert direct sum: $\mathcal{H}_f^d = \mathcal{H}_f \oplus \mathcal{P}_d(\mathbb{R}^n)$. We denote by Π_f the orthogonal projector from $\mathcal{H}_f^d$ onto $\mathcal{H}_f$, and we define on $\mathcal{H}_f^d$ the linear mapping $A_N(u) = \big(u(t_k)\big)_{1\le k\le N} \in \mathbb{R}^N$. For all $(\epsilon, z_N) \in [0,1] \times \mathbb{R}^N$, we consider the following minimal approximation problem in $\mathcal{H}_f^d$:

$$\Big(P_\epsilon(z_N)\Big) : \qquad \inf_{u\in C_\epsilon} \Big((1-\epsilon)\langle \Pi_f(u) \mid \Pi_f(u)\rangle_{\mathcal{H}_f} + \epsilon \|A_N u - z_N\|^2_{\mathbb{R}^N}\Big),$$

where

$$C_\epsilon = \begin{cases} A_N^{-1}\{z_N\}, & \text{if } \epsilon = 0 \text{ (Interpolation)},\\ \mathcal{H}_f^d, & \text{if } \epsilon \in]0,1[\text{ (Smoothing)},\\ \mathcal{P}_d(\mathbb{R}^n), & \text{if } \epsilon = 1 \text{ (Least squares)}.\end{cases}$$

The hypothesis (H1) implies that the problem $P_1(z_N)$ admits a unique solution. In the case $\epsilon \in]0,1[$ (Smoothing) we have the following theorem:

Theorem 4.1. *For all* $(\epsilon, z_N) \in]0,1[\times\mathbb{R}^N$, *the problem* $P_\epsilon(z_N)$ *admits a unique solution* σ^ϵ. *In the case* $\epsilon \in]0,1[$, *the solution* σ^ϵ *is given by*

$$\sigma^\epsilon(t) = \sum_{k=1}^{k=N} \lambda_k^\epsilon H_f(t,t_k) + \sum_{i=1}^{i=n_d} b_i^\epsilon p_i(t),$$

where the coefficients ${}^t\Lambda^\epsilon = (\lambda_1^\epsilon, \dots, \lambda_N^\epsilon) \in \mathbb{R}^N$ *and* ${}^tB^\epsilon = (b_1^\epsilon, \dots, b_{n_d}^\epsilon) \in \mathbb{R}^{n_d}$ *are the solution of the system*

$$\begin{pmatrix} H_N + \frac{\epsilon-1}{\epsilon} I_N & E \\ {}^tE & 0 \end{pmatrix} \begin{pmatrix} \Lambda^\epsilon \\ B^\epsilon \end{pmatrix} = \begin{pmatrix} z_N \\ 0 \end{pmatrix},$$

with

1) $H_N = \left(H_f(t_k,t_l)\right)_{1\le k,l\le N}$ *and* I_N *is the identity matrix,*
2) $E = \left(E_{k,i}\right)_{1\le i\le n_d}^{1\le k\le N}$ *with* $E_{k,i} = p_i(t_k)$ *and* $(p_i)_{1\le i\le n_d}$ *is a basis of* $\mathcal{P}_d(\mathbb{R}^n)$.

In particular, if there exists $p \in \mathcal{P}_d(\mathbb{R}^n)$ *such that* $\{p(t_k) = z_{N,k}, 1 \le k \le N\}$, *then* $\sigma^\epsilon = p$ *(preserving polynomials property).*

Proof: Theorem 4.1 is a consequence of general spline theory (see [1,3]):

1) A_N is continuous since $\mathcal{H}_f^d$ is a hilbertian subspace of $\mathbb{R}^\Omega$.
2) Π_f is continuous and $\Pi_f(\mathcal{H}_f^d) = \mathcal{H}_f$ is closed since Π_f is an orthogonal projector.
3) $\ker A_N \cap \ker \Pi_f = \{0\}$: derives from the hypothesis (H1) and the fact that $\ker \Pi_f = \mathcal{P}_d(\mathbb{R}^n)$.
4) $\ker A_N + \ker \Pi_f$ is closed since $\ker \Pi_f$ is a finite dimensional vector space.

□

References

1. Attéia, M., *Hilbertian Kernels and Splines Functions*, Elsevier Science, North Holland, 1992.
2. Donughue, W. F., *Distributions and Fourier Transforms*, Academic Press, 1969.
3. Laurent, P.-J., *Approximation et Optimisation*, Hermann, Paris, 1972.
4. Schwartz, L., Sous-espaces hilbertiens d'espaces vectoriels topologiques et noyaux associés, J. Analyse Math. **13** (1964), 115–256.

Benbourhim Mohammed-Najib
Université Paul Sabatier (M.I.P)
118, Route de Narbonne, 31062 Toulouse, Cedex 04, France
`bbourhim@cict.fr`

Adaptive Wavelet Galerkin Methods on Distorted Domains: Setup of the Algebraic System

Stefano Berrone and Karsten Urban

Abstract. We use the algorithm of Bertoluzza, Canuto and Urban [2] for computing integrals of products (of derivatives) of wavelets in order to solve elliptic PDEs on 2D distorted domains. We construct a variant of the original method which turns out to be more efficient. Several numerical results are presented.

§1. Introduction

Adaptive wavelet Galerkin schemes have quite recently been proven to offer great potential for numerically solving boundary value problems for partial differential equations. On the one hand, strong analytical properties such as convergence and optimal efficiency have been proven for elliptic operators [6,9]. On the other hand, first numerical tests also on non–tensor product domains indicate the applicability of such methods, [1].

However, the major obstacle so far is the efficient computation of the entries of the stiffness matrix and the right–hand side of the corresponding algebraic systems. In fact, it turns out that these entries are more expensive to compute than, e.g., in the case of adaptive Finite Element Methods. In [2], a method to adaptively approximate and compute these entries was introduced and analyzed; numerical results were given for a 1D example. In this paper, we study the application of the algorithm in [2] for 2D 'distorted' domains, which are parametric images of the unit square. This allows the study of the influence of 'realistic' parametrizations of non–tensor product domains on the assembling of the algebraic system. We incorporate some improvements over the original method in [2] to increase efficiency, and present various numerical results.

Curve and Surface Fitting: Saint-Malo 1999
Albert Cohen, Christophe Rabut, and Larry L. Schumaker (eds.), pp. 65–74.

ISBN 0-8265-1357-3.

§2. Adaptive Approximation of the Algebraic System

Given a linear boundary value problem in a bounded Lipschitz domain $\Omega \subset \mathbb{R}^n$ ($n \geq 1$), its numerical approximation by a variational method (Galerkin, Petrov–Galerkin, weighted residuals, ...) requires the computation of integrals of the form

$$\int_\Omega b_{\alpha,\beta}(x)\, D^\alpha u(x)\, D^\beta v(x)\, dx \qquad \text{or} \qquad \int_\Omega f_\beta(x)\, D^\beta v(x)\, dx, \tag{1}$$

where $\alpha, \beta \in \mathbb{N}^n$ are suitable multi–indices, $D^\alpha = \dfrac{\partial^{\|\alpha\|}}{\partial x_1^{\alpha_1} \cdots \partial x_n^{\alpha_n}}$ with $\|\alpha\| := \alpha_1 + \cdots + \alpha_n$, u and v are suitable trial and test functions belonging to $H^{\|\alpha\|}(\Omega)$ and $H^{\|\beta\|}(\Omega)$ respectively, $b_{\alpha,\beta} \in L^\infty(\Omega)$ and $f_\beta \in L^2(\Omega)$.

We consider a wavelet Galerkin method with trial and test spaces $S_\Lambda = \operatorname{span} \Psi_\Lambda$ generated by adaptively choosing a finite subset $\Psi_\Lambda = \{\psi_\lambda : \lambda \in \Lambda\}$ within a wavelet basis $\Psi = \{\psi_\lambda : \lambda \in \mathcal{J}\}$ in $L^2(\Omega)$, i.e., $\Lambda \subset \mathcal{J}$ (see, e.g., [5,10]). The wavelets are assumed to have the appropriate regularity for the above integrals to be well defined.

The construction of such wavelet bases on fairly general domains Ω is not a trivial task. However, quite recently significant progress has been made on this topic, see [3,4,8,12] and also [13] for a somewhat different approach. The main idea behind all the constructions in the first cited papers is domain decomposition and matching. The domain Ω is subdivided into N non–overlapping subdomains Ω_i. Each subdomain is mapped to the n–dimensional reference cube $\hat{\Omega} := [0,1]^n$ by means of smooth parametric mappings

$$F_i : \hat{\Omega} \to \bar{\Omega}_i, \quad \bar{\Omega}_i = F_i(\hat{\Omega}), \quad G_i := F_i^{-1}. \tag{2}$$

Then, each ψ_λ, $\lambda \in \mathcal{J}$, restricted to Ω_i is the image through F_i of a linear combination of tensor product wavelets $\hat{\psi}_\lambda$ on $\hat{\Omega}$, i.e., if $\lambda = (j,k)$ ($j =: |\lambda|$ denoting the level and k the location in space as well as the type of wavelet), then

$$\psi_\lambda(x)_{|\Omega_i} = \sum_{\hat{\lambda}' \in S(i,\lambda)} \gamma_{\hat{\lambda}',i}\, \hat{\psi}_{\hat{\lambda}'}(G_i(x)), \quad x \in \Omega_i,$$

where $S(i,\lambda)$ is a suitable set of indices of the form $\hat{\lambda}' = (j, \hat{k}')$ with $\hat{k}' \in \hat{\Omega}$ and $\gamma_{\hat{\lambda}',i}$ are suitable coefficients independent of j (see e.g. [3,4]).

2.1. Reduction to univariate integrals

We will only consider the calculation of the integral on the left–hand side of (1) which enters into the stiffness matrix. The entries for the right–hand side are treated analogously, [2]. Hence, replacing u by ψ_λ and v by ψ_μ on the left–hand side in (1) for some $\lambda, \mu \in \Lambda$, we get

$$a_{\lambda,\mu} := \int_\Omega b_{\alpha,\beta}(x)\, D^\alpha \psi_\lambda(x)\, D^\beta \psi_\mu(x)\, dx = \sum_{i=1}^N \sum_{\substack{\alpha' \in T(i,\alpha),\\ \beta' \in T(i,\beta)}} \sum_{\substack{\hat{\lambda}' \in S(i,\lambda),\\ \hat{\mu}' \in S(i,\mu)}} \gamma_{\hat{\lambda}',i}\, \gamma_{\hat{\mu}',i}$$

$$\times \int_{\hat{\Omega}} b_{\alpha,\beta}(F_i(\hat{x}))\, d_{\alpha'}(\hat{x})\, d_{\beta'}(\hat{x})\, |JF_i(\hat{x})|\, \hat{D}^{\alpha'} \hat{\psi}_{\hat{\lambda}'}(\hat{x})\, \hat{D}^{\beta'} \hat{\psi}_{\hat{\mu}'}(\hat{x})\, d\hat{x}, \tag{3}$$

where the sets $T(i,\alpha), T(i,\beta)$ are defined by the chain rule, and $d_{\alpha'}$ and $d_{\beta'}$ are smooth functions depending on G_i and its derivatives. The integrals which appear on the right–hand side of (3) take the form

$$\hat{d}_{\hat{\lambda},\hat{\mu}} := \int_{\hat{\Omega}} \hat{c}(x)\, \hat{D}^{\hat{\alpha}} \hat{\psi}_{\hat{\lambda}}(\hat{x})\, \hat{D}^{\hat{\beta}} \hat{\psi}_{\hat{\mu}}(\hat{x})\, d\hat{x}, \quad \text{where} \quad \hat{\psi}_{\hat{\lambda}}(\hat{x}) = \prod_{i=1}^{n} {}_i\hat{\theta}_{\hat{\lambda}_i}(\hat{x}_i) \tag{4}$$

and ${}_i\hat{\theta}_{\hat{\lambda}_i}$ are univariate scaling functions or wavelets on $[0,1]$. Now, we use the Two–Scale–Relation for the wavelets to express them in terms of scaling functions on the next higher level, i.e.,

$$\hat{\psi}_{\hat{\lambda}}(\hat{x}) = \sum_{\underline{\hat{\lambda}} \in \Delta_{\hat{\lambda}}} m_{\hat{\lambda},\underline{\hat{\lambda}}} \hat{\varphi}_{\underline{\hat{\lambda}}}(\hat{x}), \tag{5}$$

where $m_{\hat{\lambda},\underline{\hat{\lambda}}}$ are the refinement coefficients. Here, the index set $\Delta_{\hat{\lambda}} \subset \mathcal{I}_{|\hat{\lambda}|+1}$ is determined by the Two–Scale–Relation and $\mathcal{I}_j$ denotes the set of all scaling function indices on a level j. Hence, $\hat{d}_{\hat{\lambda},\hat{\mu}}$ becomes

$$\hat{d}_{\hat{\lambda},\hat{\mu}} = \sum_{\underline{\hat{\lambda}} \in \Delta_{\hat{\lambda}}} \sum_{\underline{\hat{\mu}} \in \Delta_{\hat{\mu}}} m_{\hat{\lambda},\underline{\hat{\lambda}}} m_{\hat{\mu},\underline{\hat{\mu}}} \int_{\hat{\Omega}} \hat{c}(x)\, \hat{D}^{\hat{\alpha}} \hat{\varphi}_{\underline{\hat{\lambda}}}(\hat{x})\, \hat{D}^{\hat{\beta}} \hat{\varphi}_{\underline{\hat{\mu}}}(\hat{x})\, d\hat{x}. \tag{6}$$

The computation of each integral on the right–hand side of (6) would be highly efficient if we could reduce it to a product of univariate integrals, but, in general, the function $\hat{c}$ is *not* a tensor product of univariate functions. However, we can expand it in an appropriate tensor product wavelet basis

$$\hat{\Theta}^* := \left\{ \hat{\theta}^*_{\hat{\nu}} : \hat{\theta}^*_{\hat{\nu}}(\hat{x}) = \prod_{i=1}^{n} {}_i\hat{\theta}^*_{\hat{\nu}_i}(\hat{x}_i),\, \hat{\nu} \in \hat{\mathcal{J}}^* \right\}, \tag{7}$$

(where ${}_i\hat{\theta}^*_{\hat{\nu}_i}$ are again univariate scaling functions and wavelets on $[0,1]$, respectively, possibly different from ${}_i\hat{\theta}_{\hat{\nu}_i}$) as follows

$$\hat{c}(\hat{x}) = \sum_{\hat{\nu} \in \hat{\mathcal{J}}^*} c_{\hat{\nu}} \hat{\theta}^*_{\hat{\nu}}(\hat{x}). \tag{8}$$

Then, we approximate $\hat{c}$ locally on $S_{\underline{\hat{\lambda}},\underline{\hat{\mu}}} := \operatorname{supp} \hat{\varphi}_{\underline{\hat{\lambda}}} \cap \operatorname{supp} \hat{\varphi}_{\underline{\hat{\mu}}}$ by a finite sum $\hat{Q}_{\Lambda^*}\hat{c}$, obtained by restricting the sum in (8) to a finite index set $\Lambda^* \subset \hat{\mathcal{J}}^*$ (depending on $\hat{c}$ as well as on $\hat{\alpha}$, $\hat{\beta}$, $\hat{\lambda}$, $\hat{\mu}$, $\underline{\hat{\lambda}}$ and $\underline{\hat{\mu}}$), whose precise definition will be given below. Correspondingly, $\hat{d}_{\hat{\lambda},\hat{\mu}}$ is approximated by

$$\hat{d}^*_{\hat{\lambda},\hat{\mu}} := \sum_{\underline{\hat{\lambda}} \in \Delta_{\hat{\lambda}}} \sum_{\underline{\hat{\mu}} \in \Delta_{\hat{\mu}}} m_{\hat{\lambda},\underline{\hat{\lambda}}} m_{\hat{\mu},\underline{\hat{\mu}}} \int_{\hat{\Omega}} \hat{Q}_{\Lambda^*}\hat{c}(\hat{x})\, \hat{D}^{\hat{\alpha}} \hat{\varphi}_{\underline{\hat{\lambda}}}(\hat{x})\, \hat{D}^{\hat{\beta}} \hat{\varphi}_{\underline{\hat{\mu}}}(\hat{x})\, d\hat{x}, \tag{9}$$

which is a finite linear combination of products of univariate integrals of the following form

$$\int_0^1 {}_i\hat{\theta}^*_{\hat{\nu}_i}(\hat{x}_i)\, {}_i\hat{\theta}^{(\hat{\alpha}_i)}_{\underline{\hat{\lambda}}_i}(\hat{x}_i)\, {}_i\hat{\theta}^{(\hat{\beta}_i)}_{\underline{\hat{\mu}}_i}(\hat{x}_i)\, d\hat{x}_i, \qquad i = 1, \ldots, n. \tag{10}$$

An algorithm for computing such integrals can be found in [2]. However, here we use biorthogonal B–spline wavelets [7,11] as trial and test functions. Due to their explicit representation, efficient direct formulas for the integrals (10) are available and have been used for the subsequent numerical experiments.

Let us mention that the above strategy slightly differs from [2] since here we approximate $\hat{c}$ locally on $S_{\underline{\hat{\lambda}},\underline{\hat{\mu}}}$, whereas in [2] this is done on the somewhat larger domain $S_{\hat{\lambda},\hat{\mu}} := \operatorname{supp} \hat{\psi}_{\hat{\lambda}} \cap \operatorname{supp} \hat{\psi}_{\hat{\mu}}$. This new method ensures automatically that only non–zero integrals are computed, avoiding a wide number of checks, which explains why the present method is more efficient than the original one.

2.2. Adaptive approximation of the stiffness matrix

Now, we are going to describe the construction of the index set Λ^* introduced above. To this end, we have to introduce some notation. Let us set

$$\mathcal{I}(\underline{\hat{\lambda}}, \underline{\hat{\mu}}) := \{\hat{\nu} \in \hat{\mathcal{J}}^* : |\operatorname{supp} \hat{\theta}^*_{\hat{\nu}} \cap \operatorname{supp} \hat{\varphi}_{\underline{\hat{\lambda}}} \cap \operatorname{supp} \hat{\varphi}_{\underline{\hat{\mu}}}| > 0\}, \tag{11}$$

and $j := \min\{|\underline{\hat{\lambda}}|, |\underline{\hat{\mu}}|\}$ as well as $J := \max\{|\underline{\hat{\lambda}}|, |\underline{\hat{\mu}}|\}$. Let $R_{\hat{\nu}}$ be the number of *zero moments* of $\hat{\theta}^*_{\hat{\nu}}$, [5,10]. Moreover, let $T_{\underline{\hat{\lambda}}}$ and $T_{\hat{\nu}}$ be the largest integers such that $\hat{\varphi}_{\underline{\hat{\lambda}}} \in W^{T_{\underline{\hat{\lambda}}},\infty}(\hat{\Omega})$ and $\hat{\theta}^*_{\hat{\nu}} \in W^{T_{\hat{\nu}},\infty}(\hat{\Omega})$, respectively. Then, we set

$$R := \min\{R_{\hat{\nu}}, T_{\underline{\hat{\lambda}}} - \|\hat{\alpha}\|, T_{\underline{\hat{\mu}}} - \|\hat{\beta}\|\}.$$

We make the following

Assumption 1. *The system $\hat{\Theta}^*$ defined in (7) allows the characterization of the Besov space $B^{\sigma}_{q,q}(\hat{\Omega})$ for indices (σ, q) in a certain range $\mathcal{S}_{\Theta^*} \subseteq \mathbb{R}^+ \times (0,1]$, i.e., the Besov seminorm $|\cdot|_{B^{\sigma}_{q,q}(\hat{\Omega})}$ has the representation*

$$|\hat{v}|_{B^{\sigma}_{q,q}(\hat{\Omega})} \sim \left(\sum_{\hat{\nu} \in \hat{\mathcal{J}}^*} 2^{|\hat{\nu}|\sigma q}\, 2^{|\hat{\nu}|n(q/2-1)}\, |\hat{v}_{\hat{\nu}}|^q \right)^{1/q}, \qquad \hat{v} \in B^{\sigma}_{q,q}(\hat{\Omega}). \tag{12}$$

The following notation will be frequently used in the sequel. For $\ell = 1, \ldots, L$, we consider (possibly different) wavelet bases ${}_\ell\Psi = \{{}_\ell\psi_{\lambda_\ell} : \lambda_\ell \in {}_\ell\mathcal{J}\}$. Then, for $\lambda_\ell \in {}_\ell\mathcal{J}$, $\ell = 1, \ldots, L$, we define

$$i(\lambda_1, \ldots, \lambda_L) := \begin{cases} 1, & \text{if } |\bigcap_{\ell=1}^{L} \operatorname{supp} {}_\ell\psi_{\lambda_\ell}| > 0, \\ 0, & \text{otherwise.} \end{cases}$$

Now, we are in a position to define the set Λ^*. Let us fix, once and for all, independently of $\underline{\hat{\lambda}}$ and $\underline{\hat{\mu}}$, a non-increasing $\ell^1(\mathbb{N}_0)$–sequence $\delta = (\delta_\ell)_{\ell\in\mathbb{N}_0}$ with strictly positive elements, whose $\ell^1(\mathbb{N}_0)$–norm is close to 1. Let $\hat{c} \in B^\sigma_{q,q}(\hat{\Omega})$ for some $(\sigma, q) \in \mathcal{S}_{\Theta^*}$. For an index $\hat{\nu} \in \hat{\mathcal{J}}^*$, we define its relevance for the computation of $\hat{c}_{\hat{\nu}} \int_{\hat{\Omega}} \hat{\theta}_{\hat{\nu}}(\hat{x})\, \hat{D}^{\hat{\alpha}} \varphi_{\underline{\hat{\lambda}}}(\hat{x})\, \hat{D}^{\hat{\beta}} \varphi_{\underline{\hat{\mu}}}(\hat{x})\, d\hat{x}$ as

$$\begin{aligned}\rho^{(\hat{\alpha},\hat{\beta})}_{\underline{\hat{\lambda}},\underline{\hat{\mu}}}(\hat{\nu}) :=& i(\underline{\hat{\lambda}}, \underline{\hat{\mu}}, \hat{\nu})\, 2^{-|\hat{\nu}|\sigma q}\, 2^{-|\hat{\nu}| n(q/2-1)}\, |\hat{c}_{\hat{\nu}}|^{1-q} \\ &\times\ 2^{-(R+n/2)(|\hat{\nu}|-J)}\, 2^{nJ/2}\, 2^{|\underline{\hat{\lambda}}|(\|\hat{\alpha}\|-s)}\, 2^{|\underline{\hat{\mu}}|(\|\hat{\beta}\|-s)} \delta^{-1}_{||\underline{\hat{\lambda}}|-|\underline{\hat{\mu}}||}.\end{aligned}$$

Finally, for any $\varepsilon > 0$, we define

$$\Lambda^* := \{\hat{\nu} \in \mathcal{I}(\underline{\hat{\lambda}}, \underline{\hat{\mu}}) : \rho^{(\hat{\alpha},\hat{\beta})}_{\underline{\hat{\lambda}},\underline{\hat{\mu}}}(\hat{\nu}) \geq \varepsilon/(m_{\hat{\lambda},\underline{\hat{\lambda}}}\, m_{\hat{\mu},\underline{\hat{\mu}}}\, \#\Delta_{\hat{\lambda}}\, \#\Delta_{\hat{\mu}}) \text{ or } |\hat{\nu}| \leq J\}, \quad (13)$$

which concludes the construction of an adaptive approximation $\hat{d}^*_{\lambda,\mu}$ of $\hat{d}_{\lambda,\mu}$.

Remark 2. *The construction of Λ^* according to (13) seems to require the explicit knowledge of all the (infinite) coefficients $\hat{c}_{\hat{\nu}}$ of $\hat{c}$. However, one can estimate a priori a level J_ε such that $\rho^{(\hat{\alpha},\hat{\beta})}_{\underline{\hat{\lambda}},\underline{\hat{\mu}}}(\nu) < \varepsilon/(m_{\hat{\lambda},\underline{\hat{\lambda}}}\, m_{\hat{\mu},\underline{\hat{\mu}}}\, \#\Delta_{\hat{\lambda}}\, \#\Delta_{\hat{\mu}})$, if $|\nu| > J_\varepsilon$. Following [2] it is easy to show that this is valid for*

$$J_\varepsilon := \left\lceil \frac{\varrho_\varepsilon}{R + \sigma + n(1 - \frac{1}{q})} \right\rceil, \quad (14)$$

where we have set

$$\begin{aligned}\varrho_\varepsilon :=& \,|\log_2 \varepsilon/(m_{\hat{\lambda},\underline{\hat{\lambda}}}\, m_{\hat{\mu},\underline{\hat{\mu}}}\, \#\Delta_{\hat{\lambda}}\, \#\Delta_{\hat{\mu}})| + (R+n)J + |\underline{\hat{\lambda}}|(|\hat{\alpha}| - s) \\ &+ |\underline{\hat{\mu}}|(|\hat{\beta}| - s) + \log_2(|\hat{c}|^{1-q}_{B^\sigma_{q,q}(\hat{\Omega})} \delta^{-1}_{||\lambda|-|\mu||}) + \log_2 \mathrm{Const}.\end{aligned}$$

Replacing $d_{\lambda,\mu}$ in the computation of $a_{\lambda,\mu}$ in (3) by $d^*_{\lambda,\mu}$ results in an adaptive approximation $a^*_{\lambda,\mu}$ of $a_{\lambda,\mu}$. As already mentioned, one can construct an adaptive approximation f^*_λ for the entry of the right–hand side $f_\lambda := \int_\Omega f_\beta(x)\, D^\beta \psi_\lambda(x)\, dx$ in the same way.

2.3. Error estimates

Let us assume that the boundary value problem we aim at approximating is elliptic of order $2s$. Let $H^s_b(\Omega)$ be the closed subspace of $H^s(\Omega)$ which accounts for the given boundary conditions. The wavelet basis Ψ introduced above is assumed to form a Riesz basis of this space. The wavelet Galerkin approximation of our problem is obtained by replacing $H^s_b(\Omega)$ by $S_\Lambda := \mathrm{span}\{\psi_\lambda\ :\ \lambda \in \Lambda\}$, where again Λ is an adaptively chosen subset of $\mathcal{J}$. The corresponding Galerkin solution will be denoted by $u_\Lambda := \sum_{\lambda\in\Lambda} u_\lambda \psi_\lambda$. The vector $\mathbf{u}_\Lambda := (u_\lambda)_{\lambda\in\Lambda}$ is obtained by solving the linear algebraic system

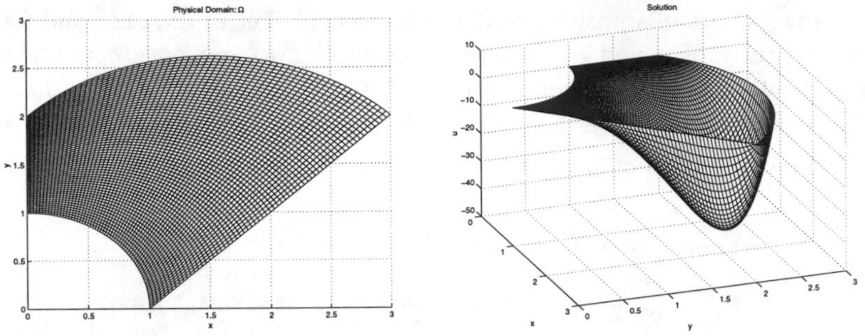

Fig. 1. Physical domain (left) and exact solution (right) for the numerical tests.

$\mathbf{A}_\Lambda \mathbf{u}_\Lambda = \mathbf{f}_\Lambda$, which is defined in a straightforward manner. Let the integrals which appear in the stiffness matrix as well as the right–hand side be computed in an approximate way, as described above. Denote the resulting matrix by $\mathbf{A}^*_\Lambda$ and the resulting vector by $\mathbf{f}^*_\Lambda$; let $\mathbf{u}^*_\Lambda$ be the solution of the modified linear system $\mathbf{A}^*_\Lambda \mathbf{u}^*_\Lambda = \mathbf{f}^*_\Lambda$ and let $u^*_\Lambda = \sum_{\lambda \in \Lambda} u^*_\lambda \psi_\lambda$.

The following estimate on the effect of the described approximation of the stiffness matrix and the right–hand side has been established in [2]. It is readily seen that it also holds for our variant of the original method in [2].

Theorem 3. *Under the above and similar assumptions for computing the right–hand side, there exists* $\varepsilon_0 > 0$ *such that for all* $0 < \varepsilon \le \varepsilon_0$*:*

$$\frac{|u_\Lambda - u^*_\Lambda|_{s,\Omega}}{|u_\Lambda|_{s,\Omega}} \lesssim \varepsilon. \qquad \square \tag{15}$$

§3. Numerical Results

In this section, we present our numerical results. We consider the Poisson problem with homogeneous Dirichlet boundary conditions on a domain Ω which is the parametric image of $\hat{\Omega}$ under a suitable transformation. The domain is displayed in Figure 1, left. The boundary of Ω consists of two straight lines and two curved parts. We computed the parametrization of the four parts of the boundary and then the parametric mapping $F : \hat{\Omega} \to \bar{\Omega}$ is determined by *transfinite interpolation*, [14].

We constructed a solution u which satisfies the boundary conditions and which has a strong layer near the upper right corner of the domain. This function is shown on the right in Figure 1. Since we have an explicit formula for u, we determined the right–hand side f by using MAPLE V.

The choice of these parameters allows us to test an interesting situation. Indeed, the parametric mapping is obviously far from being a tensor product. Hence, we can study the influence of a 'realistic' transformation. Even though this influence was studied in [2] in 1D, we face here a non–tensor product situation for the first time. Moreover, for computing the right–hand side, two

	$\|\lambda\| = \|\mu\| = j_0,$			$\|\lambda\| = j_0,\ \|\mu\| = j_0+1$			$\|\lambda\| = \|\mu\| = j_0+1$		
ε	J_ε	$j_{\max}$	#Int	J_ε	$j_{\max}$	#Int	J_ε	$j_{\max}$	#Int
0.5	5	3, 4	16, 80	8	5, 6	108, 1264	6	4, 5	196, 8150
0.25	5	4	18, 138	8	6	142, 1603	7	5	224, 13168
0.125	5	4, 5	26, 196	8	6, 7	184, 2751	7	5, 6	344, 18052
0.0625	7	4, 5	35, 260	10	7	281, 3924	8	6	442, 21924
0.03125	8	5, 6	44, 454	11	7, 8	396, 8048	9	6, 7	556, 36600
0.015625	8	6	61, 656	11	8	716, 12044	10	7	840, 49916
0.0078125	9	6, 7	82, 1318	12	8, 9	1152, 27708	11	7, 8	1172, 100464

Tab. 1. Estimated and determined maximum level and number of integrals in Λ^*.

kinds of effects are present, namely the parametric mapping and the layer near the corner. We stress that we do not intend to study any particular choice for the adaptive discretization, i.e., the choice of the set Λ. We are primarily interested in the behaviour of the adaptive approximation $\hat{Q}_{\Lambda^*}\hat{c}$ in a realistic situation.

As trial and test functions we used the biorthogonal B–spline wavelets on the interval corresponding to the parameters $d = \tilde{d} = 2$ (i.e., piecewise linear primal functions and dual functions of lowest possible order) from [11] (see also [7] for the original construction on $\mathbb{R}$). For the system $\hat{\Theta}^*$, we choose as in [2] piecewise linear interpolatory wavelets. This of course implies that the computation of the corresponding wavelet coefficients $c_{\hat{\nu}}$ can easily be performed. Moreover, since piecewise linear interpolatory wavelets are nothing else than hierarchical B–splines, the integrals in (10) actually only contain scaling functions.

We used the parameters $\sigma = 2$, $q = 3/4$, $\delta_k := (k+1)^{-2}$ as well as the corresponding parameters $\tau = 2$, $p = 3/4$ and $\bar{\delta}_k := k^{-1}$ for the right–hand side, [2].

In the 1D tests in [2], the parameter ε was chosen as the error in the H^1–norm of a corresponding uniform discretization. From a practical point of view, this is of course unrealistic. First of all, the solution is in general not known. Moreover, the ultimate goal of an adaptive scheme is to avoid a (high level) uniform discretization but to use the degrees of freedom in a more economical way. Hence, we performed various tests on the choice of the parameter ε.

Our first test concerns J_ε in Remark 2 and the number of integrals needed for computing the elements of the stiffness matrix. Our computations are performed in this way: at first, we start from the minimum level ($j_0 = 3$) for the used wavelet basis, where we fix a certain ε, then we solve the problem with scaling functions and wavelets. In Table 1 we compare the theoretical estimate J_ε on the maximum level in Λ^* with the values that were actually determined by our indicators in (13). For different portions of the stiffness matrix, we display the predicted J_ε, the detected maximum levels by the indicator as well as the minimum and maximum number of integrals needed for computing non zero–entries.

We see that the estimated J_ε is always larger than the effectively used maximum level. This is what we expected, but the efficiency of the algorithm may be reduced by an excessive over-estimate of J_ε. We also deduce that the efficiency of the method crucially depends on the choice of ε since the number of integrals strongly grows for decreasing ε. This is surely due to the low order of the interpolatory wavelets used.

Next, we present in Table 2 the average number of integrals computed for the stiffness matrix and the right–hand side for the first two levels with the same ε of Table 1. Here $J_\Lambda := \max\{|\lambda| : \lambda \in \Lambda\}$.

	ε	0.5	0.25	0.125	0.0625	$3.1e-2$	$1.6e-2$	$7.8e-3$
A	$J_\Lambda = 3$	4.04	5.47	8.57	11.05	16.99	24.15	41.62
	$J_\Lambda = 4$	111.55	48.31	234.08	305.98	489.47	738.67	1336.59
f	$J_\Lambda = 3$	9	9	11.49	17.98	31.27	55.69	95.10
	$J_\Lambda = 4$	87.4	87.4	87.94	90.17	137.56	207.72	411.61

Tab. 2. Average number of integrals per entry.

We deduce that the choice of ε not only influences the maximal and minimal number of integrals as shown in Table 1. Since the average number of integrals grows when ε decreases, the choice of ε effects the efficiency of the computation of the *whole* stiffness matrix. Moreover, the presence of the first wavelet level also increases the number of integrals.

Finally, we consider the error in the H^1–norm and the relative error

$$r_\Lambda^{\text{ex}} := \frac{|u_\Lambda^* - u|_{1,\Omega}}{|u|_{1,\Omega}}$$

for different choices of the parameter ε. In Table 3 and Figure 2 'rate' corresponds to the rate of convergence w.r.t. the exact solution in the H^1–norm for the first two levels in Λ. We see that these quantities do not depend on ε. At these levels the relative discretization error still exceeds the relative error (15). This explains why the rate of convergence is basically constant w.r.t. the choices of ε.

Hence, the choice of ε matters only if this value is at least of the same order than the relative discretization error. We remark that also for increasing ε all scaling functions on level J whose support overlap $S_{\underline{\hat{\lambda}},\underline{\hat{\mu}}}$ belong to Λ^*. This implies that the error due to the approximation of the entries of the linear system is bounded.

ε	rate	r_Λ^{ex}
0.5	1.9666	0.1335
0.25	1.9633	0.1337
0.125	1.9538	0.1344
0.0625	1.9658	0.1338
0.03125	1.9563	0.1343
0.015625	1.9591	0.1342
0.0078125	1.9568	0.1343

Tab. 3. Relative error and rate of convergence in dependence of ε.

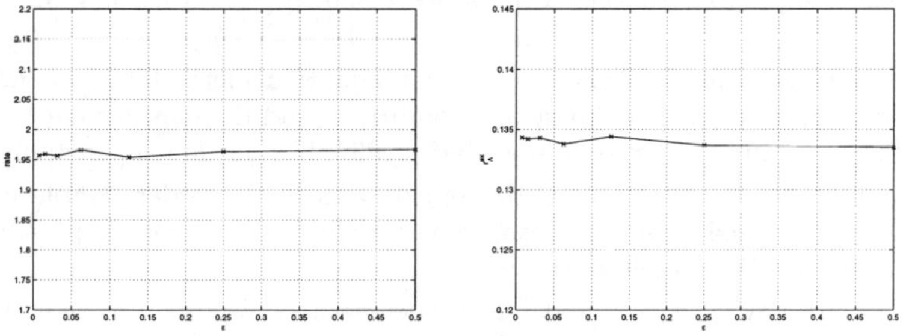

Fig. 2. Rate (left) and r_Λ^{ex} (right) of Table 3.

Acknowledgments. We thank Claudio Canuto very much for his helpful advice during the preparation of this paper. The first author is extremely grateful to the Dipartimento di Matematica of the Politecnico di Torino for using its facilities. This work was supported by the *European Commission* within the TMR project (Training and Mobility for Researchers) *Wavelets and Multiscale Methods in Numerical Analysis and Simulation*, No. ERB FMRX CT98 0184. This paper was partly written when the second author was in residence at the Istituto di Analisi Numerica del C.N.R. in Pavia, Italy.

References

1. Barinka, A., T. Barsch, P. Charton, A. Cohen, S. Dahlke, W. Dahmen, and K. Urban, Adaptive wavelet schemes for elliptic problems — Implementation and numerical experiments, RWTH Aachen, IGPM Preprint 173, 1999.
2. Bertoluzza, S., C. Canuto, and K. Urban, On the adaptive computation of integrals of wavelets, Preprint No. 1129, Istituto di Analisi Numerica del C.N.R. Pavia, 1999. To appear in Appl. Numer. Math.
3. Canuto, C., A. Tabacco, and K. Urban, The wavelet element method, part I: Construction and analysis, Appl. Comp. Harm. Anal. **6** (1999), 1–52.

4. Canuto, C., A. Tabacco, and K. Urban, The wavelet element method, part II: Realization and additional features in 2d and 3d. Preprint 1052, Istituto di Analisi Numerica del C.N.R., Pavia, 1997. To appear in Appl. Comp. Harm. Anal.
5. Cohen, A., Wavelet methods in numerical analysis, in *Handbook of Numerical Analysis*, North Holland, Amsterdam, to appear.
6. Cohen, A., W. Dahmen, and R. DeVore, Adaptive wavelet schemes for elliptic operator equations – Convergence rates, RWTH Aachen, IGPM Preprint 165, 1998.
7. Cohen, A., I. Daubechies, and J.-C. Feauveau, Biorthogonal bases of compactly supported wavelets, Comm. Pure and Appl. Math. **45** (1992), 485–560.
8. Cohen, A., and R. Masson, Wavelet adaptive method for second order elliptic problems– boundary conditions and domain decomposition, Preprint, Univ. P. et M. Curie, Paris, 1997, Numer. Math., to appear.
9. Dahlke, S., W. Dahmen, R. Hochmuth, and R. Schneider, Stable multiscale bases and local error estimation for elliptic problems, Appl. Numer. Math. **23**, No. 1 (1997), 21–48.
10. Dahmen, W., Wavelet and multiscale methods for operator equations, Acta Numerica, **6** (1997), 55–228.
11. Dahmen, W., A. Kunoth, and K. Urban, Biorthogonal spline wavelets on the interval – stability and moment conditions, Appl. Comp. Harm. Anal. **6** (1999), 132–196.
12. Dahmen, W., and R. Schneider, Composite wavelet bases for operator equations, Math. Comput. **68** (1999), 1533–1567.
13. Dahmen, W., and R. Schneider, Wavelets on manifolds I: Construction and domain decomposition, RWTH Aachen, IGPM Preprint 149, 1998. To appear in SIAM J. Math. Anal.
14. Gordon, W., and C. Hall, Transfinite element methods: blending-function interpolation over arbitrary curved element domains, Numer. Math. **21** (1973), 109–129.

Stefano Berrone
Dipartimento di Ingegneria Aeronautica e Spaziale
Politecnico di Torino
Corso Duca degli Abruzzi, 24
10129 Torino, Italy
`sberrone@calvino.polito.it`

Karsten Urban
Institut für Geometrie und Praktische Mathematik
RWTH Aachen
Templergraben 55
52056 Aachen, Germany
`urban@igpm.rwth-aachen.de`

Scattered Data Near-Interpolation with Application to Discontinuous Surfaces

Renata Besenghi and Giampietro Allasia

Abstract. This paper discusses a particular type of function approximation on scattered data in a general number of variables, and its application to surface representation with imposed conditions. If the given function values are subject to errors, it is not appropriate to interpolate the function at the data in the sense of exact matching. As a consequence, we formulate a weakened version of the classical scattered data interpolation problem, and give a simple and efficient procedure to obtain near-interpolation formulas. Near-interpolants enjoy many remarkable properties, which are very useful from both theoretical and practical points of view (shape preserving properties, operator positivity, subdivision techniques, parallel and multistage computation). Applications of near-interpolants to the representation of surfaces, in particular with faults, are discussed in detail (parameter values, localizing weights, etc.).

§1. Introduction

In many applications, the given function values are subject to errors; hence it is not appropriate to interpolate the function at the data in the sense of exact matching, but it seems more appropriate to approximate the function or, more precisely, to get a relaxed interpolation or near-interpolation. Data requiring near-interpolation by scattered data methods occur in virtually every field of science and engineering. Sources include both experimental results (experiments in chemistry, physics, engineering) and measured values of physical quantities (meteorology, oceanography, optics, geodetics, mining, geology, geography, cartography), as well as computational values (e.g., output from finite element solutions of partial differential equations).

As a consequence of this remark, we formulate a relaxed version of the classical multivariate interpolation problem at scattered data points.

Curve and Surface Fitting: Saint-Malo 1999
Albert Cohen, Christophe Rabut, and Larry L. Schumaker (eds.), pp. 75–84.

ISBN 0-8265-1357-3.

Definition 1. *Given a set of points* $S_n = \{x_i,\ i = 1, \dots, n\}$, *distinct and generally scattered, in a domain* $D \subset \mathbb{R}^s, (s \geq 1)$, *with associated values* $\{f_i,\ i = 1, \dots, n\}$, *and a linear space* $\Phi(D)$ *spanned by continuous real basis functions* $g_j(x; r)$ *with* $x \in D$, $r \geq 0$, *and* $j = 1, \dots, n$, *the multivariate near-interpolation problem at scattered data consists in finding a function* $F(x; r) \in \Phi(D)$ *such that*

$$F(x_i; r) = \sum_{j=1}^{n} a_j g_j(x_i; r) = f_i + \epsilon_i(r), \quad i = 1, \dots, n, \tag{1}$$

and

$$\lim_{r \to 0} \epsilon_i(r) = 0. \tag{2}$$

We observe that r works as a parameter, and the limit case of $F(x; r)$ when r vanishes

$$F(x) \equiv F(x; 0) = \lim_{r \to 0} F(x; r)$$

is an interpolation operator. If $F(x; r)$ is specified, then the $\epsilon_i(r)$ in (1) are known; these *near-interpolation errors at the nodes* must not be confused with the unknown errors which affect the corresponding function values f_i. However, it is reasonable to get things so that the $\epsilon_i(r)$ and the errors on f_i are quantities of the same order.

In Section 2 we give a constructive procedure to obtain a wide class of near-interpolation formulas. These enjoy many interesting properties which are listed in Section 3. A crucial point in near-interpolation is the proper choice of the parameter r in (1) and, eventually, of other parameters; the matter is discussed in Section 4. Finally, Section 5 is devoted to the application of near-interpolation to modelling faults.

§2. Construction of Near-Interpolants

To solve the classical interpolation problem, one can consider basis functions which depend on the nodes and, moreover, are cardinal. The method of cardinal basis functions involves selecting continuous *cardinal functions* $g_j : D \to \mathbb{R}$, $(j = 1, \dots, n)$, such that $g_j(x_i) = \delta_{ij}$, $(i = 1, \dots, n)$, where δ_{ij} is the Kronecker delta operator, and setting up the interpolation operator F in the form

$$F(x) = \sum_{j=1}^{n} f_j\ g_j(x).$$

The corresponding near-interpolation problem considers basis functions $g_j(x; r)$, which are no longer cardinal, but $g_j(x; r) \to g_j(x)$ for $r \to 0$. If such $g_j(x; r)$ are given, then

$$F(x; r) = \sum_{j=1}^{n} f_j\ g_j(x; r) = F(x) + \sum_{j=1}^{n} f_j\ [g_j(x; r) - g_j(x)] \tag{3}$$

represents a solution of the near-interpolation problem. In this relation the terms $\epsilon_i(r) = F(x_i; r) - f_i$, $(i = 1, \dots, n)$, are uniquely determined and satisfy (2).

As an example, let us consider the near-interpolant Shepard's formula in the product form

$$\mu(x;r) = \sum_{j=1}^{n} f_j \frac{\prod_{k=1,k\neq j}^{n}[d^2(x,x_k)+r]^\beta}{\sum_{h=1}^{n}\prod_{k=1,k\neq h}^{n}[d^2(x,x_k)+r]^\beta}, \tag{4}$$

where $d(x,y)$ is the Euclidean distance between x and y, and $\beta > 0$, or in the equivalent barycentric form

$$\mu(x;r) = \sum_{j=1}^{n} f_j \frac{[d^2(x,x_j)+r]^{-\beta}}{\sum_{h=1}^{n}[d^2(x,x_h)+r]^{-\beta}}, \quad \mu(x_i;0) = f_i, \quad i = 1,\ldots,n. \tag{5}$$

This formula no longer interpolates for $r > 0$, but $\mu(x;r) \to \mu(x)$ as $r \to 0$, where $\mu(x)$ is the well-known Shepard's formula [8, 1].

Examining the structure and the basic idea of Shepard's operator suggests a simple and efficient procedure to obtain an interpolation formula [1]. The corresponding way to obtain a near-interpolation formula is contained in

Definition 2. *Let $\alpha(x,y;r)$, with $x,y \in D$ and $r \geq 0$, be a continuous positive real function such that*

$$\lim_{r\to 0} \alpha(x,y;r) = \alpha(x,y), \tag{6}$$

where $\alpha(x,y) > 0$, if $x \neq y$ and $\alpha(x,y) = 0$, if $x = y$ for all $x,y \in D$. Define now the functions $g_j(x;r)$ by the equations

$$g_j(x;r) = \frac{\prod_{k=1,k\neq j}^{n}\alpha(x,x_k;r)}{\sum_{h=1}^{n}\prod_{k=1,k\neq h}^{n}\alpha(x,x_k;r)}, \tag{7}$$

and the near-interpolant $F(x;r)$ by

$$F(x;r) = \sum_{j=1}^{n} f_j\, g_j(x;r) = \sum_{j=1}^{n} f_j \frac{\prod_{k=1,k\neq j}^{n}\alpha(x,x_k;r)}{\sum_{h=1}^{n}\prod_{k=1,k\neq h}^{n}\alpha(x,x_k;r)}, \tag{8}$$

or equivalently by

$$F(x;r) = \sum_{j=1}^{n} f_j \frac{1/\alpha(x,x_j;r)}{\sum_{h=1}^{n} 1/\alpha(x,x_h;r)}, \quad F(x_i;0) = f_i, \quad i = 1,\ldots,n. \tag{9}$$

Many choices are possible for the function $\alpha(x,y;r)$ in (6); there are no constraints engendered by the set S_n, that is, the distribution of the nodes is irrelevant. Nevertheless, experience suggests identifying α with a radial function

$$\alpha(x,y;r) = \phi(\|x-y\|^2 + r),$$

where $\|\cdot\|$ is a convenient norm. As an example, choosing the Euclidean norm $\|\cdot\|_2$ and

$$\alpha(x,y;r) = (\|x-y\|_2^2 + r)^\beta, \qquad \beta > 0, \tag{10}$$

we obtain from (8) the near-interpolant Shepard's formula (4).

It is often convenient to consider the function $\varphi(t;r) : \mathbb{R}_{\geq 0} \to \mathbb{R}$, associated with $\phi(\|x-y\|^2 + r)$ and defined as $\varphi(t;r) = \phi(t^2 + r)$. We have just seen $\varphi(t;r) = (t^2+r)^\beta$ in (10); another possible expression that works is $\varphi(t;r) = (t^2+r)^\beta \exp(\gamma\, t^2), (\gamma > 0)$. Considering several functions φ can be useful in order to compare their behaviour and choose the most suitable for use.

§3. Properties of Near-Interpolants

Near-interpolants, as given in Definition 2, enjoy many interesting properties. We list some of them.

A) The near-interpolant $F(x;r)$ in (8) or (9) is a weighted arithmetic mean of the values f_j, $(j = 1,\ldots,n)$, since $0 \leq g_j(x;r) \leq 1$ and $\sum_{j=1}^n g_j(x;r) = 1$. As a consequence $F(x;r)$ satisfies the betweeness property $\min_i f_i \leq F(x;r) \leq \max_i f_i$, and reproduces exactly any constant function $f(x) \equiv c$, that is, if $f_i = c$, $(i = 1,\ldots,n)$, then $F(x;r) = c$. Moreover, $F(x;r)$, considered as a functional on the set of functions $f : D \to \mathbb{R}$, is linear and positive.

B) If $\alpha(x,y;r)$ is infinitely differentiable with respect to the pth component of $x = (x^{(1)},\ldots,x^{(s)})$ for all $x, y \in D$ and $p = 1,\ldots,s$, then $F(x;r)$ is also infinitely differentiable with respect to the $x^{(p)}$. For example, choosing $\alpha(x,y;r)$ as in (10), $F(x;r) = \mu(x;r)$ in (4) can be differentiated as many times as desired.

C) If α is a radial function, $F(x;r)$ enjoys some properties of invariance with respect to affine transformations. In particular, with the Euclidean norm, we have that $F(x;r)$ is invariant under translation and rotation, but not scalar invariant.

D) Subdivision techniques can be applied to near-interpolants achieving remarkable results, very well suited for parallel computation [3]. Let us make a partition of the set S_n on the domain D into q subsets S_{n_j}, so that the jth subset, $(j = 1,\ldots,q)$, consists of the nodes $x_{j1}, x_{j2},\ldots,x_{jn_j}$, with $n_1 + n_2 + \cdots + n_q = n$, and the values $f_{jk_j}, (j = 1,\ldots,q; k_j = 1,\ldots,n_j)$, correspond to the nodes x_{jk_j}. The indexing of the nodes in the subsets may not depend on the indexing in the set, provided the biunivocity is saved.

Given $S_{n_j} = \{x_{j1}, x_{j2},\ldots,x_{jn_j}\}$, $(j = 1,\ldots,q)$, let $S_n = S_{n_1} \cup S_{n_2} \cup \cdots \cup S_{n_p}$ and $S_{n_q} \cap S_{n_r} = \emptyset$ for $q \neq r$, then $F(x;r)$ in (9) can be rewritten in the form

$$F_{S_n}(x;r) = \sum_{j=1}^{q} F_{S_{n_j}}(x;r)\,\frac{A_j}{\sum_{j=1}^q A_j}\,, \quad \text{where } A_j = \sum_{k_j=1}^{n_j} 1/\alpha(x, x_{jk_j}; r). \tag{11}$$

E) As a consequence of (11) the following multistage procedure works very well. In the first stage, a given set of nodes $S_{n_1} = \{x_i, i = 1, \ldots, n_1\}$ is considered, and the corresponding near-interpolant $F_{S_{n_1}}(x;r)$ is evaluated. In the second stage, it is required to enlarge the considered set S_{n_1}, taking the union of it and another set of nodes $S_{n_2} = \{x_j, j = 1, \ldots, n_2\}$. Now the near-interpolating function referred to the union set, i.e., $F_{S_{n_1} \cup S_{n_2}}(x;r)$, with $S_{n_1} \cap S_{n_2} = \emptyset$, can be obtained simply by evaluating the near-interpolant $F_{S_{n_2}}(x;r)$, corresponding to the added set S_{n_2}, and using the relation

$$F_{S_{n_1} \cup S_{n_2}}(x;r) = \frac{F_{S_{n_1}}(x;r)A_1 + F_{S_{n_2}}(x;r)A_2}{A_1 + A_2}, \tag{12}$$

where $A_1 = \sum_{i=1}^{n_1} 1/\alpha(x, x_i; r)$, $A_2 = \sum_{j=1}^{n_2} 1/\alpha(x, x_j; r)$, and A_1 is known. The procedure can be repeated as many times as required.

F) Near-interpolants have the remarkable property that an additional node, say x_{n+1}, can be added to the interpolation set S_n by simply combining an extra term with the original formula. The goal is achieved by using a particular case of (12), that is, the recurrence relation

$$F_{S_{n+1}}(x;r) = \frac{F_{S_n}(x;r)A_n + f_{n+1}\ 1/\alpha(x, x_{n+1}; r)}{A_n + 1/\alpha(x, x_{n+1}; r)},$$

where $A_n = \sum_{k=1}^{n} 1/\alpha(x, x_k; r)$.

G) It is often convenient to extend (8), or better (9), in the following way:

$$F_1(x;r) = \sum_{j=1}^{n} f_j \frac{\tau(x, x_j; \gamma)\ 1/\alpha(x, x_j; r)}{\sum_{h=1}^{n} \tau(x, x_h; \gamma)\ 1/\alpha(x, x_h; r)}, \tag{13}$$

where $\tau(x, y; \gamma)$, with $x, y \in D$ and $\gamma \geq 0$, is a continuous positive real function. Choosing suitably $\tau(x, y; \gamma)$, one can modify the weights in (13) in order either to cancel a useless characteristic, or to introduce a new feature. In particular, it is possible to localize the method considering a factor $\tau(x, y; \gamma)$ rapidly decreasing with distance [2]. The formulas obtained in this way maintain, in general, the analytical and computational properties of the corresponding original ones.

The use of the exponential-type function

$$\tau(x, y; \gamma) = \exp(-\gamma \|x - y\|_2^2) \tag{14}$$

is suggested by McLain [4] for Shepard's formula; he observes that much more accurate results can be obtained in this way. The use of exponential-type weights increases the computational effort, but generally this drawback can be tolerated.

The value of the parameter in the mollifying function $\tau(x, y; \gamma)$ may depend on the nodes, as happens in the popular case [4]

$$\tau(x, x_j; \rho_j) = \left(1 - \frac{\|x - x_j\|_2}{\rho_j}\right)_+^2, \tag{15}$$

where ρ_j is the radius of the circle of support at the point x_j, and $(u)_+ > 0$ if $u > 0$, $(u)_+ = 0$ if $u \leq 0$.

H) The precision of the operator $F(x;r)$ can be increased, considering the Taylor expansion for the function f in each node x_j instead of the function value f_j. This leads to the following extension of (9).

If $f \in C^m(D)$ and $T_j(x)$ is the truncated Taylor expansion for f up to derivatives of order m evaluated at the point x_j and referred to the displacement $h_j = x - x_j$, with $h_j \subset D$, then the operator

$$F_2(x;r) = \sum_{j=1}^{n} T_j(x)\; g_j(x;r), \tag{16}$$

near-interpolates to $T_j(x)$ at $x = x_j$. In this form, $F_2(x;r)$ reproduces exactly algebraic polynomials of degree $\leq m$.

Combining the modifications in (13) and (16), we have

$$F_3(x;r) = \sum_{j=1}^{n} T_j(x)\, \frac{\tau(x,x_j;\gamma)\; 1/\alpha(x,x_j;r)}{\sum_{h=1}^{n}\tau(x,x_h;\gamma)\; 1/\alpha(x,x_h;r)}\,. \tag{17}$$

Obviously, the technique calls for additional derivative values that are not normally available as data. A more practical solution is discussed below.

K) For simplicity, we refer here to an Euclidean radial function $\alpha(x,y;r) = \phi(\|x-y\|_2^2 + r)$, because in this case the procedure is well established. The primary modifications required involve using $\tau(x,y;\gamma)$ to localize the overall approximation, and replacing f_j with a suitable "local approximation" to the surface. To carry out the approximation (17), a practical way is to get, in a first stage, local approximants $M_j(x)$ to $f(x)$ at the points $x_j, (j = 1,\ldots,n)$, obtained by means of the moving weighted least-squares method using weight functions with reduced compact support. Then, in a second stage, the near-interpolating operator is expressed as a convex combination of the local approximants

$$F_4(x;r) = \sum_{j=1}^{n} M_j(x)\;\; \frac{\tau(x,x_j;\gamma)\; 1/\phi(\|x-x_j\|_2^2 + r)}{\sum_{h=1}^{n}\tau(x,x_h;\gamma)\; 1/\phi(\|x-x_h\|_2^2 + r)}\,. \tag{18}$$

In particular, by (18), (10), and (14),

$$\mu_1(x;r) = \sum_{j=1}^{n} M_j(x)\, \frac{\exp(-\gamma\|x-x_j\|_2^2)\; (\|x-x_j\|_2^2 + r)^{-\beta}}{\sum_{h=1}^{n}\exp(-\gamma\|x-x_h\|_2^2)\; (\|x-x_h\|_2^2 + r)^{-\beta}}\,, \tag{19}$$

which extends (5).

Very good performance is achieved by a version of (18) which uses quadratic approximations for $M_j(x)$, and mollifying functions given by (15). This method has been developed by Franke and Nielson [4], and Renka [7] for Shepard's operator.

§4. Determining Parameter Values

The near-interpolating operator (19) works very well in a large variety of cases. Our attention is here focused on finding values of the parameters r, β and γ, which can be regarded as "optimal" from a practical viewpoint. The considerations which follow are mainly based on experiments.

A relatively small increase of the parameter r in (19) in some right neighbourhood of zero has a considerable effect on the behaviour of $\mu_1(x;r)$. In fact, if r is small, x fixed and near to the node x_{j^*}, then the value of $g_{j^*}(x;r)$ equals nearly one; but, if r increases, $g_{j^*}(x;r)$ decreases. Since $\sum_{j=1}^{n} g_j(x;r) = 1$, diminishing of $g_{j^*}(x;r)$ makes the other weight values $g_j(x;r)$, $j \neq j^*$, increase. Summing up, if the weight attributed to f_{j^*} in $\mu_1(x;r)$ decreases, then $\mu_1(x_{j^*};r) = f_{j^*} + \epsilon_{j^*}(r)$ diverges from f_{j^*}, namely $\epsilon_{j^*}(r)$ increases and reduces the accuracy of $\mu_1(x_{j^*};r)$.

Introducing the parameter r in (19), and in particular in (5), has the effect that, in general, the gradient of the rendered surface is not zero at the nodes. As a consequence, the surface is considerably smoother than for $r = 0$. However, if r is too small, the first derivatives of $\mu_1(x;r)$ are highly oscillating and their values are nearly zero. Clearly, the goal is to choose an "optimal" value of r, such that $\mu_1(x;r)$ does not exhibit the characteristic irregularities of the basic Shepard's formula, but at the same time, it maintains a sufficient computational accuracy, in particular at the nodes.

The search for the optimal value of r can be done by many applications of (19) with different values of the parameter, and then by choosing that value which minimizes the global root mean square error. Although this is currently considered in the literature, the estimate of r is not a simple matter; in a sense, it can be compared with the analogous difficult problem of computing the optimal value of the parameter in multiquadric interpolants.

The optimal value of the parameter β has been determined with particular attention to computational accuracy. The performance analysis on some test functions proposed by Franke leads to prefer the value $\beta = 3/2$.

As for the optimal value of the parameter γ in the strongly localizing function (14), McLain has proposed $\gamma = 1.62n/\text{diam}(D)$, where n is the number of nodes and $\text{diam}(D)$ is the diameter of D. However, this value is, in general, too large, whereas it is sufficient to consider for γ a value of the order of tens.

§5. Application to Modelling Faults

Using the near-interpolating operator $\mu_1(x;r)$ of (19), with a suitable value of the parameter r, instead of the corresponding interpolating operator $\mu_1(x;0)$, increases considerably the performance of the approximation in a rich variety of applications, because it permits consideration of supplementary information connected with the characteristics of the examined problem. A typical case occurs with surface discontinuities, in particular faults, which are frequently met when modelling geological surfaces.

Following Shepard [8], we observe that, if some physical barrier such as a fault separates the set of nodes, the relationship between nodes on the opposite sides of the barrier may be attenuated. Through the inclusion of barriers, a user may specify discontinuities in the metric space in which the distance between two points is calculated to simulate this attenuation. Suppose a "detour" of length $b(x, x_k)$ were required to go over the barrier between x, the current near-interpolation point, and the node x_k. The quantity $b(x, x_k)$ is considered the strength of the barrier, and an effective distance between x and x_k is given by

$$d^*(x, x_k) = \sqrt{[d(x, x_k)]^2 + [b(x, x_k)]^2}.$$

This definition is general so that if no barrier separates x and x_k, then $b(x, x_k) = 0$ and $d^*(x, x_k) = d(x, x_k)$. Because of the discontinuity in effective distance as the near-interpolation point x crosses the barrier, the rendered surface will be discontinuous at the barrier.

Since extensive tests [4] have shown that the modified quadratic Shepard's method performs very well for a variety of data sets, Franke and Nielson [5] have chosen it as a basis to investigate the problem of simulating faults. Our approach uses instead the near-interpolant (19), with significant differences in distance penalty, localizing functions, fault forms, etc., as compared to Franke and Nielson.

The possibility of having to model faults can occur in different ways [5]; to save space, we limit our attention to the following case: there is a known fault line $\Gamma \subset D \subset \mathbb{R}^2$, in a known location, with a known jump. More complicated situations (see, e.g., [5,6]) require extensive considerations that will be discussed in a further work.

As a first step, it is convenient to focus on the basic situation in which the fault line Γ is a segment l and, moreover, the jump is constant along l. Then, a known polygonal curve can be considered as a fault line; in fact, the reduction to the case of a fault line segment is straightforward by using the subdivision procedure considered in Section 3. In principle, any curve can be considered as a fault line, provided it is well approximated by a polygonal. Another extension consists in considering a jump varying along the fault line. Also the reduction to the basic case is now possible, subdividing the fault line into a convenient number of segments, and using a mean value of the jump for each segment.

To deal with the basic case, we modify the value of the parameter r in (19) in order to take the jump into account. Let x be the near-interpolation point, x_k a node and l^* the segment joining x and x_k. Then for $x, x_k \notin l$ we set

$$r = \begin{cases} r_{opt}\,, & \text{if} \quad l \cap l^* = \emptyset, \\ b(x, x_k), & \text{if} \quad l \cap l^* \neq \emptyset, \end{cases}$$

where the quantity r_{opt} is the optimal value obtained for r in (19) on the opposite sides of the fault and $b(x, x_k)$ represents the "effort" required to go over the barrier, due to the discontinuity dividing the two points. If the jump

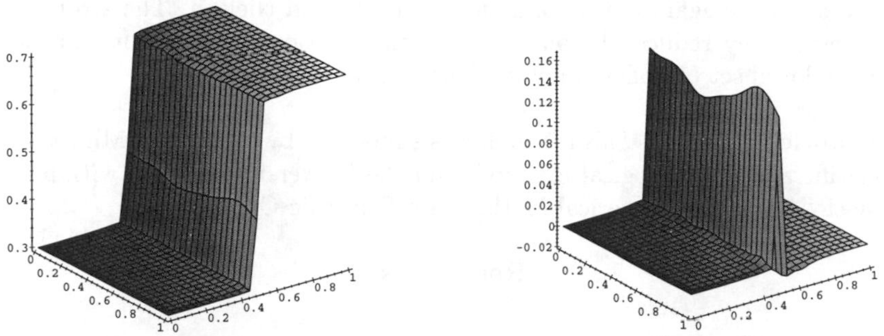

Fig. 1. Function $f_1(x, y)$: near-interpolation and signed error surfaces.

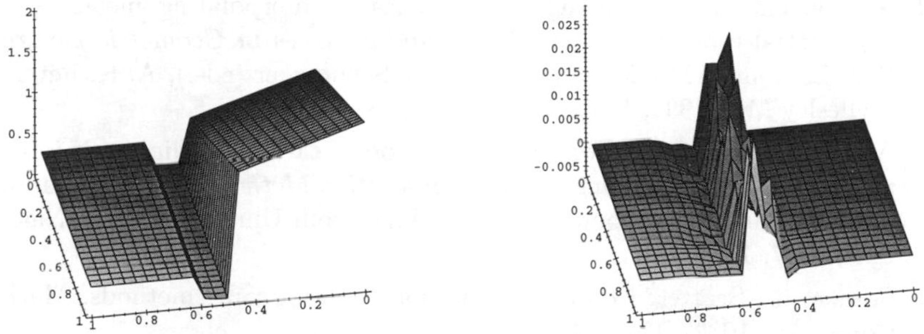

Fig. 2. Function $f_2(x, y)$: near-interpolation and signed error surfaces.

is constant or almost constant along l, it is possible to simply set $b(x, x_k) = h$, where h is the jump size.

Formula (19), after these adjustments in the parameter r, gives results quite good both for the appearance of the graphic representation and the accuracy in computation. Comparing the rendered surface with the one obtained by the modified quadratic Shepard's formula shows that the introduction of the parameter r gives a smoother surface which is closer to the approximated function.

Our procedure has been used to fit the test function proposed by Franke and Nielson [5] using their set of nodes. Numerous tests were also made on other surfaces. We present two examples of the rendered surfaces and the signed error surfaces for the functions

$$f_1(x, y) = \begin{cases} 0.3, & \text{if } 0 \le x < 0.5, \\ 0.7, & \text{if } x \ge 0.5; \end{cases} \qquad f_2(x, y) = \begin{cases} y - x + 1, & \text{if } 0 \le x < 0.5, \\ 0, & \text{if } 0.5 \le x < 0.6, \\ 0.3, & \text{if } x \ge 0.6, \end{cases}$$

defined on the unit square (see Fig. 1 and Fig. 2). We used the parameter values $r = 0.0036, \beta = 1.5, \gamma = 24$, and $r = 0.0025, \beta = 1.5, \gamma = 30$ respectively, and once again the set of nodes of Franke and Nielson. The errors can be considerably reduced by adding more information on the faults; in fact, the employed set of nodes is not obviously an ad hoc choice.

Acknowledgments. This research was supported by the Italian Ministry of Scientific and Technological Research and the University of Turin within the project "Advanced Numerical Methods for Scientific Computing".

References

1. Allasia, G., A class of interpolating positive linear operators: theoretical and computational aspects, in *Recent Developments in Approximation Theory, Wavelets and Applications*, S. P. Singh (ed.), Kluwer, Boston, 1995, 1–36.
2. Allasia, G., and R. Besenghi, Properties of interpolating means with exponential-type weights, in *Curves and Surfaces in Geometric Design*, P.-J. Laurent, A. Le Méhauté, and L. L. Schumaker (eds.), A. K. Peters, Wellesley MA, 1994, 1–8.
3. Allasia, G., and P. Giolito, Fast evaluation of cardinal radial basis interpolants, in *Surface Fitting and Multiresolution Methods*, A. Le Méhauté, C. Rabut, and L. L. Schumaker (eds.), Vanderbilt University Press, Nashville, 1997, 1–8.
4. Franke, R., Scattered data interpolation: tests of some methods, Math. Comp. **38** (1982), 181–200.
5. Franke, R., and G. M. Nielson, Surfaces approximation with imposed condition, in *Surfaces in CAGD*, R. E. Barnhill and W. Boehm (eds.), North-Holland, Amsterdam, 1983, 135–146.
6. Parra, M. C., M. C. López de Silanes and J. J. Torrens, Vertical fault detection from scattered data, J. Comp. Appl. Math. **73** (1996), 225–239.
7. Renka, R. J., Multivariate interpolation of large sets of scattered data, ACM Trans. Math. Softw. **14** (2) (1988), 139–148.
8. Shepard, D., A two-dimensional interpolation function for irregularly spaced data, *Proc. 23rd Nat. Conf. ACM*, 1968, 517–524.

Renata Besenghi, Giampietro Allasia
Dipartimento di Matematica
Università di Torino
I-10123 Torino, Italy
`besenghi@dm.unito.it`, `allasia@dm.unito.it`

On a Method of Numerical Differentiation

Mira Bozzini and Milvia Rossini

Abstract. In this paper we present a method for the numerical differentiation of two-dimensional functions when scattered data are given. The method is based on a regularization of the given sample.

§1. Introduction

In this paper we present a method for the numerical differentiation of two-dimensional functions when scattered data are given. The problem of numerical differentiation is very important when dealing with function approximation. In fact, a satisfactory recovery of a function given in a sampled form needs some knowledge of the derivatives.

It is also well known that this problem is ill-conditioned. Consider for instance, one-dimensional equispaced data with $h_1 = 10^l$ and $h_2 = 2^p$. On a computer, because of the base change, numerical differentiation gives considerable errors in the first case, and a more accurate solution in the second case. Moreover, it is strongly influenced by the data position.

The literature on scattered data, includes the papers [7,8,10], their improvements [3,5], and some experiments on their use [9]. These papers provide the gradient approximation at the sampled points (see [7,10]) or the approximation of the gradient function (see [8]). This is done by triangulation or local and global moving least square interpolation. In some of them, asymptotic bounds for the error are also supplied.

Generally, these methods provide a satisfactory approximation inside the domain in which the data are given, but they may give large errors at the boundary (see Figs. 1–4 below).

In this paper we present a method based on a regularization of the sample which gives an error with an uniform behaviour on the whole domain in which the data are assigned. The regularization is done by constructing a new set on a regular grid. Namely, taking into account the previous observations, we have considered dyadic grids. Obviously the construction of this new set will

Curve and Surface Fitting: Saint-Malo 1999
Albert Cohen, Christophe Rabut, and Larry L. Schumaker (eds.), pp. 85–94.

ISBN 0-8265-1357-3.

generate errors that can be thought of as causal errors depending on the sample. Then (see [4]), to smooth the data, we perform a wavelet decomposition of the signal. Using the smoothed data, we approximate the gradient at the grid points by the classical finite centered difference formulas, and finally we construct a smooth function approximating the unknown gradient.

The paper is organised as follows. In §2 the method is described and the gradient estimator is constructed. In §3 the convergence properties are considered. Finally in §4 we discuss some questions related to the numerical aspects of the problem and we provide some numerical examples.

§2. The Method

Suppose we are given a set of scattered points in a domain $Q \subset \mathbb{R}^2$

$$S = \{P_i^*(x_i, y_i) | \, P_i^* \in Q, \, i = 1, \dots, N\},$$

and the set of functional data

$$F = \{(P_i^*, f(P_i^*)), \, i = 1, \dots, N\},$$

where $f(x, y)$ is an unknown function defined on Q. Without loss of generality, we suppose $Q = [0,1] \times [0,1]$.

The first step of our method consists in generating, from F, a new set of functional values, say $\hat{F}$, located on a dyadic lattice T of Q,

$$T = \{\bar{P}_{\mathbf{k}}, \, \mathbf{k} = (k_1, k_2) \in \mathbb{Z}^2, \, \mathbf{k} = 1, \dots, 2^{\bar{n}}\},$$

that is

$$\hat{F} = \{(\bar{P}_{\mathbf{k}}, \hat{f}(\bar{P}_{\mathbf{k}})), \quad \bar{P}_{\mathbf{k}} \in T\}.$$

It is clear that the new values are affected by errors

$$\hat{f}(\bar{P}_{\mathbf{k}}) = f(\bar{P}_{\mathbf{k}}) + e(\bar{P}_{\mathbf{k}}).$$

Therefore we need to construct $\hat{F}$ by an efficient computational method such that the error is less or of the same order as that generated by the derivative approximation we will use. A possible strategy is to interpolate the data by a local method of a suitable order m, $m \leq \alpha$ (for instance a moving least squares technique, [6, 8]) which gives a smooth interpolating function $\hat{f}(x, y) \in C^m(Q)$ with an error $e(x, y) = O(h^m)$, where h is a local parameter depending on the distribution of the points P_i^* in Q (usually $h = 1/\sqrt{N}$).

The new set $\hat{F}$ can be thought of as a sample coming from a stochastic process depending on the points $P_i^* \in S$. Then we can use the method described in [4] for noisy data. In the next section, we will perform a wavelet decomposition in order to smooth the errors e, and we will define an estimator $\hat{g}_{j(\bar{n})}(x, y)$ of the underlying function $f(x, y)$. Then we will approximate the unknown gradient using the function estimator $\hat{g}_{j(\bar{n})}(x, y)$, and the classical approach of central finite differences.

In this paper we assume that

i) $f(x,y)$ belongs to an Hölder space of order $\alpha > 2$ on $Q^* \supset Q$, say $C^\alpha(Q^*)$.

ii) We consider a multiresolution analysis of $L^2(\mathbb{R}^2)$ given by the tensor product of two one-dimensional s-regular multiresolution analysis such that the scaling function ϕ is a coiflet, that is a compactly supported orthonormal function such that the scaling function $\phi(x)$ and the wavelet $\psi(x)$ have $L-1$ and L, vanishing moments respectively. Moreover, we assume that $L > [\alpha] + 1$ and $s > \alpha$.

2.1. The function estimator

As mentioned above, for approximating the gradient of $f(x,y)$, we will use the function $\hat{g}_{j(\bar{n})}(x,y)$ defined as in [4]. In this section we briefly describe its definition and the motivations that lead to consider this function.

It is known that when we perform a MRA using data given on a subset of $\mathbb{R}^2$, we need to take into account the problem of reducing the boundary errors. To this end, we consider a function $g(x,y) \in C^\alpha(\mathbb{R}^2)$ compactly supported on $Q^* \supset Q$ such that

$$g(x,y) = f(x,y), \quad \forall (x,y) \in Q,$$

and a new dyadic lattice T^* on Q^* of dimension $2^n \times 2^n$, such that $T^* = \{T \cup \{\text{ points sampled in } Q^* \setminus Q\}\}$. Moreover, the advantage of the nested structure of a MRA is that to provide an efficient tree-structure algorithm for the decomposition of functions in $\boldsymbol{V}_n$ for which the smoothing coefficients $\langle g, \Phi_{n,\boldsymbol{k}} \rangle$ are given.

In applications, a function is given in sampled form, and it is therefore necessary to approximate the projection $P_{\boldsymbol{V}_n}$ on the space $\boldsymbol{V}_n$, by some operator Π_n, and to derive a reasonable estimator of Π_n in terms of the sampled values. The choice of Π_n and of its estimator is suggested by the following facts (see [1,4]).

The set of nonzero coefficients $\langle g, \Phi_{n,\boldsymbol{k}} \rangle$ has cardinality equivalent to $O(2^{2n})$. Moreover,

$$\left| \langle g, \Phi_{n,\boldsymbol{k}} \rangle - 2^{-n} g(P_{\boldsymbol{k}}) \right| \le C 2^{-n} 2^{-n\alpha}, \tag{1}$$

where C is a constant depending on the smoothing function $\Phi(x,y)$ and on $g(x,y)$.

As a consequence, we define

$$(\Pi_n g)(x,y) = 2^{-n} \sum_{P_{\boldsymbol{k}} \in T^*} g(P_{\mathbf{k}}) \Phi_{n,\mathbf{k}}(x,y). \tag{2}$$

Since we have data corrupted by the interpolation errors, we consider the estimator of Π_n

$$(\hat{\Pi}_n g)(x,y) = 2^{-n} \{ \sum_{\bar{P}_{\boldsymbol{k}} \in T} \hat{f}(\bar{P}_{\boldsymbol{k}}) \Phi_{n,\boldsymbol{k}}(x,y) + \sum_{P_{\mathbf{k}} \in T^* \setminus T} g(P_{\mathbf{k}}) \Phi_{n,\mathbf{k}}(x,y) \}. \tag{3}$$

This estimator may lead to an oscillatory solution bearing too much fidelity to the data. Since, for numerical differentiation, we need to correctly smooth the data, we have to associate to each sample of size $2^{\bar{n}} \times 2^{\bar{n}}$ a resolution $j(\bar{n}) < \bar{n}$, and to consider the orthogonal projection of $(\hat{\Pi}_n g)(x, y)$ onto $\boldsymbol{V}_{j(\bar{n})}$, that is

$$\hat{g}_{j(\bar{n})}(x, y) = (P_{\boldsymbol{V}_{j(\bar{n})}} \hat{\Pi}_n g)(x, y). \tag{4}$$

The parameter $j(\bar{n})$ governs the smoothness of our estimator, and it is important to choose it in the right way because it controls the tradeoff between the fidelity to the data and the smoothness of the resulting solution. From a theoretical point of view, the smoothing parameter must tend to infinity at the correct rate, as the amount of information in the data grows to infinity.

2.2. The gradient estimator

We now consider the construction of the gradient estimator. When dealing with gridded data, it is natural to approximate the gradient using the usual centered difference formulas. Let

$$(grad\, f)(x, y) = \left(f^{(x)}(x, y), f^{(y)}(x, y)\right),$$

be the gradient of $f(x, y)$, and let D_r^x, D_r^y be the centered difference operators which use r equispaced points in the x or y direction respectively. If $r \leq [\alpha]$, we know that

$$(D_r^t f)(\bar{P}_{\boldsymbol{k}}) = f^{(t)}(\bar{P}_{\boldsymbol{k}}) + O(2^{-n(r-1)}), \quad t = \{x, y\}.$$

Then, at each point of the lattice T, we approximate $(grad\, f)(\bar{P}_{\boldsymbol{k}})$ by

$$(\hat{grad}\, f)(\bar{P}_{\mathbf{k}}) = \left((D_r^x \hat{g}_{j(\bar{n})})(\bar{P}_{\mathbf{k}}), (D_r^y \hat{g}_{j(\bar{n})})(\bar{P}_{\mathbf{k}})\right). \tag{5}$$

Using the data (5), it is possible to define, at each point of Q, an estimator of $f^{(t)}(x, y)$, $t = \{x, y\}$, by the operator $\hat{\Pi}_n$. Namely, we consider the function $g(x, y)$, and we define

$$(\hat{grad}\, f)(x, y) = \left((\hat{\Pi}_n g^{(x)})(x, y), (\hat{\Pi}_n g^{(y)})(x, y)\right), \tag{6}$$

where

$$(\hat{\Pi}_n g^{(t)})(x, y) = 2^{-n}\{ \sum_{\bar{P}_{\mathbf{k}} \in T} (D_r^t \hat{g}_{j(\bar{n})})(\bar{P}_{\mathbf{k}}) \Phi_{n,\mathbf{k}}(x, y)$$
$$+ \sum_{P_{\mathbf{k}} \in T^* \setminus T} g^{(t)}(P_{\mathbf{k}}) \Phi_{n,\mathbf{k}}(x, y)\}.$$

§3. Asymptotic Properties

Under the assumptions of Section 2, and taking into account the properties of the projection $P_{\mathbf{V}_l}$, we know that

$$|(P_{\mathbf{V}_l}g)(x,y) - g(x,y)| = O(2^{-l\alpha}), \quad \forall (x,y) \in Q^*. \tag{7}$$

Moreover, from (1) we have

$$|(\Pi_l g)(x,y) - (P_{\mathbf{V}_l}g)(x,y)| = O(2^{-l\alpha}) \quad \forall (x,y) \in Q^*, \tag{8}$$

then

$$|(\Pi_l g)(x,y) - f(x,y)| = O(2^{-l\alpha}) \quad \forall (x,y) \in Q. \tag{9}$$

Using relations (7), (8), (9) and the results stated in [4], we have proved

Proposition 1. *If assumptions i) and ii) of Section 2 hold, we have*

$$|\hat{g}_{j(\bar{n})}(x,y) - f(x,y)| = O(2^{-j(\bar{n})\alpha}) + O(h^m),$$

for every $(x,y) \in Q$.

Remark 1. *This result points out how the choice of* $j(\bar{n})$ *depends on the sample dimension* N *and on* m. *In fact it has to be chosen so that* $2^{-j(\bar{n})\alpha}$ *is less or of the same order of* h^m.

Proposition 2. *If assumptions i), ii) of Section 2 hold, the approximation (6) of the gradient satisfies, asymptotically, the following bound*

$$\begin{aligned}|(\hat{\Pi}_n g^{(t)})(x,y) - f^{(t)}(x,y)| =& O(2^{-j(\bar{n})\alpha}) + O(2^{-n(r-1)}) \\ &+ O(h^m) + O(2^{-n(\alpha-1)}),\end{aligned}$$

for every $(x,y) \in Q$.

§4. Numerical Results

In this section, we discuss some questions related to the computational costs and to the numerical implementation of the method we have studied. We also present some numerical results.

4.1. Computational costs

The computational costs are essentially given by the wavelet decomposition and by the construction of the gridded data set $\hat{F}$ of dimension 2^{2n}. For the wavelet decomposition, they are at the most of the same order of the sample dimension, that is $O(2^{2n})$. For the construction of $\hat{F}$, if we use a local method of order $m \ll N$, they are given by the solution of N linear systems of dimension $2m+1$. Then the computational costs will be $O\big((\frac{2m+1}{3})^3 N\big) + O(2^{2n})$.

4.2. Numerical implementation

In this section we discuss some questions related to the construction of $\hat{g}_{j(\bar{n})}(x,y)$ and $(\hat{\Pi}_n g^{(t)})(x,y)$. Following the idea of §3.1, we need to extend the given signal to a suitable square $Q^* \supset Q$

$$Q^* = [-2^{-\bar{n}}K, 1+2^{-\bar{n}}K] \times [-2^{-\bar{n}}K, 1+2^{-\bar{n}}K],$$

forcing it to be zero at the Q^* boundary. This is necessary in order to avoid undesirable behaviour at the boundary. The extension can be done following a method proposed in [2]. Note that K is related to the number of points we consider outside Q. On one hand, it cannot be chosen too small, otherwise undesirable boundary oscillations could occur. On the other hand, it depends on the resolution level $j(\bar{n})$. In fact, it has to be chosen so that the discrete wavelet decomposition can be performed.

In the numerical examples we have used the coiflets with $L = 6$ vanishing moments. For the pointwise gradient approximation, we have used the central finite difference of order 4 ($r = 5$).

4.3. Numerical examples

In this paper we present the results achieved for the test functions

$$\begin{aligned} f_1(x,y) &= 0.75\exp[-((9x-2)^2+(9y-2)^2)/4] \\ &\quad + 0.75\exp[-((9x+1)^2/49+(9y+1)^2/10)] \\ &\quad + 0.5\exp[-((9x-7)^2+(9y-3)^2)/4] \\ &\quad - 0.2\exp[-((9x-4)^2+(9y-7)^2)], \\ f_2(x,y) &= \frac{1}{\sqrt{(1+2\exp(-2\sqrt{100x^2+100y^2}-6.7)}}, \end{aligned}$$

defined on the unit square $[0,1]\times[0,1]$. We have considered N scattered points on Q, and have constructed a new gridded data set $\hat{F}$ of size $2^{\bar{n}} \times 2^{\bar{n}}$ using the modified quadratic Shepard method. The smoothing parameter $j(\bar{n})$ has been chosen taking into account Remark 1 of Section 3. Therefore, having used the modified quadratic Shepard method with $\bar{n} = 5$, a possible choice is $j(\bar{n}) = 4$.

We now present our results, and compare them with those obtained with the method (L-method) proposed in [8] which, among those we find in the literature, we belive is preferable both for its theoretical aspects and numerical performance.

The following examples show how the proposed method provides an approximation which seems to have the same behaviour for the functions considered, both for graphical results and for errors. For the sake of brevity, we show only the approximations of one gradient component, but give the relative errors for both of them.

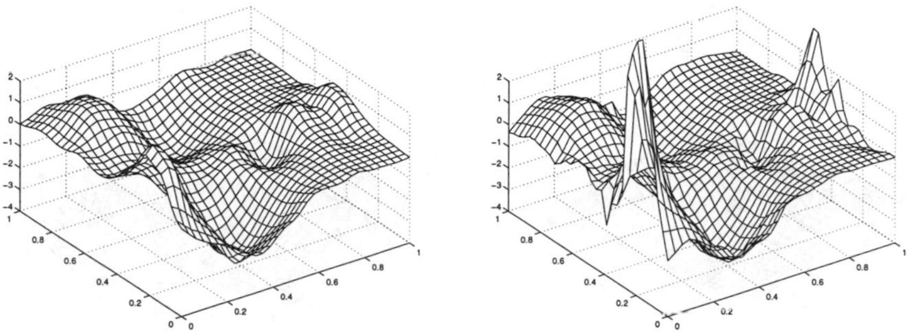

Fig. 1. Example 1, $\boldsymbol{N} = \mathbf{100}$. On the left-hand side, the result of our method. On the right-hand side, the result of the L-method.

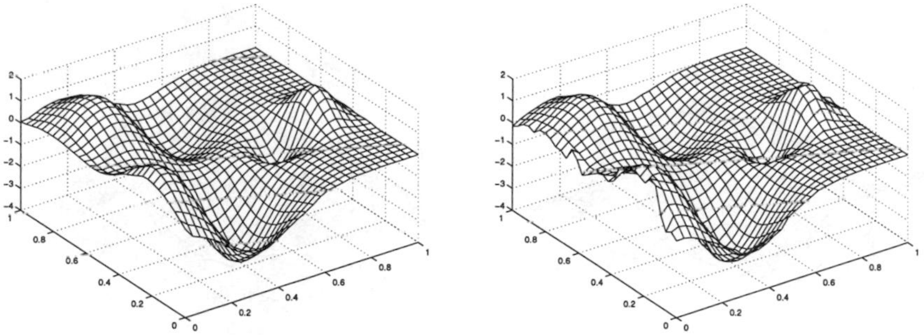

Fig. 2. Example 1. $\boldsymbol{N} = \mathbf{300}$. On the left-hand side, the results of our method. On the right-hand side, the result of the L-method.

Example 1. Consider $N = 100$ and $N = 300$ values of $f_1(x, y)$. In Figs. 1 and 2 we present the y-partial derivative approximations. The following table lists er:= relative error of our method and erL:= relative error of the L-method:

f_1	$N = 100$		$N = 300$	
	er	erL	er	erL
f^x	24%	55%	6%	12%
f^y	43%	226%	36%	36%

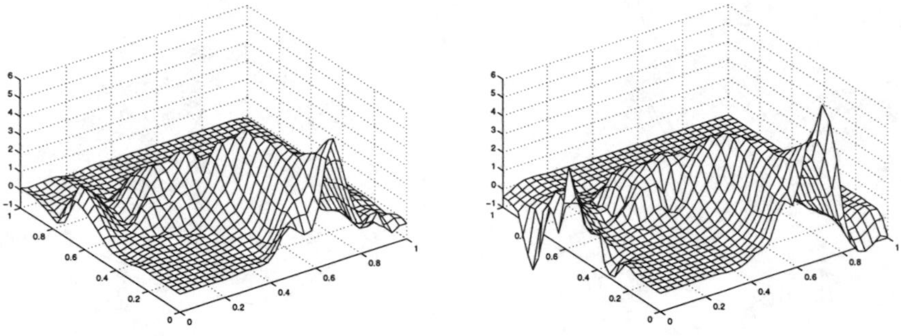

Fig. 3. Example 2. $\boldsymbol{N = 100}$. On the left-hand side, the result of our method. On the right-hand side, the result of the L-method.

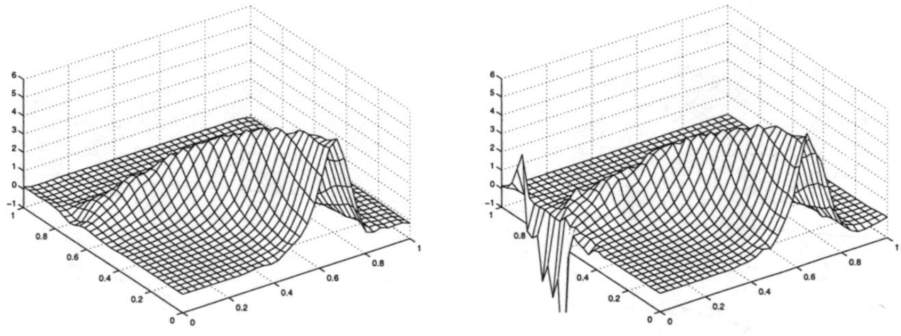

Fig. 4. Example 2. $\boldsymbol{N = 300}$. On the left-hand side, the result of our method. On the right-hand side, the result of the L-method.

Example 2. Consider $N = 100$ and $N = 300$ values of $f_2(x, y)$. In Figs. 3 and 4 we present the results achieved for the y-partial derivative, and in Fig. 5 the error functions for $N = 300$. The following table lists er:= relative error of our method, and erL:= relative error of the L-method:

f_2	$N = 100$		$N = 300$	
	er	erL	er	erL
f^x	35%	400%	6%	54%
f^y	27%	64%	6%	135%

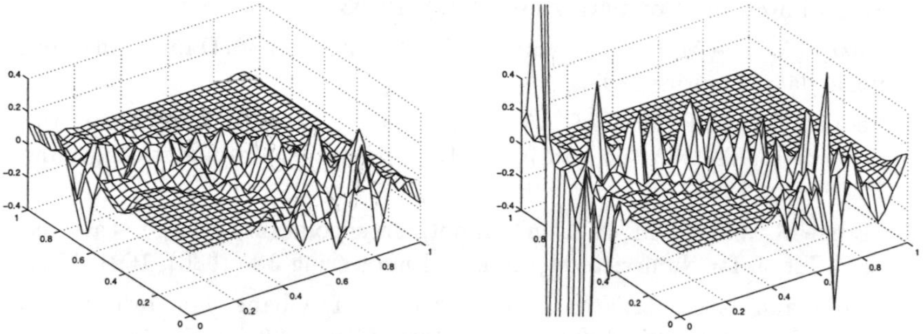

Fig. 5. Example 2. $\boldsymbol{N = 300}$. On the left-hand side, the error function of our method. On the right-hand side, the error function of L-method.

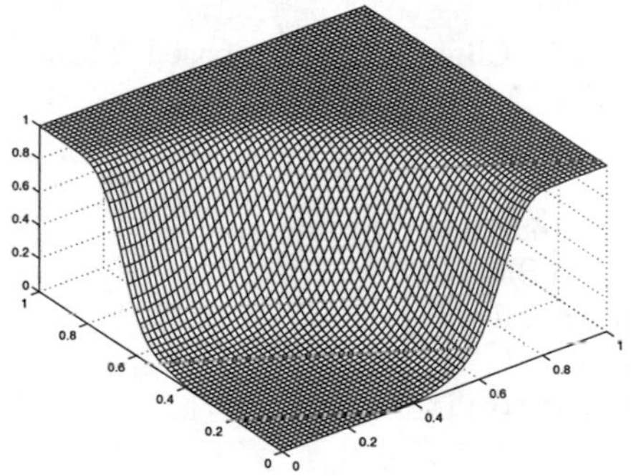

Fig. 6. Example 2. $\boldsymbol{N} = 300$. The approximation of $f_2(x, y)$. The absolute maximum error is 0.015.

Finally, as is usual in the literature, we consider an application to function recovery which shows the goodness of the gradient approximation. We recover $f_2(x, y)$ by Hermite interpolation at 16×16 nodes, where we interpolate the data coming from the estimator $\hat{g}_{j(\bar{n})}(x, y)$ and from the approximated gradient (Fig. 6).

References

1. Antoniadis A., Wavelet methods for smoothing noisy data, in *Curves and Surfaces in Geometric Design*, P.-J. Laurent, A. Le Méhauté, and L. L. Schumaker (eds.), A. K. Peters, Wellesley MA, 1994, 485–492.
2. Bozzini M. and M. Rossini, Approximating surfaces with discontinuity, Mathematical and Computer Modelling, to appear.

3. De Marchi S., On Computing derivatives for C^1 interpolating schemes: an optimization. Computing **60** (1998), 29–53.
4. Bozzini M. and M. Rossini, Numerical differentiation of 2D functions from noisy data, preprint 1999.
5. Feraudi F. and F. De Tisi, A local Method for shape preserving partial derivatives, Technical Report of the Department of Mathematics, Milano University, 1999.
6. Franke R. and G. M. Nielson, Smooth interpolation of large scattered data, Int. J. for Numerical Methods in Engineering **38** (1980), 1691–1704.
7. Goodmann T. N. T., H. B. Said, and L. H. T. Chang, Local derivative estimation for scattered data, App. Math. Comp. **68** (1995), 41–50.
8. Levin D., The approximation power of moving least-squares, Math. Comp. **67**, n. 224 (1998), 1517–1531.
9. Moriconi S., Sull'approssimazione numerica delle derivate e loro utilizzo nella ricostruzione di superfici da dati sparsi. Math. Degree Thesis, University of Milano, 1997.
10. Renka R. J. and A. K. Cline, A triangle based C^1 interpolation method, Rocky Mountain J. of Mathematics **14** (1984), 223–237.

Mira Bozzini
Dipartimento di Matematica e Applicazioni
Università di Milano Bicocca
via Bicocca degli Arcimboldi 8
20126, Milano Italy
`bozzini@matapp.unimib.it`

Milvia Rossini
Dipartimento di Matematica e Applicazioni
Università di Milano Bicocca
via Bicocca degli Arcimboldi 8
20126, Milano Italy
`rossini@matapp.unimib.it`

Ridgelets and their Derivatives: Representation of Images with Edges

Emmanuel J. Candès

Abstract. This paper reviews the development of several recent tools from computational harmonic analysis. These new systems are presented under a coherent perspective, namely, the representation of bivariate functions that are singular along smooth curves (edges). First, the representation of functions that are smooth away from straight edges is presented, and ridgelets will be shown to provide near optimal nonlinear approximations to these objects. Motivated by the limitations of the ridgelet methodology, new representation systems, namely, monoscale ridgelets and curvelets – both of which use the ridgelet transform as a building block – will be introduced. Curvelets are shown to provide concrete and constructive optimal nonlinear approximations to smooth functions with twice differentiable singularities. In addition, these approximations are obtained simply by thresholding the curvelet series.

§1. Introduction

Throughout the sciences, sparse representations of classes of objects are often sought because of the well-known applications of sparsity to problems ranging from data compression and statistical estimation to feature detection. Indeed, finding sparse representations together with rapid algorithms to compute them is one of the main objectives of a rapidly growing field, computational harmonic analysis (CHA). In this paper, we will argue that CHA has not really addressed the problem of efficiently representing smooth multivariate functions with sharp discontinuities, like smooth images with edges. Motivated by this gap in the literature, we present a collection of new representation tools that efficiently represent smooth functions that are singular along curves. Here, the tone is expository; details may be found in the cited references. In this paper, attention is restricted to the two-dimensional situations although extensions to higher dimensions exist, or are anticipated.

Curve and Surface Fitting: Saint-Malo 1999
Albert Cohen, Christophe Rabut, and Larry L. Schumaker (eds.), pp. 95–104.

ISBN 0-8265-1357-3.

The wavelet miracle

One of the most appealing features of wavelet systems is their ability to provide efficient representations of spatially inhomogeneous functions, i.e., functions that may be discontinuous, spiky, etc. In Mallat's words "bases of smooth wavelets are the best bases for representing objects composed of singularities, when there may be an arbitrary number of singularities, which may be located in all possible spatial positions" [8]. For instance, on the unit interval define

$$f(t) = H(t - t_0)\, g(t), \quad t \in [0, 1], \tag{1}$$

where H is the Heavyside $H(t) = 1_{\{t>0\}}$ and g is a smooth arbitrary function with compact support and finite Sobolev norm $\|g\|_{W_2^s}$ (see [1] for the classical definition of L_2 Sobolev norms). Then, the number of Fourier coefficients of f exceeding $1/n$ in absolute value is bounded below by $c \cdot n$, regardless of the degree of smoothness of f away from the singular point t_0. This means that a lot of different terms are needed to obtain good partial reconstructions; keeping the n largest terms in the Fourier series gives only an L_2 error of approximation of order $n^{-1/2}$. (Throughout the paper, it will always be implicit that the error is measured in the L_2 norm.) In contrast, the sparsity of the wavelet coefficient sequence of f is in some sense the same as if f were not singular. In effect, the number of wavelet coefficients exceeding $1/n$ is bounded by $C\, n^{2/(2s+1)}$ giving rates of approximation of order n^{-s} corresponding to the nonlinear bandwidth of W_2^s Sobolev balls. This remarkable adaptivity property is what we call the "wavelet miracle."

The curse

Unfortunately, wavelets can deal with point-like singularities, but are seriously challenged by line-like singularities in dimension two. Let us for instance consider the object

$$f(x_1, x_2) = H(x_1 \cos\theta_0 + x_2 \sin\theta_0 - t_0)\, g(x_1, x_2), \quad (x_1, x_2) \in [0, 1]^2, \tag{2}$$

where, again, g is a bivariate function taken from the Sobolev space W_2^s; f is singular on the line $x_1 \cos\theta_0 + x_2 \sin\theta_0 = t_0$, but smooth otherwise. Then, the number of wavelet coefficients exceeding $1/n$ is now of the order n. Hence, partial n-term wavelet reconstructions will only converge at a rate $n^{-1/2}$, regardless of the almost everywhere degree s of smoothness. The edge limits the speed of convergence. This result is intuitively not very surprising as wavelet bases are made of local isotropic oscillatory bumps at various scales, and are not adapted to represent long elongated structures like edges.

This clearly raises an important question: in two dimensions (and, more generally, in arbitrary d dimensions) can we develop a representation enjoying the same adaptivity features as wavelets in dimension one?

§2. Ridgelets and Linear Singularities

In [3], Candès introduced a new tiling of the frequency plane that led to the construction of ridgelet frames. We say that a collection (φ_n) is a frame of a Hilbert space H if there exist two constants $A, B > 0$ such that for any element of H, we have

$$A \|f\|_H^2 \leq \sum_n |\langle f, \varphi_n \rangle_H|^2 \leq B \|f\|_H^2.$$

When $A = B$, the frame is said to be tight. A collection (φ_n) that verifies the frame property is of course complete and there is a very concrete way to reconstruct f from the datum of its coefficients $(\langle f, \varphi_n \rangle_H)$. Generalities about frames can be found in [11].

Let ψ be a univariate oscillatory function and $\psi_{j,k}(t) = 2^{j/2}\psi(2^j t - k)$. The ridgelet frame $\psi_{j,\ell,k}$ is a collection of ridge functions given by

$$\hat{\psi}_{j,\ell,k}(\xi) = \hat{\psi}_{j,k}(|\xi|)\delta(\theta - 2\pi\, 2^{-j}\ell) + \hat{\psi}_{j,k}(-|\xi|)\delta(\theta + \pi - 2\pi\, 2^{-j}\ell)$$

in the frequency domain [3] (δ denotes the dirac distribution).

Donoho [9] modified the ridgelet construction by essentially replacing the discretization of the angular variable with a periodic wavelet transform resulting in an orthonormal basis. He called these new basis elements orthonormal ridgelets. In the remainder of this paper, we make the choice of the orthonormal ridgelets, although all the results and constructions that follow would hold true if one were to use 'pure ridgelets.'

As stated in [9], such a system can be defined as follows: let $(\psi_{j,k}(t))_{j,k \in \mathbb{Z}}$ be an orthonormal basis of Meyer wavelets for $L^2(\mathbb{R})$ [12], and let $(w^0_{i_0,\ell}(\theta),\ \ell = 0, \ldots, 2^{i_0} - 1;\ w^1_{i,\ell}(\theta),\ i \geq i_0,\ \ell = 0, \ldots, 2^i - 1)$ be an orthonormal basis for $L^2[0, 2\pi)$ made of periodized Lemarié scaling functions $w^0_{i_0,\ell}$ at level i_0 and periodized Meyer wavelets $w^1_{i,\ell}$ at levels $i \geq i_0$. (We suppose a particular normalization of these functions.) Let $\hat{\psi}_{j,k}(\omega)$ denote the Fourier transform of $\psi_{j,k}(t)$, and define ridgelets $\rho_\lambda(x)$, $\lambda = (j, k; i, \ell, \varepsilon)$ as functions of $x \in \mathbb{R}^2$ using the frequency-domain definition

$$\hat{\rho}_\lambda(\xi) = |\xi|^{-\frac{1}{2}}(\hat{\psi}_{j,k}(|\xi|)w^\varepsilon_{i,\ell}(\theta) + \hat{\psi}_{j,k}(-|\xi|)w^\varepsilon_{i,\ell}(\theta + \pi))/2. \tag{3}$$

Here the indices run as follows: $j, k \in \mathbb{Z}$, $\ell = 0, \ldots, 2^{i-1} - 1$; $i \geq i_0$, $i \geq j$. Notice the restrictions on the range of ℓ and on i. Let λ denote the set of all such indices λ. It turns out that $(\rho_\lambda)_{\lambda \in \Lambda}$ is a complete orthonormal system for $L^2(\mathbb{R}^2)$. Hence, we have a new decomposition of the form

$$f = \sum_\lambda \langle f, \rho_\lambda \rangle \rho_\lambda.$$

Ridgelets turn out to be optimal for representing functions with linear singularities. Indeed, let us consider the template (2). The following theorem is proved in [4].

Theorem 1. *Let $g \in W_2^s(\mathbb{R}^2)$ and $f(x_1, x_2) = H(x_1 \cos\theta_0 + x_2 \sin\theta_0 - t_0)\, g(x_1, x_2)$. Then the sequence $(\alpha_\lambda = \langle f, \rho_\lambda \rangle)$ of orthonormal ridgelet coefficients of f satisfies*

$$\#\{|\alpha_\lambda| \geq 1/n\} \leq C\, n^{1/(s+1)}\, \|g\|_{W_2^s}$$

for some constant C not depending on f. As a consequence, the n-term approximation f_n – obtained by keeping the terms corresponding to the n largest coefficients in the ridgelet expansion – satisfies

$$\|f - f_n\| \leq C\, n^{-s/2}\, \|g\|_{W_2^s}.$$

Hence, the theorem states that we obtain a rate of approximation as if the object were not singular, simply by thresholding the orthonormal ridgelet expansion. Whereas the singularity caused partial wavelet reconstructions to converge very slowly, its effect on the approximation rate of truncated ridgelet series is 'harmless.'

§3. Ridgelets and Curved Edges.

Theorem 1 considered linear singularities and it seems natural to ask whether similar results will hold if one replaces the singularity along a straight line with one along an arbitrary curve γ. To simplify our exposition, consider the simple case of a singular function defined on the unit square by

$$f(x_1, x_2) = g(x_1, x_2)\, 1_{\{x_2 \leq \gamma(x_1)\}}, \tag{4}$$

where g is a smooth function and γ is smooth curve. Then the ridgelet coefficient sequence of such an object is in general not sparse:

$$\#\{\lambda, |\alpha_\lambda| \geq 1/n\} \geq c\, n.$$

Thus, the speed of convergence of the best n-term ridgelet approximation is only of order $n^{-1/2}$. It is interesting to observe that the degree of approximation of both wavelet and ridgelet partial reconstructions is the same, although they correspond to radically different systems of representation. Ridgelets are elongated and directional, whereas wavelets are isotropic and local.

The limitations that we presented in this section motivate the refinements and new tools that we are about to introduce.

§4. Monoscale Ridgelets

The approach developed in this section builds on Theorem 1. The idea here is to take advantage of the optimal representation of linear singularities by localizing the ridgelets. A detailed exposition is provided in [5].

For an integer $s \geq 0$ and integers k_1, k_2, we let Q be the dyadic square defined by $Q = [k_1/2^s, (k_1+1)/2^s) \times [k_2/2^s, (k_2+1)/2^s)$. The collection of all dyadic squares at scale s will be denoted by $\mathcal{Q}_s$. The idea is to smoothly localize the function f we wish to represent near each of the dyadic squares of $\mathcal{Q}_s$. We choose an orthonormal partition of unity w_Q; that is, a collection of windows such that w_Q^2 is a partition of unity

$$\sum_{Q \in \mathcal{Q}_s} w_Q^2 = 1.$$

The following details a way of making up such an orthonormal partition: take a C^∞ univariate window ν supported in $[-3/4, 3/4]$ such that $\nu(t) = 1$ on $[-1/2, 1/2]$; define $v_Q = \nu(2^s x_1 - k_1)\, \nu(2^s x_2 - k_2)$; and renormalize the windows v_Q with

$$w_Q = v_Q / (\sum_{Q \in \mathcal{Q}_s} v_Q^2)^{1/2}.$$

It is then clear that the w_Q's obey the desired condition.

Define the rescaling operator $T_Q g$ by

$$T_Q g = 2^s g(2^s x_1 - k_1, 2^s x_2 - k_2),$$

which is an isometry of L_2. Throughout this section, s is arbitrary but fixed. Monoscale ridgelets are defined as follows: let ρ_λ be an orthonormal ridgelet basis and define

$$\psi_{Q,\lambda}(x_1, x_2) = w_Q(x_1, x_2)(T_Q \rho_\lambda)(x_1, x_2);$$

the collection

$$\{\psi_{Q,\lambda}, Q \in \mathcal{Q}_s, \lambda \in \Lambda\} \tag{5}$$

is what we call the monoscale ridgelet dictionary.

It is easy to check that the monoscale ridgelet dictionary is a tight frame of $L_2(\mathbb{R}^2)$ as we have a Parseval relationship

$$\|f\|_2^2 = \sum_{Q \in \mathcal{Q}_s} \sum_\lambda \langle f, \psi_{Q,\lambda} \rangle^2.$$

Standard arguments show that we then have the decomposition

$$f = \sum_{Q \in \mathcal{Q}_s} \sum_\lambda \langle f, \psi_{Q,\lambda} \rangle \psi_{Q,\lambda}, \tag{6}$$

with equality holding in an L_2 sense.

We add an "extra layer of coarse scale coefficients" to eliminate various artifacts. Consider a standard multiresolution analysis that is adapted to the unit square [7] so that the set of translates $\{2^s \varphi(2^s \cdot - k)\}$, $k = (k_1, k_2)$, $k_i = 0, 1, \ldots, 2^s - 1$ is orthonormal. Let P_0 be the orthogonal projector onto V_s, the span of the $\varphi_{s,k}$'s; i.e.,

$$P_0 f := \sum_k \langle f, \varphi_{s,k} \rangle \varphi_{s,k} := \sum_k \beta_{s,k} \varphi_{s,k}.$$

The following Pythagorean relationship holds:

$$\|f\|_2^2 = \|P_0 f\|_2^2 + \|(I - P_0)f\|_2^2. \tag{7}$$

Finally, define the coefficients

$$\alpha_{s,\mu} = \langle Rf, \psi_{Q,\lambda}\rangle \quad \mu = (Q, \lambda), \quad Q \in \mathcal{Q}_s, \lambda \in \Lambda. \tag{8}$$

Definition 1. *The* monoscale ridgelet transform with base scale s *is the mapping from functions* $f \in L_2(\mathbb{R}^2)$ *to the amalgamation of coefficients* $(\beta_{s,k})$ *and* $(\alpha_{s,\mu})$.

Note that we again have a partial isometry

$$\|f\|_2^2 = \sum_k |\beta_{s,k}|^2 + \sum_\mu |\alpha_{s,\mu}|^2,$$

thanks to the Pythagorean relationship (7).

Let us return now to the main theme of this paper, and study the efficiency of monoscale ridgelets to represent objects that are singular along curves. Suppose that one is interested in constructing an n-term approximation of the function f in (4). Without loss of generality, we will suppose that n is of the form $n = 2^{2J+1}$. We simply expand f in the monoscale ridgelet dictionary (5) *with* $s = J$ *as a choice of base scale*; that is, we define the n-term approximation by

$$f_n = P_0 f + R_{n/2} f, \tag{9}$$

where $R_{n/2}f$ is the partial reconstruction of the residual Rf obtained by keeping the terms corresponding to the $n/2 = 2^{2J}$ largest coefficients $\alpha_{J,\mu}$.

It is interesting to observe that the choice of the base scale s of the monoscale dictionary depends on the number n of terms we wish to keep in the approximant. We have the following result [5]:

Theorem 2. *Let* $g \in W_2^s(\mathbb{R}^2)$ *and* $f(x) = g(x)\, 1_{\{x_2 \le \gamma(x_1)\}}$, *with* γ *being three times differentiable. Let* f_n *be the* n*-term approximation defined by (9). Then,*

$$\|f - f_n\|_2 \le C \max(n^{-s/2}, n^{-3/4}).$$

This simple approximation scheme provides optimal rates of convergence as long as $s \le 3/2$; that is, approximation bounds as if f were not singular. In some sense, one is allowed to say that unlike wavelets, ridgelets can be adapted to provide efficient representations of curved singularities. There is a critical value $s = 3/2$ of the smoothness parameter, however, beyond which the method saturates; as s increases, the approximation rate is blocked at $n^{-3/4}$. Nevertheless, this represents already a substantial improvement over wavelet approximations whose convergence rates are blocked at $n^{-1/2}$.

Better results are theoretically possible. For instance, let $\mathcal{F}(C)$ be a model of smooth images with twice differentiable edges defined as follows:

$$\mathcal{F}(C) = \{f : f \text{ satisfies (4) with } \|g\|_{W_2^2} \le C \text{ and } \|\gamma\|_{\dot{C}^2} \le C\}.$$

The condition $\|\gamma\|_{\dot{C}^2} \le C$ states that the homogeneous Hölder norm of order 2 is bounded by C. In other words, γ is differentiable and its first derivative satisfies the Lipschitz condition $|\gamma'(u) - \gamma'(v)| \le C\,|u - v|$. For this class of objects, it can be shown that there are reasonable ways of constructing approximations converging at the rate $n^{-1}\log n$. Monoscale ridgelets do not attain this optimal rate.

§5. Curvelets and Curved Singularities

The curvelet transform – introduced by Candès and Donoho in [6] – is the last of the representation tools that we will review. Whereas the monoscale ridgelet transform involved taking ridgelet coefficients with a fixed base scale s, the curvelet transform spans all possible scales $s \ge 0$. A useful slogan is that the curvelet transform is obtained by filtering and then applying a multiscale ridgelet transform. The multiscale ridgelet dictionary is the collection of the monoscale dictionaries at all possible scales $s \ge 0$; i.e.,

$$\{\psi_\mu := \psi_{Q,\lambda},\ s \ge 0, Q \in \mathcal{Q}_s, \lambda \in \Lambda\}. \tag{10}$$

The curvelet transform requires the use of a sequence of filters that we now describe. Let Φ_0 and Ψ_{2s}, $s = 0, 1, 2, \ldots$ satisfy the following properties:

- Φ_0 is a lowpass filter and is concentrated at frequencies $|\xi| \le 2$;
- Ψ_{2s} is bandpass and concentrated at frequencies $|\xi| \in [2^{2s-1}, 2^{2s+3}]$;
- the filters satisfy

$$|\hat{\Phi}_0(\xi)|^2 + \sum_{s\ge0} |\hat{\Psi}_{2s}(\xi)|^2 = 1.$$

Existence and constructions of such filters are well-known. The last relationship implies that the transformation of f into a bank of functions

$$f \mapsto (P_0 f = \Phi_0 * f, \Delta_0 f = \Psi_0 * f, \Delta_1 f = \Psi_1 * f, \ldots, \Delta_s f = \Psi_{2s} * f, \ldots)$$

is a partial isometry in the sense that

$$\|f\|_2^2 = \|P_0 f\|_2^2 + \sum_{s\ge0} \|\Delta_s * f\|_2^2.$$

Equipped with both a multiscale ridgelet dictionary and a sequence of filters, define the *curvelet coefficient* α_μ of f by

$$\alpha_\mu = \langle \Delta_s f, \psi_{Q,\lambda}\rangle, \quad Q \in \mathcal{Q}_s, \lambda \in \Lambda. \tag{11}$$

Thus, the coefficient α_μ is interpreted as the multiscale ridgelet coefficient of a piece of f containing information at frequencies near 2^{2s}. We would like to point out that there is a quadratic scaling relationship between the scale 2^s of the multiscale ridgelet and the frequency content, localized around the corona of radius 2^{2s}, of the piece that is analyzed. This relationship is the key feature of the curvelet transform.

We proceed a little bit differently for the piece of f containing information at low frequencies $P_0 f$. Recall the orthogonal collection of Lemarié-Meyer scaling functions $V_k(x_1, x_2) = V(x_1 - k_1, x_2 - k_2)$, for $k = (k_1, k_2) \in \mathbb{Z}^2$. We make the choice of a base scale so that $\hat{V}_0(\xi) = 1$ for $|\xi| \leq 4/3$; and we make sure that the span of the translates V_k contains the range of the projector $P_0 f$. We define the coarse scale curvelet coefficients by

$$\beta_k = \langle P_0 f, V_k \rangle, \quad k \in \mathbb{Z}^2.$$

It will be more convenient to use a single notation to index the set of curvelet coefficients; the notation M' will stand for the union of M and $k \in \mathbb{Z}^2$. When $\mu \in M' \setminus M$, we let $\alpha_\mu = \beta_k$.

Definition 2. *The* curvelet transform *is the mapping that associates the coefficients sequence α_μ, $\mu \in M'$ to an arbitrary square integrable function f.*

We will call curvelets those elements σ_μ defined by

$$\sigma_\mu = \Delta_s \psi_{Q,\lambda}, \quad Q \in \mathcal{Q}_s, \lambda \in \Lambda, \tag{12}$$

with an obvious modification for the piece corresponding to the low frequencies, $\sigma_\mu = P_0 V_k$.

The collection of curvelets is then a tight frame for $L_2(\mathbb{R}^2)$

$$\|f\|_2^2 = \sum_{\mu \in M'} \langle f, \sigma_\mu \rangle^2, \tag{13}$$

and, of course, we have the decomposition

$$f = \sum_{\mu \in M'} \langle f, \sigma_\mu \rangle \sigma_\mu \tag{14}$$

with equality in an L_2 sense.

Let f_n be the truncated n-term curvelet series

$$f_n = \sum_{\mu \in M'} \alpha_\mu 1_{\{|\alpha_\mu| \geq |\alpha|_{(n)}\}} \sigma_\mu. \tag{15}$$

The following theorem is proved in [6].

Theorem 3. *Let $g \in W_2^s(\mathbb{R}^2)$ and $f(x) = g(x)\, 1_{\{x_2 \leq \gamma(x_1)\}}$, with γ being two times differentiable. Let f_n be the n-term approximation (15). Then,*

$$\|f - f_n\|_2 \leq C\, n^{-1} (\log n)^{1/2}.$$

Again, we have a very concrete procedure that achieves rates of approximation that cannot be fundamentally improved. A detailed discussion about the optimality of this result is in [6].

§6. Conclusion

In this paper, we presented a connected set of ideas originating in the ridgelet transform and culminating in the curvelet transform. We have shown how these representations provide efficient representations of objects that are singular along curves. These tools, however, may have several other potential applications.

Because of space limitations, we set aside questions related to the practicability of these new methods. We would like to point out that fast algorithms have been developed to implement the ridgelet, monoscale ridgelet and curvelet transform. We will report on the numerical aspects of these transforms in a separate paper.

Acknowledgments. The author is especially grateful to David Donoho for many fruitful discussions. This research was supported by National Science Fundation grant DMS 98–72890 (KDI) and grant DMS 95–05151 and by AFOSR MURI 95–P49620–96–1–0028.

References

1. Adams, R. A., *Sobolev Spaces*, New York, Academic Press, 1975.
2. Candès, E. J., Ridgelets: theory and applications, Ph.D. thesis, Department of Statistics, Stanford University, 1998.
3. Candès, E. J., Harmonic analysis of neural networks, Appl. Comput. Harmonic Anal. **6** (1999) 197–218.
4. Candès, E. J., On the representation of mutilated Sobolev functions, Tech. Report, Department of Statistics, Stanford University, 1999.
5. Candès, E. J., Monoscale ridgelets for the representation of images with edges, Tech. Report, Department of Statistics, Stanford University, 1999.
6. Candès, E. J., and D. L. Donoho, Curvelets, Tech. Report, Department of Statistics, Stanford University, 1999.
7. Cohen, A., I. Daubechies, and P. Vial, Wavelets on the interval and fast wavelet transforms, Appl. Comput. Harmonic Anal. **1** (1993), 54–81.
8. Donoho, D. L., Unconditional bases are optimal bases for data compression and for statistical estimation, Appl. Comput. Harmonic Anal. **1** (1993), 100–115.
9. Donoho, D. L., Orthonormal ridgelets and linear singularities, Tech. Report, Department of Statistics, Stanford University, 1998.
10. Daubechies, I., Orthonormal bases of compactly supported wavelets, Commun. Pure Appl. Math. **41** (1988), 909–996.
11. Daubechies, I., *Ten Lectures on Wavelets*, Society for Industrial and Applied Mathematics, Philadelphia, PA, 1992.
12. Lemarié, P. G., and Y. Meyer, Ondelettes et bases Hilbertiennes, Rev. Mat. Iberoamericana **2** (1986), 1–18.
13. Meyer Y., *Ondelettes et Opérateurs*, Hermann, 1990.

Department of Statistics
Stanford University
Stanford, CA 94305-4065, USA
`emmanuel@stat.stanford.edu`
`http://www-stat.stanford.edu/~emmanuel/`

Curvelets: A Surprisingly Effective Nonadaptive Representation for Objects with Edges

Emmanuel J. Candès and David L. Donoho

Abstract. It is widely believed that to efficiently represent an otherwise smooth object with discontinuities along edges, one must use an adaptive representation that in some sense 'tracks' the shape of the discontinuity set. This folk-belief — some would say folk-theorem — is incorrect. At the very least, the possible quantitative advantage of such adaptation is vastly smaller than commonly believed. We have recently constructed a tight frame of curvelets which provides stable, efficient, and near-optimal representation of otherwise smooth objects having discontinuities along smooth curves. By applying naive thresholding to the curvelet transform of such an object, one can form m-term approximations with rate of L^2 approximation rivaling the rate obtainable by complex adaptive schemes which attempt to 'track' the discontinuity set. In this article we explain the basic issues of efficient m-term approximation, the construction of efficient adaptive representation, the construction of the curvelet frame, and a crude analysis of the performance of curvelet schemes.

§1. Introduction

In many important imaging applications, images exhibit edges – discontinuities across curves. In traditional photographic imaging, for example, this occurs whenever one object occludes another, causing the luminance to undergo step discontinuities at boundaries. In biological imagery, this occurs whenever two different organs or tissue structures meet.

In image *synthesis* applications, such as CAD, there is no problem in dealing with such discontinuities, because one knows where they are and builds the discontinuities into the representation by specially adapting the representation — for example, inserting free knots, or adaptive refinement rules.

In image *analysis* applications, the situation is different. When working with real rather than synthetic data, one of course doesn't 'know' where these edges are; one only has a digitized pixel array, with potential imperfections caused by noise, by blurring, and of course by the unnatural pixelization of the underlying continuous scene. Hence the typical image analyst only

Curve and Surface Fitting: Saint-Malo 1999
Albert Cohen, Christophe Rabut, and Larry L. Schumaker (eds.), pp. 105–120.

ISBN 0-8265-1357-3.

has recourse to representations which don't 'know' about the existence and geometry of the discontinuities in the image.

The success of discontinuity-adapting methods in CAD and related image synthesis fields creates a temptation for an image analyst – a temptation to spend a great deal of time and effort importing such ideas into image analysis. Almost everyone we know has yielded to this temptation in some form, which creates a possibility for surprise.

Oracles and ideally-adapted representation

One could imagine an ideally-privileged image analyst who has recourse to an *oracle* able to reveal the positions of all the discontinuities underlying the image formation. It seems natural that this ideally-privileged analyst could do far better than the normally-endowed analyst who knows nothing about the position of the discontinuities in the image.

To elaborate this distinction, we introduce terminology borrowed from fluid dynamics, where 'edges' arise in the form of fronts or shock fronts.

A Lagrangian representation is constructed using full knowledge of the intrinsic structure of the object and adapting perfectly to that structure.

- In fluid dynamics this means that the fluid flow pattern is known, and one constructs a coordinate system which 'flows along with the particles', with coordinates mimicking the shape of the flow streamlines.
- In image representation this could mean that the edge curves are known, and one constructs an image representation adapted to the structure of the edge curves. For example, one might construct a basis with discontinuities exactly where the underlying object has discontinuities.

An Eulerian representation is fixed, constructed once and for all. It is non-adaptive – having nothing to do with the known or hypothesized details of the underlying object.

- In fluid dynamics, this would mean a usual euclidean coordinate system, one that does not depend in any way on the fluid motion.
- In image representation, this could mean that the representation is some fixed coordinate representation, such as wavelets or sinusoids, which does not change depending on the positions of edges in the image.

It is quite natural to suppose that the Lagrangian perspective, when it is available, is much more powerful that the Eulerian one. Having the privilege of 'inside information' about the position of important geometric characteristics of the solution seems *a priori* rather valuable. In fact, this position has rather a large following. Much recent work in computational harmonic analysis (CHA) attempts to find bases which are optimally adapted to the specific object in question [7,10,11]; in this sense much of the ongoing work in CHA is based on the presumption that the Lagrangian viewpoint is best.

In the setting of edges in images, there has, in fact, been considerable interest in the problem of developing representations which are adapted to the structure of discontinuities in the object being studied. The (equivalent) concepts of probing and minimum entropy segmentation are old examples of this: wavelet systems which are specifically constructed to allow discontinuities in the basis elements at specific locations [8,9]. More recently, we are aware of much informal unpublished or preliminary work attempting to build $2D$ edge-adapted schemes; we give two examples.

- *Adaptive triangulation* aims to represent a smooth function by partitioning the plane into a sequence of triangular meshes, refining the meshes at one stage to create finer meshes at the next stage. One represents the underlying object using piecewise linear functions supported on individual triangles. It is easy to see how, in an *image synthesis* setting, one can in principle develop a triangulation where the triangles are arranged to track a discontinuity very faithfully, with the bulk of refinement steps allocated to refinements near the discontinuity, and one obtains very effective representation of the object. It is not easy to see how to do this in an *image analysis* setting, but one can easily be persuaded that the development of adaptive triangulation schemes for noisy, blurred data is an important and interesting project.
- In an *adaptively warped wavelet representation*, one deforms the underlying image so that the object being analyzed has all its discontinuities aligned purely horizontal or vertical. Then one analyzes the warped object in a basis of tensor-product wavelets where elements take the form $\psi_{j,k}(x_1) \cdot \psi_{j',k'}(x_2)$. This is very effective for objects which are smooth apart from purely horizontal and purely vertical discontinuities. Hence, the warping deforms the singularities to render the the tensor product scheme very effective. It is again not easy to see how adaptive warping could work in an *image analysis* setting, but one is easily persuaded that development of adaptively warped representations for noisy, blurred data is an important and interesting project.

Activity to build such adaptive representations is based on an article of faith: namely, that *Eulerian approaches are inferior, that oracle-driven Lagrangian approaches are ideal*, and that one should, in an image analysis setting, mimic Lagrangian approaches, attempting empirically to estimate from noisy, blurred data the information that an oracle would supply, and build an adaptive representation based on that information.

Quantifying rates of approximation

In order to get away from articles of faith, we now quantify performance, using an asymptotic viewpoint.

Suppose we have an object supported in $[0,1]^2$ which has a discontinuity across a nice curve Γ, and which is otherwise smooth. Then using a standard Fourier representation, and approximating with $\tilde{f}_m^F$ built from the best m nonzero Fourier terms, we have

$$\|f - \tilde{f}_m^F\|_2^2 \asymp m^{-1/2}, \qquad m \to \infty. \tag{1}$$

This rather slow rate of approximation is improved upon by wavelets. The approximant $\tilde{f}_m^W$ built from the best m nonzero wavelet terms satisfies

$$\|f - \tilde{f}_m^W\|_2^2 \asymp m^{-1}, \qquad m \to \infty. \tag{2}$$

This is better than the rate of Fourier approximation, and, until now, is the best published rate for a fixed non-adaptive method (i.e. best published result for an 'Eulerian viewpoint').

On the other hand, we will discuss below a method which is adapted to the object at hand, and which achieves a much better approximation rate than previously known 'nonadaptive' or 'Eulerian' approaches. This adaptive method selects terms from an overcomplete dictionary and is able to achieve

$$\|f - \tilde{f}_m^A\|_2^2 \asymp m^{-2}, \qquad m \to \infty. \tag{3}$$

Roughly speaking, the terms in this dictionary amount to triangular wedges, ideally fitted to approximate the shape of the discontinuity.

Owing to the apparent trend indicated by (1)-(3) and the prevalence of the puritanical belief that 'you can't get something for nothing', one might suppose that inevitably would follow the

Folk-Conjecture/[Folk-Theorem]. *The result (3) for adaptive representations far exceeds the rate of m-term approximation achievable by fixed non-adaptive representations.*

This conjecture appeals to a number of widespread beliefs:

- the belief that adaptation is very powerful,
- the belief that the way to represent discontinuities in image analysis is to mimic the approach in image synthesis,
- the belief that wavelets give the best fixed nonadaptive representation.

In private discussions with many respected researchers we have many times heard expressed views equivalent to the purported Folk-Theorem.

The surprise

It turns out that performance almost equivalent to (3) can be achieved by a *non*adaptive scheme. In other words, the Folk-Theorem is effectively false.

There is a tight frame, fixed once and for all nonadaptively, which we call a frame of curvelets, which competes surprisingly well with the ideal adaptive rate (3). A very simple m-term approximation – summing the m biggest terms in the curvelet frame expansion – can achieve

$$\|f - \tilde{f}_m^C\|_2^2 \le C \cdot m^{-2}(\log m)^3, \qquad m \to \infty, \tag{4}$$

which is nearly as good as (3) as regards asymptotic order. In short, *in a problem of considerable applied relevance, where one would have thought that adaptive representation was essentially more powerful than fixed nonadaptive representation, it turns out that a new fixed nonadaptive representation is essentially as good as adaptive representation, from the point of view of asymptotic m-term approximation errors.* As one might expect, the new nonadaptive representation has several very subtle and distinctive features.

Contents

In this article, we would like to give the reader an idea of why (3) represents the ideal behavior of an adaptive representation, of how the curvelet frame is constructed, and of the key elements responsible for (4). We will also attempt to indicate why curvelets perform for singularities along curves the task that wavelets perform for singularities at points.

§2. A Precedent: Wavelets and Point Singularities

We mention an important precedent – a case where a nonadaptive scheme is roughly competitive with an ideal adaptive scheme. Suppose we have a piecewise polynomial function f on the interval $[0, 1]$, with jump discontinuities at several points.

An obvious adaptive representation is to fit a piecewise polynomial with breakpoints at the discontinuities. If there are P pieces and each polynomial is of degree $\leq D$, then we need only keep $P \cdot (D + 1)$ coefficients and $P - 1$ breakpoints to exactly represent this function. Common sense tells us that this is the natural, and even, the ideal representation for such a function.

To build this representation, we need to know locations of the discontinuities. If the measurements are noisy or blurred, and if we don't have recourse to an oracle, then we can't necessarily build this representation.

A less obvious but much more robust representation is to take a nice wavelet transform of the object, and keep the few resulting nonzero wavelet coefficients. If we have an N-point digital signal $f(i/N)$, $1 \leq i \leq N$, and we use Daubechies wavelets of compact support, then there are no more than $C \cdot \log_2(N) \cdot P \cdot (D + 1)$ nonzero wavelet coefficients for the digital signal.

In short, the nonadaptive representation needs only to keep a factor $C \log_2(N)$ more data to give an equally faithful representation.

We claim that this phenomenon is at least partially responsible for the widespread success of wavelet methods in data compression settings. One can build a single fast transform and deal with a wide range of different f, with different discontinuity sets, without recourse to an oracle.

In particular, since one almost never has access to an oracle, the natural first impulse of one committed to the adaptive viewpoint would be to 'estimate' the break points – i.e. to perform some sort of edge detection. Unfortunately this is problematic when one is dealing with noisy blurred data. Edge detection is a whole topic in itself which has thousands of proposed solutions and (evidently, as one can see from the continuing rate of publication in this area) no convincing solution.

In using wavelets, one does not need edge detectors or any other problematic schemes, one simply extracts the big coefficients from the transform domain, and records their values and positions in an organized fashion.

We can lend a useful perspective to this phenomenon by noticing that the discontinuities in the underlying f are *point singularities*, and we are saying that wavelets need in some sense at most $\log(n)$ coefficients to represent a point singularity out to scale $1/n$.

It turns out that even in higher dimensions, wavelets have a near-ideal ability to represent objects with point singularities.

The two-dimensional object $f_\beta(x_1, x_2) = 1/((x_1 - 1/2)^2 + (x_2 - 1/2)^2)^\beta$ has, for $\beta < 1/2$, a square-integrable singularity at the point $(1/2, 1/2)$ and is otherwise smooth. At each level of the $2D$ wavelet pyramid, there are effectively only a few wavelets which 'feel' the point singularity, other coefficients being effectively negligible. In approximation out to scale $1/n$, only about $O(\log(n))$ coefficients are required.

Another approach to understanding the representation of singularities, which is not limited by scale, is to consider rates of decay of the countable coefficient sequence. Analysis of wavelet coefficients of f_β shows that for any desired rate ρ, the N-th largest coefficient can be bounded by $C_\rho N^{-\rho}$ for all N. In short, the wavelet coefficients of such an object are very sparse.

Thus we have a slogan: *wavelets perform very well for objects with point singularities in dimensions* 1 *and* 2.

§3. Failure of Wavelets on Edges

We now briefly sketch why wavelets, which worked surprisingly well in representing point discontinuities in dimension 1, are less successful dealing with 'edge' discontinuities in dimension 2.

Suppose we have an object f on the square $[0,1]^2$ and that f is smooth away from a discontinuity along a C^2 curve Γ. Let's look at the number of substantial wavelet coefficients.

A grid of squares of side 2^{-j} by 2^{-j} has order 2^j squares intersecting Γ. At level j of the two-dimensional wavelet pyramid, each wavelet is localized near a corresponding square of side 2^{-j} by 2^{-j}. There are therefore $O(2^j)$ wavelets which 'feel' the discontinuity along Γ. Such a wavelet coefficient is controlled by

$$|\langle f, \psi_{j,k_1,k_2}\rangle| \le \|f\|_\infty \cdot \|\psi_{j,k_1,k_2}\|_1 \le C \cdot 2^{-j};$$

and in effect no better control is available, since the object f is not smooth within the support of ψ_{j,k_1,k_2} [14]. Therefore there are about 2^j coefficients of size about 2^{-j}. In short, the N-th largest wavelet coefficient is of size about $1/N$. The result (2) follows.

We can summarize this by saying that in dimension 2, discontinuities across edges are spatially distributed; because of this they can interact rather extensively with many terms in the wavelet expansion, and so the wavelet representation is not sparse.

In short, wavelets do well for point singularities, and not for singularities along curves. The success of wavelets in dimension 1 derived from the fact that all singularities in dimension 1 are point singularities, so wavelets have a certain universality there. In higher dimensions there are more types of singularities, and wavelets lose their universality.

For balance, we need to say that wavelets *do* outperform classical methods. If we used sinusoids to represent an object of the above type, then we have the result (1), which is far worse than that provided by wavelets. For completeness, we sketch the argument. Suppose we use for 'sinusoids' the complex exponentials on $[-\pi, \pi]^2$, and that the object f tends smoothly to zero at the boundary of the square $[0,1]^2$, so that we may naturally extend it to a function living on $[-\pi, \pi]^2$. Now typically the Fourier coefficients of an otherwise smooth object with a discontinuity along a curve decay with wavenumber as $|k|^{-3/2}$ (the very well-known example is f = indicator of a disk, which has a Fourier transform described by Bessel functions). Thus there are about R^2 coefficients of size $\geq c \cdot R^{-3/2}$, meaning that the N-th largest is of size $\geq c \cdot N^{-3/4}$, from which (1) follows.

In short: neither wavelets nor sinusoids really sparsify two-dimensional objects with edges (although wavelets are better than sinusoids).

§4. Ideal Representation of Objects with Edges

We now consider the optimality result (3), which is really two assertions. On the one hand, no reasonable scheme can do better than this rate. On the other hand, there is a certain adaptive scheme, with intimate connections to adaptive triangulation, which achieves it. For more extensive discussion see [10,11,13].

In talking about adaptive representations, we need to define terms carefully, for the following reason. For any f, there is always an adaptive representation of f that does very well: namely the orthobasis $\Psi = \{\psi_0, \psi_1, \ldots\}$ with first element $\psi_0 = f/\|f\|_2$! This is, in a certain conception, an 'ideal representation' where each object requires only one nonzero coefficient. In a certain sense it is a useless one, since all information about f has been hidden in the definition of representation, so actually we haven't learned anything. Most of our work in this section is in setting up a notion of adaptation that will free us from fear of being trapped at this level of triviality.

Dictionaries of atoms

Suppose we are interested in approximating a function in $L^2(T)$, and we have a countable collection $\mathcal{D} = \{\phi\}$ of atoms in $L^2(T)$; this could be a basis, a frame, a finite concatenation of bases or frames, or something even less structured.

We consider the problem of m-term approximation from this dictionary, where we are allowed to select m terms $\phi_1, \ldots, \phi_m$ from $\mathcal{D}$ and we approximate f from the L^2-closest member of the subspace they span:

$$\tilde{f}_m = Proj\{f|\text{span}(\phi_1, \ldots, \phi_m)\}.$$

We are interested in the behavior of the m-term approximation error

$$e_m(f; \mathcal{D}) = \|f - \tilde{f}_m\|_2^2,$$

where in this *provisional* definition, we assume $\tilde{f}_m$ is a best approximation of this form after optimizing over the selection of m terms from the dictionary.

However, to avoid a trivial result, we impose regularity on the selection process. Indeed, we allow rather arbitrary dictionaries, including ones which enumerate a dense subset of $L^2(T)$, so that in some sense the trivial result $\phi_1 = f/\|f\|_2$ $e_m = 0$, $\forall m$ is always a lurking possibility. To avoid this possibility we forbid arbitrary selection rules. Following [10] we propose

Definition. *A sequence of selection rules $(\sigma_m(\cdot))$ choosing m terms from a dictionary $\mathcal{D}$,*

$$\sigma_m(f) = (\phi_1, \ldots, \phi_m),$$

is said to implement polynomial depth search *if there is a single fixed enumeration of the dictionary elements and a fixed polynomial $\pi(t)$ such that terms in $\sigma_m(f)$ come from the first $\pi(m)$ elements in the dictionary.*

Under this definition, the trivial representation based on a countable dense dictionary is not generally available, since in any fixed enumeration, a decent 1-term approximation to typical f will typically be so deep in the enumeration as to be unavailable for polynomial-depth selection. (Of course, one can make this statement quantitative, using information-theoretic ideas).

More fundamentally, our definition not only forbids trivialities, but it allows us to speak of optimal dictionaries and get meaningful results. Starting now, we think of dictionaries as *ordered*, having a first element, second element, etc., so that different enumerations of the same collection of functions are *different* dictionaries. We define the m-optimal approximation number for dictionary $\mathcal{D}$ and limit polynomial π as

$$e_m(f; \mathcal{D}; \pi) = \|f - \tilde{f}_m\|_2^2,$$

where $\tilde{f}_m$ is constructed by optimizing the choice of m atoms among the first $\pi(m)$ in the fixed enumeration. Note that we use *squared error* for comparison with (1)-(3) in the Introduction.

Approximating classes of functions

Suppose we now have a class $\mathcal{F}$ of functions whose members we wish to approximate. Suppose we are given a countable dictionary $\mathcal{D}$ and polynomial depth search delimited by polynomial $\pi(\cdot)$.

Define the error of approximation by this dictionary over this class by

$$e_m(\mathcal{F}; \mathcal{D}, \pi) = \max_{f \in \mathcal{F}} e_m(f; \mathcal{D}, \pi).$$

We may find, in certain examples, that we can establish bounds

$$e_m(\mathcal{F}; \mathcal{D}, \pi) = O(m^{-\rho}), \qquad m \to \infty,$$

for all $\rho < \rho^*$. At the same time, we may have available an argument showing that for every dictionary and every polynomial depth search rule delimited by $\pi(\cdot)$,

$$e_m(\mathcal{F}; \mathcal{D}, \pi) \geq c m^{-\rho^*}, \qquad m \geq m_0(\pi).$$

Then it seems natural to say that ρ^* is the optimal rate of m-term approximation from any dictionary when polynomial depth search delimited by $\pi(\cdot)$.

Starshaped objects with C^2 boundaries

We define Star-Set$^2(C)$, a class of star-shaped sets with C^2-smooth boundaries, by imposing regularity on the boundaries using a kind of polar coordinate system. Let $\rho(\theta) : [0, 2\pi) \to [0, 1]$ be a radius function and $b_0 = (x_{1,0}, x_{2,0})$ be an origin with respect to which the set of interest is star-shaped. With $\delta_i(x) = x_i - x_{i,0}$, $i = 1, 2$, define functions $\theta(x_1, x_2)$ and $r(x_1, x_2)$ by

$$\theta = \arctan(-\delta_2/\delta_1); \qquad r = ((\delta_1)^2 + (\delta_2)^2)^{1/2}.$$

For a starshaped set, we have $(x_1, x_2) \in B$ iff $0 \le r \le \rho(\theta)$. Define the class Star-Set$^2(C)$ of *sets* by

$$\{B : B \subset [\frac{1}{10}, \frac{9}{10}]^2, \ \frac{1}{10} \le \rho(\theta) \le \frac{1}{2}, \quad \theta \in [0, 2\pi), \quad \rho \in C^2, |\ddot{\rho}(\theta)| \le C\},$$

and consider the corresponding *functional class*

$$\mathrm{Star}^2(C) \ = \ \{f \ = \ 1_B : \ \ B \in \text{Star-Set}^2(C)\}.$$

The following lower rate bound should be compared with (3).

Lemma. *Let the polynomial $\pi(\cdot)$ be given. There is a constant c so that, for every dictionary $\mathcal{D}$,*

$$e_m(\mathrm{Star}^2(C); \mathcal{D}, \pi) \ge c\frac{1}{m^2 \log(m)}, \qquad m \to \infty.$$

This is proved in [10] by the technique of hypercube embedding. Inside the class Star$^2(C)$ one can embed very high-dimensional hypercubes, and the ability of a dictionary to represent *all* members of a hypercube of dimension n by selecting $m \ll n$ terms from a subdictionary of size $\pi(m)$ is highly limited if $\pi(m)$ grows only polynomially.

For each f, a corresponding orthobasis is adaptively constructed in [13] which achieves the rate (3). It tracks the boundary of B at increasing accuracy using a sequence of polygons; in fact these are n-gons connecting equispaced points along the boundary of B, for $n = 2^j$. The difference between n-gons for $n = 2^j$ and $n = 2^{j+1}$ is a collection of thin triangular regions obeying *width* $\approx$ *length*^{2}; taking the indicators of each region as a term in a basis, one gets an orthonormal basis whose terms at fine scales are thin triangular pieces. Estimating the coefficient sizes by simple geometric analysis leads to the result (3). In fact, [13] shows how to do this under the constraint of polynomial-depth selection, with polynomial Cm^7.

Although space constraints prohibit a full explanation, our polynomial-depth search formalism also makes perfect sense in discussing the warped wavelet representations of the Introduction. Consider the noncountable 'dictionary' of all wavelets in a given basis, with all continuum warpings applied. Notice that for wavelets at a given fixed scale, warpings can be quantized with a certain finite accuracy. Carefully specifying the quantization of the warping, one obtains a countable collection of warped wavelets, for which polynomial depth search constraints make sense, and which is as effective as adaptive triangulation, but *not more so.* Hence (3) applies to (properly interpreted) deformation methods as well.

§5. Curvelet Construction

We now briefly describe the curvelet construction. It is based on combining several ideas, which we briefly review:

- Ridgelets, a method of analysis suitable for objects with discontinuities across straight lines.
- Multiscale Ridgelets, a pyramid of windowed ridgelets, renormalized and transported to a wide range of scales and locations.
- Bandpass Filtering, a method of separating an object into a series of disjoint scales.

We briefly describe each idea in turn, and then their combination.

Ridgelets

The theory of ridgelets was developed in the Ph.D. Thesis of Emmanuel Candès (1998). In that work, Candès showed that one could develop a system of analysis based on ridge functions

$$\psi_{a,b,\theta}(x_1, x_2) = a^{-1/2}\psi((x_1\cos(\theta) + x_2\sin(\theta) - b)/a). \tag{5}$$

He introduced a continuous ridgelet transform $R_f(a,b,\theta) = \langle \psi_{a,b,\theta}(x), f\rangle$ with a reproducing formula and a Parseval relation. He also constructed frames, giving stable series expansions in terms of a special discrete collection of ridge functions. The approach was general, and gave ridgelet frames for functions in $L^2[0,1]^d$ in all dimensions $d \geq 2$ – For further developments, see [3,5].

Donoho [12] showed that in two dimensions, by heeding the sampling pattern underlying the ridgelet frame, one could develop an orthonormal set for $L^2(\mathbb{R}^2)$ having the same applications as the original ridgelets. The orthonormal ridgelets are convenient to use for the curvelet construction, although it seems clear that the original ridgelet frames could also be used. The ortho-ridgelets are indexed using $\lambda = (j, k, i, \ell, \epsilon)$, where j indexes the ridge scale, k the ridge location, i the angular scale, and ℓ the angular location; ϵ is a gender token. Roughly speaking, the ortho-ridgelets look like pieces of ridgelets (5) which are windowed to lie in discs of radius about 2^i; $\theta_{i,\ell} = \ell/2^i$ is roughly the orientation parameter, and 2^{-j} is roughly the thickness.

A formula for ortho-ridgelets can be given in the frequency domain

$$\hat{\rho}_\lambda(\xi) = |\xi|^{-\frac{1}{2}}(\hat{\psi}_{j,k}(|\xi|)w^\epsilon_{i,\ell}(\theta) + \hat{\psi}_{j,k}(-|\xi|)w^\epsilon_{i,\ell}(\theta + \pi))/2 .$$

Here the $\psi_{j,k}$ are Meyer wavelets for $\mathbb{R}$, $w^\epsilon_{i,\ell}$ are periodic wavelets for $[-\pi, \pi)$, and indices run as follows: $j, k \in \mathbb{Z}$, $\ell = 0, \ldots, 2^{i-1} - 1$; $i \geq 1$, and, if $\epsilon = 0$, $i = \max(1, j)$, while if $\epsilon = 1$, $i \geq \max(1, j)$. We let Λ be the set of such λ. The formula is an operationalization of the ridgelet sampling principle:

- Divide the frequency domain in dyadic coronae $|\xi| \in [2^j, 2^{j+1}]$.
- In the angular direction, sample the j-th corona at least 2^j times.
- In the radial frequency direction, sample behavior using local cosines.

The sampling principle can be motivated by the behavior of Fourier transforms of functions with singularities along lines. Such functions have Fourier transforms which decay slowly along associated lines through the origin in the frequency domain. As one traverses a constant radius arc in Fourier space, one encounters a 'Fourier ridge' when crossing the line of slow decay. The ridgelet sampling scheme tries to represent such Fourier transforms by using wavelets in the angular direction, so that the 'Fourier ridge' is captured neatly by one or a few wavelets. In the radial direction, the Fourier ridge is actually oscillatory, and this is captured by local cosines. A precise quantitative treatment is given in [4].

Multiscale ridgelets

Think of ortho-ridgelets as objects which have a "length" of about 1 and a "width" which can be arbitrarily fine. The multiscale ridgelet system renormalizes and transports such objects, so that one has a system of elements at all lengths and all finer widths.

In a light mood, we may describe the system impressionistically as "brush strokes" with a variety of lengths, thicknesses, orientations and locations.

The construction employs a nonnegative, smooth partition of energy function w, obeying $\sum_{k_1,k_2} w^2(x_1 - k_1, x_2 - k_2) \equiv 1$. Define a transport operator, so that with index Q indicating a dyadic square $Q = (s, k_1, k_2)$ of the form $[k_1/2^s, (k_1+1)/2^s) \times [k_2/2^s, (k_2+1)/2^s)$, by $(T_Q f)(x_1, x_2) = f(2^s x_1 - k_1, 2^s x_2 - k_2)$. The Multiscale Ridgelet with index $\mu = (Q, \lambda)$ is then

$$\psi_\mu = 2^s \cdot T_Q(w \cdot \rho_\lambda).$$

In short, one transports the normalized, windowed ortho-ridgelet.

Letting $\mathcal{Q}_s$ denote the dyadic squares of side 2^{-s}, we can define the subcollection of Monoscale Ridgelets at scale s:

$$\mathcal{M}_s = \{(Q, \lambda) : Q \in \mathcal{Q}_s, \lambda \in \Lambda\}.$$

Orthonormality of the ridgelets implies that each system of monoscale ridgelets makes a tight frame, in particular obeying the Parseval relation

$$\sum_{\mu \in \mathcal{M}_s} \langle \psi_\mu, f \rangle^2 = \|f\|_{L^2}^2.$$

It follows that the dictionary of multiscale ridgelets at all scales, indexed by

$$\mathcal{M} = \cup_{s \geq 1} \mathcal{M}_s,$$

is not frameable, as we have energy blow-up:

$$\sum_{\mu \in \mathcal{M}} \langle \psi_\mu, f \rangle^2 = \infty. \tag{6}$$

The Multiscale Ridgelets dictionary is simply too massive to form a good analyzing set. It lacks inter-scale orthogonality – $\psi_{(Q,\lambda)}$ is not typically orthogonal to $\psi_{(Q',\lambda')}$ if Q and Q' are squares at different scales and overlapping locations. In analyzing a function using this dictionary, the repeated interactions with all different scales causes energy blow-up (6).

Our construction of curvelets solves (6) by disallowing the full richness of the Multiscale Ridgelets dictionary. Instead of allowing 'brushstrokes' of all different 'lengths' and 'widths', we allow only those where *width* $\approx$ *length*^{2}.

Subband filtering

Our solution to the 'energy blow-up' (6) is to decompose f into subbands using standard filterbank ideas. Then we assign one specific monoscale dictionary $\mathcal{M}_s$ to analyze one specific (and specially chosen) subband.

We define coronae of frequencies $|\xi| \in [2^{2s}, 2^{2s+2}]$, and subband filters Δ_s extracting components of f in the indicated subbands; a filter P_0 deals with frequencies $|\xi| \leq 1$. The filters decompose the energy exactly into subbands:

$$\|f\|_2^2 = \|P_0 f\|_2^2 + \sum_s \|\Delta_s f\|_2^2.$$

The construction of such operators is standard [15]; the coronization oriented around powers 2^{2s} is nonstandard – and essential for us. Explicitly, we build a sequence of filters Φ_0 and $\Psi_{2s} = 2^{4s}\Psi(2^{2s}\cdot)$, $s = 0, 1, 2, \ldots$ with the following properties: Φ_0 is a lowpass filter concentrated near frequencies $|\xi| \leq 1$; Ψ_{2s} is bandpass, concentrated near $|\xi| \in [2^{2s}, 2^{2s+2}]$; and we have

$$|\hat{\Phi}_0(\xi)|^2 + \sum_{s\geq 0} |\hat{\Psi}(2^{-2s}\xi)|^2 = 1, \quad \forall \xi.$$

Hence, Δ_s is simply the convolution operator $\Delta_s f = \Psi_{2s} * f$.

Definition of curvelet transform

Assembling the above ingredients, we are able to sketch the definition of the Curvelet transform. We let M' consist of M merged with the collection of integral pairs (k_1, k_2) indexing unit-side squares in the plane.

The curvelet transform is a map $L^2(\mathbb{R}^2) \mapsto \ell^2(M')$, yielding Curvelet coefficients $(\alpha_\mu : \mu \in M')$. These come in two types. At coarse scale we have wavelet scaling function coefficients

$$\alpha_\mu = \langle \phi_{k_1,k_2}, P_0 f \rangle, \qquad \mu = (k_1, k_2) \in M' \backslash M,$$

where ϕ_{k_1,k_2} is the Lemarié scaling function of the Meyer basis, while at *fine scale* we have Multiscale Ridgelets coefficients of the bandpass filtered object:

$$\alpha_\mu = \langle \Delta_s f, \psi_\mu \rangle, \qquad \mu \in M_s, s = 1, 2, \ldots.$$

Note well that each coefficient associated to scale 2^{-s} derives from the subband filtered version of f – $\Delta_s f$ – and not from f. Several properties are immediate:

- Tight Frame:

$$\|f\|_2^2 = \sum_{\mu \in M'} |\alpha_\mu|^2.$$

- Existence of Coefficient Representers (Frame Elements):

$$\alpha_\mu \equiv \langle f, \gamma_\mu \rangle.$$

- L^2 Reconstruction Formula:

$$f = \sum_{\mu \in M'} \langle f, \gamma_\mu \rangle \gamma_\mu.$$

- Formula for Frame Elements:

$$\gamma_\mu = \Delta_s \psi_\mu, \qquad \mu \in \mathcal{Q}_s.$$

  In short, the curvelets are obtained by bandpass filtering of Multiscale Ridgelets with *passband* is rigidly linked to the *scale* of spatial localization

- Anisotropy Scaling Law: Linking the filter passband $|\xi| \approx 2^{2s}$ to the spatial scale 2^{-s} imposes that (1) most curvelets are negligible in norm (most multiscale ridgelets do not survive the bandpass filtering Δ_s); (2) the non-negligible curvelets obey *length* $\approx 2^{-s}$ while *width* $\approx 2^{-2s}$. So the system obeys approximately the scaling relationship

$$\textit{width} \approx \textit{length}^2.$$

It is here that the 2^{2s} coronization scheme comes into play.

§6. Why Should This Work?

The curvelet decomposition can be equivalently stated in the following form:

- *Subband Decomposition.* The object f is filtered into subbands:

$$f \mapsto (P_0 f, \Delta_1 f, \Delta_2 f, \ldots).$$

- *Smooth Partitioning.* Each subband is smoothly windowed into "squares" of an appropriate scale:

$$\Delta_s f \mapsto (w_Q \Delta_s f)_{Q \in \mathcal{Q}_s}.$$

- *Renormalization.* Each resulting square is renormalized to unit scale

$$g_Q = 2^{-s} (T_Q)^{-1} (w_Q \Delta_s f), \qquad Q \in \mathcal{Q}_s.$$

- *Ridgelet Analysis.* Each square is analyzed in the ortho-ridgelet system

$$\alpha_\mu = \langle g_Q, \rho_\lambda \rangle, \qquad \mu = (Q, \lambda).$$

We can now give a crude explanation of why the main result (4) holds. Effectively, the bandpass images $\Delta_s f$ are almost vanishing at x far from the edges in f. Along edges, the bandpass images exhibit *ridges* of width $\approx 2^{-2s}$ – the width of the underlying bandpass filter.

The partitioned bandpass images are broken into squares of side $2^{-s} \times 2^{-s}$. The squares which do not intersect edges have effectively no energy, and we ignore them. The squares which do intersect edges result from taking a nearly-straight ridge and windowing it. Thus the squares which 'matter' exhibit tiny *ridge fragments* of aspect ratio 2^{-s} by 2^{-2s}. After renormalization, the resulting g_Q exhibits a ridge fragment of about unit length and of width 2^{-s}. The ridge fragment is then analyzed in the ortho-ridgelet system, which should (we hope) yield only a few significant coefficients.

In fact, simple arguments of size and order give an idea how the curvelet coefficients roughly behave. We give an extremely loose description.

First, at scale 2^{-s}, there are only about $O(2^s)$ squares $Q \in \mathcal{Q}_s$ that interact with the edges. Calculating the energy in such a square using the size of the support and the height of the ridge leads to

$$(\text{length} \cdot \text{width})^{1/2} \cdot \text{height} \approx (2^{-s} \times 2^{-2s})^{1/2} \times 1.$$

Indeed, the height of the ridge is bounded by

$$\|\Delta_s f\|_\infty = \|\Psi_{2s} * f\|_\infty \leq \|\Psi_{2s}\|_1 \|f\|_\infty = \|\Psi\|_1 \|f\|_\infty.$$

Since we are interested in uniformly bounded functions f, the height is thus bounded by a constant C. The calculation of the norm $\|g_Q\|_2 \approx 2^{-3/2}$ follows immediately (because of the renormalization, the height of the ridge g_Q is now 2^{-s}) . Now *temporarily* suppose that for some fixed K not depending on Q,

$$\text{each ridge fragment } g_Q \text{ is a sum of at most K ortho-ridgelets.} \qquad (7)$$

This would imply that at level s we have a total number of coefficients

$$O(2^s) \text{ squares which 'matter'} \times K \text{coefficients/square},$$

while the norm estimate for g_Q and the orthonormality of ridgelets give

$$\text{coefficient amplitude} \leq C \cdot 2^{-3s/2}.$$

The above assumptions imply that the N-th largest curvelet coefficient is of size $\leq C \cdot N^{-3/2}$. Letting $|\alpha|_{(N)}$ denote the N-th coefficient amplitude, the tail sum of squares would obey

$$\sum_{N>m} |\alpha|^2_{(N)} \leq C \cdot m^{-2}. \qquad (8)$$

This coefficient decay leads to (4) as follows. Let $\mu_1, \ldots, \mu_m$ enumerate indices of the m largest curvelet coefficients. Build the m-term approximation

$$\tilde{f}_m^C = \sum_{i=1}^{m} \alpha_{\mu_i} \gamma_{\mu_i}.$$

By the tight frame property,

$$\|f - \tilde{f}_m^C\|^2 \leq \sum_{N=m+1}^{\infty} |\alpha|_{(N)}^2, \leq C \cdot m^{-2},$$

where the last step used (8). This of course would establish (4) – in fact something even stronger, something fully as good as (3).

However, we have temporarily assumed (7) – which is not true. Each ridge fragment generates a countable number of nonzero ridgelet coefficients in general. The paper [6] gets (4) using much more subtle estimates.

§7. Discussion

Why call these things curvelets?

The visual appearance of curvelets does not match the name we have given them. The curvelets waveforms look like brushstrokes; brushlets would have been an appropriate name, but it was taken already, by F. Meyer and R. Coifman, for an unrelated scheme (essentially, Gabor Analysis).

Our point of view is simply that curvelets exemplify a certain curve scaling law, *width* = *length*^{2}, which is naturally associated to curves.

A deeper connection between curves and curvelets was alluded to in our talk at *Curves and Surfaces '99*. Think of a curve in the plane as a distribution supported on the curve, in the same way that a point in the plane can be thought of as a Dirac distribution supported on that point. The curvelets scheme can be used to represent that distribution as a superposition of functions of various lengths and widths obeying the scaling law *width* = *length*^{2}. In a certain sense this is a near-optimal representation of the distribution.

The analogy and the surprise

Sections 2 and 3 showed that wavelets do surprisingly well in representing *point singularities*. Without attempting an explicit representation of 'where the bad points are', wavelets do essentially as well as ideal adaptive schemes in representing the point singularities.

Sections 4–6 showed that the non-adaptive curvelets representation can do nearly as well in representing objects with discontinuities along curves as adaptive methods that explicitly track the shape of the discontinuity and use a special adaptive representation dependent on that tracking.

We find it surprising and stimulating that the curvelet representation can work so well despite the fact that it never constructs or depends on the existence of any 'map' of the discontinuity set.

We also find it interesting that there is a system of analysis which plays the role for curvilinear singularities that wavelets play for point singularities.

Acknowledgments. This research was supported by National Science Foundation grants DMS 98–72890 (KDI), DMS 95–05151, and by AFOSR MURI 95–P49620–96–1–0028.

References

1. Candès, E. J., Harmonic analysis of neural networks, Appl. Comput. Harmon. Anal. **6** (1999), 197–218.
2. Candès, E. J., Ridgelets: theory and applications, Ph.D. Thesis, Statistics, Stanford, 1998.
3. Candès, E. J., Monoscale ridgelets for the representation of images with edges, Technical Report, Statistics, Stanford, 1999.
4. Candès, E. J., On the representation of mutilated Sobolev functions, Technical Report, Statistics, Stanford, 1999.
5. Candès, E. J., and D. L. Donoho, Ridgelets: The key to high-dimensional intermittency? Phil. Trans. R. Soc. Lond. A. **357** (1999), 2495–2509.
6. Candès, E. J., and D. L. Donoho, Curvelets, Manuscript, 1999.
7. Coifman, R. R., and M. V. Wickerhauser, Entropy-based algorithms for best basis selection, IEEE Trans. Inform. Theory **38** (1992), 1713–1716.
8. Deng, B., B. Jawerth, G. Peters, and W. Sweldens, Wavelet probing for compression-based segmentation, in Proc. SPIE Symp. Math. Imaging: Wavelet Applications in Signal and Image Processing, 1993. Proceedings of SPIE conference July 1993, San Diego.
9. Donoho, D. L., Minimum entropy segmentation, in *Wavelets: Theory, Algorithms and Applications*, C. K. Chui, L. Montefusco and L. Puccio (eds.), Academic Press, San Diego, 1994, 233–270.
10. Donoho, D. L., and I. M. Johnstone, Empirical Atomic Decomposition, Manuscript, 1995.
11. Donoho, D. L., Wedgelets: nearly minimax estimation of edges, Ann. Statist. **27** (1999), 859–897.
12. Donoho, D. L., Orthonormal ridgelets and linear singularities, Technical Report, Statistics, Stanford, 1998, SIAM J. Math. Anal., to appear.
13. Donoho, D. L., Sparse components analysis and optimal atomic decomposition, Technical Report, Statistics, Stanford, 1998.
14. Meyer, Y., *Wavelets and Operators*, Cambridge University Press, 1992.
15. Vetterli, M., and J. Kovacevic, *Wavelets and Subband Coding*, Prentice Hall, 1995.

Department of Statistics
Stanford University
Stanford, CA 94305-4065, USA
`{emmanuel,donoho}@stat.stanford.edu`
`http://www-stat.stanford.edu/{~emmanuel,~donoho}/`

Bases in Function Spaces on Compact Sets

Zbigniew Ciesielski

Abstract. This is a brief history, covering the twentieth century, of spline bases on cubes, and an exposition of constructing bases in classical function spaces over compact smooth finite dimensional manifolds.

§1. Introduction

The aim of this paper is to present an overview on some twentieth century developments in the theory of spline bases. We start by recalling some of the relevant notions on bases in Banach spaces (for more details see e.g. [1,30]). For simplicity we are going to stay within the real Banach spaces. An abstract Banach space X with the norm $\|\cdot\|_X$ is denoted as $[X, \|\cdot\|_X]$. The sequence $(x_n, n = 0, 1, \ldots)$ in $[X, \|\cdot\|_X]$ is called a basis in X if to each $x \in X$ there is a unique sequence of scalars $\underline{a} = (a_n, n = 0, 1, \ldots)$ such that

$$x = \sum_{n=0}^{\infty} a_n\, x_n. \tag{1}$$

There are unique linear functionals $(x_n^*) \subset X^*$ such that $a_n = x_n^*(x)$. The system $(x_0, x_1, \ldots; x_0^*, x_1^*, \ldots)$ is biorthogonal i.e. $x_k^*(x_i) = \delta_{k,i}$. The basis (x_n) is unconditional if for each $x \in X$ the series in the right hand side of (1) converges unconditionally. Now, denote by $\mathcal{A}$ the set of all $\underline{a}$ appearing in (1) while x is running through X. The linear space $\mathcal{A}$ becomes a Banach space linearly isomorphic to X with the norm

$$\|\underline{a}\|_{\mathcal{A}} = \sup_{n\ge 0} \|\sum_{i=0}^{n} a_i\, x_i\|_X. \tag{2}$$

The Banach space $[\mathcal{A}, \|\cdot\|_{\mathcal{A}}]$ is customarily called the coefficient space. Introducing the basis constant, which by the Banach-Steinhaus theorem is finite,

$$\beta = \sup_{\|x\|_X \le 1} \sup_{n\ge 0} \|\sum_{i=0}^{n} a_i\, x_i\|_X, \tag{3}$$

Curve and Surface Fitting: Saint-Malo 1999
Albert Cohen, Christophe Rabut, and Larry L. Schumaker (eds.), pp. 121–134.

ISBN 0-8265-1357-3.

we obtain the equivalence of norms

$$\|x\|_X \leq \|a\|_{\mathcal{A}} \leq \beta \|x\|_X. \tag{4}$$

Thus, *every Banach space with a basis is linearly isomorphic to a sequence space.* However, depending on the space X and on the particular basis, the corresponding sequence space may be of little use. Sometimes it helps to consider equivalent basis in X. Two basis $(x_n) \subset X$ and $(x'_n) \subset X'$ are said to be equivalent if $\mathcal{A} = \mathcal{A}'$. In case of equivalent bases we conclude that $\|\underline{a}\|_{\mathcal{A}} \sim \|\underline{a}\|_{\mathcal{A}'}$ for $\underline{a} \in \mathcal{A}$. Now, we may describe the program of the paper. For a given compact C^∞ finite dimensional manifold M and for given order of smoothness m, we are going to describe the construction of biorthogonal system of functions of class C^m over M such that the system itself is a basis in $VMO(M)$, $H_1(M)$ and and in the whole scale of Sobolev $W_p^k(M)$, $-m \leq k \leq m$, and of Besov spaces $B_{p,q}^s(M)$, $-m < s < m$, with $1 \leq p, q \leq \infty$. At the same time the dual system is going to be a basis in the same scale of function spaces with the corresponding spaces VMO, H_1, W and B replaced by $V\overset{\circ}{M}O, \overset{\circ}{H}_1, \overset{\circ}{W}$ and $\overset{\circ}{B}$, respectively. The constructed system of functions (or its dual) is always an unconditional basis whenever the space admits an unconditional basis. Moreover, for the constructed basis, we are able to describe the coefficient spaces in case of the BMO and Besov spaces. The duality questions will be treated at the same time. The main idea of the general construction was announced by T. Figiel and the author at the Gdańsk 1979 conference: *Approximation and function spaces* (cf. [14]), and then carried out in the subsequent papers [10,11,15,16,17].

The material is arranged as follows: Sections 2 presents historical remarks on the Haar, Faber-Schauder, Franklin and spline systems; Section 3 treats function spaces and bases with boundary conditions on the cube; Section 4 describes the reduction of function spaces and bases from manifolds to the cubes with boundary conditions.

It is encouraging, that in recent years, the ideas of the constructions from [16,17] stimulated works on modifications of the decomposition of the function spaces on smooth compact manifolds into standard spaces, and also on constructing new bases in the standard spaces. The new investigations of W. Dahmen and R. Schneider as they were presented at this Saint Malo conference (see also [19]) are very promising as they show that these constructions can be applied to treat singular operators on manifolds both theoretically and numerically.

§2. The History of Haar, Faber-Schauder, Franklin and Spline Systems

At the very origin there is the construction of A. Haar (1909) [25] of a simple ONC (orthonormal and complete) system $\underline{\chi} = (\chi_n, n = 1, \ldots)$ on $I = [0, 1]$. The system $\underline{\chi}$ has the nice property that each continuous function has its Fourier-Haar series uniformly convergent on I. Here and later on, unless otherwise stated, the orthogonality is understood with respect to the Lebesgue measure. The orthonormal Haar functions over I can be defined by means of a single function h, where

$$h(t) = \begin{cases} 1 & \text{for } -\frac{1}{2} < t \leq 0, \\ -1 & \text{for } 0 < t \leq \frac{1}{2}, \\ 0 & \text{otherwise.} \end{cases}$$

Define for $j \geq 0$, $1 \leq k \leq 2^j$ and $n = 2^j + k$

$$\chi_1 = 1, \quad \text{and} \quad \chi_n(t) = \chi_{j,k}(t) = 2^{j/2} h(2^j(t - \frac{2k-1}{2^{j+1}})). \tag{5}$$

The Haar functions are piecewise constant and left continuous i.e. they are splines of order $r = 1$ (of degree $r - 1 = 0$). For later convenience for the support of χ_n we introduce the symbol $(n) = [\frac{k-1}{2^j}, \frac{k}{2^j}]$ (and let $t_n = \frac{2k-1}{2^{j+1}}$ denote the middle point). It was pointed out by J. Schauder (1927) [29] that the Haar system $\underline{\chi}$ is a basis in the Lebesgue space $L_p(I)$, $1 \leq p < \infty$, with the basis constant equal to 1. Much more involved was the proof of R.E.A.C. Paley (1932) [27] (see also J. Marcinkiewicz (1937) [26]) that the Haar system is an unconditional basis in each $L_p(I)$, $1 < p < \infty$. For a real variable proof of this property, we refer e.g. to Ch. Watari (1964) [31]. The unconditional basic constant for the Haar system in $L_p(I)$ appears to be equal to $\max(p, p') - 1$ where $1/p + 1/p' = 1$ (see e.e. D.L. Burkholder [5]). The extensively investigated martingale theory covers many results on the Haar system, but it is not very related to our subject, and will not be discussed here (see e.g. [23]).

To construct the Faber-Schauder, Franklin and more general spline systems, it is convenient to introduce the following operations on sequences of functions. For a given sequence $\underline{\psi} = (\psi_n, n = 1, 2, \ldots)$ of integrable functions on I, we define

$$\mathbf{G}\underline{\psi} = (1, \mathrm{G}\psi_n, n = 1, 2, \ldots) \quad \text{and} \quad \mathbf{G}_0\underline{\psi} = (\mathrm{G}\psi_n, n = 1, 2, \ldots),$$

where $\mathrm{G}f(t) = \int_0^t f(s)\,ds$. If in addition, the functions in $\underline{\psi}$ are linearly independent, then the result of the Gram orthogonalization process applied to $\underline{\psi}$ is denoted by $\mathbf{O}\underline{\psi}$. It is assumed in this definition that so obtained orthogonal set is normalized in $L^2(I)$.

The Faber-Schauder system can now be obtained from the Haar system by the operation $\mathbf{G}$

$$\underline{\phi} = (\phi_n, n \geq 0) = \mathbf{G}\underline{\chi}. \tag{6}$$

The Faber-Schauder functions are continuous splines of order 2. It was proved by G. Faber (1910) [21] (see also J. Schauder (1927) [29]) that this system is a basis in the space of continuous functions $[C(I), \|\cdot\|_\infty]$. In this case the basis constant is again equal to 1 and the basis itself is interpolating at dyadic points of I.

The orthonormal set constructed by Ph. Franklin (1928) [22] can now be defined as the result of application of the operation **O** to the Schauder system

$$\underline{f} = (f_n, n = 0, 1, \ldots) = \mathbf{O}\underline{\phi}. \tag{7}$$

These functions are again continuous splines of order $r = 2$. Ph. Franklin proved in [22] that $\underline{f}$ is a basis in $[C(I), \|\cdot\|_\infty]$. For an elegant proof that the Franklin system is a basis in $C(I)$ and in $L_p(I)$, $1 \le p < \infty$, we refer to [6]. Using the same idea as in [31], S. V. Botchkarev (1974, 1975) [3, 4] proved the unconditionality of the Franklin system in each $L_p(I)$, $1 < p < \infty$. There is an extensive literature on the pointwise behavior of the Franklin series, but we mention only the expository article by G. G. Gevorkyan [24].

The operation **G** increases the order r of splines and the order of their smoothness by 1, and **O** preserves these orders. We may repeat this two step process starting now with the orthonormal Franklin system and then repeat it again and again. In general, for $r \ge 1$ we use the notation

$$\underline{f}^{(r)} = (f_n^{(r)}, n > 1 - r) \quad \text{and} \quad \underline{\phi}^{(r,1)} = (\phi_n^{(r,1)}, n > -r) = \mathbf{G}\underline{f}^{(r)}$$

and

$$\underline{f}^{(r+1)} = \mathbf{O}\underline{\phi}^{(r,1)}.$$

Consequently, we have for the order $r \ge 1$ the following inductive formula for the spline ONC system on I

$$\underline{f}^{(r+1)} = \mathbf{O} \circ \mathbf{G}\underline{f}^{(r)} \quad \text{with} \quad \underline{f}^{(1)} = \underline{\chi}. \tag{8}$$

In particular in this notation $\underline{\phi}^{(1,1)} = \underline{\phi}$ and $\underline{f} = \underline{f}^{(2)}$. It was proved in [7] that $\underline{\phi}^{(2,1)}$ is a basis in $[C(I), \|\cdot\|_\infty]$ and by J. Radecki (1970) [28] that $\underline{f}^{(3)}$ is a basis in $[C(I), \|\cdot\|_\infty]$ and in each $L_p(I)$, $1 \le p < \infty$. The proof that $\underline{f}^{(r)}$ for arbitrary $r \ge 1$ is a basis in $C(I)$ and in $L_p(I)$ follows from the work of J. Domsta (1972) [20] (see also [12]).

From the construction of the ONC system $\underline{f}^{(r)}$ it follows that its first r elements $f_{2-r}^{(r)}, \ldots, f_1^{(r)}$ are simply the orthonormal Legendre polynomials on I; the degree of $f_i^{(r)}$ is $i + r - 2$. Now, with each $r \ge 1$ we associate a family of spline systems

$$\underline{f}^{(r,k)} = (f_n^{(r,k)}, n > |k| + 1 - r) \quad \text{with} \quad -r \le k < r, \tag{9}$$

where

$$f_n^{(r,k)} = \begin{cases} \mathrm{D}^k f_n^{(r)} & \text{for } 0 \le k < r, \\ \mathrm{H}^{-k} f_n^{(r)} & \text{for } -r \le k < 0, \end{cases} \tag{10}$$

and $\mathrm{D}f(t) = \frac{d}{dt}f(t)$ and $\mathrm{H}f(t) = \int_t^1 f(s)\,ds$. Since D is inverse to G and H is adjoint to G in $L^2(I)$, it follows that for $|k| < r$,

$$(f_n^{(r,k)}, f_m^{(r,-k)}) = \delta_{n,m} \quad \text{for} \quad n, m > |k| + 1 - r. \tag{11}$$

Equally important are spline ON systems defined by formula similar to (8) with $\mathbf{G}$ replaced by $\mathbf{G}_0$, i.e.

$$\underline{g}^{(r+1)} = \mathbf{O} \circ \mathbf{G}_0 \underline{g}^{(r)} \ \text{ with } \ \underline{g}^{(1)} = \underline{\chi}. \tag{12}$$

Here again, with each $r \geq 1$ we associate a family of spline systems

$$\underline{g}^{(r,k)} = (g_n^{(r,k)}, n \geq 1) \quad \text{with} \quad -r \leq k < r, \tag{13}$$

where

$$g_n^{(r,k)} = \begin{cases} \mathrm{D}^k g_n^{(r)} & \text{for } 0 \leq k < r, \\ \mathrm{H}^{-k} g_n^{(r)} & \text{for } -r \leq k < 0; \end{cases} \tag{14}$$

and as before we have for $|k| < r$,

$$(g_n^{(r,k)}, g_m^{(r,-k)}) = \delta_{n,m} \quad \text{for} \quad n, m \geq 1. \tag{15}$$

In what follows in this section we denote by $\underline{F}^{(r)}$ either $\underline{f}^{(r)}$ or $\underline{g}^{(r)}$. Since the family $\{\underline{F}^{(r,k)}, -r \leq k < r\}$ of spline systems is the main ingredient in the construction presented in the next section, it is natural to recall now its basic properties. Notice that the elements of $\underline{F}^{(r,k)}$ are indexed by $n \geq n(k, F)$, where $n(k, f) = |k| + 2 - r$ and $n(k, g) = 1$.

For given r and k such that $|k| < r$ and for given $n \geq n(k, F)$, we have the kernel corresponding to the partial sum operator with index n

$$K_n^{(r,k)}(s,t) = \sum_{\nu = n(k,F)}^{n} F_\nu^{(r,-k)}(s) \cdot F_\nu^{(r,k)}(t) \quad \text{for} \quad s, t \in I. \tag{16}$$

The following exponential estimates (cf. [8,13,17]) play a fundamental role in our construction. There are two constants: $C = C_r < \infty$ and $q = q_r$, $0 < q < 1$, such that for $|k| < r$ we have

$$|K_n^{(r,k)}(s,t)| \leq C \cdot (n+r) \cdot q^{(n+r)|s-t|} \quad \text{for} \quad s, t \in I, \tag{17}$$

and

$$|F_n^{(r,k)}(t)| \leq C \cdot (n+r)^{k+\frac{1}{2}} \cdot q^{(n+r)|t-t_n|} \quad \text{for} \quad t \in I, \tag{18}$$

where t_n has been defined earlier as the middle point of (n). Now, the biorthogonality (11), (15) and (18) imply for $1 \leq p \leq \infty$, $|k| < r$, and for any real sequence $(a_n, 2^j < n \leq 2^{j+1})$, $(j \geq 0)$, the equivalence

$$\| \sum_{2^j < n \leq 2^{j+1}} a_n \cdot F_n^{(r,k)} \|_p \sim 2^{j(k+\frac{1}{2}-\frac{1}{p})} \cdot (\sum_{2^j < n \leq 2^{j+1}} |a_n|^p)^{\frac{1}{p}}. \tag{19}$$

Moreover, it follows that

$$\| \sum_{2^j < n \leq 2^{j+1}} a_n \cdot F_n^{(r,k)} \|_p \sim \| \sum_{2^j < n \leq 2^{j+1}} |a_n \cdot F_n^{(r,k)}| \|_p, \tag{20}$$

where the positive constants in the equivalences $\sim$ in (19) and (20) depend on r only.

Now, as one of the consequences of (17) and (18), we obtain

Theorem 1. *For given r, k, $|k| < r$ and p, $1 \le p < \infty$, the system $\underline{F}^{(r,k)}$ is a basis in $L_p(I)$. Moreover, for $1 < p < \infty$, each of the systems is an unconditional basis in $L_p(I)$, and all of them are equivalent bases in this space. Moreover, $\underline{F}^{(r,k)}$ is a basis in $C(I)$ for each k, $0 \le k \le r-2$.*

§3. The Standard Spaces over Cubes

We start with general setup which will be needed in the following sections. Let the dimension d be fixed, and let M be a compact C^∞ d-dimensional manifold (d-manifold). For simplicity, we assume here that M has no boundary. We denote by μ one of the measures which locally is of the form $d\mu = h\,dx$ where h is positive C^∞ function. A closed set $Q \subset M$ is said to be a d-cube if it is diffeomorphic to the standard cube $[0,1]^d$. A compact set $K \subset M$ or $K \subset R^d$ is said to be proper if it can be viewed locally as an epigraph of a lipschitzian function of $d-1$ variables (cf. Def. 3.1 in [16]). We are going to discuss function spaces $\mathcal{F}(K)$ over a proper subsets K, in particular the Sobolev spaces with $\mathcal{F} = W_p^m$ and the Besov spaces with $\mathcal{F} = B_{p,q}^\alpha$. In the Sobolev space $W_p^m(K)$, $K \subset R^d$, $1 \le p \le \infty$, $m \ge 0$, we shall use the norm

$$\|f\|_p^{(m)}(K) = \sum_{|\alpha| \le m} \|D^\alpha f\|_p(K). \tag{21}$$

Clearly, $W_p^0(K) = L_p(K)$ and we denote by $W_\infty^m(K)$ the space $C^m(K)$. Moreover, the space of equivalence classes of measurable functions over K equipped with the topology of convergence in measure is denoted by $L_0(K)$. In order to define $W_p^k(K)$ for $k < 0$, we introduce $\overset{\circ}{W}{}_p^m(K)$ for each $m \ge 0$ as the closure in the norm (21) of smooth functions f such that $\overline{\operatorname{supp} f} \subset \operatorname{int} K$. For $1 \le p \le \infty$, $k < 0$ and for $g \in W_{p'}^0(K)$ put

$$\|g\|_{p'}^{(k)}(K) = \sup\left\{\left|\int_K fg\,dx\right| : \|f\|_p^{(-k)}(K) \le 1,\ f \in \overset{\circ}{W}{}_p^{-k}(K)\right\}, \tag{22}$$

where $p' = p/(p-1)$ for $1 < p < \infty$ and $1' = \infty$, $\infty' = 1$. Now, the completion of $W_{p'}^0(K)$ in the norm (22) defines the space $W_{p'}^k(K)$.

Let now $I = [0,1]$, $Q = I^d$ and let $\boldsymbol{Z}$ be a boundary set i.e. a set which is a union of $(d-1)$-dimensional faces of Q. To each pair $\{\mathcal{F}(Q), \boldsymbol{Z}\}$ we associate a subspace of $\mathcal{F}(Q)$ of functions which are vanishing on $\boldsymbol{Z} \subset \partial Q$ in the sense described below. To each $\boldsymbol{Z}$ there are unique $Z_i \subset \partial I$, $i = 1, \ldots, d$, such that

$$\boldsymbol{Z} = Q \setminus (I \setminus Z_1) \times \ldots \times (I \setminus Z_d). \tag{23}$$

Now, define for each $\boldsymbol{Z}$ the parallelepiped

$$Q_{\boldsymbol{Z}} = I_{Z_1} \times \ldots \times I_{Z_d}, \tag{24}$$

where

$$I_Z = \begin{cases} [0,1] & \text{for } Z = \emptyset, \\ [-1,1] & \text{for } Z = \{0\}, \\ [0,2] & \text{for } Z = \{1\}, \\ [-1,2] & \text{for } Z = \{0,1\}. \end{cases} \tag{25}$$

If $f \in L_0(Q)$, we denote by $f_{\boldsymbol{Z}}$ the element of $L_0(Q_{\boldsymbol{Z}})$ such that $f_{\boldsymbol{Z}}|_Q = f$ and $f_{\boldsymbol{Z}} = 0$ on $Q_{\boldsymbol{Z}} \setminus Q$.

Definition 2. *For given integer k and $1 \le p \le \infty$, put*

$$\|f\|_p^{(k)}(Q)_{\boldsymbol{Z}} = \|f_{\boldsymbol{Z}}\|_p^{(k)}(Q_{\boldsymbol{Z}}).$$

Now, if $k \ge 0$, define

$$W_p^k(Q)_{\boldsymbol{Z}} = \{f \in W_p^0(Q) : \ f_{\boldsymbol{Z}} \in W_p^k(Q_{\boldsymbol{Z}})\},$$

and if $k < 0$ then introducing $W_0 = \{f \in W_p^0(Q) : \ f_{\boldsymbol{Z}} \in W_p^k(Q_{\boldsymbol{Z}})\}$, define

$$W_p^k(Q)_{\boldsymbol{Z}} = \textit{completion of } W_0 \textit{ in the norm } \|f\|_p^{(k)}(Q)_{\boldsymbol{Z}}.$$

The spaces $[W_p^k(Q)_{\boldsymbol{Z}}, \|\cdot\|_p^{(k)}(Q)_{\boldsymbol{Z}}]$ are called standard.

Notice that for $k \ge 0$ the set $\{f_{\boldsymbol{Z}} : \ f \in W_p^k(Q)_{\boldsymbol{Z}}\}$ is a closed subspace of $W_p^k(Q_{\boldsymbol{Z}})$, and by Definition 2 the map $f \mapsto f_{\boldsymbol{Z}}$ is an isometry. Now let $k < 0$. In this case the map $f \mapsto f_{\boldsymbol{Z}}$ extends to an isometry of $W_p^k(Q)_{\boldsymbol{Z}}$ into $W_p^k(Q_{\boldsymbol{Z}})$. Thus $W_p^k(Q)_{\boldsymbol{Z}}$ is always complete, and the image of the map $f \mapsto f_{\boldsymbol{Z}}$ is a closed subspace of $W_p^k(Q_{\boldsymbol{Z}})$. We have constructed in [16], using the formulae (23) and (24) and the generalized Hestenes extension operators, a bounded projection onto this subspace.

Proposition 3. *Let $m \ge 1$ and the boundary set $\boldsymbol{Z} \subset \partial Q$ be given. Then there are a continuous linear operator P in $L_0(Q_{\boldsymbol{Z}})$ and $C < \infty$ such that P projects $L_0(Q_{\boldsymbol{Z}})$ onto $\{f : \ f = 0$ a.e. on $Q_{\boldsymbol{Z}} \setminus Q\}$ and for $1 \le p \le \infty$ we have*

$$\|Pf\|_p^{(k)}(Q_{\boldsymbol{Z}}) \le C\,\|f\|_p^{(k)}(Q_{\boldsymbol{Z}}) \ \textit{for } f \in W_p^k(Q_{\boldsymbol{Z}}),\ |k| \le m. \tag{26}$$

Thus, P projects $W_p^k(Q_{\boldsymbol{Z}})$ onto a subspace which is via the map $f \mapsto f_{\boldsymbol{Z}}$ linearly isomorphic to $W_p^k(Q)_{\boldsymbol{Z}}$.

Now, for $k \le 0$ and $1 \le p \le \infty$ we define the bilinear form

$$g^*(f) = \int_Q fg\,dx \quad \text{for} \quad g^* \in (W_{p'}^{-k}(Q)_{\boldsymbol{Z}'})^*,\ g \in W_p^0(Q). \tag{27}$$

Proposition 4. *Let $k \le 0$ and $1 \le p \le \infty$ be given. Then the map*

$$g \mapsto g^* : W_p^0(Q) \to (W_{p'}^{-k}(Q)_{\boldsymbol{Z}'})^*$$

defined in (27) extends to a linear isomorphism of $W_p^k(Q)_{\boldsymbol{Z}}$ onto a subspace of $(W_{p'}^{-k}(Q)_{\boldsymbol{Z}'})^*$.

Now suppose real s and $1 \le p, q \le \infty$ are given. Moreover, let K be a proper set. Then for any integers k, l such that $l < s < k$, we have the real interpolation formula for the Besov space (with $\theta = (s-l)/(k-l)$)

$$B_{p,q}^s(K) = \big(W_p^l(K), W_p^k(K)\big)_{\theta,q}. \tag{28}$$

For $f \in B_{p,q}^s(K)$ the norm is denoted by $\|f\|_{p,q}^{(s)}(K)$. The Besov space over Q with $l = 0 < s < k$ and corresponding to the boundary set $\boldsymbol{Z} \subset \partial Q$ is now defined by the formula

$$B_{p,q}^s(Q)_{\boldsymbol{Z}} = \{f \in W_p^0(Q) :\ f_{\boldsymbol{Z}} \in B_{p,q}^s(Q_{\boldsymbol{Z}})\}. \tag{29}$$

Moreover, let us define for $f \in B_{p,q}^s(Q)_{\boldsymbol{Z}}$

$$\|f\|_{p,q}^{(s)}(Q)_{\boldsymbol{Z}} = \|f_{\boldsymbol{Z}}\|_{p,q}^{(s)}(Q_{\boldsymbol{Z}}). \tag{30}$$

The Besov space $[B_{p,q}^s(Q)_{\boldsymbol{Z}}, \|\cdot\|_{p,q}^{(s)}(Q)_{\boldsymbol{Z}}]$ will be called standard as well. Notice, that $B_{p,q}^s(Q)_{\boldsymbol{Z}} \subseteq B_{p,q}^s(Q)$, but it may be not a closed subset of $B_{p,q}^s(Q)$.

Proposition 5. *Let the parameters l, k, θ, s, p, q be given as for (28), and let*

$$\mathcal{F}(Q)_{\boldsymbol{Z}} = (\mathcal{F}_0(Q)_{\boldsymbol{Z}}, \mathcal{F}_1(Q)_{\boldsymbol{Z}})_{\theta,q} \text{ where } \mathcal{F}_0 = W_p^l,\ \mathcal{F}_1 = W_p^k. \tag{31}$$

Then, $\mathcal{F}(Q)_{\boldsymbol{Z}} = B_{p,q}^s(Q)_{\boldsymbol{Z}}$ for $s > 0$ and for $s < 0$ the space $\mathcal{F}(Q)_{\boldsymbol{Z}}$ is naturally identified with the closure of W_p^0 in $(B_{p',q'}^{-s}(Q)_{\boldsymbol{Z}'})^*$.

The proof is based on the existence of the projection P in Proposition 3, and on the general properties of the real interpolation spaces (see [2, 16]).

Corollary 6. *Suppose we are given real numbers s, $1 \le p, q \le \infty$, an integer k, and a boundary set $\boldsymbol{Z} \subseteq \partial Q$ of the cube Q. Then the standard spaces $\mathcal{F}(Q)_{\boldsymbol{Z}}$ are well defined for $\mathcal{F} = W_p^k$ or $B_{p,q}^s$. Moreover, if $l < s < k$, then formula (31) takes place.*

In the last part of Section 3 we are going to present a construction of spline bases in the $\mathcal{F}(Q)_{\boldsymbol{Z}}$ spaces. Actually, according to (31), it is sufficient to do it for the Sobolev spaces $W_p^k(Q)_{\boldsymbol{Z}}$.

We start with the case of dimension $d = 1$. To each $Z \subset \partial I$ and for an integer $m = r - 2 \ge 0$ a spline system is defined as follows:

$$F_n^{(m)}(t; Z) = \begin{cases} f_n^{(2r,r)}(t) & \text{if } Z = \emptyset \text{ and } n \ge 2 - r, \\ f_n^{(2r,-r)}(t) & \text{if } Z = \{0,1\} \text{ and } n \ge 2 - r, \\ g_n^{(2r,r)}(t) & \text{if } Z = \{0\} \text{ and } n \ge 1, \\ g_n^{(2r,-r)}(t) & \text{if } Z = \{1\} \text{ and } n \ge 1; \end{cases} \tag{32}$$

where the $\underline{f}^{(r,l)}$ and $\underline{g}^{(r,l)}$ are given as in (10) and (14), respectively. Moreover, let

$$n(Z) = n(Z,0) \quad \text{and} \quad n(Z,k) = \begin{cases} |k| + 2 - r & \text{if } Z = \emptyset, \\ |k| + 2 - r & \text{if } Z = \{0,1\}, \\ 1 & \text{if } Z = \{0\}, \\ 1 & \text{if } Z = \{1\}. \end{cases} \tag{33}$$

For simplicity let us write $\underline{F}^{(m)}(Z) = (F_n^{(m)}(\cdot\,;Z), n \geq n(Z))$. Notice that $n(Z,k) = n(Z', -k)$, and that the two systems $\underline{F}^{(m)}(Z)$ and $\underline{F}^{(m)}(Z')$, where $Z' = \partial I \setminus Z$, are dual, i.e. they are biorthogonal in the $L^2(I)$ scalar product

$$(F_i^{(m)}(\cdot\,;Z), F_j^{(m)}(\cdot\,;Z')) = \delta_{i,j} \quad \text{for} \quad i,j \geq n(Z). \tag{34}$$

Now, we introduce related family of biorthogonal systems indexed by k with $|k| \leq m$. Namely, for $j \geq n(Z,k)$, let

$$F_j^{(m,k)}(\cdot\,;Z) = \begin{cases} \mathrm{D}^k F_j^{(m)}(\cdot\,;Z) & \text{for } 0 \leq k \leq m, \\ \mathrm{H}^{-k} F_j^{(m)}(\cdot\,;Z) & \text{for } -m \leq k < 0; \end{cases} \tag{35}$$

and the biorthogonality for $|k| \leq m$ is as follows

$$(F_i^{(m,k)}(\cdot\,;Z), F_j^{(m,-k)}(\cdot\,;Z')) = \delta_{i,j} \quad \text{for} \quad i,j \geq n(Z,k). \tag{36}$$

Theorem 7. *For each $Z \subset \partial I$, $1 \leq p \leq \infty$, the system $\underline{F}^{(m)}(Z)$ is in $W_p^m(I)_Z$ and it is a basis (an unconditional basis if $1 < p < \infty$) in each $W_p^k(I)_Z$ for $k = 0, \ldots, m$. This means that it is a simultaneous basis (simultaneous unconditional basis if $1 < p < \infty$) in $[W_p^m(I)_Z, \|\cdot\|_p^{(m)}]$.*

Proof: To see how the proof works, let

$$P_n f(x;Z) = \sum_{n(Z) \leq j \leq n} (f, F_j^{(m)}(\cdot\,;Z'))F_j^{(m)}(x;Z) \quad \text{for} \quad f \in L_p(I). \tag{37}$$

Then we find that for $0 \leq k \leq m$,

$$D^k P_n f(x;Z) = P_n^{(k)}(D^k f)(x;Z) \quad \text{for} \quad f \in W_p^{(k)}(I)_Z, \tag{38}$$

where for $g \in W_p^0(I)$

$$P_n^{(k)}(g)(x;Z) = \sum_{j=n(Z,k)}^{n} (g, F_j^{(m,-k)}(\cdot\,;Z'))F_j^{(m,k)}(x;Z). \tag{39}$$

Now, Theorem 7 follows immediately from Theorem 1 by (38) and (39). □

Now we consider the case of dimension $d > 1$, with $Q = I^d$. Suppose we are given $\boldsymbol{Z} \subset \partial Q$. Then according to (23) the $Z_i \subset \partial I$, for $i = 1, \ldots, d$, are determined. We are ready now to construct the tensor product basis corresponding to the boundary set $\boldsymbol{Z}$. Each function of the basis under construction is determined by an integer vector $\boldsymbol{j} = (j_1, \ldots, j_d)$ satisfying the inequality $\boldsymbol{j} \geq \boldsymbol{n}(\boldsymbol{Z})$ with $\boldsymbol{n}(\boldsymbol{Z}) = (n(Z_1), \ldots, n(Z_d))$ i.e. $j_i \geq n(Z_i)$ for $i = 1, \ldots, d$.

Given the order of smoothness $m \geq 0$, we now define the $\boldsymbol{j}$'s function as follows:

$$F_{\boldsymbol{j}}^{(m)}(\boldsymbol{x}; \boldsymbol{Z}) = F_{j_1}^{(m)}(x_1; Z_1) \times \cdots \times F_{j_d}^{(m)}(x_d; Z_d), \tag{40}$$

where $\boldsymbol{x} = (x_1, \ldots, x_d)$. The indices $\boldsymbol{j}$ are ordered in the rectangular way (cf. [12], p. 221).

Theorem 8. *The system* $(F_{\boldsymbol{j}}^{(m)}(\cdot\,; \boldsymbol{Z}), \boldsymbol{j} \geq \boldsymbol{n}(\boldsymbol{Z}))$ *in the rectangular ordering is a basis in* $W_p^m(Q)_{\boldsymbol{Z}}$ *for* $1 \leq p \leq \infty$, *and in addition it is unconditional in these spaces if* $1 < p < \infty$.

Our next goal is to modify the basis (40) in such a way that the elements of the new basis will be concentrated around the corresponding dyadic points in Q. To this end let us introduce in dimension one the following finite dimensional spline spaces:

$$S_\mu^m(Z) = \operatorname{span}\{F_j^{(m)}(\cdot\,; Z) : \ n(Z) \leq j \leq 2^\mu\} \quad \text{where} \quad \mu \geq 1. \tag{41}$$

Now, without going into details, we accept the Definition 10.17 of [17] of the new spline basis in $S_\mu^m(Z)$, i.e. of $F_{\mu,j}^{(m)}(\cdot\,; Z)$ with $n(Z) \leq j \leq 2^\mu$. The new basic functions for the standard space $W_p^m(Q)_{\boldsymbol{Z}}$ are now defined as follows. For convenience, let $D = \{1, \ldots, d\}$, and let

$$N_0(\boldsymbol{Z}) = \{\boldsymbol{j} : \ n(Z_i) \leq j_i \leq 1 \text{ for } i \in D\},$$

and for every $e \subset D$, $e \neq \emptyset$, $\mu \geq 1$, let

$$N_{e,\mu}(\boldsymbol{Z}) = \{\boldsymbol{j} : \ 2^{\mu-1} < j_i \leq 2^\mu \text{ for } i \in e, \ n(Z_i) \leq j_i \leq 2^{\mu-1} \text{ for } i \in D \setminus e\}.$$

We also introduce

$$N_\mu(\boldsymbol{Z}) = \bigcup_{\emptyset \neq e \subset D} N_{e,\mu}(\boldsymbol{Z}).$$

Definition 9. *Let*

$$G_{\boldsymbol{j}}^{(m)}(\cdot\,; \boldsymbol{Z}) = F_{\boldsymbol{j}}^{(m)}(\cdot\,; \boldsymbol{Z}) \quad \text{for} \quad \boldsymbol{j} \in N_0(\boldsymbol{Z}),$$

and

$$G_{\boldsymbol{j}}^{(m)}(\cdot\,; \boldsymbol{Z}) = \bigotimes_{i \in e} F_{j_i}^{(m)}(\cdot\,; Z_i) \otimes \bigotimes_{i \in D \setminus e} F_{\mu, j_i}^{(m)}(\cdot\,; Z_i) \quad \text{for} \quad \boldsymbol{j} \in N_{e,\mu}(\boldsymbol{Z})$$

for any $e \subset D$, $e \neq \emptyset$, *and* $\mu \geq 1$.

Any ordering $\prec$ of the set of indices $\{\boldsymbol{j} : \boldsymbol{j} \geq \boldsymbol{n}(\boldsymbol{Z})\}$ is said to be regular if $\boldsymbol{j} \prec \boldsymbol{j}'$ for any $\boldsymbol{j} \in N_\mu$ and $\boldsymbol{j}' \in N_{\mu'}$ whenever $\mu < \mu'$ (cf. [10]). We also have the biorthogonality relation

$$(G_{\boldsymbol{j}}^{(m)}(\cdot\,;\boldsymbol{Z}), G_{\boldsymbol{j}'}^{(m)}(\cdot\,;\boldsymbol{Z}')) = \delta_{\boldsymbol{j}',\boldsymbol{j}}. \tag{42}$$

We can now state the result on 'universal basis' in standard spaces (cf. [10, 16])

Theorem 10. *Let $m \geq 0$ be a given order of smoothness. The system $(G_{\boldsymbol{j}}^{(m)}(\cdot\,;\boldsymbol{Z}), \boldsymbol{j} \geq \boldsymbol{n}(\boldsymbol{Z}))$ in the regular ordering is a basis in all the spaces $W_p^k(Q)_{\boldsymbol{Z}}$ for $0 \leq k \leq m$, $1 \leq p \leq \infty$, and in addition it is unconditional if $1 < p < \infty$. Moreover, for $\mu \geq 1$, $1 \leq p \leq \infty$, we have*

$$\|\sum_{\boldsymbol{j}\in N_\mu} a_{\boldsymbol{j}}\, G_{\boldsymbol{j}}^{(m)}(\cdot\,;\boldsymbol{Z})\|_p \sim 2^{\mu(1/2-1/p)d}\Big(\sum_{\boldsymbol{j}\in N_\mu} |a_{\boldsymbol{j}}|^p\Big)^{1/p}, \tag{43}$$

where the constants in the relation $\sim$ depend only on d and m.

Corollary 11. *The system $(G_{\boldsymbol{j}}^{(m)}(\cdot\,;\boldsymbol{Z}), \boldsymbol{j} \geq \boldsymbol{n}(\boldsymbol{Z}))$ in the regular ordering is a basis in all the spaces $B_{p,q}^s(Q)_{\boldsymbol{Z}}$ with $1 \leq p, q \leq \infty$, $0 < s < m$. Moreover, for*

$$f(\cdot) = \sum_{\boldsymbol{j}\geq \boldsymbol{n}(\boldsymbol{Z})} a_{\boldsymbol{j}}\, G_{\boldsymbol{j}}^{(m)}(\cdot\,;\boldsymbol{Z}),$$

letting $\sigma = s/d + 1/2 - 1/p$, we have

$$\|f\|_{p,q}^{(s)}(Q)_{\boldsymbol{Z}} \sim \left\{\sum_{\mu=0}^{\infty}\left[2^{\mu\sigma d}\Big(\sum_{\boldsymbol{j}\in N_\mu}|a_{\boldsymbol{j}}|^p\Big)^{1/p}\right]^q\right\}^{1/q}.$$

The constants in the relation $\sim$ depend on m, s and on d.

For an arbitrary d-cube Q, the function space $\mathcal{F}(Q)_{\boldsymbol{Z}}$ is defined as the image of $\mathcal{F}(I^d)_{\boldsymbol{Z}}$ under the linar mapping induced by the diffeomorphism between Q and I^d.

§4. Decomposition of Function Spaces over Smooth Manifolds

Let us start with the decomposition of M without boundary (for M with boundary cf. [16]). We say that M admits decomposition into d-cubes if for some N there are d-cubes $Q_1, \ldots, Q_N \subset M$ such that $\cup_{j\leq N} Q_j = M$ and if Φ_j is a diffeomorphism of $[0,1]^d$ onto Q_j, $1 \leq j \leq N$, then the set $\Phi_j^{-1}(\cup_{i<j} Q_i)$ is the union of some $(d-1)$-dimensional faces of $[0,1]^d$. The decomposition $Q_1, \ldots Q_N$, is said to be proper if the sets $\cup_{i\leq j} Q_i$ are proper for $j = 0, \ldots, N$. Now we have the following result whose proof depends very much on Morse theory (cf. Theorem 3.3 in [16]):

Theorem 12. *Let M be a compact d-manifold. Then M admits a proper decomposition.*

For the Sobolev $W_p^k(M)$ and Besov $B_{p,q}^s(M)$ spaces, we have the real interpolation formula for any integers l, r, real $s, l < s < r$, and for $1 \le p, q \le \infty$ (cf. [15])

$$B_{p,q}^s(M) = \big(W_p^l(M), W_P^r(M)\big)_{\theta,q}, \tag{44}$$

where $s = (1-\theta)l + \theta r$. We recall that for the standard Besov spaces, we have similar formula (cf. Proposition 5). Having a proper decomposition of M (Theorem 12), we would like to obtain a corresponding decomposition of $\mathcal{F}(M)$ into a direct sum of standard spaces $\mathcal{F}(Q_j)_{\boldsymbol{Z}_j}$. Let $L_0(M)$ denote the space of all measurable functions (of equivalence classes) with the topology of convergence in measure.

Proposition 13. *Let $Q_1, \dots, Q_N$ be a proper decomposition of M into d-cubes as in Theorem 12. Let μ be a smooth measure on M. Then, for any $m \ge 1$, one can construct continuous linear operators $P_1, \dots, P_N$ in the space $L_0(M)$ with the following properties for $f \in L_0(M)$:*

$$\sum_{i \le N} P_i f = f, \tag{45}$$

$$P_i P_j f = 0, \qquad \text{if } 1 \le i \ne j \le N, \tag{46}$$

$$\chi_{Q_i} P_j f = P_i \chi_{Q_j} f, \qquad \text{if } 1 \le i < j \le N, \tag{47}$$

there is $C < \infty$ such that for all spaces $W_p^k(M)$, $0 \le k \le m, 1 \le p \le \infty$, for all $g \in W_p^k(M)$ and for $1 \le i \le N$ we have

$$\|P_i g\|_p^{(k)}(M) \le C \|g\|_p^{(k)}(M), \tag{48}$$

the adjoint operators (in the Hilbert space $L_2(M, \mu)$) $P_1^*, \dots, P_N^*$ satisfy the analog of (48).

Proposition 13 implies the main result on decomposing the function spaces $\mathcal{F}(M)$ (see [16]), i.e.

Theorem 14. *Let $Q_1, \dots, Q_N$ be a proper decomposition of M as in Theorem 12, and let $P_1, \dots, P_N$ be the linear operators from Proposition 13. Then the formulae*

$$T_0 f = \sum_{i \le N} \chi_{Q_i} P_i f, \quad V_0 f = \sum_{i \le N} \chi_{Q_i} P_i^* f$$

define linear isomorphism of $L_0(M)$ onto itself, the inverse maps being, respectively,

$$S_0 f = \sum_{i \le N} P_i \chi_{Q_i} f, \quad U_0 f = \sum_{i \le N} P_i^* \chi_{Q_i} f.$$

Moreover, if $\mathcal{F}$ denotes W_p^k, $0 \le k \le m$, $1 \le p \le \infty$, then T_0, V_0 induce linear topological isomorphism

$$T : \mathcal{F}(M) \to \bigoplus_{i \le N} \mathcal{F}(Q_i)_{\boldsymbol{Z}_i}, \quad V : \mathcal{F}(M) \to \bigoplus_{i \le N} \mathcal{F}(Q_i)_{\boldsymbol{Z}'_i},$$

where $\boldsymbol{Z}_i = Q_i \cup_{i<j} Q_j$ and $\boldsymbol{Z}'_i = Q_i \cup_{j>i} Q_j$.

Corollary 15. *The assertions of Theorem 14 remain true for* $\mathcal{F} = W_p^k,\ B_{p,q}^s$ *with* $|k| \le m, |s| < m, 1 \le p, q \le \infty$. *Moreover, there is now an obvious extension of Theorem 10 and Corollary 11 to* $\mathcal{F}(M)$ *for all these spaces.*

Acknowledgments. The author is very much indebted to the Program and Organizing Committees for the invitation.

References

1. Banach, S., *Théorie des Opérations Linéaires*, Warszawa, Monografje Matematyczne, 1932.
2. Bergh, J. and J. Löfström, *Interpolation Spaces*, Springer-Verlag, Berlin – Heidelberg – New York, 1976.
3. Bochkarev, S. V., Existence of a basis in the space of functions analytic in the disk, and some properties of Franklin's system, Mat. Sb. **95 (137)** (1974), 3–18; English transl. in Math. USSR-Sb. **24** (1974).
4. Bochkarev, S. V., Some inequalities for Franklin series, Analysis Mathematica **1** (1975), 249–257.
5. Burkholder, D. L., A proof of Pelczynski conjecture for the Haar system, Studia Math. **91** (1988), 79–83.
6. Ciesielski, Z., Properties of the orthonormal Franklin system, Studia Math. **23** (1963), 141–157.
7. Ciesielski, Z., A construction of basis in $C^{(1)}(I^2)$, Studia Math. **33** (1969), 243–247.
8. Ciesielski, Z., Equivalence, unconditionality and convergence a.e. of spline bases in L_p spaces, in *Approximation Theory*, Banach Center Publications, Vol. 4, PWN – Polish Scientific Publishers, Warszawa, 1979, 55–68.
9. Ciesielski, Z., Bases and K-functionals for Sobolev spaces over C^∞ compact manifolds, in *Recent Trends in Mathematics, Reinhardsbrunn 1982*, Taubner Texte zur Mathematik, Band 50, Leipzig 1983, 47–55. Trudy Inst. Steklova **164** (1983), 197–202.
10. Ciesielski, Z., Spline bases in classical function spaces on C^∞ compact manifolds. Part III, in *Constructive Theory of Functions 84*, Sofia 1984, 214–223.
11. Ciesielski, Z., Bases in function spaces, in *Approximation Theory V*, C. Chui, L. Schumaker, and J. Ward (eds.), Academic Press, New York, 1986, 31–54.
12. Ciesielski, Z. and J. Domsta, Construction of an orthonormal basis in $C^m(I^d)$ and $W_p^m(I^d)$, Studia Math. **41** (1972), 211–224.
13. Ciesielski, Z. and J. Domsta, Estimates for the spline orthonormal functions and for their derivatives, Studia Math. **46** (1972), 315–320.
14. Ciesielski, Z. and T. Figiel, Construction of Schauder bases in function spaces on smooth compact manifolds, in *Proc. International Conference "Approximation and Function Spaces, Gdańsk August 27–31, 1979"*, PWN and North Holland, Warszawa 1981, 217–232.

15. Ciesielski, Z. and T. Figiel, Spline approximations and Besov spaces on compact manifolds, Studia Math. **75** (1982), 13–36.

16. Ciesielski, Z. and T. Figiel, Spline bases in classical function spaces on compact C^∞ manifolds. Part I, Studia Math. **76** (1983), 1–58.

17. Ciesielski, Z. and T. Figiel, Spline bases in classical function spaces on compact C^∞ manifolds. Part II, Studia Math. **76** (1983), 95–136.

18. Ciesielski, Z., P. Simon, and P. Sjölin, Equivalence of Haar and Franklin bases in L_p spaces, Studia Math. **60** (1977), 195–210.

19. Dahmen, W., Wavelet methods for PDEs – Some recent developments, J. Comput. Appl. Math., to appear.

20. Domsta, J., A theorem on B-splines, Studia Math. **41** (1972), 291–314.

21. Faber, G., Über die Orthogonalfunktionen des Herrn Haar, Jahresber. Deutsch. Math. Verein **19** (1910), 104–112.

22. Franklin, Ph., A set of continuous orthogonal functions, Math. Ann. **100** (1928), 522–529.

23. Garcia, A., *Martingale Inequalities. Seminar Notes on Recent Progresses*, W. A. Benjamin, Inc. Reading, 1973.

24. Gevorkyan, G. G., Some theorems on unconditional convergence and the majorant of Franklin series and their application to the spaces $Re(H^p)$, Trudy Mat. Inst. Steklova **190** (1989), 49–74; English transl. in Proc. Steklov Inst. Math. (1992), **no. 1(190)**, 49–76.

25. Haar, A., Zur Theorie der Orthogonalen Funktionensysteme, Math. Ann. **69** (1910), 331–371.

26. Marcinkiewicz, J., Quelques théorèmes sur les séries orthogonales, Ann. Soc. Polon. Math. **16** (1937), 85–96.

27. Paley, R.E.A.C., A remarkable series of orthogonal functions I, II, Proc. London Math. Soc. (2) **34** (1932), 241–264, 265–279.

28. Radecki, J., Orthonormal basis in the space $C_1[0,1]$, Studia Math. **35** (1970), 123–163.

29. Schauder, J., Zur Theorie stetiger Abbildungen in Funktionalräumen, Math. Z. **26** (1927), 47–65.

30. Singer, I., *Bases in Banach Spaces I*, Springer-Verlag, Berlin – Heidelberg – New York, 1970.

31. Watari, Ch., Mean convergence of Walsh Fourier series, Tôhoku Math. J. (2) **16** (1964), 183–188.

Zbigniew Ciesielski
Instytut Matematyczny PAN
ul. Abrahama 18
81-825 Sopot, Poland
`Z.Ciesielski@impan.gda.pl`

A Note on Convolving Refinable Function Vectors

C. Conti and K. Jetter

Abstract. When convolving two refinable function vectors which give rise to convergent subdivision schemes, the convolved scheme is again convergent. Moreover, the conditions on the mask symbols which characterize the approximation order of the associated shift invariant spaces show that the order of the convolved space is, essentially, the sum of the order of the two spaces originating from the convolution factors.

§1. Kronecker Convolved Function Vectors

Our previous paper [3] deals with special subdivision schemes associated with a shift invariant space of bivariate spline functions, where the "generators" of the shift invariant space are produced through convolving lower order splines of small support. The present paper gives a more detailed and more systematic analysis of this convolution process. In this way it is possible to prove that (i) the convolution of convergent subdivision schemes yields a scheme which is convergent as well, and (ii) essentially, the approximation power of a convolved shift invariant space is at least the sum of the approximation powers of the two convolution factors.

We recall that a vector $\mathbf{\Phi} = (\phi_1, \phi_2, \ldots, \phi_n)^T$ of (continuous, compactly supported) d-variate functions is called refinable, if it satisfies a refinement equation

$$\mathbf{\Phi} = \sum_{\alpha \in \mathbb{Z}^d} A_\alpha \, \mathbf{\Phi}(2 \cdot - \alpha) \, . \tag{1.1}$$

Here, the refinement matrix mask $\mathbf{A} = \left(A_\alpha\right)_{\alpha \in \mathbb{Z}^d}$ is a matrix sequence with each 'coefficient' A_α being a real $(n \times n)$-matrix. We allow only masks of finite support, *i.e.*, $A_\alpha = 0$ except for finitely many $\alpha \in \mathbb{Z}^d$.

Curve and Surface Fitting: Saint-Malo 1999
Albert Cohen, Christophe Rabut, and Larry L. Schumaker (eds.), pp. 135–142.

ISBN 0-8265-1357-3.

Given another refinable function vector $\mathbf{\Psi} = (\psi_1, \psi_2, \ldots, \psi_m)^T$ of d-variate functions, satisfying the refinement equation

$$\mathbf{\Psi} = \sum_{\beta \in \mathbb{Z}^d} B_\beta \, \mathbf{\Psi}(2 \cdot - \beta) \tag{1.2}$$

with corresponding matrix mask $\mathbf{B} = \left(B_\beta\right)_{\beta \in \mathbb{Z}^d}$, which is a matrix sequence of $(m \times m)$-matrices B_β, we use the following Kronecker type notion of convolving the two vectors:

$$\mathbf{\Theta} := \mathbf{\Phi} * \mathbf{\Psi} := \begin{pmatrix} \phi_1 * \mathbf{\Psi} \\ \phi_2 * \mathbf{\Psi} \\ \vdots \\ \phi_n * \mathbf{\Psi} \end{pmatrix} .$$

Here, the convolution of a scalar function ϕ_i with the vector function $\mathbf{\Psi}$ is taken componentwise. This operation produces a vector function $\mathbf{\Theta} = (\theta_1, \theta_2, \ldots, \theta_{mn})^T$ with $m \times n$ components of type $\theta_{(i-1)m+j} := \phi_i * \psi_j$. It is not too hard to see that $\mathbf{\Theta}$ is refinable again,

$$\mathbf{\Theta} = \sum_{\gamma \in \mathbb{Z}^d} C_\gamma \, \mathbf{\Theta}(2 \cdot - \gamma) \, , \tag{1.3}$$

where the refinement mask $\mathbf{C} = \left(C_\gamma\right)_{\gamma \in \mathbb{Z}^d}$, which is a matrix sequence of $(nm \times nm)$-matrices, is computed as follows:

$$C_\gamma = \frac{1}{2^d} \sum_{\alpha \in \mathbb{Z}^d} A_\alpha \otimes B_{\gamma - \alpha} \, , \quad \alpha \in \mathbb{Z}^d \, . \tag{1.4}$$

Here, the symbol $\otimes$ denotes the Kronecker product of matrices.

For the definition and some properties of this Kronecker product, we refer to [4, Section 4.2]. It should be noted at least that neither the Kronecker product of matrices nor the Kronecker type convolution is commutative.

§2. Convergence of the Convolved Subdivision Scheme

Let us also recall that a refinable function vector gives rise to a vector-valued subdivision scheme as follows: In the situation of (1.1), the subdivision operator associated with the refinable function vector $\mathbf{\Phi} = (\phi_1, \phi_2, \ldots, \phi_n)^T$ is defined as

$$\begin{aligned} S_A \, &: \, (\ell(\mathbb{Z}^d))^n \to (\ell(\mathbb{Z}^d))^n \, , \\ (S_A \, \mathbf{\Lambda})_\alpha &:= \sum_{\beta \in \mathbb{Z}^d} A^T_{\alpha - 2\beta} \, \Lambda_\beta, \quad \alpha \in \mathbb{Z}^d \, , \end{aligned} \tag{2.1}$$

where $\ell(\mathbb{Z}^d)$ denotes the linear space of sequences indexed by $\mathbb{Z}^d$. The complete (stationary) subdivision scheme consists in the iterates of S_A, namely:

$$\begin{aligned} &\text{For a given initial vector sequence } \mathbf{\Lambda} \in (\ell(\mathbb{Z}^d))^n \\ &\text{Put } \mathbf{\Lambda}^{(0)} := \mathbf{\Lambda} \text{ and} \\ &\text{Compute } \mathbf{\Lambda}^{(k+1)} := S_A \, \mathbf{\Lambda}^{(k)}, \; k = 0, 1, \ldots \end{aligned} \tag{2.2}$$

Hence, the iterate $\mathbf{\Lambda}^{(k)} = (\Lambda_\alpha^{(k)})_{\alpha\in\mathbb{Z}^d}$ has components

$$\Lambda_\alpha^{(k)} := \sum_{\beta\in\mathbb{Z}^d} (A_{\alpha-2^k\beta}^{(k)})^T \,\Lambda_\beta^{(0)}, \quad \alpha \in \mathbb{Z}^d , \tag{2.3}$$

with the iterated matrices $\mathbf{A}^{(k)} = (A_\alpha^{(k)})_{\alpha\in\mathbb{Z}^d}$ defined by $\mathbf{A}^{(1)} := \mathbf{A}$ and

$$A_\alpha^{(k)} := \sum_{\beta\in\mathbb{Z}^d} A_\beta^{(k-1)} A_{\alpha-2\beta} , \quad \alpha \in \mathbb{Z}^d, \quad \text{for } k > 1. \tag{2.4}$$

Following [2, Section 2.4] we say that the subdivision scheme converges for $\mathbf{\Lambda} = (\lambda^1, \ldots, \lambda^n)^T \in (\ell^\infty(\mathbb{Z}^d))^n$ if there exists a continuous function $f_\Lambda : \mathbb{R}^d \to \mathbb{R}$ such that

$$\lim_{k\to\infty} \left\| f_\Lambda(\tfrac{\cdot}{2^k})\mathbf{e} - \mathbf{\Lambda}^{(k)} \right\|_\infty = 0 \quad \text{for} \quad \mathbf{e} = (\underbrace{1\;1\;\cdots\;1}_{n})^T. \tag{2.5}$$

Here, $\|\cdot\|_\infty$ denotes the sup-norm of the vector sequence $\Lambda = (\lambda^1, \ldots, \lambda^n)^T$ given by

$$\left\|\Lambda\right\|_\infty := \max_{i=1,\ldots,n} \|\lambda^i\|_\infty .$$

The symbol $f_\Lambda(\frac{\cdot}{2^k})$ is short for the scalar-valued sequence $\left(f_\Lambda(\frac{\alpha}{2^k})\right)_{\alpha\in\mathbb{Z}^d}$. It should be noted that, since for any convergent scheme the limit function f_Λ is given by

$$f_\Lambda = \sum_{\alpha\in\mathbb{Z}^2} \Lambda_\alpha^T \,\mathbf{\Phi}(\cdot - \alpha),$$

we can recover the components ϕ_i, $i = 1, \ldots, n$, as follows: choose the initial sequence $\mathbf{\Lambda} = (\lambda^1, \ldots, \lambda^n)^T$ in the following way that λ^i is a delta sequence (*i.e.*, $\lambda_\alpha^i = \delta_\alpha$ for $\alpha \in \mathbb{Z}^d$) while all other sequences λ^j, $j \neq i$, are null sequences. This special initial vector sequence and its iterates will be denoted by $\mathbf{E}_i$ and $\mathbf{E}_i^{(k)}$, $k \geq 0$, respectively. It then turns out that

$$\lim_{k\to\infty} \left\| \phi_i(\tfrac{\cdot}{2^k})\mathbf{e} - \mathbf{E}_i^{(k)} \right\|_\infty = 0 \quad \text{for} \quad i = 1, \ldots, n. \tag{2.6}$$

Theorem 2.1. *Given two convergent subdivision schemes associated with the refinable function vectors* $\mathbf{\Phi}$ *and* $\mathbf{\Psi}$, *the (Kronecker type) convolved scheme associated with* $\mathbf{\Theta} = \mathbf{\Phi} * \mathbf{\Psi}$ *is again convergent.*

Proof: Let $\mathbf{f} := (\underbrace{1\;1\;\cdots\;1}_{m})^T$, and let $\mathbf{F}_j$ denote the vector sequence (composed by m sequences) with the delta sequence at position j and the null sequence at all other positions. If $\mathbf{F}_j^{(k)}$, $k \geq 0$, are the iterated vectors with respect to the $\mathbf{\Psi}$-subdivision, we have

$$\lim_{k\to\infty} \left\| \psi_j(\tfrac{\cdot}{2^k})\mathbf{f} - \mathbf{F}_j^{(k)} \right\|_\infty = 0 \quad \text{for} \quad j = 1, \ldots, m, \tag{2.7}$$

in addition to (2.6). In order to prove the theorem, we will show that

$$\lim_{k\to\infty} \left\| \left((\phi_i * \psi_j)(\frac{\cdot}{2^k})\right)(\mathbf{e}\otimes \mathbf{f}) - \left(\mathbf{E}_i * \mathbf{F}_j\right)^{(k)} \right\|_\infty = 0 \tag{2.8}$$

for $i = 1,\ldots,n$ and $j = 1,\ldots,m$. Here, the convolution of two vector sequences is defined by

$$\mathbf{\Lambda} * \mathbf{\Gamma} := \begin{pmatrix} \lambda^1 \\ \lambda^2 \\ \vdots \\ \lambda^n \end{pmatrix} * \begin{pmatrix} \gamma^1 \\ \gamma^2 \\ \vdots \\ \gamma^m \end{pmatrix} := \begin{pmatrix} \lambda^1 * \mathbf{\Gamma} \\ \lambda^2 * \mathbf{\Gamma} \\ \vdots \\ \lambda^n * \mathbf{\Gamma} \end{pmatrix},$$

where the convolution of a scalar sequence λ^i with the vector sequence $\mathbf{\Gamma}$ is taken componentwise, *i.e.*, $\lambda^i * \mathbf{\Gamma} = (\lambda^i * \gamma^1, \lambda^i * \gamma^2, \ldots, \lambda^i * \gamma^m)^T$. We will also use the estimate

$$\left\|\mathbf{\Lambda} * \mathbf{\Gamma}\right\|_\infty \le \min \{ \left\|\mathbf{\Lambda}\right\|_1 \left\|\mathbf{\Gamma}\right\|_\infty , \left\|\mathbf{\Lambda}\right\|_\infty \left\|\mathbf{\Gamma}\right\|_1 \},$$

where $\left\|\mathbf{\Lambda}\right\|_1 := \max_{i=1,\ldots,n} \|\lambda^i\|_1$, with the usual 1-norm of scalar sequences.

Now, the iterated matrices $\mathbf{C}^{(k)}$ for the $\mathbf{\Theta}$-subdivision can be expressed by the iterated matrices $\mathbf{A}^{(k)}$ and $\mathbf{B}^{(k)}$ for the $\mathbf{\Phi}$- and $\mathbf{\Psi}$-schemes as follows:

$$C_\alpha^{(k)} = \frac{1}{2^{d\cdot k}} \sum_{\delta\in\mathbb{Z}^d} A_\delta^{(k)} \otimes B_{\alpha-\delta}^{(k)}, \quad \alpha \in \mathbb{Z}^d. \tag{2.9}$$

Thus, taking $\mathbf{E}_i * \mathbf{F}_j$ as a starting vector for the $\mathbf{\Theta}$-subdivision, the iterated vectors are given by

$$\left(\mathbf{E}_i * \mathbf{F}_j\right)^{(k)} = \frac{1}{2^{d\cdot k}} \mathbf{E}_i^{(k)} * \mathbf{F}_j^{(k)}, \quad k \ge 0, \tag{2.10}$$

whence

$$\begin{aligned}
&\left((\phi_i * \psi_j)(\frac{\cdot}{2^k})\right)(\mathbf{e}\otimes \mathbf{f}) - \left(\mathbf{E}_i * \mathbf{F}_j\right)^{(k)} \\
&= \left((\phi_i * \psi_j)(\frac{\cdot}{2^k})\right)(\mathbf{e}\otimes \mathbf{f}) - \frac{1}{2^{d\cdot k}}\left(\phi_i(\frac{\cdot}{2^k})\mathbf{e}\right) * \left(\psi_j(\frac{\cdot}{2^k})\mathbf{f}\right) \\
&\quad + \frac{1}{2^{d\cdot k}} \mathbf{E}_i^{(k)} * \left(\psi_j(\frac{\cdot}{2^k})\mathbf{f} - \mathbf{F}_j^{(k)}\right) \\
&\quad - \frac{1}{2^{d\cdot k}}\left(\mathbf{E}_i^{(k)} - \phi_i(\frac{\cdot}{2^k})\mathbf{e}\right) * \left(\psi_j(\frac{\cdot}{2^k})\mathbf{f}\right).
\end{aligned}$$

We estimate the three vector sequences in the preceding three lines as follows: the first term is a vector sequence where each component of the vector consists of the sequence $(e_\alpha^{(k)})_{\alpha\in\mathbb{Z}^d}$ with

$$e_\alpha^{(k)} := \int_{\mathbf{R}^d} \left(\phi_i(x)\right) \left(\psi_j(\frac{\alpha}{2^k} - x)\right) dx - \frac{1}{2^{d\cdot k}} \sum_{\delta\in\mathbb{Z}^d} \left(\phi_i(\frac{\delta}{2^k})\right) \left(\psi_j(\frac{\alpha-\delta}{2^k})\right).$$

This is the error of a tensor product rectangular rule applied to the convolution integral, and due to the continuity and the compact support of ϕ_i and ψ_j we get

$$\lim_{k\to\infty} \sup_{\alpha\in\mathbb{Z}^d} |e_\alpha^{(k)}| = 0.$$

For the third term,

$$\frac{1}{2^{d\cdot k}}\Big\|\Big(\mathbf{E}_i^{(k)} - \phi_i(\tfrac{\cdot}{2^k})\mathbf{e}\Big) * \Big(\psi_j(\tfrac{\cdot}{2^k})\mathbf{f}\Big)\Big\|_\infty \le \Big\|\mathbf{E}_i^{(k)} - \phi_i(\tfrac{\cdot}{2^k})\mathbf{e}\Big\|_\infty \frac{1}{2^{d\cdot k}}\Big\|\psi_j(\tfrac{\cdot}{2^k})\mathbf{f}\Big\|_1,$$

and the bound tends to zero as $k \to \infty$, since

$$\frac{1}{2^{d\cdot k}}\Big\|\psi_j(\tfrac{\cdot}{2^k})\mathbf{f}\Big\|_1 = \frac{1}{2^{d\cdot k}} \sum_{\alpha\in\mathbb{Z}^d} \Big|\psi_j(\tfrac{\alpha}{2^k})\Big| \to \int_{\mathbf{R}^d} |\psi_j(x)|\,dx.$$

Finally, for the second term

$$\frac{1}{2^{d\cdot k}}\Big\|\mathbf{E}_i^{(k)} * \Big(\psi_j(\tfrac{\cdot}{2^k})\mathbf{f} - \mathbf{F}_j^{(k)}\Big)\Big\|_\infty \le c_k \ \Big\|\psi_j(\tfrac{\cdot}{2^k})\mathbf{f} - \mathbf{F}_j^{(k)}\Big\|_\infty \to 0$$

as $k \to \infty$. This follows from

$$c_k := \frac{1}{2^{d\cdot k}}\Big\|\mathbf{E}_i^{(k)}\Big\|_1 \le \frac{1}{2^{d\cdot k}}\Big\|\phi_i(\tfrac{\cdot}{2^k})\mathbf{e}\Big\|_1 + \frac{1}{2^{d\cdot k}}\Big\|\phi_i(\tfrac{\cdot}{2^k})\mathbf{e} - \mathbf{E}_i^{(k)}\Big\|_1,$$

since the first term on the right-hand side converges to $\int_{\mathbf{R}^d} |\phi_i(x)|\,dx$, and the second term is bounded due to (2.6), the uniform continuity of the compactly supported function ϕ_i, and the fact that for some compact set $K \subset \mathbb{R}^d$ (in fact we can take the support of ϕ_i):

$$(\mathbf{E}_i^{(k)})_\alpha = \mathbf{0} \quad \text{for } \frac{\alpha}{2^k} \notin K.$$

This latter property is a consequence of the compact support of the matrix mask $\mathbf{A}$, since we start the iteration with a 'delta'-sequence $\mathbf{E}_i$. In conclusion, (2.8) holds true. □

§3. Approximation Order

Approximation orders of shift invariant spaces have been studied quite intensively in the past few years. Concerning definitions and notation we refer to the recent survey [5]. A characterization of approximation power in terms of the mask symbol is also given there. In case of the refinement equation (1.1), this mask symbol is defined by

$$\mathbf{H}_\Phi(\xi) := \frac{1}{2^d} \sum_{\alpha\in\mathbb{Z}^d} A_\alpha \, e^{-2\pi i\alpha\cdot\xi}. \tag{3.1}$$

The result is as follows: For given $k \in \mathbb{N}$ we say that $\mathbf{H}_\Phi$ *satisfies condition* (Z_k) if there exists a row vector $\mathbf{v} = (\tau_1, \ldots, \tau_n)$ of trigonometric polynomials such that

$$\mathbf{v}(0)\ \hat{\mathbf{\Phi}}(0) \neq 0, \quad \mathbf{v}(0)\ \mathbf{H}_\Phi(0) = \mathbf{v}(0), \tag{3.2.a}$$

and

$$D^\mu\Big(\mathbf{v}(\cdot)\mathbf{H}_\Phi(\tfrac{\cdot}{2})\Big)\Big|_\beta = 0 \quad \text{for } |\mu| < k \text{ and } \beta \in E_0', \tag{3.2.b}$$

where E_0' is short for the set of corners of the cube $[0,1]^d$ with the origin removed. [5, Theorem 3.2.8.(i)] then asserts that *condition* (Z_k) *implies that* S_Φ *has* L_2*- approximation order* k, *for* $f \in W_2^k(\mathbb{R}^d)$.

Lemma 3.1. *If the mask symbol* $\mathbf{H}_\Phi$ *satisfies condition* (Z_k) *(with the row vector* $\mathbf{v}$*) and the mask symbol* $\mathbf{H}_\Psi$ *satisfies condition* (Z_ℓ) *(with the row vector* $\mathbf{w}$*) then the mask symbol* $\mathbf{H}_\Theta$ *of the convolved function vector* $\mathbf{\Theta} := \mathbf{\Phi} * \mathbf{\Psi}$ *satisfies condition* $Z_{k+\ell}$ *(with the row vector* $\mathbf{z} := \mathbf{v} \otimes \mathbf{w}$*).*

Proof: We apply the identity

$$\mathbf{H}_\Theta = \mathbf{H}_\Phi \otimes \mathbf{H}_\Psi \tag{3.3}$$

several times. Condition (3.2.a) holds for the convolved function vector, since

$$(\mathbf{v} \otimes \mathbf{w})(0)\ (\widehat{\mathbf{\Phi} * \mathbf{\Psi}})(0) = \mathbf{v}(0)\hat{\mathbf{\Phi}}(0)\mathbf{w}(0)\hat{\mathbf{\Psi}}(0) \neq 0,$$

and by (3.3),

$$\mathbf{z}(0)\mathbf{H}_\Theta(0) = (\mathbf{v}(0)\mathbf{H}_\Phi(0)) \otimes (\mathbf{w}(0)\mathbf{H}_\Psi(0)) = \mathbf{v}(0) \otimes \mathbf{w}(0) = \mathbf{z}(0).$$

In order to verify condition (3.2.b),

$$D^\mu\Big(\mathbf{z}(\cdot)\mathbf{H}_\Theta(\tfrac{\cdot}{2})\Big)\Big|_\beta = 0 \quad \text{for } |\mu| < k + \ell \text{ and } \beta \in E_0',$$

we make use of the Leibniz-type formula

$$D^\mu\Big((\mathbf{v} \otimes \mathbf{w})(\cdot)(\mathbf{H}_\Phi \otimes \mathbf{H}_\Psi)(\tfrac{\cdot}{2})\Big) = \sum_{0 \le |\gamma| \le |\mu|} \binom{\mu}{\gamma} D^\gamma\Big(\mathbf{v}(\cdot)\mathbf{H}_\Phi(\tfrac{\cdot}{2})\Big) \otimes D^{\mu-\gamma}\Big(\mathbf{w}(\cdot)\mathbf{H}_\Psi(\tfrac{\cdot}{2})\Big)$$

at every point $\beta \in E_0'$. □

In order to derive an approximation order result from this lemma, we refer to the precise statement of [5, Theorem 3.2.8]. A sample result is as follows: *If the shift invariant spaces associated with* $\mathbf{\Phi}$ *and* $\mathbf{\Psi}$ *have approximation orders* k *and* ℓ, *respectively, and if the Gramians* $\mathbf{G}_\Phi$ *and* $\mathbf{G}_\Psi$ *satisfy the regularity condition given in [5, Theorem 3.2.8.(ii)], then the 'convolved' shift invariant space has (at least) approximation order* $k + \ell$. However, in general the 'convolved' Gramian $\mathbf{G}_\Theta$ does not satisfy this regularity condition.

§4. An Example: Bivariate C^1 Cubics on a 3-directional Mesh

In [3], we have given an example of piecewise C^1-cubics on the four-directional mesh. Following [1], we consider piecewise C^1-cubics on the three-directional mesh generated by the lines $x = k$, $y = l$, with $k, l \in \mathbb{Z}$ when adding the diagonals $x - y = m$, $m \in \mathbb{Z}$. A basis of this space is given by the two functions $\theta_1 = B_{111} * \chi_{T_1}$ and $\theta_2 = B_{111} * \chi_{T_2}$, with B_{111} the linear three-directional box-spline (or "Courant" element), and χ_{T_1}, χ_{T_2} the characteristic functions of the two triangles T_1, T_2 obtained by cutting the unit square $[0,1]^2$ by the 'north-east' diagonal. Thus, $\mathbf{\Theta} = \mathbf{\Phi} * \mathbf{\Psi}$ with $\mathbf{\Phi} := (B_{111})$ and $\mathbf{\Psi} := (\chi_{T_1}, \chi_{T_2})^T$.

Now, $\mathbf{\Phi} = (B_{111})$ satisfies a scalar refinement equation (1.1) with refinement mask

$$\mathbf{A} = (A_\alpha)_{\alpha \in \mathbb{Z}^2} = \begin{pmatrix} & \vdots & \vdots & \vdots & \vdots & \vdots & \\ \cdots & 0 & 0 & 0 & 0 & 0 & \cdots \\ \cdots & 0 & 0 & \frac{1}{2} & \frac{1}{2} & 0 & \cdots \\ \cdots & 0 & \frac{1}{2} & \mathbf{1} & \frac{1}{2} & 0 & \cdots \\ \cdots & 0 & \frac{1}{2} & \frac{1}{2} & 0 & 0 & \cdots \\ \cdots & 0 & 0 & 0 & 0 & 0 & \cdots \\ & \vdots & \vdots & \vdots & \vdots & \vdots & \end{pmatrix},$$

and $\mathbf{\Psi} = (\chi_{T_1}, \chi_{T_1})^T$ satisfies the refinement equation (1.2) with matrix mask

$$\mathbf{B} = (B_\alpha)_{\alpha \in \mathbb{Z}^2} = \begin{pmatrix} & \vdots & \vdots & \vdots & \vdots & \vdots & \\ \cdots & 0 & 0 & 0 & 0 & 0 & \cdots \\ \cdots & 0 & 0 & \begin{pmatrix} 0 & 0 \\ 1 & 1 \end{pmatrix} & \begin{pmatrix} 1 & 0 \\ 0 & 1 \end{pmatrix} & 0 & \cdots \\ \cdots & 0 & 0 & \begin{pmatrix} \mathbf{1} & \mathbf{0} \\ \mathbf{0} & \mathbf{1} \end{pmatrix} & \begin{pmatrix} 1 & 1 \\ 0 & 0 \end{pmatrix} & 0 & \cdots \\ \cdots & 0 & 0 & 0 & 0 & 0 & \cdots \\ & \vdots & \vdots & \vdots & \vdots & \vdots & \end{pmatrix}.$$

Here, the indexing of the 'coefficients' is such that the **boldface** entry is at position $\alpha = (0,0)$.

It follows that $\mathbf{\Theta} = \mathbf{\Phi} * \mathbf{\Psi}$ satisfies the refinement equation (1.3), with the matrix mask $\mathbf{C}$ as displayed on the following page, and Theorem 2.1 provides the convergence property for the associated subdivision scheme.

Concerning approximation order, Lemma 3.1 can be applied by putting $k = 2$, $\ell = 1$, and $\mathbf{z} := (1) \otimes (1\ 1)$. As a consequence, the approximation order for bivariate C^1-cubics on the the three-directional mesh is at least 3, and this is the precise approximation order (as was proved in [1]).

$$
\mathbf{C} = \begin{pmatrix}
 & \vdots & \vdots & \vdots & \vdots & \vdots & \vdots & \\
\cdots & 0 & 0 & 0 & 0 & 0 & 0 & \cdots \\
\cdots & 0 & 0 & \begin{pmatrix} 0 & 0 \\ \frac{1}{8} & \frac{1}{8} \end{pmatrix} & \begin{pmatrix} \frac{1}{8} & 0 \\ \frac{1}{8} & \frac{1}{4} \end{pmatrix} & \begin{pmatrix} \frac{1}{8} & 0 \\ 0 & \frac{1}{8} \end{pmatrix} & 0 & \cdots \\
\cdots & 0 & \begin{pmatrix} 0 & 0 \\ \frac{1}{8} & \frac{1}{8} \end{pmatrix} & \begin{pmatrix} \frac{1}{4} & 0 \\ \frac{1}{4} & \frac{1}{2} \end{pmatrix} & \begin{pmatrix} \frac{1}{2} & \frac{1}{8} \\ \frac{1}{8} & \frac{1}{2} \end{pmatrix} & \begin{pmatrix} \frac{1}{4} & \frac{1}{8} \\ 0 & \frac{1}{8} \end{pmatrix} & 0 & \cdots \\
\cdots & 0 & \begin{pmatrix} \frac{1}{8} & 0 \\ \frac{1}{8} & \frac{1}{4} \end{pmatrix} & \begin{pmatrix} \mathbf{\frac{1}{2}} & \mathbf{\frac{1}{8}} \\ \mathbf{\frac{1}{8}} & \mathbf{\frac{1}{2}} \end{pmatrix} & \begin{pmatrix} \frac{1}{2} & \frac{1}{4} \\ 0 & \frac{1}{4} \end{pmatrix} & \begin{pmatrix} \frac{1}{8} & \frac{1}{8} \\ 0 & 0 \end{pmatrix} & 0 & \cdots \\
\cdots & 0 & \begin{pmatrix} \frac{1}{8} & 0 \\ 0 & \frac{1}{8} \end{pmatrix} & \begin{pmatrix} \frac{1}{4} & \frac{1}{8} \\ 0 & \frac{1}{8} \end{pmatrix} & \begin{pmatrix} \frac{1}{8} & \frac{1}{8} \\ 0 & 0 \end{pmatrix} & 0 & 0 & \cdots \\
\cdots & 0 & 0 & 0 & 0 & 0 & 0 & \cdots \\
 & \vdots & \vdots & \vdots & \vdots & \vdots & \vdots &
\end{pmatrix}.
$$

References

1. de Boor, C., R. A. DeVore, and A. Ron, Approximation orders of FSI spaces in $L_2(\mathbb{R}^d)$, Constr. Approximation **14** (1998), 411–427.
2. Cavaretta, A., W. Dahmen, and C. A. Micchelli, *Stationary Subdivision*, Memoirs of the Amer. Math. Soc., vol. 93, no. 453, Providence, RI, 1991.
3. Conti, C., and K. Jetter, A new subdivision method for bivariate splines on the four-directional mesh, J. Comp. Appl. Math., to appear.
4. Horn, R. A., and C. R. Johnson, *Topics in Matrix Analysis*, Cambridge University Press, 1994.
5. Jetter, K., and G. Plonka, A survey on L_2-approximation order from shift-invariant spaces, in *Proceedings of an International Conference on Multivariate Approximation and Interpolation with Applications in CAGD, Signal and Image Processing, Eilat*, submitted.

C. Conti
Dipartimento di Energetica
Università di Firenze
Via C. Lombroso 6/17
I–50134 Firenze, Italy
`costanza@sirio.de.unifi.it`

K. Jetter
Institut für Angew. Math. und Statistik
Universität Hohenheim
D–70593 Stuttgart, Germany
`kjetter@uni-hohenheim.de`

Interpolating Polynomial Macro-Elements with Tension Properties

Paolo Costantini and Carla Manni

Abstract. In this paper we present the construction of nine–parameter polynomial macro elements, based on the classical Powell–Sabin split, which can be connected to form a C^1 surface. Variable degrees, which act as independent tension parameters, are associated with any vertex of the triangulation, i.e. to any interpolation point.

§1. Introduction

Among the several approaches for avoiding extraneous inflection points in interpolating functions, the so called tension methods are the oldest and probably the most famous ones. Basically they consist of C^k, $k \geq 1$, piecewise functions depending on a set of parameters which are selected in a local or global way to control the shape of the interpolants, stretching their patches between data points.

Piecewise polynomial splines with variable degrees have turned out to be a useful alternative to classical (exponential or rational) tension methods, and have been successfully applied both in free-form design and in constrained interpolation of spatial data [1]. On the other hand, a limited number of methods for constrained interpolation of bivariate scattered data are at present available, and very few of them offer the possibility of controlling the shape via tension parameters (see for example [2,3] and references quoted therein).

Let a set of scattered points (x_i, y_i, f_i), $i = 0, 1, \cdots, N$, be given, and suppose they have been associated with a proper triangulation $\mathcal{T}$. We are interested in local methods where the polynomial pieces of a spline are determined one triangle at time using only local data. Such methods are called macro–element methods. The aim of the paper is to describe a new class of variable degree polynomial triangular macro–elements, and to show their tension properties. These polynomial patches, which are based on the classical Powell–Sabin split of a triangle, can be connected to form a C^1 interpolating function.

Curve and Surface Fitting: Saint-Malo 1999
Albert Cohen, Christophe Rabut, and Larry L. Schumaker (eds.), pp. 143–152.

ISBN 0-8265-1357-3.

Typically, the main drawback of macro-element methods is in the strong influence of the triangulation on the shape of the interpolating surface, and, as far as we know, it is still not clear how to construct a "good" triangulation. Given that any interpolating scheme based on triangular macro–elements cannot be completely independent of the triangulation, we can nevertheless try to reduce this dependence. From this point of view, the main advantage over existing methods is in the fact that we have a variable degree associated to each vertex of the triangulation; therefore the modification in the shape of the interpolant is similar for all the patches around the same common vertex (see Fig. 8 right). In other words, although constructed over $\mathcal{T}$, the tension parameters are more related to the interpolation points than to the edges of the adopted triangulation.

It is worthwhile to anticipate that the possibly high degrees we use for our construction act only as tension parameters, and do not modify the basic structure of the macro–element. In other words, we always use a nine–parameter macro–element, and the computational complexity is almost independent of the size of the degrees used.

The scheme, being local, requires that the gradients are also known at the vertices of the triangulation. If this information is unavailable, gradients can be recovered from the data points. In the presented numerical test, gradients have been computed from the data according to the classical least square strategy.

The paper is divided into six sections. In the next section we introduce some notations. Sections 3 and 4 are devoted to the construction of the control points defining the macro–element, whose properties are briefly discussed in Section 5. We end with Section 6 where a graphical example is presented.

§2. Notation and Preliminaries

In this section we introduce some notation. To aid in comprehension, we notice that points and vectors in $\mathbb{R}^2$ and in $\mathbb{R}^3$ have been denoted by bold–faced characters unless classical notations (as for gradients) have been used. As usual, we describe the polynomial macro–element in terms of its Bernstein–Bézier control points. Let $\mathbf{P}_r$, $r = 1, 2, 3$, be three non–collinear points in $\mathbb{R}^2$, and let T denote the triangle they form. An n–degree Bernstein polynomial has the form

$$b(x, y; n) = b(u, v, w; n) := \sum_{i+j+k=n} \frac{n!}{i!j!k!} \, l_{ijk} \, u^i v^j w^k,$$

where, setting $\mathbf{P} = [x, \; y]^T \in \mathbb{R}^2$, $u = u(x, y)$, $v = v(x, y)$, $w = w(x, y)$, are the barycentric coordinates of $\mathbf{P}$ with respect to the vertices of T, that is

$$\mathbf{P} = u\mathbf{P}_1 + v\mathbf{P}_2 + w\mathbf{P}_3; \quad u + v + w = 1,$$

and l_{ijk}, $i + j + k = n$, are the Bernstein ordinates of $b(., ., .; n)$.

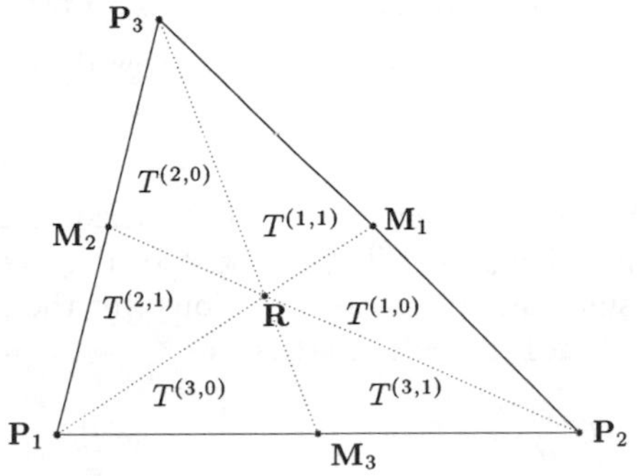

Fig. 1. Powell–Sabin split of a triangle.

Setting $x = x(u,v,w),\ y = y(u,v,w)$, the points in $\mathbb{R}^3$

$$\mathbf{L}_{ijk} := \begin{bmatrix} x(\frac{i}{n},\frac{j}{n},\frac{k}{n}) \\ y(\frac{i}{n},\frac{j}{n},\frac{k}{n}) \\ l_{ijk} \end{bmatrix}, \quad i,j,k \geq 0\ ,\ i+j+k=n,$$

are called control points [4].

We finally recall the so-called Powell–Sabin split ([5]) of a given triangle, T, which consists of dividing $T := \mathbf{P}_1\mathbf{P}_2\mathbf{P}_3$ into six *mini–triangles* (Fig. 1):

$$T^{(1,0)} := \mathbf{P}_2\mathbf{M}_1\mathbf{R},\ T^{(2,0)} := \mathbf{P}_3\mathbf{M}_2\mathbf{R},\ T^{(3,0)} := \mathbf{P}_1\mathbf{M}_3\mathbf{R},$$

$$T^{(1,1)} := \mathbf{M}_1\mathbf{P}_3\mathbf{R},\ T^{(2,1)} := \mathbf{M}_2\mathbf{P}_1\mathbf{R},\ T^{(3,1)} := \mathbf{M}_3\mathbf{P}_2\mathbf{R},$$

where

$$\mathbf{R} = \beta_1\mathbf{P}_1 + \beta_2\mathbf{P}_2 + \beta_3\mathbf{P}_3\ ,\ \beta_1+\beta_2+\beta_3 = 1,$$

is a point internal to T and

$$\mathbf{M}_i = (1-\alpha_i)\mathbf{P}_{i+1} + \alpha_i\mathbf{P}_{i+2},\ 0 < \alpha_i < 1,$$

is a point internal to the edge of T opposite to $\mathbf{P}_i$. Here, and in the following, indices will be considered modulus 3.

We will denote by $\mathbf{L}_{ijk}^{(p,q)}$, $\overline{\mathbf{L}}_{ijk}^{(p,q)}$ the control points of Bernstein polynomials in the mini triangle $T^{(p,q)}$ (see below). Let the data

$$(\mathbf{P}_i, f_i = f(\mathbf{P}_i), \nabla f_i = \nabla f(\mathbf{P}_i)), \mathbf{P}_i \in \mathbb{R}^2,\quad i = 1,2,3, \tag{1}$$

be given. As mentioned in the introduction, our goal is to construct a C^1 polynomial macro–element on T interpolating the data and having tension properties.

The macro–element will be obtained considering a Powell–Sabin split of T, and constructing in each mini triangle $T^{(p,q)}$ a Bernstein polynomial via suitable control points, $\overline{\mathbf{L}}_{ijk}^{(p,q)}$. To obtain the final control points (FCP) $\overline{\mathbf{L}}_{ijk}^{(p,q)}$, we follow two basic steps: first we construct the basic control points (BCP) $\mathbf{L}_{ijk}^{(p,q)}$, then we modify them to reach the required smoothness of the macro–element.

§3. Defining the Basic Control Points

For each vertex $\mathbf{P}_i$, let us consider an associated given degree

$$n_i \geq 3, \quad n_i \in \mathbb{N}.$$

Let us now describe the construction of the BCP considering for the sake of simplicity only the mini triangle $T^{(3,0)}$. See Fig. 2 for the role of indices.

First of all we assume interpolation conditions (for the position and the gradient) at $\mathbf{P}_1$ (see "$*$" in Fig. 3 left) that is

$$\begin{aligned} l^{(3,0)}_{n_1,0,0} &:= f_1 \ , \\ l^{(3,0)}_{n_1-1,1,0} &:= f_1 + \frac{1}{n_1}\langle \nabla f_1, \mathbf{P}_1\mathbf{M}_3 \rangle \ , \\ l^{(3,0)}_{n_1-1,0,1} &:= f_1 + \frac{1}{n_1}\langle \nabla f_1, \mathbf{P}_1\mathbf{R} \rangle. \end{aligned} \tag{2}$$

In order to define the BCP around $\mathbf{M}_3$, let us consider the univariate piecewise linear function, l_3, defined along the edge $\mathbf{P}_1\mathbf{P}_2$ having breakpoints at

$$\mathbf{P}_1, \ \ \mathbf{P}_1 + \frac{1}{n_1}\mathbf{P}_1\mathbf{M}_3, \ \ \mathbf{P}_2 - \frac{1}{n_2}\mathbf{M}_3\mathbf{P}_2, \ \ \mathbf{P}_2,$$

interpolating f and its derivatives at the extremes of the edge. We define (see "$\otimes$" in Fig. 3 left)

$$\begin{aligned} l^{(3,0)}_{0,n_1,0} &:= l_3(\mathbf{M}_3), \\ l^{(3,0)}_{1,n_1-1,0} &:= l_3\left(\mathbf{M}_3 - \frac{1}{n_1}\mathbf{P}_1\mathbf{M}_3\right), \\ l^{(3,0)}_{0,n_1-1,1} &:= l_3(\mathbf{M}_3) + \frac{1}{n_1}d_3, \end{aligned} \tag{3}$$

where

$$d_3 := \langle (1-\alpha_3)\nabla f(\mathbf{P}_1) + \alpha_3 \nabla f(\mathbf{P}_2), \mathbf{M}_3\mathbf{R} \rangle \, .$$

Moreover, we require that $\mathbf{L}^{(3,0)}_{1,n_1-2,1}$ (see "$\odot$" in Fig. 3 left) belongs to the plane through

$$\mathbf{L}^{(3,0)}_{1,n_1-1,0}, \ \mathbf{L}^{(3,0)}_{0,n_1,0}, \ \mathbf{L}^{(3,0)}_{0,n_1-1,1}.$$

Concerning the remaining control points, we assume that the not yet defined control points of the first two rows in the mini triangle parallel to $\mathbf{P}_1\mathbf{P}_2$ (see "o" in Fig. 3 left) belong to the straight lines through the above defined control points, that is

$$\begin{aligned} \mathbf{L}^{(3,0)}_{i,n_1-i,0} &= \frac{i-1}{n_1-2}\mathbf{L}^{(3,0)}_{n_1-1,1,0} + \frac{(n_1-2)-(i-1)}{n_1-2}\mathbf{L}^{(3,0)}_{1,n_1-1,0} \ , \\ \mathbf{L}^{(3,0)}_{i,n_1-i-1,1} &= \frac{i-1}{n_1-2}\mathbf{L}^{(3,0)}_{n_1-1,0,1} + \frac{(n_1-2)-(i-1)}{n_1-2}\mathbf{L}^{(3,0)}_{1,n_1-2,1} \ , \\ i &= n_1-2, \ldots, 2 \ . \end{aligned} \tag{4}$$

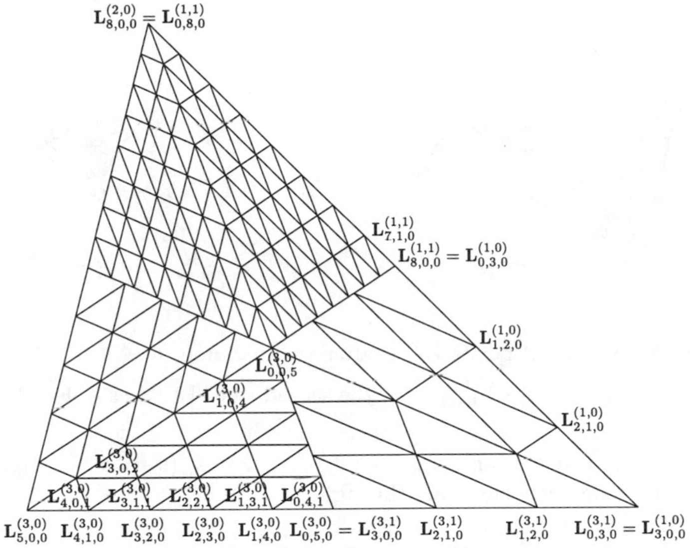

Fig. 2. The role of indices of the control points (projection onto the x, y plane) for the split of Figure 1 with $n_1 = 5, n_2 = 3, n_3 = 8$.

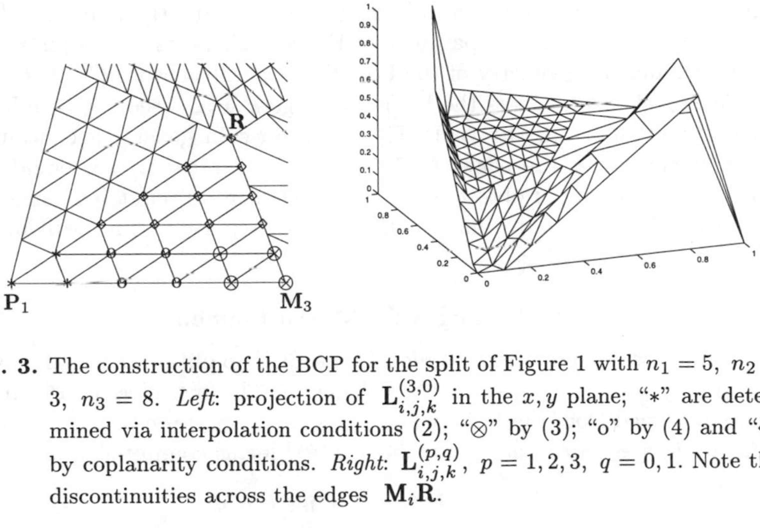

Fig. 3. The construction of the BCP for the split of Figure 1 with $n_1 = 5$, $n_2 = 3$, $n_3 = 8$. *Left*: projection of $\mathbf{L}^{(3,0)}_{i,j,k}$ in the x, y plane; "$*$" are determined via interpolation conditions (2); "$\otimes$" by (3); "o" by (4) and "$\diamond$" by coplanarity conditions. *Right*: $\mathbf{L}^{(p,q)}_{i,j,k}$, $p = 1, 2, 3,\ q = 0, 1$. Note the discontinuities across the edges $\mathbf{M}_i\mathbf{R}$.

Moreover, we require that conditions for C^1 continuity across the edge $\mathbf{P}_1\mathbf{R}$ hold [4]. Then, in particular,

$$\mathbf{L}^{(2,1)}_{1,n_1-2,1}, \mathbf{L}^{(2,1)}_{0,n_1-1,1}, \mathbf{L}^{(2,1)}_{0,n_1-2,2} = \mathbf{L}^{(3,0)}_{n_1-2,0,2},\ \mathbf{L}^{(3,0)}_{n_1-1,0,1}, \mathbf{L}^{(3,0)}_{n_1-2,1,1},$$

lie onto the same plane.

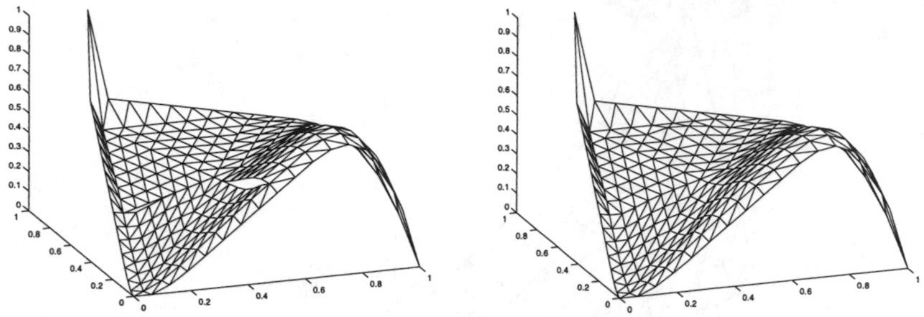

Fig. 4. *Left*: the control points $\overline{\mathbf{L}}_{i,j,k}^{(p,q)}$ after the first step, $n = 8$. *Right*: The final control points $\overline{\mathbf{L}}_{i,j,k}^{(p,q)}$ after the second step. $\mathbf{L}_{i,j,k}^{(p,q)}$ as in Fig. 3.

Finally, the central control points $\mathbf{L}_{i,j,k}^{(3,0)}$, $k \geq 2$, (see "$\diamond$" in Fig. 3 left) are assumed to lie onto the plane through

$$\mathbf{L}_{n_1-2,0,2}^{(3,0)},\ \mathbf{L}_{n_2-2,0,2}^{(1,0)},\ \mathbf{L}_{n_3-2,0,2}^{(2,0)}.$$

Similarly we define the BCP in the other mini triangles.

The above construction provides a polynomial macro–element which turns out to be of class C^1 across the interior edges $\mathbf{P}_i\mathbf{R}$.

Moreover, we assume that the points $\mathbf{M}$ lying on the internal edges of the initial triangulation $\mathcal{T}$ also lie on the straight lines joining the $\mathbf{R}$ points of the triangles which those edges separate (see Fig. 8). This classical requirement, (2), (3), (4), and the geometry of the Powell–Sabin split ensure C^1 continuity of two macro–elements across the boundary edge $\mathbf{P}_i\mathbf{P}_{i+1}$ (see Fig. 3 right).

On the other hand, the BCP do not produce in general a continuous macro–element across the edges $\mathbf{M}_i\mathbf{R}$ unless the degrees n_j are equal (see Fig. 3 right). In order to obtain a C^1 macro–element without imposing any conditon on the degrees, we modify the constructed BCP. This will be described in the next section.

§4. Obtaining a C^1 Macro Element

In this section we describe how to modify the BCP to obtain the final control points (FCP) producing a C^1 macro–element. The modified FCP will be basically obtained from the BCP via the degree-raising process in two steps.

As a first step, for each mini triangle $T^{(p,q)}$ let us compute

$$\overline{\mathbf{L}}_{i,j,k}^{(p,q)},\ i,j,k \geq 0\ ,\ i+j+k = n,$$

the control points obtained from $\mathbf{L}_{i,j,k}^{(p,q)}$, $i,j,k \geq 0$, $i+j+k = n_{p+q+1}$, by the degree-raising process ([4]) from the degree n_{p+q+1} (that is the degree associated with the mini triangle $T^{(p,q)}$) to the degree $n := \max\{n_1, n_2, n_3\}$ (see Fig. 4 left). The control points $\overline{\mathbf{L}}_{i,j,k}^{(p,q)}$ allow us to express the polynomial of degree n_{p+q+1} defined by $\mathbf{L}_{i,j,k}^{(p,q)}$ as a Bernstein polynomial of degree n.

We emphasize that, due to the geometry of the split, the control points $\overline{\mathbf{L}}^{(p,q)}_{i,j,k}$, define a macro–element which is of class C^1 across the interior edges $\mathbf{M}_p\mathbf{R}$ if and only if the control points

$$\overline{\mathbf{L}}^{(p,0)}_{1,n-1-k,k},\ \overline{\mathbf{L}}^{(p,0)}_{0,n-k,k} = \overline{\mathbf{L}}^{(p,1)}_{n-k,0,k},\ \overline{\mathbf{L}}^{(p,1)}_{n-k-1,1,k},\quad k = 0,\cdots,n-1,$$

are collinear. Moreover, we notice that, due to the construction of the BCP around $\mathbf{M}_p$ and near $\mathbf{R}$ and to the properties of the degree-raising process, the control points

$$\overline{\mathbf{L}}^{(p,0)}_{1,n-1-k,k},\ \overline{\mathbf{L}}^{(p,0)}_{0,n-k,k} = \overline{\mathbf{L}}^{(p,1)}_{n-k,0,k},\ \overline{\mathbf{L}}^{(p,1)}_{n-k-1,1,k},\quad k = 0,1,n-1,$$

lie on the same plane, then they are collinear due to the geometry of the split (see Fig. 4 left). Then in order to obtain C^1 continuity across the edge $\mathbf{M}_p\mathbf{R}$, we simply consider a second step in which we modify

$$\overline{\mathbf{L}}^{(p,0)}_{0,n-k,k}, \overline{\mathbf{L}}^{(p,1)}_{n-k,0,k},\quad k = 2,\cdots,n-2,\ p = 1,2,3, \tag{5}$$

imposing that (see Fig. 4 right) they lie on the segment through

$$\overline{\mathbf{L}}^{(p,0)}_{1,n-k-1,k}, \overline{\mathbf{L}}^{(p,1)}_{n-k-1,1,k},\quad k = 2,\cdots,n-2,\ p = 1,2,3.$$

§5. Properties of the Macro–Element

In this section we analyze the interpolation, smoothness and tension properties of the macro–element defined by the FCP constructed in Section 4. First of all, we notice that the construction of the macro–element is completely local: it only depends on the data at the vertices of T and on the degrees n_j associated with the vertices which are given input parameters.

Theorem 1. *The polynomial macro–element defined by the control points*

$$\overline{\mathbf{L}}^{(p,q)}_{i,j,k},\ p = 1,2,3, q = 0,1,\ i,j,k \geq 0\ ,\ i+j+k = n,$$

interpolates the data (1) and is of class $C^1(T)$. Moreover, let $\mathcal{T}$ be a given triangulation equipped with a classical Powell–Sabin split, and with a fixed degree associated with any vertex. Then the collection of the macro–elements corresponding to the triangles of $\mathcal{T}$ produces an interpolating surface of class C^1.

Proof: The BCP defined in Section 3 produce an interpolating macro–element which is of class C^1 across the interior edges $\mathbf{P}_i\mathbf{R}$ and the boundary edges $\mathbf{P}_i\mathbf{P}_{i+1}$. After the degree-raising process, only the control points in (5) are modified in order to obtain C^1 continuity across $\mathbf{M}_i\mathbf{R}$. Since the C^1 continuity across one edge only depends on the control points lying on the

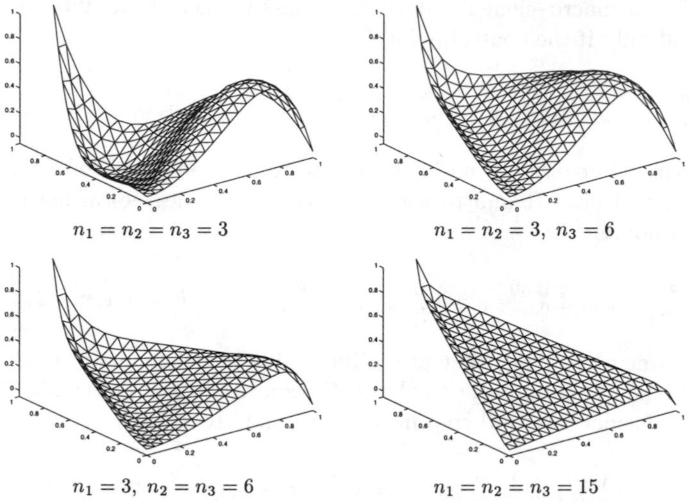

Fig. 5. The local tension effect of the degrees: vertices are numbered counterclockwise from the origin.

first two rows parallel to that edge ([4]), the second step in Section 4 does not affect C^1 continuity across $\mathbf{P}_i\mathbf{R}$ and $\mathbf{P}_i\mathbf{P}_{i+1}$. □

As mentioned before, the degrees n_j are free input parameters. From the construction described in Section 3, it is clear that increasing their values causes the BCP to approach the plane interpolating $(\mathbf{P}_i, f_i),\ i = 1, 2, 3$. Similarly, the FCP approach the same plane because they have been obtained via a degree-raising process, that is via a convex combination of the BCP. Therefore, the same property is shared by the macro–element, due to the convex hull property of the Bernstein representation. We summarize the tension properties of our macro–element with the following theorem.

Theorem 2. *If* $n_1, n_2, n_3 \to +\infty$, *then the polynomial macro–element defined by the control points*

$$\overline{\mathbf{L}}^{(p,q)}_{i,j,k},\ p = 1, 2, 3, q = 0, 1,\ i, j, k \geq 0\ ,\ i + j + k = n,$$

approaches the plane through $(\mathbf{P}_i, f_i),\ i = 1, 2, 3$.

We end this section by emphasizing that the degrees act as *local* tension parameters: each degree affects the shape of the interpolating surface only around the associated vertex (see Fig. 8 right), and, as previously said, this is the main feature of this method. The increase of a degree pushes the surface to the piecewise linear interpolant around the corresponding point, giving it a cuspidal appearance. This local tension effect is clearly shown in Fig. 5.

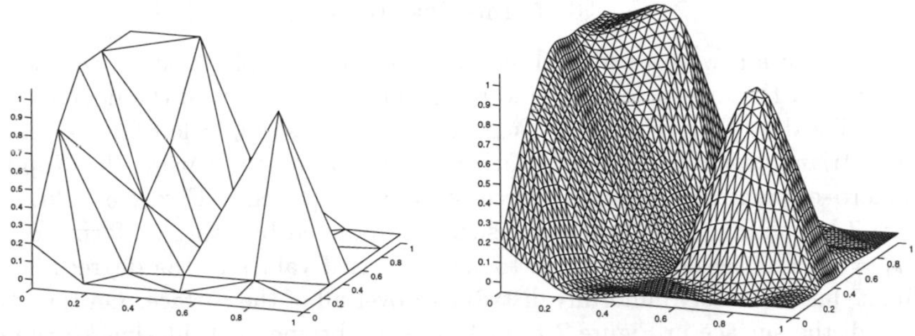

Fig. 6. *Left*: the Ritchie's Hill data. *Right*: the interpolating surface with uniform degrees $n_j = 3$.

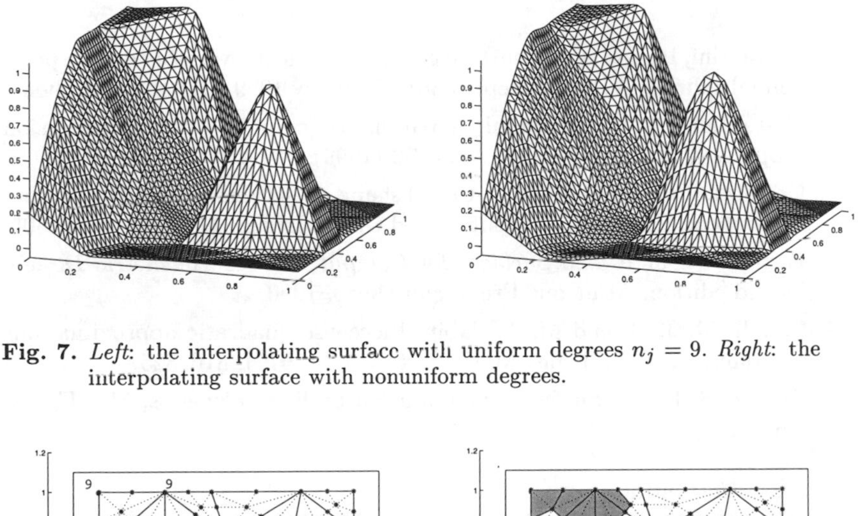

Fig. 7. *Left*: the interpolating surface with uniform degrees $n_j = 9$. *Right*: the interpolating surface with nonuniform degrees.

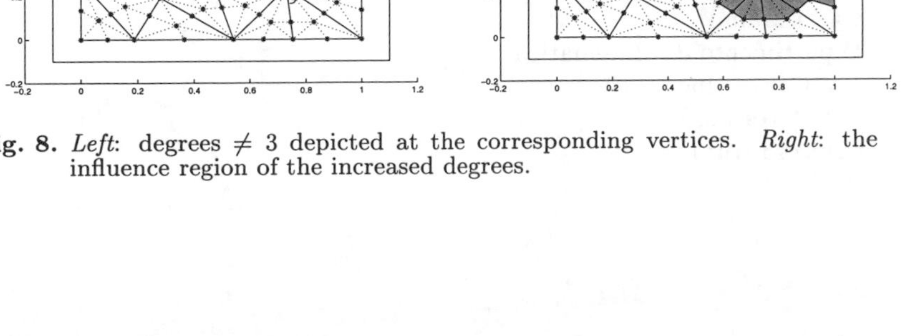

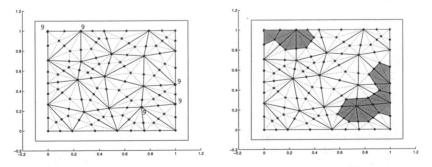

Fig. 8. *Left*: degrees $\neq 3$ depicted at the corresponding vertices. *Right*: the influence region of the increased degrees.

§6. Numerical Results

In this section we present a classical graphical example to show the performances of the macro–element and the local tension effect of the degrees.

The data (Fig. 6 left) are taken from [6] (see also Fig. 8 left for the considered triangulation). The interpolating surfaces obtained by using the proposed macro–element with uniform degrees $n_j = 3$ is depicted in Figure 6 right. Figure 7 left shows the interpolating surfaces obtained by using uniform degrees $n_j = 9$. The tension effect due to the increased values of the degrees is evident, but obviously uniformly distributed over all of the surface. On the other hand, the surface in Figure 7 right has been obtained considering all the degrees equal to 3, except those associated with the five vertices as depicted in Figure 8 left. The local tension effect of the degrees is clear. The influence region of the increased degrees can be also seen in Figure 8 right.

References

1. Costantini, P., Curve and surface construction using variable degree polynomial splines, Technical Report 361, February 1999, Universitá di Siena.
2. Costantini, P. and C. Manni, A parametric cubic element with tension properties, SIAM J. Numer. Anal. **36** (1999), 607–628.
3. Costantini, P. and C. Manni, A local shape-preserving interpolation scheme for scattered data, Comput. Aided Geom. Design **16** (1999), 385–405.
4. Farin, G., *Curves and Surfaces for Computer Aided Geometric Design*, second edition, Academic Press, San Diego, 1990.
5. Powell, M. J. D. and M. A. Sabin, Piecewise quadratic approximations on triangles, ACM Trans. Math. Software **3** (1977), 316–325.
6. Ritchie, S. I. M. , *Surface representation by finite elements*, Ms. Thesis, Univ. Calgary, 1978.

Paolo Costantini
Dipartimento di Matematica " Roberto Magari"
Via del Capitano 15
53100 Siena, Italy
costantini@unisi.it

Carla Manni
Dipartimento di Matematica
Via Carlo Alberto 10
10123 Torino, Italy
manni@dm.unito.it

Quantized Frame Decompositions

M. Craizer, D. A. Fonini, Jr., and E. A. B. da Silva

Abstract. In this paper, we consider a certain type of decomposition of vectors in frames, in which the coefficients are already quantized and thus are ready for coding. This decomposition is a generalization for vectors of the usual binary expansion of real numbers, and the algorithm for obtaining it can be seen as a quantized version of the matching pursuits algorithm. We show that, in several cases, applying this algorithm is better than first finding the frame coefficients and then quantizing them.

§1. Introduction

Let $\mathcal{F} = \{\boldsymbol{e}_1, \boldsymbol{e}_2, \ldots, \boldsymbol{e}_p\}$ be a collection of unit vectors generating $\mathbb{R}^N$. This means that every $\boldsymbol{x} \in \mathbb{R}^N$ can be expressed as

$$\boldsymbol{x} = \sum_{i=1}^{p} a_i \boldsymbol{e}_i.$$

The vectors $\{\boldsymbol{e}_1, \boldsymbol{e}_2, \ldots, \boldsymbol{e}_p\}$ may or may not be linearly independent. In the case that they are linearly dependent, the set $\mathcal{F}$ is called a frame or an over-complete basis. In this paper, we shall call $\mathcal{F}$ a frame even if the vectors $\{\boldsymbol{e}_1, \boldsymbol{e}_2, \ldots, \boldsymbol{e}_p\}$ are linearly independent. More on frame expansions can be found in [7].

Let $q = 2p$ and, for $1 \le i \le p$, let $\boldsymbol{v}_i = \boldsymbol{e}_i$ and $\boldsymbol{v}_{i+p} = -\boldsymbol{e}_i$. We shall call the set $\mathcal{D} = \{\boldsymbol{v}_1, \ldots, \boldsymbol{v}_q\}$ a codebook or a dictionary. Let α be a real number in the interval $(0, 1)$. A representation of a vector $\boldsymbol{x} \in \mathbb{R}^N$ in the form

$$\boldsymbol{x} = \sum_{i=0}^{\infty} \alpha^i \boldsymbol{v}_{k_i}$$

with $\boldsymbol{v}_{k_i} \in \mathcal{D}$, will be called an $(\alpha, \mathcal{D})$-expansion. When the dictionary being used is clear by the context, we shall call this representation simply an α-expansion. Observe that the $(\alpha, \mathcal{D})$-expansion of a vector $\boldsymbol{x}$ can be seen as a decomposition of $\boldsymbol{x}$ in the frame $\mathcal{F}$.

Curve and Surface Fitting: Saint-Malo 1999
Albert Cohen, Christophe Rabut, and Larry L. Schumaker (eds.), pp. 153–160.

ISBN 0-8265-1357-3.

Define the $n^{\underline{th}}$ residual of a vector $\boldsymbol{x}$ by

$$r_n(\boldsymbol{x}) = \begin{cases} r_0(\boldsymbol{x}) = \boldsymbol{x}, & \text{if } n = 0, \\ \boldsymbol{x} - \sum_{i=0}^{n-1} \alpha^i \boldsymbol{v}_{k_i}, & \text{if } n > 0. \end{cases}$$

Given $\boldsymbol{x}$, the sequence $(k_0, k_1, \ldots)$ can be obtained recursively by the relation

$$\langle \boldsymbol{r}_n(\boldsymbol{x}), \boldsymbol{v}_{k_n} \rangle = \max_k \langle \boldsymbol{r}_n(\boldsymbol{x}), \boldsymbol{v}_k \rangle .$$

We shall call this algorithm the nearest point algorithm and it may be seen as a quantized version of the matching pursuits algorithm [4].

Denote by $\Lambda_\alpha = \Lambda_\alpha(\mathcal{D})$ the set of points of $\mathbb{R}^N$ that can be represented as an α-expansion of vectors that belong to $\mathcal{D}$, and by $\Lambda_\alpha^0 = \Lambda_\alpha^0(\mathcal{D})$ the subset of $\Lambda_\alpha(\mathcal{D})$ whose α-expansion can be obtained by the nearest point algorithm. In order for the α-expansion or the nearest point algorithm to be a suitable scheme for quantized frame decomposition, we must choose α such that Λ_α or Λ_α^0, respectively, contain an open set of $\mathbb{R}^N$. In Section 2, we shall give conditions on α that guarantee these facts.

At this point, a question arises: is it worthwhile to decompose a vector in a frame using the α-expansion, or is it better to decompose it in the usual way and then quantize the coefficients in a second step (see [6,9])? We shall answer this question by considering the rate-distortion characteristic of each scheme. We show in Section 3 that the first scheme is better, in an asymptotic sense, if and only if we can choose α satisfying

$$\frac{\log_2(2p)}{\log_2 \frac{1}{\alpha}} < p.$$

We shall also give examples where this inequality holds.

Take $\boldsymbol{x} = (x_1, \ldots, x_N) \in \mathbb{R}^N$. We can quantize $\boldsymbol{x}$ by taking the n-term binary representation of each coordinate x_i. This procedure can be considered as an n-term α-expansion using the dictionary $\mathcal{B}_N$ whose code vectors are the corners of the hypercube $[-1, 1]^N$ and $\alpha = \frac{1}{2}$. So the α-expansion in an arbitrary dictionary $\mathcal{D}$ can be considered as a generalization of the usual binary expansion for vectors. The relevant question is whether there is any dictionary $\mathcal{D}$ that is better in some sense than $\mathcal{B}_N$. It is worthy of note that some special dictionaries, related to the sphere packing problem [2], have already been used for image coding, yielding better results than $\mathcal{B}_N$ ([3,8]).

§2. Theory of Alpha-Expansions

General representation

Let $\mathcal{D} = \{\boldsymbol{v}_1, \ldots, \boldsymbol{v}_q\}$ be a collection of vectors that generates all $\mathbb{R}^N$, and $0 < \alpha < 1$ be a parameter. Denote by Λ_α the set of points $\boldsymbol{x} \in \mathbb{R}^N$ that can be written as

$$\boldsymbol{x} = \sum_{i=0}^{\infty} \alpha^i \boldsymbol{v}_{k_i}$$

with $\boldsymbol{v}_{k_i} \in \mathcal{D}$. Let P_α be the convex hull of the vectors $\{\frac{1}{1-\alpha}\boldsymbol{v}_k\}_{k=1,\ldots,q}$. Observe that any $\boldsymbol{x} \in \Lambda_\alpha$ is a linear combination of the vectors $\{\boldsymbol{v}_k\}_{k=1,\ldots,q}$ with coefficients whose sum is not larger than $\frac{1}{1-\alpha}$, and therefore $\Lambda_\alpha \subset P_\alpha$.

Define the contracting maps $f_k = f_{k,\alpha}$ by $f_k(\boldsymbol{x}) = \alpha\boldsymbol{x} + \boldsymbol{v}_k$ for $k \in \{1,\ldots,q\}$. We observe that f_k is a homotety of center $\frac{1}{1-\alpha}\boldsymbol{v}_k$, which implies that $f_k(P_\alpha) \subset P_\alpha$. Therefore the set $\{f_1,\ldots,f_q\}$ forms an iterated function system (IFS) [1] on P_α. It is not difficult to show that the attractor of this system is exactly Λ_α, i.e.,

$$\Lambda_\alpha = \bigcap_{n=0}^{\infty} F^n(P_\alpha),$$

where $F = F_\alpha$ is the function of sets defined by

$$F(A) = f_1(A) \cup \cdots \cup f_q(A).$$

Example 1. *Let $\mathcal{D} = \mathcal{B}_N$, the dictionary whose code vectors are the corners of the hypercube $[-1,1]^N$. For any $\alpha \geq \frac{1}{2}$, we have that $F(P_\alpha) = P_\alpha$ and therefore $\Lambda_\alpha = P_\alpha$.*

We are interested in finding the smallest value of α such that Λ_α contains an open set. In the above example, this occurs for $\alpha = \frac{1}{2}$, when in fact $\Lambda_\alpha = P_\alpha$.

Remark 1. *One can show that for any dictionary, if $\alpha \geq \frac{N}{N+1}$, then $\Lambda_\alpha = P_\alpha$, which shows that the smallest value of α such that Λ_α contains an open set is smaller than $\frac{N}{N+1}$.*

In Example 1, the smallest α such that Λ_α contains an open set satisfies also $\Lambda_\alpha = P_\alpha$. But this is not a general fact, as the following example shows.

Example 2. *Let $\mathcal{D} = \mathcal{B}_3 \cup \{(1,0,0),(-1,0,0)\}$, where $\mathcal{B}_3$ is the dictionary whose code vectors are the corners of the cube $[-1,1]^3$. If we consider $\alpha = \frac{1}{2}$, then $[-1,1]^3 \subset \Lambda_\alpha$, but Λ_α is strictly contained in P_α. This fact can be seen by observing that the centroid of the face of P_α whose vertices are $(1,0,0)$, $\frac{1}{\sqrt{3}}(1,1,1)$ and $\frac{1}{\sqrt{3}}(1,1,-1)$ are not contained in $F(P_\alpha)$, which implies that $F(P_\alpha) \neq P_\alpha$, and thus $\Lambda_\alpha \neq P_\alpha$.*

In all examples that we have considered, we observe that if Λ_α contains an open set of $\mathbb{R}^N$, then it also contains the convex hull of some of the points of the dictionary. We don't know whether this is always true, so we formulate it as a question:

Question 1. *If, for some $0 < \alpha < 1$, $\Lambda_\alpha(\mathcal{D})$ contains an open set of $\mathbb{R}^N$, then will a subdictionary $\mathcal{D}_1 \subset \mathcal{D}$ always exist such that $\Lambda_\alpha(\mathcal{D}) \supseteq P_\alpha(\mathcal{D}_1)$?*

Basic algorithm

How do we obtain the sequence of indexes $(k_0, k_1, \ldots)$ that represent a given vector $\boldsymbol{x} \in \Lambda_\alpha$? In general, the representation of a vector $\boldsymbol{x}$ is not unique. In order to define which of the sequences representing the vector x we shall look for, we consider a choice function $K : F(P_\alpha) \to \{1, \ldots, q\}$ with the following properties:

1) $f_k(V_k) \subset V_k$, for any $k \in \{1, \ldots, q\}$, where $V_k = K^{-1}(k)$.
2) If $K(\boldsymbol{x}) = k$, then $\boldsymbol{x} \in f_k(P_\alpha)$, for any $\boldsymbol{x} \in F(P_\alpha)$.

It can be shown that such a function always exists. This choice function K determines a function $g : F(P_\alpha) \to P_\alpha$ given by

$$g(\boldsymbol{x}) = \frac{\boldsymbol{x} - \boldsymbol{v}_{K(\boldsymbol{x})}}{\alpha}.$$

It is not difficult to show that

$$\Lambda_\alpha = \cap_{n=1}^{\infty} g^{-n}(P_\alpha).$$

This implies that if $\boldsymbol{x} \in \Lambda_\alpha$, then $g^i(\boldsymbol{x}) \in \Lambda_\alpha$, for any $i \geq 0$.

By the last paragraph, given $\boldsymbol{x} \in \Lambda_\alpha$, we can choose the sequence $(k_0, k_1, \ldots)$ by the relation $k_i = K(g^i(\boldsymbol{x}))$. We shall call this the *basic algorithm.* This algorithm always works, but it is computationally expensive. So we propose another algorithm, computationally feasible, called the nearest point algorithm.

Nearest point algorithm

The nearest point algorithm is used for obtaining a sequence $(k_0, k_1, \ldots)$ representing a vector $\boldsymbol{x} \in \Lambda_\alpha$. It can be seen as a quantized version of the matching pursuits algorithm.

It is determined by the choice function K_0 defined by the property that $K_0(\boldsymbol{x})$ is the code vector in $\mathcal{D}$ nearest to $\boldsymbol{x}$. We denote by $V_{0,k}$ the set $K_0^{-1}(k)$ and by g_0 the function

$$g_0(\boldsymbol{x}) = \frac{\boldsymbol{x} - \boldsymbol{v}_{K_0(\boldsymbol{x})}}{\alpha}.$$

This choice function certainly satisfies Property 1 above, but Property 2 can fail. It is not difficult to see that Property 2 holds if and only if $V_{0,k} \subset f_k(P_\alpha)$, for every $k \in \{1, \ldots, q\}$.

Let $\Lambda_\alpha^0(\mathcal{D}) = \cap_{n=1}^{\infty} g_0^{-n}(P_\alpha)$. We have that $\Lambda_\alpha^0(\mathcal{D}) \subset \Lambda_\alpha(\mathcal{D})$, but they are not necessarily equal. One can verify that $\Lambda_\alpha^0(\mathcal{D}) = \Lambda_\alpha(\mathcal{D})$ if and only if the choice function K_0 satisfies Property 2 above.

Example 3. *Let $\mathcal{D} = \{(1,0), (0,1), (-1,0)\}$. For any $0 < \alpha < 1$, the segment $(0, \delta), 0 < \delta < 1$ is not contained in $f_2(P_\alpha)$. Therefore $g_0(0, \delta)$ is not in P_α, which implies that this segment is not in $\Lambda_\alpha^0(\mathcal{D})$. On the other hand, if $\alpha \geq \frac{2}{3}$, $\Lambda_\alpha(\mathcal{D}) = P_\alpha$.*

We have also observed in examples that if $\Lambda_\alpha^0(\mathcal{D})$ contains an open set of $\mathbb{R}^N$, then it must contain the convex hull of some code vectors. This prompts the following question:

Question 2. *If, for some* $0 < \alpha < 1$, $\Lambda_\alpha^0(\mathcal{D})$ *contains an open set of* $\mathbb{R}^N$, *then will there be a subdictionary* $\mathcal{D}_1 \subset \mathcal{D}$ *such that* $\Lambda_\alpha^0(\mathcal{D}) \supseteq P_\alpha(\mathcal{D}_1)$?

§3. Comparison between Alpha-Expansions and the Decompose-Quantize Procedure

In this section we shall compare the α-expansion in a frame with the 2-step procedure of first decomposing in the frame and then quantizing the coefficients so obtained in a second step. We shall do this by comparing the rate-distortion functions of each scheme.

Let $\mathcal{F} = \{e_1, e_2, \ldots, e_p\}$ be a frame in $\mathbb{R}^N$, and take $x \in \mathbb{R}^N$ with $\|x\| \leq M$. We shall assume that the coefficients $(a_1, a_2, \ldots, a_p)$ of the decomposition x in the frame $\mathcal{F}$ satisfy $|a_i| \leq C_1 M$, for some constant C_1 that depends only on $\mathcal{F}$. We shall consider here the quantization of these coefficients by binary expansions. If each coefficient is represented by n bits, the total number of bits used is $R = np$, and the maximum square error per coefficient is given by $\left[C_1 M \left(\frac{1}{2}\right)^{n-1}\right]^2$. If we multiply this by p, we obtain the total maximum square distortion D. Therefore, we can write the rate-distortion relation

$$R = \frac{p}{2} \log_2 \left(\frac{4C_1^2 M^2}{D} \right).$$

Let us consider now the α-expansion procedure. If we approximate $x \in \Lambda_\alpha$, $\|x\| \leq M$, by its n-term α-expansion $(v_{i_0}, v_{i_1}, \ldots, v_{i_{n-1}})$, the maximum square distortion is given by $D = [C_2 M \alpha^n]^2$, where C_2 is a constant that depends only on $\mathcal{F}$. The number of bits necessary to code this sequence is $R = n \log_2(2p)$, and thus we have the rate-distortion relation

$$R = \frac{\log_2(2p)}{2\log_2\left(\frac{1}{\alpha}\right)} \log_2 \left(\frac{C_2^2 M^2}{D} \right).$$

We conclude that asymptotically, the α-expansion is better than the decompose-quantize procedure if we can choose α such that Λ_α contains an open set and

$$\frac{\log_2(2p)}{\log_2\left(\frac{1}{\alpha}\right)} \leq p \tag{1}$$

Example 4. *Let* $F = \left\{ (\frac{\sqrt{3}}{2}, \frac{1}{2}), (1, 0), (-\frac{\sqrt{3}}{2}, \frac{1}{2}) \right\}$ *and take* $\alpha = \frac{1}{2}$. *Then* $\Lambda_\alpha = P_\alpha$ *(which in this case is a hexagon) and*

$$\frac{\log_2(2p)}{\log_2 \frac{1}{\alpha}} = \log_2 6 < 3,$$

which implies that in this case the α*-expansion is better than the decompose-quantize procedure.*

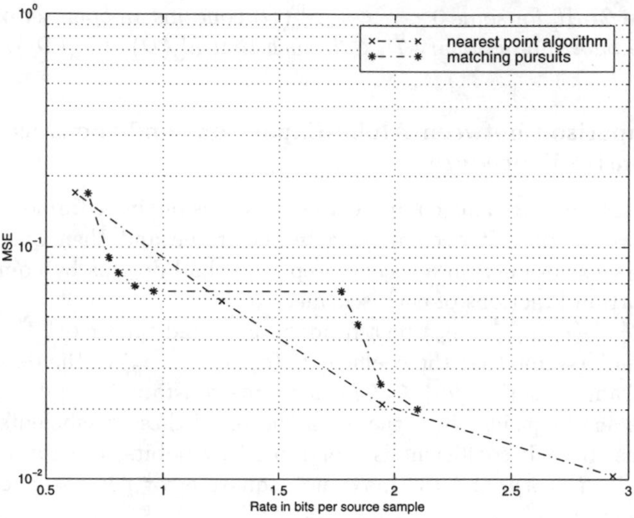

Fig. 1. The 8-dimensional hypercube as codebook.

Remark 2. *If for a given frame $\mathcal{F}$, Λ_α contains an open set, then this remains true for any other frame $\mathcal{F}_1$ obtained from $\mathcal{F}$ by adding some more vectors. Hence, even if relation (1) does not hold for $\mathcal{F}$, it will hold for highly redundant frames $\mathcal{F}_1$ containing $\mathcal{F}$.*

Remark 3. *By Remark 1, $\alpha \geq \frac{N}{N+1}$ implies that $\Lambda_\alpha = P_\alpha$. Therefore, if we consider frames satisfying*

$$\frac{\log_2(2p)}{p} \leq \log_2\left(\frac{N+1}{N}\right),$$

the α-expansion scheme will be better than the decompose-quantize procedure.

Experimental results

In order to directly compare our method (the nearest point algorithm) with some established results, we look at some examples presented in [6] (3.4.2, pp. 41–45). To this end, we used a similar source and the same codebook.

A zero-mean gaussian AR source with correlation coefficient $\rho = 0.9$ was used to provide the data points. Vectors were formed by blocks of N samples. Rate was measured as the first order entropy of the index stream produced by the algorithm – similarly to [6].

In Fig. 1, we used the vertices of the 8-dimensional hypercube as the codebook. In this case, α was set to 0.501 and $\| \boldsymbol{v}_k \| = 1.4142$ (that is $\frac{\sqrt{8}}{2}$).

As can be seen from this data, our method does give some performance benefits on low bit-rates.

Although our algorithm can be very expensive in terms of computational effort, so are other greedy algorithms like matching pursuits. But there are some well structured codebooks which lend themselves to fast calculation of the steps involved – like the one used in this example.

§4. Conclusions

In this paper we have further developed the theory of α-expansions and applied it in the context of quantized frame expansions.

We have shown that α-expansions perform asymptotically better, in a rate-distortion sense, than the decompose-quantize method. In addition, preliminary experimental results indicate that this method also compares favorably to the decompose-quantize method in practical cases. This was verified by direct comparison between our method and quantized matching pursuits from [6].

References

1. Barnsley, M. F. and L. P. Hard, *Fractals Image Compression*, Wellesley, MA, AKPeters Ltd, 1992.
2. Conway, J. H. and N. J. A. Sloane, *Sphere Packings, Lattices and Groups*, Springer-Verlag, 1988.
3. Craizer, M., da Silva, E. A. B., and E. G. Ramos, Convergent algorithms for successive approximation vector quantization with applications to wavelet image compression, IEEE Proceedings - Vision, Image and Signal Processing, to appear.
4. Davis, G. M., S. Mallat, and M. Avellaneda, Greedy adaptive approximations, J. of Constr. Approx. **13** (1997), 57–98.
5. Gersho, A. and R. M. Gray, *Vector Quantization and Signal Compression*, Kluwer Academic Publishers, 1992.
6. Goyal, V. K., Quantized overcomplete expansions: Analysis, synthesis and algorithms, MsC Thesis, University of California, Berkeley, 1995.
7. Mallat, S., *A Wavelet Tour of Signal Processing*, Academic Press, 1997.
8. da Silva, E. A. B., D. G. Sampson, and M. Ghanbari, A successive approximation vector quantizer for wavelet transform image coding, IEEE Transactions on Image Processing, Special Issue on Vector Quantization **5** (1996), 299–310.
9. Thao, N. T., and M. Vetterli, Deterministic analysis of oversampled a/d conversion and decoding improvement based on consistent estimates, IEEE Trans. Signal Proc. **42** (1994), 519–531.

Marcos Craizer
Departamento de Matemática
Pontifícia Universidade Católica do Rio de Janeiro
R. Marquês de São Vicente, 225 - Gávea
Rio de Janeiro, RJ
22453-900 Brazil
craizer@mat.puc-rio.br

Eduardo Antônio Barros da Silva
DEL/PEE/EE/COPPE
Universidade Federal do Rio de Janeiro, Brazil
eduardo@lps.ufrj.br

Décio Angelo Fonini Jr.
PEE/COPPE
Universidade Federal do Rio de Janeiro, Brazil
fonini@lps.ufrj.br

Cubic Spline Interpolation on Nested Polygon Triangulations

Oleg Davydov, Günther Nürnberger, and Frank Zeilfelder

Abstract. We develop an algorithm for constructing Lagrange and Hermite interpolation sets for spaces of cubic C^1-splines on general classes of triangulations built up of nested polygons whose vertices are connected by line segments. Additional assumptions on the triangulation are significantly reduced compared to the special class given in [4]. Simultaneously, we have to determine the dimension of these spaces, which is not known in general. We also discuss the numerical aspects of the method.

§1. Introduction

In contrast to univariate splines, it is a non-trivial problem to construct even one single set of interpolation points for bivariate spline spaces. Such interpolation sets for $S_q^r(\triangle)$, the space of splines of degree q and smoothness r, were constructed for crosscut-partitions $\triangle$ (see the survey [9] and the references therein). For general triangulations $\triangle$, interpolation sets were constructed for $S_q^1(\triangle)$, $q \geq 4$ in [3].

The case $q = 3$ is much more complicated given that not even the dimension of $S_3^1(\triangle)$ is known for arbitrary triangulations $\triangle$. It is an open question whether the dimension of $S_3^1(\triangle)$ is equal to Schumaker's lower bound [12]. The aim of this paper is to investigate interpolation by $S_3^1(\triangle)$ for general classes of triangulations $\triangle$ consisting of nested polygons whose vertices are connected by line segments. Following a general principle of locally choosing interpolation points for $S_3^1(\triangle)$ by passing from triangle to triangle, we describe an inductive method for constructing point sets that admit unique Lagrange (respectively Hermite) interpolation by $S_3^1(\triangle)$ under certain assumptions on $\triangle$. Moreover, we prove that the dimension of these spaces is equal to Schumaker's lower bound.

In this way we obtain a class of triangulations $\triangle$ which is significantly larger than the special class described in [4]. Moreover, the methods of proof in this paper are different from those in [4]. It is important to note that

Curve and Surface Fitting: Saint-Malo 1999
Albert Cohen, Christophe Rabut, and Larry L. Schumaker (eds.), pp. 161–170.

ISBN 0-8265-1357-3.

triangulations of this type can be constructed starting from any given points in the plane, see [11].

The numerical examples (with up to 100,000 interpolation points) show that in order to obtain good approximations, it is desirable to subdivide some of the triangles. Our method of constructing interpolation points also works for these modified triangulations.

We note that our interpolation method can be used for the construction of smooth surfaces without involving any derivative data. For scattered data fitting, the needed Lagrange data are approximately computed by local methods. In contrast to the finite element methods for cubic splines, we do not need to subdivide all triangles by a Clough-Tocher split or use derivatives.

§2. Preliminaries

Let $\triangle$ be a regular triangulation of a simply connected polygonal domain Ω in $\mathbb{R}^2$. We denote by $S_3^1(\triangle) = \{s \in C^1(\Omega) : s|_T \in \Pi_3, T \in \triangle\}$ the space of bivariate splines of degree 3 and smoothness 1 (with respect to $\triangle$). Here $\Pi_3 = \operatorname{span}\{x^\nu y^\mu : \nu, \mu \geq 0, \nu + \mu \leq 3\}$ denotes the space of bivariate polynomials of total degree 3.

We investigate the following interpolation problem. Construct a set $\{z_1, \ldots, z_N\}$ in Ω, where $N = \dim S_3^1 = (\triangle)$, such that for each function $f \in C(\Omega)$, a unique spline $s \in S_3^1(\triangle)$ exists such that $s(z_i) = f(z_i)$, $i = 1, \ldots, N$. Such a set $\{z_1, \ldots, z_N\}$ is called a Lagrange interpolation set for $S_3^1(\triangle)$. If also partial derivatives of f are involved, then we speak of a Hermite interpolation set for $S_3^1(\triangle)$.

In contrast to [4], we will use *Bernstein-Bézier techniques* [2,5]. Given a spline $s \in S_3^1(\triangle)$, we consider the following representation of the polynomial pieces $p = s|_T \in \Pi_3$ on the triangle $T \in \triangle$ with vertices v_1, v_2, v_3,

$$p(x,y) = \sum_{\nu+\mu+\sigma=3} a_{\nu,\mu,\sigma}^{[T]} \tfrac{3!}{\nu!\mu!\sigma!} \Phi_1^\nu(x,y) \Phi_2^\mu(x,y) \Phi_3^\sigma(x,y), \ (x,y) \in T, \qquad (1)$$

where $\Phi_l \in \Pi_1$, $l = 1,2,3$, is uniquely defined by $\Phi_l(v_k) = \delta_{k,l}$, $k = 1,2,3$. This representation of p is called the Bernstein-Bézier representation of p, the real numbers $a_{\nu,\mu,\sigma}^{[T]}$ are called the Bernstein-Bézier coefficients of p, and $\Phi_l(x,y)$, $l = 1,2,3$, are the barycentric coordinates (w.r.t. T) of $(x,y) \in T$.

Definition 1. *A set $A \subset \{(\nu,\mu,\sigma,T) : \nu + \mu + \sigma = 3, T \in \triangle\}$ is called an* admissible set *for $S_3^1(\triangle)$ if for every choice of coefficients $a_{\nu,\mu,\sigma}^{[T]}$, $(\nu,\mu,\sigma,T) \in A$, a unique spline $s \in S_3^1(\triangle)$ exists with these coefficients in the above Bernstein-Bézier representation.*

The above Bernstein-Bézier form can be used to express smoothness conditions of polynomial pieces on adjacent triangles T_1, T_2 with vertices v_1, v_2, v_3, respectively v_1, v_2, v_4 (cf. [2,5]).

Theorem 2. *Let s be a piecewise cubic polynomial function defined on $T_1 \cup T_2$. Then $s \in S_3^1(\{T_1, T_2\})$ iff $a_{\nu,\mu,0}^{[T_2]} = a_{\nu,\mu,0}^{[T_1]}$, $\nu + \mu = 3$, and $a_{\nu,\mu,1}^{[T_2]} = a_{\nu+1,\mu,0}^{[T_1]} \Phi_1(v_4) + a_{\nu,\mu+1,0}^{[T_1]} \Phi_2(v_4) + a_{\nu,\mu,1}^{[T_1]} \Phi_3(v_4)$, $\nu + \mu = 2$.*

For later use, we also mention here the following relations between the Bernstein-Bézier coefficients of a cubic polynomial p in the representation (1) and its partial derivatives at v_1 in direction of a unit vector parallel to the edge $e = [v_1, v_2]$, denoted by $\frac{\partial}{\partial e}$.

$$\begin{aligned}
a^{[T]}_{3,0,0} &= p(v_1), \quad a^{[T]}_{2,1,0} = p(v_1) + \frac{1}{3}\frac{\partial p(v_1)}{\partial e}\|v_1 - v_2\|_2, \\
a^{[T]}_{1,2,0} &= p(v_1) + \frac{2}{3}\frac{\partial p(v_1)}{\partial e}\|v_1 - v_2\|_2 + \frac{1}{6}\frac{\partial^2 p(v_1)}{\partial e^2}\|v_1 - v_2\|_2^2, \\
\frac{\partial p(v_1)}{\partial e} &= \frac{3(a^{[T]}_{2,1,0} - a^{[T]}_{3,0,0})}{\|v_1 - v_2\|_2}, \quad \frac{\partial^2 p(v_1)}{\partial e^2} = \frac{6(a^{[T]}_{1,2,0} - 2a^{[T]}_{2,1,0} + a^{[T]}_{3,0,0})}{\|v_1 - v_2\|_2^2}.
\end{aligned} \tag{2}$$

§3. Main Results

In this section, we state our main results on $S_3^1(\triangle)$, where $\triangle$ consists of nested polygons whose vertices are connected by line segments. We first define this class of triangulations. Then, we determine the dimension and construct interpolation sets for the corresponding spline space. Moreover, we show that this dimension is equal to Schumaker's lower bound [12]. Finally, we discuss a property of $\triangle$ which is essential for the local construction of interpolation points.

First, we describe triangulations of nested polygons and decompose the domain into finitely many subsets needed in our construction of interpolation points.

Triangulations of nested polygons. We consider the following general type of triangulation $\triangle$. Let $P_0, P_1, \ldots, P_k$ be a sequence of closed simple polygonal lines, and let Ω_μ be the closed (not necessarily convex) bounded polygon with boundary P_μ. Suppose that the polygons Ω_μ are *nested*, *i.e.*, $\Omega_{\mu-1} \subset \Omega_\mu$, $\mu = 0, \ldots, k$. The vertices of $\triangle$ are the vertices of P_μ, $\mu = 0, \ldots, k$, and one vertex inside P_0. The edges of $\triangle$ are the edges of P_μ, $\mu = 0, \ldots, k$, and additional line segments connecting the vertices of P_μ with the vertices of $P_{\mu+1}$, $\mu = 0, \ldots, k-1$. The resulting triangulation $\triangle$ of $\Omega := \Omega_k$ does not have vertices in the interior of $\Omega_{\mu+1} \setminus \Omega_\mu$, $\mu = 0, \ldots, k-1$, and does not have edges connecting two vertices of P_μ other than the edges of P_μ, see Figure 1.

Decomposition of the domain. We decompose the domain Ω into finitely many sets $V_0 \subset V_1 \subset \cdots \subset V_m = \Omega$, where each set V_i, is the union of closed triangles of $\triangle$, $i = 0, \ldots, m$. Let V_0 be an arbitrary closed triangle of $\triangle$ in Ω_0. We define the sets $V_1 \subset \cdots \subset V_m$ by induction. Assuming V_{i-1} is defined, we choose a vertex v_i of $\triangle$ such that there exists at least one triangle of $\triangle$ with vertex v_i and a common edge with V_{i-1}. Let $T_{i,1}, \ldots, T_{i,n_i}$, $n_i \geq 1$, be all such triangles. We set $V_i = V_{i-1} \cup \overline{T}_{i,1} \cup \cdots \cup \overline{T}_{i,n_i}$, and denote by $\triangle_i = \{T \in \triangle : T \subset V_i\}$ the subtriangulation which corresponds to the set V_i.

The vertices v_i, $i = 1, \ldots, m$, are chosen as follows. After choosing V_0 to be an arbitrary closed triangle of $\triangle$ in Ω_0, we pass through the vertices of P_0 in clockwise order by applying the above rule. (It is clear that the choice of

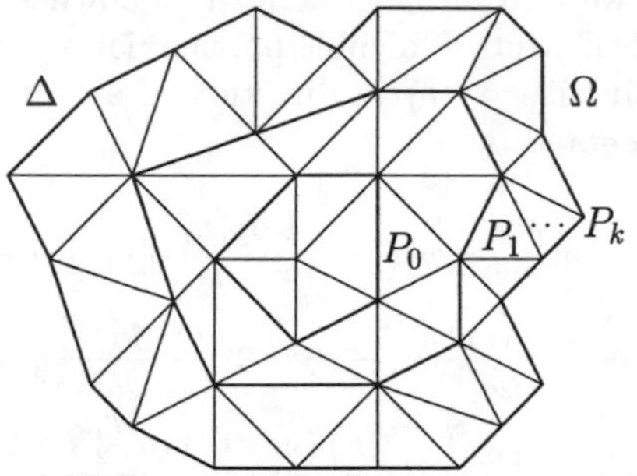

Fig. 1. Triangulation of nested polygons.

these vertices is unique after fixing the first vertex.) Now, we assume that we have passed through the vertices of $P_{\mu-1}$. We fix a vertex w_μ of P_μ that is connected with at least two vertices of $P_{\mu+1}$. Then w.r.t. clockwise order, we choose the first vertex of P_μ greater than w_μ which is connected with at least two vertices of $P_{\mu-1}$. Then we pass through the vertices of P_μ in clockwise order until w_μ^-, and pass through the vertices of P_μ in anticlockwise order until w_μ^+ by applying the above rule. (Here w_μ^+ denotes the vertex next to w_μ in clockwise order and w_μ^- denotes the vertex next to w_μ in anticlockwise order.) Finally, we choose the vertex w_μ. (It is clear that after fixing w_μ, the choice of the vertices on P_μ is unique.)

The construction of an admissible set for $S_3^1(\Delta)$ and the choice of interpolation points depend on the following properties of the triangulation Δ.

Definition 3. *(1) An interior edge e with vertex v of the triangulation Δ is called* degenerate *at v if the edges with vertex v adjacent to e lie on a line. (2) An interior vertex v of Δ is called* singular *if v is a vertex of exactly four edges and these edges lie on two lines. (3) An interior vertex v of Δ on the boundary of a given subtriangulation Δ' of Δ is called* semi-singular of type 1 *w.r.t. Δ' if exactly one edge with endpoint v is not contained in Δ' and this edge is degenerate at v. (4) An interior vertex v of Δ on the boundary of a given subtriangulation Δ' of Δ is called* semi-singular of type 2 *w.r.t. Δ' if exactly two edges with endpoint v are not contained in Δ' and these edges are degenerate at v. (5) A vertex v of Δ is called* semi-singular *w.r.t. Δ' if v satisfies (3) or (4).*

In the following, we construct an admissible set and interpolation sets for $S_3^1(\Delta)$, where Δ is a nested-polygon triangulation.

Construction of an admissible set. First, we choose $\mathcal{A}_0 = \{(\nu, \mu, \sigma, V_0) : \nu+\mu+\sigma = 3\}$ and then, proceeding by induction, we successively add admissible points on $V_i \setminus V_{i-1}$, $i = 1, \ldots, m$. Assuming that an admissible set $\mathcal{A}_{i-1}$ on V_{i-1} has been constructed, we choose admissible points on $V_i \setminus V_{i-1}$ as follows. By the above decomposition of Ω, $V_i \setminus V_{i-1}$ is the union of consecutive triangles $T_{i,1}, \ldots, T_{i,n_i}$ with vertex v_i and common edges with V_{i-1}. We denote the consecutive endpoints of these edges by $v_{i,0}, v_{i,1}, \ldots, v_{i,n_i}$, and the piecewise polynomials in the representation (1) on $T_{i,j}$ by $p_{i,j} \in \Pi_3$, $j = 1, \ldots n_i$, where the vertices of $T_{i,j}$ are ordered as follows: $v_i, v_{i,j}, v_{i,j+1}$. Furthermore, we denote by $e_{i,j}$ the edges $[v_{i,j}, v_i]$, $j = 0, \ldots, n_i$.

We need the following properties of the subtriangulation $\triangle_i = \{T \in \triangle : T \subset V_i\}$ at the vertices $v_{i,0}, \ldots, v_{i,n_i}$:

(a) $e_{i,j}$ is non-degenerate at $v_{i,j}$,

(b) $v_{i,j}$ is semi-singular w.r.t. $\triangle_i$. (This latter property is only relevant if $v_{i,j}$ lies on the boundary of $\triangle_i$, *i.e.*, for $j \in \{0, n_i\}$.)

For $j \in \{1, \ldots, n_i - 1\}$, we set $c_{i,j} = 1$ if (a) holds, and $c_{i,j} = 0$ otherwise. For $j \in \{0, n_i\}$, we set $c_{i,j} = 1$ if both (a) and (b) hold, and $c_{i,j} = 0$ otherwise. Moreover, we set $c_i = \sum_{j=0}^{n_i} c_{i,j}$, and assume $c_i \leq 3$, $i = 1, \ldots, n$.

Now, we construct the following admissible points on $V_i \setminus V_{i-1}$. If $c_i = 3$, then no point is chosen. If $c_i = 2$, then we choose $(3, 0, 0, T_{i,1})$. If $c_i = 1$, then we choose $(3, 0, 0, T_{i,1})$ and $(2, 0, 1, T_{i,j})$, where $e_{i,j}$ is an edge with $c_{i,j} = 0$. If $c_i = 0$, then we choose $(3, 0, 0, T_{i,1}), (2, 0, 1, T_{i,1})$ and $(2, 1, 0, T_{i,1})$. The admissible set $\mathcal{A}_i$ on V_i is obtained by adding these points to $\mathcal{A}_{i-1}$.

Construction of interpolation sets. We choose interpolation points in V_0 and then in $V_i \setminus V_{i-1}$, $i = 1, \ldots, m$, successively. In the first step, we choose 10 different points in V_0 (respectively 10 Hermite interpolation conditions) which admit unique Lagrange interpolation (respectively Hermite interpolation) by the space Π_3. For example, for Lagrange interpolation, we may choose four parallel line segments l_ν in V_0 and ν different points on each $l_\nu, \nu = 1, 2, 3, 4$. Assuming that the interpolation points in V_{i-1} have already been chosen, we proceed to $V_i \setminus V_{i-1}$ as follows.

For Lagrange interpolation, we choose the following points in $V_i \setminus V_{i-1}$. If $c_i = 3$, then no point is chosen. If $c_i = 2$, then we choose v_i. If $c_i = 1$, then we choose v_i and one further point on some edge $e_{i,j}$ with $c_{i,j} = 0$. If $c_i = 0$, then we choose v_i and two further points on two different edges.

For Hermite interpolation, we require the following interpolation conditions for $s \in S_3^1(\triangle)$ at the vertex v_i. If $c_i = 3$, then no interpolation condition is required at v_i. If $c_i = 2$, then we require $s(v_i) = f(v_i)$. If $c_i = 1$, then we require $s(v_i) = f(v_i)$ and $\frac{\partial s}{\partial e_{i,j}}(v_i) = \frac{\partial f}{\partial e_{i,j}}(v_i)$, where $e_{i,j}$ is some edge with $c_{i,j} = 0$. If $c_i = 0$, then we require $s(v_i) = f(v_i)$, $\frac{\partial s}{\partial x}(v_i) = \frac{\partial f}{\partial x}(v_i)$ and $\frac{\partial s}{\partial y}(v_i) = \frac{\partial f}{\partial y}(v_i)$.

By the above construction, we obtain a set of points for Lagrange interpolation respectively a set of Hermite interpolation conditions.

Theorem 4. *Let $\triangle$ be a triangulation of nested polygons. If for all $i \in \{1, \ldots, m\}$, $c_i \leq 3$ and no vertex v_i is simultanously semi-singular (of type 2) w.r.t $\triangle_i$ and non-singular, then a unique spline in $S_3^1(\triangle)$ exists which satisfies the above Lagrange (respectively Hermite) interpolation conditions. In particular, the total number of interpolation conditions is equal to the dimension of $S_3^1(\triangle)$.*

Proof: First, we prove that the set constructed above is an admissible set for $S_3^1(\triangle)$. To this end, we show by induction that $\mathcal{A}_i$ is an admissible set for $S_3^1(\triangle)|_{\triangle_i} = \{s|_{\triangle_i} : s \in S_3^1(\triangle)\}$. This is clear for $i = 0$. Now, we assume that $\mathcal{A}_{i-1}$ is an admissible set for $S_3^1(\triangle)|_{\triangle_{i-1}}$, where $i \in \{1, \ldots, m\}$, and consider V_i. For simplicity, we omit here the index i for $v_i, v_{i,j}, e_{i,j}, p_{i,j}, T_{i,j}$ and n_i. It follows from the induction hypothesis and Theorem 2 that the coefficients $a^{[T_j]}_{\nu,3-\nu-\sigma,\sigma}$, $\sigma = 0, \ldots, 3-\nu$, $\nu = 0, 1$, of $p_j \in \Pi_3$, $j = 1, \ldots, n$, on T_j, are uniquely determined. Moreover, if $c_{i,j} = 1$ for some $j \in \{1, \ldots, n-1\}$, then it follows from Theorem 2 that the coefficient $a^{[T_j]}_{2,0,1}$ is uniquely determined.

In the following, we show that if $c_{i,0} = 1$, then the coefficient $a^{[T_1]}_{2,1,0}$ is uniquely determined. Let us consider the case where v_0 is semi-singular of type 2 w.r.t. $\triangle_i$. (The case that v_0 is semi-singular of type 1 w.r.t. $\triangle_i$ is analogous.) We denote by $\tilde{T}_l \in \triangle$, $l = 1, \ldots, 3$, the triangles with vertex v_0 not contained in $\triangle_i$ in anticlockwise order, and by $\tilde{e}_l$ the common edge of $\tilde{T}_l$ and $\tilde{T}_{l+1}$, $l = 1, 2$. Since $\tilde{T}_3$ has a common edge with $\triangle_{i-1}$, it follows from Theorem 2 that the coefficient $\tilde{a}^{[\tilde{T}_3]}_{1,1,1}$ of $\tilde{p}_3 \in \Pi_3$ on $\tilde{T}_3$ is uniquely determined. Moreover, since $\tilde{e}_2$ and $\tilde{e}_1$ are degenerate at v_0, the coefficients $\tilde{a}^{[\tilde{T}_l]}_{1,1,1}$ of $\tilde{p}_l \in \Pi_3$ on $\tilde{T}_l$, $l = 1, 2$, are uniquely determined. Since e_0 is non-degenerate at v_0, it follows from Theorem 2 that the coefficient $a^{[T_1]}_{2,1,0}$ is uniquely determined. We note that since $\triangle$ is a nested-polygon triangulation, at least two triangles with vertex v_0 not contained in $\triangle_i$ exist. Therefore, if $c_{i,0} = 0$, then the coefficient $a^{[T_1]}_{2,1,0}$ is not yet determined.

Analogously as above, it can be shown that the coefficient $a^{[T_n]}_{2,0,1}$ is uniquely determined if $c_{i,n} = 1$. Otherwise, this coefficient is not yet determined.

Now, we consider the vertex v. The arguments below will show that we may assume that v is an interior point of $\triangle$. We denote by $T_{n+l} \in \triangle$, $l = 1, \ldots, r$, $r \geq 3$, the triangles with vertex v not contained in $\triangle_i$ in anticlockwise order. Moreover, let the piecewise polynomials $p_{n+l} \in \Pi_3$, $l = 1, \ldots, r$, on T_{n+l} in the representation (1) be given such that the first barycentric coordinate always corresponds to v. The above arguments show that exactly $c_i \leq 3$ coefficients of the set $\mathcal{C}_1 = \{a^{[T_l]}_{\nu,3-\nu-\sigma,\sigma} : \sigma = 0, \ldots, 3-\nu, \ \nu = 2, 3, \ l = 1, \ldots, n+r\}$ are uniquely determined. On the other hand, we construct $3 - c_i$ additional admissible points from $\mathcal{C}_1$ on $V_i \setminus V_{i-1}$. Now, it follows from the C^1-property at v and Theorem 2 that all coefficients from $\mathcal{C}_1$ are uniquely determined. By our method of passing through the vertices of $\triangle$, v is not semi-singular of type 1 w.r.t. $\triangle_i$. In particular, if $v = w_\mu$ for some $\mu \in \{0, \ldots, k\}$. Moreover, by assumption v can be semi-singular of type 2 w.r.t. $\triangle_i$ only if v is singular. In this case, we have $r = 3$, and it follows from Theorem 3.3 in [13] that the coefficient $a^{[T_3]}_{1,1,1}$ is uniquely determined. Otherwise, if $r \geq 4$, then for some $l \in \{1, \ldots, r-1\}$ one common edge of T_{n+l} and T_{n+l+1} is non-degenerate at v, and we can also proceed with our arguments.

Since all relevant differentiability conditions at the edges with endpoint v, respectively v_j, were involved, the above shows that $\mathcal{A}_i$ is an admissible set for $S_3^1(\triangle)|_{\triangle_i}$. Thus, the set $\mathcal{A}_m$ is an admissible set for $S_3^1(\triangle)$.

Therefore, the cardinality of $\mathcal{A}_m$ is equal to the dimension of $S_3^1(\triangle)$. By construction, it is evident that the number of Lagrange interpolation points, respectively the number of Hermite interpolation conditions coincides with this cardinality.

By an inductive argument, it follows from (2) that the Hermite interpolation conditions at v determine the Bernstein-Bézier coefficients of the admissible points chosen on $V_i \setminus V_{i-1}$. Analogously, the Lagrange interpolation conditions uniquely determine the interpolating spline on the edges of $V_i \setminus V_{i-1}$. Therefore, the interpolating spline is uniquely determined on all of $V_i \setminus V_{i-1}$. This completes the proof of Theorem 4. □

For arbitrary triangulations, Schumaker [12] gave the following lower bound $L(\triangle)$ for the dimension of $S_3^1(\triangle)$,

$$L(\triangle) = 3V_B(\triangle) + 2V_I(\triangle) + \sigma(\triangle) + 1. \tag{3}$$

Here, $V_B(\triangle)$ is the number of boundary vertices of $\triangle$, $V_I(\triangle)$ is the number of interior vertices of $\triangle$ and $\sigma(\triangle)$ is the number of singular vertices of $\triangle$. For bounds on the dimension of bivariate spline spaces see also Manni [6].

Theorem 5. *If a triangulation $\triangle$ of nested polygons satisfies the hypotheses of Theorem 4, then the dimension of $S_3^1(\triangle)$ is equal to $L(\triangle)$.*

Proof: We have to show that the cardinality of $\mathcal{A}_m$ is equal to $L(\triangle)$. We prove this by induction. We set $\mathcal{S}(\triangle_0) = \emptyset$ and for $i \in \{1, \ldots m\}$, we denote by $\mathcal{S}(\triangle_i)$ the set of boundary vertices w of $\triangle_i$ such that $w = v_{l,0}$ and $c_{l,0} = 1$ (respectively $w = v_{l,n_l}$ and $c_{l,n_l} = 1$) for some $l \in \{1, \ldots, i\}$. Moreover, let $\tilde{\sigma}_i$ be the cardinality of $\mathcal{S}(\triangle_i)$ and a_i be the cardinality of $\mathcal{A}_i$. We will show that

$$L(\triangle_i) = a_i + \tilde{\sigma}_i, \qquad i = 0, \ldots, m. \tag{4}$$

This is evident for $i = 0$. We assume that $L(\triangle_{i-1}) = a_{i-1} + \tilde{\sigma}_{i-1}$ for some $i \in \{1, \ldots, m\}$ and consider V_i. We have $V_B(\triangle_i) = V_B(\triangle_{i-1}) - n_i + 2$, $V_I(\triangle_i) = V_I(\triangle_{i-1}) + n_i - 1$, $\sigma(\triangle_i) = \sigma(\triangle_{i-1}) + \gamma_i$, where γ_i is the number of singular vertices from the set $\{v_{i,j} : j = 1, \ldots, n_i - 1\}$. Since $a_i = a_{i-1} + 3 - c_i$, it follows from the induction hypothesis and some elementary computations that

$$L(\triangle_i) = a_i + \tilde{\sigma}_{i-1} + c_i + \gamma_i - n_i + 1.$$

By our method of passing through the vertices of $\triangle$, it is evident that if $v_{i,0} = v_{i-1} \in \mathcal{S}(\triangle_i)$, then $v_{i,0} \notin \mathcal{S}(\triangle_{i-1})$. In the following, we show that if $v_{i,n_i} \in \mathcal{S}(\triangle_i)$, then $v_{i,n_i} \notin \mathcal{S}(\triangle_{i-1})$. First, let us assume that $v_{i,n_i} = v_{l,0}$ for some $l \in \{1, \ldots, i-1\}$. If v_{i,n_i} is semi-singular of type 2 w.r.t. $\triangle_i$, then at least three edges of $\triangle$ not contained in $\triangle_l$ are attached to $v_{l,0}$. Hence, $c_{l,0} = 0$. If v_{i,n_i} is semi-singular of type 1 w.r.t. $\triangle_i$, then the edge e_{i,n_i} is non-degenerate at v_{i,n_i}, since $c_{i,n_i} = 1$. Therefore, $v_{l,0}$ is not semi-singular of type 2 w.r.t. $\triangle_l$. Again, $c_{l,0} = 0$ holds. The remaining case $v_{i,n_i} = v_{i-1,n_{i-1}}$, where $n_i = 1$, follows by the same arguments.

Now, we show for $j \in \{1,\ldots,n_i-1\}$ that every non-singular vertex $v_{i,j}$ such that $e_{i,j}$ is degenerate at $v_{i,j}$ lies in $\mathcal{S}(\triangle_{i-1})$. First, we consider the case $j=1$. Set $v_{i_1}=v_{i,1}$ and let $\tilde{e}_0$ be the edge that connects $v_{i,0}$ and $v_{i,1}$. We have to consider two cases.

Case 1. (The vertices $v_{i,1}$ and v_{i-2} are connected by an edge e.) If $\tilde{e}_0$ is non-degenerate at $v_{i,1}$ then $c_{i-1,n_{i-1}}=1$. (In this case $v_{i,1}$ is semi-singular of type 1 w.r.t. $\triangle_{i-1}$.) Otherwise, since $v_{i,1}$ is non-singular, e is non-degenerate at $v_{i,1}$. Thus, $c_{i-2,n_{i-2}}=1$. (In this case $v_{i,1}$ is semi-singular of type 2 w.r.t. $\triangle_{i-2}$.) We note that $v_{i,1}$ is not semi-singular w.r.t. $\triangle_{i_1+1}$, since at least three edges of $\triangle$ not contained in $\triangle_{i_1+1}$ are attached to $v_{i,1}$.

Case 2. (The vertices $v_{i,1}$ and v_{i-2} are not connected by an edge.) If $\tilde{e}_0$ is non-degenerate at $v_{i,1}$ then we also have $c_{i-1,n_{i-1}}=1$. (In this case $v_{i,1}$ is semi-singular of type 1 w.r.t. $\triangle_{i-1}$.) We note that $v_{i,1}$ is not semi-singular w.r.t. $\triangle_{i_1+1}$, since $\tilde{e}_0$ is non-degenerate at $v_{i,1}$. Otherwise, let e be the edge that connects $v_{i,1}$ with v_{i_1+1}. Since $v_{i,1}$ is non-singular, e is non-degenerate at $v_{i,1}$. Thus, $c_{i_1+1,0}=1$. (In this case $v_{i,1}$ is semi-singular of type 2 w.r.t. $\triangle_{i_1+1}$.) We note that in this case $v_{i,1}$ is semi-singular of type 1 w.r.t. $\triangle_{i-1}$, but $c_{i-1,n_{i-1}}=0$.

Now, we consider the remaining case $j \in \{2,\ldots,n_i-1\}$. Set $v_{i_j}=v_{i,j}$ and let e be the edge that connects $v_{i,j}$ with v_{i_j+1}. Since $v_{i,j}$ is non-singular, it follows that $v_{i,j}$ is not semi-singular of type 2 w.r.t. to $\triangle_{i_j}$. Therefore, e is non-degenerate at $v_{i,j}$. Hence, $c_{i_j+1,0}=1$. (In this case $v_{i,j}$ is semi-singular of type 1 w.r.t. $\triangle_{i_j+1}$.) We note that in the case $j \in \{2,\ldots,n_i-1\}$, by our method of passing through the vertices of $\triangle$, the value $c_{i_j-1,n_{i_j-1}}$ is not influenced by the geometrical properties of $\triangle$ at $v_{i,j}$.

The above proof now implies $\tilde{\sigma}_i=\tilde{\sigma}_{i-1}+c_i+\gamma_i-n_i+1$, and therefore, (4) holds. Since $\tilde{\sigma}_m=0$, we get $L(\triangle)=a_m$. This proves the theorem. □

In Theorem 4 we assume that for all $i \in \{1,\ldots,m\}$, no vertex v_i is simultaneously semi-singular (of type 2) w.r.t. $\triangle_i$ and non-singular. In the following, we show that this assumption is essential for the local construction of interpolation points.

Example 6. Let $v=v_i=(0,0)$, $v_5=v_0=v_{i,0}=(\gamma,0)$, $\gamma<0$, $v_1=v_{i,1}=(\tau,m\tau)$, $\tau<0$, $m>0$, $v_2=v_{i,2}=(0,\delta)$, $\delta<0$, and set $v_3=(\alpha,0)$, $\alpha>0, v_4=(0,\beta)$, $\beta>0$. Let v be connected with v_3 and v_4 and v_{l-1} be connected with v_l, $l=1,\ldots,5$. Then v is simultaneously semi-singular (of type 2) w.r.t. $\triangle_i$ and non-singular. Furthermore, we denote by T_l the triangle with vertices v, v_{l-1}, v_l and by $p_l \in \Pi_3$ the polynomial pieces on T_l, $l=1,\ldots 5$, in the representation (1). We consider the set $\mathcal{C}_2=\{a^{[T_l]}_{\nu,3-\nu-\sigma,\sigma},\ \sigma=0,\ldots,3-\nu, \nu=1,\ldots,3,\ l=1,\ldots,5\}$. For C^1-splines, it follows from Theorem 3.3 in [13] that each subset of $\mathcal{C}_2$ that uniquely determines all coefficients of $\mathcal{C}_2$ has cardinality 8 and contains the coefficients $a^{[T_l]}_{1,0,2}$, $l=3,4$. By the proof of Theorem 4, the coefficients $a^{[T_1]}_{1,2-\sigma,\sigma}$, $\sigma = 0,1,2$, and $a^{[T_2]}_{1,2-\sigma,\sigma}$, $\sigma = 1,2$, are uniquely determined. If $e_{i,1}$ is non-degenerate at v_1, then in addition $a^{[T_1]}_{2,0,1}$ is uniquely determined. Otherwise, this coefficient is not determined. Hence, if $e_{i,1}$ is non-degenerate at v_1, then we have to choose exactly one additional coefficient to determine all coefficients of $\mathcal{C}_2$, and otherwise, we have to choose exactly two additional coefficients. We claim that in the latter case every choice of exactly two additional coefficients from the set $\{a^{[T_1]}_{3,0,0}, a^{[T_1]}_{2,1,0}, a^{[T_1]}_{2,0,1}, a^{[T_2]}_{2,0,1}\}$ fails to determine all coefficients of $\mathcal{C}_2$.

Proof: Suppose that $e_{i,1}$ is degenerate at v_1 and choose, for example, $a^{[T_1]}_{2,1,0}$ and $a^{[T_1]}_{2,0,1}$. For simplicity, we set $a_1 = a^{[T_3]}_{2,1,0}$, $a_2 = a^{[T_3]}_{3,0,0}$, $a_3 = a^{[T_4]}_{2,0,1}$, $a_4 = a^{[T_4]}_{1,1,1}$, $a_5 = a^{[T_4]}_{2,1,0}$, $a_6 = a^{[T_3]}_{1,1,1}$, and assume that the remaining coefficients in $\mathcal{C}_2$ are zero. By Theorem 2,

$$a_3 = (\frac{\tau-\gamma}{m\tau\gamma}\beta + 1)a_2, \quad a_4 = (1-\frac{\alpha}{\gamma})a_3, \quad a_1 = (1-\frac{\delta}{\beta})a_2 + \frac{\delta}{\beta}a_3,$$
$$a_6 = (1-\frac{\delta}{\beta})a_5 + \frac{\delta}{\beta}a_4, \; a_5 = (1-\frac{\alpha}{\gamma})a_2, \; 0 = ((-\tau)(\frac{m}{\delta}+\frac{1}{\alpha})+1)a_1 + \frac{\tau}{\alpha}a_6.$$

Eliminating a_j, $j \in \{3,4,5\}$, yields $a_1 = (1+\delta\frac{\tau-\gamma}{m\tau\gamma})a_2$, $a_6 = (1-\frac{\alpha}{\gamma})(1+\delta\frac{\tau-\gamma}{m\tau\gamma})a_2$. By some elementary computations, we obtain for the determinant D of the corresponding system

$$D = \frac{(-1)(\tau(m\gamma+\delta)-\delta\gamma)^2}{m\tau\delta\gamma^2},$$

and it is easy to verify that $D = 0$ iff $e_{i,1}$ is degenerate at v_1. Other choices of exactly two additional coefficients from the set $\{a^{[T_1]}_{3,0,0}, a^{[T_1]}_{2,1,0}, a^{[T_1]}_{2,0,1}, a^{[T_2]}_{2,0,1}\}$ can be examined in the same way, which proves our claim. □

Note that if $e_{i,1}$ is non-degenerate at v_1, then every choice of exactly one additional coefficient in the set $\{a^{[T_1]}_{3,0,0}, a^{[T_1]}_{2,1,0}, a^{[T_2]}_{2,0,1}\}$ determines all coefficients in $\mathcal{C}_2$.

We finally discuss some numerical aspects of our scheme. A method for constructing nested polygon triangulations $\triangle$ of given points in the plane which satisfy the conditions of Theorem 4 was developed in [11]. Our numerical tests show that in order to obtain good approximations, it is necessary to subdivide some of the triangles (for details see [10,11]). Meanwhile, we have computed such examples with a high number of interpolation conditions. We only mention here that, for example, Lagrange respectively Hermite interpolation of Franke's test function by cubic C^1-splines with 118,822 interpolation conditions yields an error of $4.66902 * 10^{-6}$ in the uniform norm.

References

1. Alfeld, P., B. Piper, and L. L. Schumaker, An explicit basis for C^1 quartic bivariate splines, SIAM J. Numer. Anal. **24** (1987), 891–911.
2. de Boor, C., B-form basics, in G. Farin (ed.), *Geometric Modeling*, SIAM, Philadelphia, 1987, 131–148.
3. Davydov, O., and G. Nürnberger, Interpolation by C^1 splines of degree $q \geq 4$ on triangulations, J. Comput. Appl. Math., to appear.
4. Davydov, O., G. Nürnberger, and F. Zeilfelder, Interpolation by cubic splines on triangulations, in C.K. Chui and L.L. Schumaker (eds.) *Approximation Theory IX*, Vanderbilt University Press, Nashville, 1998, 17–24.
5. Farin, G., Triangular Bernstein-Bézier patches, Comput. Aided Geom. Design **3** (1986) 83–127.
6. Manni, C., On the dimension of bivariate spline spaces over rectilinear partitions, Approx. Theory Appl. **7**, no. 2 (1991), 23–34.
7. Morgan, J., and R. Scott, A nodal basis for C^1 piecewise polynomials of degree $n \geq 5$, Math. Comp. **29** (1975), 736–740.
8. Nürnberger, G., *Approximation by Spline Functions*, Springer-Verlag, Berlin, Heidelberg, New York, 1989.
9. Nürnberger, G., O. V. Davydov, G. Walz, and F. Zeilfelder, Interpolation by bivariate splines on crosscut partitions, in G. Nürnberger, J. W. Schmidt and G. Walz (eds.), *Multivariate Approximation and Splines*, Birkhäuser Verlag, Basel, 1997, 189–204.
10. Nürnberger, G., and F. Zeilfelder, Lagrange interpolation by splines on triangulations, in *Proceedings of the Morningside Institute*, R.H. Wang (ed.), Peking, 1998.
11. Nürnberger, G., and F. Zeilfelder, Interpolation by spline spaces on classes of triangulations, J. Comput. Appl. Math., to appear.
12. Schumaker, L. L., Bounds on the dimension of spaces of multivariate piecewise polynomials, Rocky Mountain J. Math. **14** (1984), 251–264.
13. Schumaker, L.L., Dual bases for spline spaces on a cell, Comput. Aided Geom. Design **5** (1987), 277–284.

Oleg Davydov
Justus-Liebig-Universität Giessen
Mathematisches Institut
35392 Giessen, Germany
`oleg.davydov@math.uni-giessen.de`

Günther Nürnberger and Frank Zeilfelder
Universität Mannheim
Fakultät für Mathematik und Informatik, A5
68131 Mannheim, Germany
`nuern@euklid.math.uni-mannheim.de`
`zeilfeld@fourier.math.uni-mannheim.de`

Stable Local Nodal Bases for C^1 Bivariate Polynomial Splines

Oleg Davydov and Larry L. Schumaker

Abstract. We give a stable construction of local nodal bases for spaces of C^1 bivariate polynomial splines of degree $d \geq 5$ defined on arbitrary triangulations. The bases given here differ from recently constructed locally linearly independent bases, and in fact we show that stability and local linear independence cannot be achieved simultaneously.

§1. Introduction

Given a regular triangulation $\triangle$, let

$$\mathcal{S}_d^r(\triangle) := \{s \in C^r(\Omega) : \ s|_T \in \mathcal{P}_d \ \text{ for all triangles } T \in \triangle\},$$

where $\mathcal{P}_d$ is the space of polynomials of degree d, and Ω is the union of the triangles in $\triangle$. In this paper we focus on the case $r = 1$ and $d \geq 5$. The main result of the paper is a construction of a basis $\mathcal{B} := \{B_i\}_{i=1}^n$ for $\mathcal{S}_d^1(\triangle)$ with the following properties:

P1) The basis $\mathcal{B}$ is *local* in the sense that for each $1 \leq i \leq n$, the support of B_i is contained in star(v_i) (see the end of this section) for some vertex v_i,

P2) The set $\mathcal{B}$ is *stable* in the sense that there exist constants K_1 and K_2 dependent only on the smallest angle $\theta_\triangle$ in $\triangle$ such that

$$K_1 \|c\|_\infty \ \leq \ \|\sum_{i=1}^n c_i B_i\|_\infty \ \leq \ K_2 \|c\|_\infty \tag{1}$$

for all choices of the coefficient vector $c = (c_1, \ldots, c_n)$.

Bases for $\mathcal{S}_d^1(\triangle)$ satisfying property P1 were constructed in [14] using nodal techniques, but they fail to satisfy property P2 for triangulations with near

Curve and Surface Fitting: Saint-Malo 1999
Albert Cohen, Christophe Rabut, and Larry L. Schumaker (eds.), pp. 171–180.

ISBN 0-8265-1357-3.

singular vertices or near degenerate edges, even if the smallest angle in the triangulation is controlled.

For convenience, we recall the definitions of some of the terminology used above. Suppose v is a vertex of a triangulation which is connected to v_1, v_2, v_3 in counter-clockwise order. Then the edge $e := \langle v, v_2 \rangle$ is said to be near-degenerate at v (degenerate at v) provided that the edges $\langle v, v_1 \rangle$ and $\langle v, v_3 \rangle$ are near-collinear (collinear). The vertex v is called near-singular (singular) if there are exactly four near-degenerate (degenerate) edges attached to it. Given a vertex v of $\triangle$, $\mathrm{star}(v) = \mathrm{star}^1(v)$ is the set of triangles sharing v, and $\mathrm{star}^\ell(v)$ is defined recursively as the union of the stars of the vertices of $\mathrm{star}^{\ell-1}(v)$.

§2. Nodal Determining Sets and Nodal Bases

Suppose s is a spline in $\mathcal{S}_d^1(\triangle)$, and that v is a point in Ω. In this paper we are interested in certain *linear functionals* defined on $\mathcal{S}_d^1(\triangle)$ in terms of values and derivatives of s at points v in Ω. Such functionals are called nodal functionals. There are three types of nodal functionals of interest here:

1) the value $s(v)$,
2) the directional derivative $D_w^m s(v)$, where w is a given vector and m is a positive integer,
3) the mixed derivative $D_{w_1}^{m_1} D_{w_2}^{m_2} s(v)$ at a vertex v of $\triangle$, where w_1 and w_2 are two noncollinear vectors which point into a common triangle T of $\triangle$.

Definition 1. *A collection $\mathcal{M} := \{\lambda_i\}_{i=1}^n$ of nodal functionals is called a* minimal nodal determining set *for $\mathcal{S}_d^1(\triangle)$ provided they form a basis for the dual space $(\mathcal{S}_d^1(\triangle))^*$. If $\mathcal{M}$ is such a set, then there exist unique splines $\mathcal{B} := \{B_i\}_{i=1}^n$ in $\mathcal{S}_d^1(\triangle)$ such that*

$$\lambda_i B_j = \delta_{ij}, \qquad i, j = 1, \ldots, n. \tag{2}$$

We call $\mathcal{B}$ a nodal basis *for $\mathcal{S}_d^1(\triangle)$.*

In this paper we will concentrate on nodal functionals which involve derivatives D_e along edges $e := \langle v_1, v_2 \rangle$ of the triangulation $\triangle$, or perpendicular to such edges. Denoting the Cartesian coordinates of a point v by (v^x, v^y), we see that the derivative along the edge e is given by

$$D_e s(v) := \frac{(v_2^x - v_1^x) D_x s(v) + (v_2^y - v_1^y) D_y s(v)}{\sqrt{(v_2^x - v_1^x)^2 + (v_2^y - v_1^y)^2}},$$

while the derivative perpendicular to the edge e is given by

$$D_{e^\perp} s(v) := \frac{(v_2^y - v_1^y) D_x s(v) - (v_2^x - v_1^x) D_y s(v)}{\sqrt{(v_2^x - v_1^x)^2 + (v_2^y - v_1^y)^2}}.$$

Note that

$$D_{\langle v_1, v_2 \rangle} s(v) = -D_{\langle v_2, v_1 \rangle} s(v), \qquad D_{\langle v_1, v_2 \rangle^\perp} s(v) = -D_{\langle v_2, v_1 \rangle^\perp} s(v).$$

§3. Smoothness Conditions Between Polynomial Pieces

It is well-known how to describe smoothness between polynomials defined on adjoining triangles in terms of the Bernstein-Bézier coefficients of the two polynomials. Here we need similar conditions in terms of nodal information. Suppose $T = \langle v_1, v_2, v_3\rangle$ and $\widetilde{T} = \langle v_1, v_2, \tilde{v}_3\rangle$ are two adjacent triangles with a common edge $e = \langle v_1, v_2\rangle$. We set $\theta_1 = \angle v_3v_1v_2$, $\theta_2 = \angle v_3v_2v_1$, $\tilde{\theta}_1 = \angle \tilde{v}_3v_1v_2$, $\tilde{\theta}_2 = \angle \tilde{v}_3v_2v_1$. Suppose

$$\begin{aligned} v_1 < v_1^{e,0} < \cdots v_{d-5}^{e,0} < v_2 \\ v_1 < v_1^{e,1} < \cdots v_{d-4}^{e,1} < v_2 \end{aligned} \tag{3}$$

are given points lying in the interior of the edge e.

Lemma 2. *Let $p, \tilde{p}$ be polynomials of degree $d \geq 5$ defined on adjoining triangles T and $\widetilde{T}$ as above. Then p and $\tilde{p}$ join together with smoothness C^1 across the edge $e := \langle v_1, v_2\rangle$ if and only if the difference $g = p - \tilde{p}$ satisfies*

$$g(v_i) = D_e g(v_i) = D_{e^\perp} g(v_i) = D_e^2 g(v_i) = 0,, \quad i = 1, 2, \tag{4}$$

$$\begin{aligned} g(v_i^{e,0}) = 0, \quad i = 1, \ldots, d-5, \\ D_{e^\perp} g(v_i^{e,1}) = 0, \quad i = 1, \ldots, d-4, \end{aligned} \tag{5}$$

and

$$\begin{aligned} \hat{\sigma}_1 D^2_{\langle v_1, v_2\rangle} p(v_1) = \tilde{\sigma}_1 D_{\langle v_1,v_2\rangle} D_{\langle v_1,v_3\rangle} p(v_1) + \sigma_1 D_{\langle v_1,v_2\rangle} D_{\langle v_1,\tilde{v}_3\rangle} \tilde{p}(v_1), \\ \hat{\sigma}_2 D^2_{\langle v_2, v_1\rangle} p(v_2) = \tilde{\sigma}_2 D_{\langle v_2,v_1\rangle} D_{\langle v_2,v_3\rangle} p(v_2) + \sigma_2 D_{\langle v_2,v_1\rangle} D_{\langle v_2,\tilde{v}_3\rangle} \tilde{p}(v_2), \end{aligned} \tag{6}$$

where $\sigma_i := \sin\theta_i$, $\tilde{\sigma}_i := \sin\tilde{\theta}_i$, $\hat{\sigma}_i := \sin(\theta_i + \tilde{\theta}_i)$, $i = 1, 2$.

Proof: We follow the method of proof of the main result in [14]. Concerning necessity, we first observe that if p and $\tilde{p}$ join with C^1 continuity across e, then

$$g(v) = D_w g(v) = 0, \qquad \text{for all } v \in e, \tag{7}$$

where w is any unit vector noncollinear with the edge e. This implies (5) and the conditions on g, $D_e g$ and $D_{e^\perp} g$ in (4). The conditions on the second derivatives are easily obtained by differentiating the identities (7) along the edge e and using the fact that

$$\begin{aligned} \hat{\sigma}_1 D_{\langle v_1,v_2\rangle} p(v_1) = \tilde{\sigma}_1 D_{\langle v_1,v_3\rangle} p(v_1) + \sigma_1 D_{\langle v_1,\tilde{v}_3\rangle} p(v_1), \\ \hat{\sigma}_2 D_{\langle v_2,v_1\rangle} p(v_2) = \tilde{\sigma}_2 D_{\langle v_2,v_3\rangle} p(v_2) + \sigma_2 D_{\langle v_2,\tilde{v}_3\rangle} p(v_2). \end{aligned}$$

To prove sufficiency, suppose that p and $\tilde{p}$ satisfy (4)–(6). Then the univariate polynomial $g|_e$ is of degree at most d and satisfies $d+1$ homogeneous Hermite interpolation conditions on e. Therefore $g(v) \equiv 0$ for $v \in e$. This shows that p and $\tilde{p}$ join continuously. We now consider the cross-derivative $q := D_{e^\perp} g|_e$ which is a univariate polynomial of degree at most

$d-1$. By (4)–(5), q has $d-2$ zeros $v_1, v_1^{e,1}, \ldots, v_{d-4}^{e,1}, v_2$ on e. Moreover, by (6), $D_e q(v_1) = D_e q(v_2) = 0$, as is easy to check by expressing $D_e D_{e^\perp} p(v_1)$ as a linear combination of $D_e^2 p(v_1)$ and $D_e D_{\langle v_1, v_3 \rangle} p(v_1)$ and expressing $D_e D_{e^\perp} \tilde{p}(v_1)$ in terms of $D_e^2 \tilde{p}(v_1)$ and $D_e D_{\langle v_1, \tilde{v}_3 \rangle} \tilde{p}(v_1)$, and similarly for v_2. Therefore, $q \equiv 0$, and we have shown that p and $\tilde{p}$ join with C^1-smoothness. □

For a different set of nodal smoothness conditions, see [5].

§4. Construction of a Stable Local Nodal Basis for $\mathcal{S}_d^1(\triangle)$

In this section we begin by defining a spanning set $\mathcal{N}_\triangle$ of nodal functionals for $(\mathcal{S}_d^1(\triangle))^*$. Then we choose an appropriate linearly independent subset $\mathcal{M}$ which forms a basis for $(\mathcal{S}_d^1(\triangle))^*$. This will involve analysing the linear dependencies between elements of $\mathcal{N}_\triangle$ (*i.e.*, the smoothness conditions). The corresponding nodal basis determined by the duality conditions (2) will be the desired stable local basis for $\mathcal{S}_d^1(\triangle)$. Given a triangle $T := \langle v_1, v_2, v_3 \rangle$, let

$$v_{ijk} := \frac{iv_1 + jv_2 + kv_3}{d}, \qquad i + j + k = d.$$

Given an edge e of $\triangle$, let $v_i^{e,0}$ and $v_i^{e,1}$ be the points defined in (3). We define

$$\mathcal{C}^T := \{\lambda_{ijk}^T s = s(v_{ijk}^T) : \; i + j + k = d, \;\; 2 \le i, j, k\},$$

$$\mathcal{E}(e) := \{\lambda_i^{e,0} s = s(v_i^{e,0}) : \; i = 1, \ldots, d-5\}$$
$$\cup \{\lambda_i^{e,1} s = |e| D_{e^\perp} s(v_i^{e,1}) : \; i = 1, \ldots, d-4\},$$

where $|e|$ denotes the length of e.

Given a vertex v in $\triangle$, suppose the vertices connected to v are $v_1, \ldots, v_n$ in counterclockwise order (with v_1 a boundary vertex if v lies on the boundary), and let $T^{[i]} = \langle v, v_i, v_{i+1} \rangle$, $e_i = \langle v, v_i \rangle$, $\theta_i = \angle e_i e_{i+1}$, where if v is an interior vertex, we identify $v_{\ell+n} = v_\ell$, $e_{\ell+n} = e_\ell$. Denote by $|\text{star}\,(v)|$ the diameter of star (v). Let

$$\mathcal{D}_1(v) := \{\lambda_{ij}^v s = |\text{star}\,(v)|^{i+j} D_x^i D_y^j s(v) : \; 0 \le i + j \le 1\}$$

$$\mathcal{R}_2(v) := \{\lambda_{i,p}^v s = \tfrac{1}{\sin\theta_i \sin\theta_{i-1}} |\text{star}\,(v)|^2 D_{e_i}^2 s(v) : \; i = 1, \ldots, n\}$$
$$\cup \{\lambda_{i,m}^v s = \tfrac{1}{\sin\theta_i} |\text{star}\,(v)|^2 D_{e_i} D_{e_{i+1}} s(v) : \; i = 1, \ldots, n\}$$

if v is an interior vertex, and

$$\mathcal{R}_2(v) := \{\lambda_{i,p}^v s = \tfrac{1}{\sin\theta_i \sin\theta_{i-1}} |\text{star}\,(v)|^2 D_{e_i}^2 s(v) : \; i = 2, \ldots, n-1\}$$
$$\cup \{\lambda_{i,p}^v s = |\text{star}\,(v)|^2 D_{e_i}^2 s(v) : \; i = 1, n\}$$
$$\cup \{\lambda_{i,m}^v s = \tfrac{1}{\sin\theta_i} |\text{star}\,(v)|^2 D_{e_i} D_{e_{i+1}} s(v) : \; i = 1, \ldots, n-1\}$$

if v is a boundary vertex. Let

$$\mathcal{N}_\triangle := \bigcup_{T \in \triangle} \mathcal{C}^T \cup \bigcup_{e \in \triangle} \mathcal{E}(e) \cup \bigcup_{v \in \triangle} [\mathcal{D}_1(v) \cup \mathcal{R}_2(v)].$$

The Markov inequality implies that for all $s \in \mathcal{S}_d^1(\triangle)$ and all $\lambda \in \mathcal{N}_\triangle$,

$$|\lambda s| \le K \|s\|_\infty, \tag{8}$$

for some constant depending only on d and the smallest angle $\theta_\triangle$ in $\triangle$.

Lemma 3. *The set $\mathcal{N}_\triangle$ is a spanning set for $(\mathcal{S}_d^1(\triangle))^*$. Moreover, the only linear dependencies between elements of $\mathcal{N}_\triangle$ are given by*

$$\lambda_{i,m}^v + \lambda_{i-1,m}^v = \sin(\theta_i + \theta_{i-1})\lambda_{i,p}^v \tag{9}$$

for every vertex v and every interior edge e_i attached to v.

Proof: Let $s \in \mathcal{S}_d^1(\triangle)$. If $\lambda s = 0$ for all $\lambda \in \mathcal{N}_\triangle$, then on each triangle $T \in \triangle$ there are exactly $\binom{d+2}{2}$ homogeneous Hermite interpolation conditions on s, and it is easy to see that they force s to be zero. It follows that $\mathcal{N}_\triangle$ is a spanning set for $(\mathcal{S}_d^1(\triangle))^*$. The second statement follows immediately from Lemma 2. □

Algorithm 4. *(Construction of a stable local nodal basis for $\mathcal{S}_d^1(\triangle)$.) Let $\{B_i\}_{i=1}^n$ be the set of splines determined by the duality conditions (2) corresponding to the following set $\mathcal{M} := \{\lambda_i\}_{i=1}^n$ of nodal functionals:*

1) *For each triangle T, choose the $\binom{d-4}{2}$ nodal functionals λ_{ijk}^T in $\mathcal{C}^T$.*
2) *For each edge $e = \langle v_1, v_2\rangle$, choose the $2d-9$ nodal functionals $\lambda_i^{e,0}$ and $\lambda_i^{e,1}$ in $\mathcal{E}(e)$.*
3) *For each vertex v, choose the three nodal functionals λ_{ij}^v in $\mathcal{D}_1(v)$.*
4) *For each vertex v, choose the following nodal functionals in $\mathcal{R}_2(v)$:*
   a) *one of the functionals $\lambda_{i,m}^v$ corresponding to the first mixed derivative at v, and*
   b) *all functionals $\lambda_{i,p}^v$ corresponding to the pure second derivatives at v, with one exception: if v is a nonsingular interior vertex, the functional $\lambda_{i_0,p}^v$ is omitted, where i_0 is chosen such that*

$$|\sin(\theta_{i_0} + \theta_{i_0-1})| \ge |\sin(\theta_i + \theta_{i-1})|, \qquad \textit{for all } i = 1, \ldots, n. \tag{10}$$

Theorem 5. *The set $\mathcal{M}$ of Algorithm 4 is a minimal nodal determining set for $\mathcal{S}_d^1(\triangle)$, and the nodal basis $\{B_1, \ldots, B_N\}$ for $\mathcal{S}_d^1(\triangle)$ defined in (2) is local and stable, i.e., it satisfies both conditions P1 and P2.*

Proof: The fact that $\mathcal{M}$ is a basis for $(\mathcal{S}_d^1(\triangle))^*$ follows easily from Lemma 3. To construct a typical basis spline B_j, we set $\lambda_i B_j = \delta_{ij}$ for all $i = 1, \ldots, n$. Then the remaining nodal values λB_j, $\lambda \in \mathcal{N}_\triangle \setminus \mathcal{M}$ are computed from the smoothness conditions (9). It is easy to see that the support of the resulting spline is at most the star of a vertex. This shows that P1 is satisfied.

It remains to show that the B_j form a stable basis. This follows from (8) by a standard argument [12], provided we can show that

$$\|B_j\|_\infty \le K, \qquad 1 \le j \le n, \tag{11}$$

where K is a constant depending only on d and the smallest angle $\theta_\triangle$ in $\triangle$. This clearly holds if

$$|\lambda B_j| \le K, \qquad \text{for all } \lambda \in \mathcal{N}_\triangle,$$

for a similar constant K. By construction, $|\lambda B_j| \le 1$ for all $\lambda \in \mathcal{M}$. Since $\mathcal{N}_\triangle \setminus \mathcal{M} \subset \bigcup_v \mathcal{R}_2(v)$, let us take an arbitrary vertex v of $\triangle$ and notice that if $\lambda \in \mathcal{R}_2(v)$, then λB_j can be nonzero only if the corresponding λ_j lies in $\mathcal{R}_2(v)$. Therefore, it will be sufficient to show that $|\lambda B_j| \le K$ for all j such that $\lambda_j \in \mathcal{R}_2(v)$ and all $\lambda \in \mathcal{R}_2(v) \setminus \mathcal{M}$. We distinguish four cases.

Case 1: (v is a boundary vertex.) In this case, $\mathcal{R}_2(v) \setminus \mathcal{M} = \{\lambda^v_{i,m} : i = 1, \ldots, n-1,\ i \ne i_1\}$, where $\lambda^v_{i_1,m}$ is the functional included in $\mathcal{M}$ in step 4a) of Algorithm 4. Without loss of generality we assume that $i_1 = 1$. For any $s \in \mathcal{S}^1_d(\triangle)$, (9) implies

$$\begin{aligned}
\lambda^v_{2,m}s &= -\lambda^v_{1,m}s + \sigma_2\lambda^v_{2,p}s\\
\lambda^v_{3,m}s &= \lambda^v_{1,m}s - \sigma_2\lambda^v_{2,p}s + \sigma_3\lambda^v_{3,p}s\\
&\ \vdots\\
\lambda^v_{n-1,m}s &= (-1)^n\lambda^v_{1,m}s + \sum_{i=2}^{n-1}(-1)^{n+i-1}\sigma_i\lambda^v_{i,p}s,
\end{aligned}$$

where we set

$$\sigma_i := \sin(\theta_i + \theta_{i-1}).$$

If we take s to be the basis spline B_j corresponding to a $\lambda_j \in \mathcal{R}_2(v)$, then all but one of the values on the right-hand side of the expression for $\lambda^v_{i,m}B_j$ vanishes, and thus

$$|\lambda^v_{i,m}B_j| \le |\lambda_j B_j| = 1, \qquad i = 2, \ldots, n-1,$$

which proves the assertion.

Case 2: (v is an interior vertex with $n \ne 4$.) In this case, $\mathcal{R}_2(v) \setminus \mathcal{M} = \{\lambda^v_{i,m} : i = 1, \ldots, n,\ i \ne i_1\} \cup \{\lambda^v_{i_0,p}\}$. For $\lambda^v_{i,m}s$, $i = 1, \ldots, n$, $i \ne i_1$, the same calculation as in Case 1 applies: we start from $\lambda^v_{i_1,m}s$ and calculate $\lambda^v_{i,m}s$ consecutively counterclockwise until $\lambda^v_{i_0-1,m}s$, and then also clockwise until $\lambda^v_{i_0,m}s$. For $\lambda^v_{i_0,p}s$, we have by (9),

$$\lambda^v_{i_0,p}s = \sigma^{-1}_{i_0}(\lambda^v_{i_0,m}s + \lambda^v_{i_0-1,m}s). \tag{12}$$

Therefore, our claim will be established if we show that

$$|\sigma^{-1}_{i_0}| = |\sin^{-1}(\theta_{i_0} + \theta_{i_0-1})| \le K_3 \qquad \text{if } n \ne 4, \tag{13}$$

where K_3 is a constant dependent only on $\theta_\triangle$. This is obvious for $n = 3$. Assuming $n \ge 5$, we have $|\theta_1 + \theta_2 + \theta_3 + \theta_4 - 2\pi| \ge \theta_\triangle$. Hence,

$$|\theta_{i_0} + \theta_{i_0-1} - \pi| \ge \max\{|\theta_1 + \theta_2 - \pi|, |\theta_3 + \theta_4 - \pi|\} \ge \theta_\triangle/2,$$

and (13) follows.

Case 3: (v is a singular vertex.) In this case, $\mathcal{R}_2(v)\backslash\mathcal{M} = \{\lambda^v_{i,m} : i = 2,3,4\}$ (where we assume for simplicity that $i_1 = 1$). Since $\sigma_1 = \cdots = \sigma_4 = 0$ for a singular vertex, (9) now reduces to

$$\lambda^v_{i,m}s + \lambda^v_{i-1,m}s = 0, \qquad i = 1,2,3,4.$$

Therefore,

$$\lambda^v_{i,m}s = (-1)^{i+1}\lambda^v_{1,m}s, \qquad i = 2,3,4,$$

and the assertion follows.

Case 4: (v is a nonsingular interior vertex with $n = 4$.) We proceed as in Case 2, but calculate $\lambda^v_{i_0,p}s$ differently. At first glance it may seem that (10) does not guarantee stability since $|\sigma_{i_0}|$ may be arbitrary small (in the case of near-singularity), while $\lambda^v_{i_0,p}s$ is to be computed from (12). However, the complete system of equations (9) for $\mathcal{R}_2(v)$ is

$$\sigma_i\lambda^v_{i,p}s = \lambda^v_{i,m}s + \lambda^v_{i-1,m}s, \quad i = 1,2,3,4.$$

Taking the sum with alternating signs, we get

$$\sum_{i=1}^{4}(-1)^i\sigma_i\lambda^v_{i,p}s = 0,$$

and hence

$$|\lambda^v_{i_0,p}s| \le \sum_{i\ne i_0}\frac{|\sigma_i|}{|\sigma_{i_0}|}|\lambda^v_{i,p}s| \le 1$$

for every $s = B_j$, with $\lambda_j \in \mathcal{R}_2(v)$. This completes the proof of (11), and the theorem has been established. □

§5. Stability vs. LLI

We recall (cf. [2,4,6,8,9]) that a set $\mathcal{B}$ of basis splines in $\mathcal{S}^1_d(\triangle)$ is called locally linearly independent (LLI) provided that for every $T \in \triangle$, the splines $\{B_i : i \in \Sigma_T\}$ are linearly independent on T, where

$$\Sigma_T := \{i : T \subset \operatorname{supp} B_i\}. \tag{14}$$

A star-supported LLI nodal basis was constructed for $\mathcal{S}^1_d(\triangle)$ in [4]. We now establish the following surprising result.

Theorem 6. *For $d \ge 5$, it is impossible to construct a basis for $\mathcal{S}^1_d(\triangle)$ which satisfies both conditions P2 and (14) simultaneously.*

Proof: Suppose $\{B_1, \ldots, B_n\}$ is a locally linearly independent basis for $\mathcal{S}^1_d(\triangle)$ on a triangulation $\triangle$ which contains an interior near-singular vertex. Suppose v is connected to $v_1, \ldots, v_4$ in counterclockwise order, and let e_i be the edge $\langle v, v_i\rangle$, T_i the triangle $\langle v_i, v_{i+1}, v\rangle$, and θ_i the angle between e_i and e_{i+1}.

Suppose that none of e_i is degenerate at v. For each $1 \le j \le 4$, let s_j be the unique spline in $\mathcal{S}_d^1(\triangle)$ such that

$$\lambda_{i,m}^v s_j = \delta_{ij}, \qquad i,j = 1,2,3,4,$$
$$\lambda s_j = 0, \qquad \text{for all } \lambda \in \mathcal{N}_\triangle \setminus \mathcal{R}_2(v)$$

Clearly,

$$\operatorname{supp} s_j = T_{j-1} \cup T_j \cup T_{j+1},$$

and we can write (see [2,9])

$$s_j = \sum_{i \in I_j} c_i^{[j]} B_i,$$

where $I_j := \{i : \operatorname{supp} B_i \subset \operatorname{supp} s_j\}$. We now consider the spline

$$\hat{s} = -s_1 + s_2 - s_3 + s_4 = -\sum_{i \in I_1} c_i^{[1]} B_i + \sum_{i \in I_2} c_i^{[2]} B_i - \sum_{i \in I_3} c_i^{[3]} B_i + \sum_{i \in I_4} c_i^{[4]} B_i$$
$$= \sum_{i \in I_1 \cup I_2 \cup I_3 \cup I_4} a_i B_i.$$

Using the smoothness conditions (9), it is easy to see that

$$\lambda_{i,p}^v \hat{s} = 0, \quad \lambda_{i,m}^v \hat{s} = (-1)^i, \qquad i = 1,2,3,4,$$
$$\lambda \hat{s} = 0, \qquad \text{for all } \lambda \in \mathcal{N}_\triangle \setminus \mathcal{R}_2(v).$$

and thus $\|\hat{s}\|_\infty \le K_4$, where K_4 depends only on d. If the basis $\{B_1, \dots, B_n\}$ satisfies P2, we get

$$\|a\|_\infty \le K_1^{-1} \|\hat{s}\|_\infty \le K_4/K_1.$$

Moreover, since $\lambda_{2,m}^v B_i \ne 0$ only if $T_1 \cup T_2 \cup T_3 \subset \operatorname{supp} B_i$, we have

$$1 = \lambda_{2,m}^v \hat{s} = \sum_{i \in \tilde{I}_2} a_i \lambda_{2,m}^v B_i \le \#\tilde{I}_2 \|a\|_\infty \max_i |\lambda_{2,m}^v B_i|,$$

with $\tilde{I}_2 := \{i : \operatorname{supp} B_i = T_1 \cup T_2 \cup T_3\}$. Clearly, $\#\tilde{I}_2 < 3\binom{d+2}{2}$, and hence there exists $i_0 \in \tilde{I}_2$ such that

$$|\lambda_{2,m}^v B_{i_0}| \ge K_5 > 0,$$

where K_5 depends only on $\theta_\triangle$. However, $\lambda_{1,m}^v B_{i_0} = 0$, so that by (9) we have

$$|\lambda_{2,p}^v B_{i_0}| = \frac{1}{|\sin(\theta_1 + \theta_2)|} |\lambda_{2,m}^v B_{i_0}| \ge \frac{K_5}{|\sin(\theta_1 + \theta_2)|},$$

which is unbounded as $\theta_1 + \theta_2 \to \pi$. In view of the Markov inequality, it follows that $\|B_{i_0}\|_\infty$ is unbounded. But then the basis $\{B_1, \dots, B_n\}$ cannot be stable. □

§6. Remarks

Remark 1. Stable local bases are important for both theoretical and practical purposes. For example, it can be shown (see [12]) that if a spline space has such a basis, then it has full approximation power. Applications where stable bases are useful include data fitting and the numerical solution of boundary-value problems.

Remark 2. For $d \geq 5$, stable local bases for certain superspline subspaces of $\mathcal{S}_d^1(\triangle)$, can be constructed using classical finite elements, see [15]. However, it is also important to have such bases for the full spaces $\mathcal{S}_d^1(\triangle)$, since in contrast to supersplines, they are nested, *i.e.*, $\mathcal{S}_d^1(\triangle_1) \subseteq \mathcal{S}_d^1(\triangle_2)$ whenever $\triangle_2$ is a refinement of $\triangle_1$. This is important for multiresolution applications, see [3,13].

Remark 3. Algorithm 4 is a modification of the algorithm used in [14] to construct a star-supported basis for $\mathcal{S}_d^1(\triangle)$. The only change is in the choice of nodal functionals in step 4b) where i_0 was taken to be any index such that e_{i_0} is nondegenerate at v. To get stability, we have to choose i_0 more carefully. The basis constructed in Algorithm 4 is not locally linearly independent. To get an LLI basis, step 4) has to be modified in a different way, see [4].

Remark 4. Star-supported bases were constructed for general spline spaces $\mathcal{S}_d^r(\triangle)$ for $d \geq 4r+1$ in [1], and for $d \geq 3r+2$ in [10,11]. The constructions were based on Bernstein-Bézier techniques, and are not stable for triangulations that contain near-degenerate edges and/or near-singular vertices.

Remark 5. In [7] we use Bernstein-Bézier techniques to construct stable local bases for general spline spaces $\mathcal{S}_d^r(\triangle)$ and their superspline subspaces for all $d \geq 3r+2$. In a related work [6], we also used Bernstein-Bézier techniques to construct locally linearly independent bases for the same range of spline spaces and superspline spaces. For more on LLI spaces, including applications to almost interpolation, see [2,4,6,8,9].

Remark 6. Following the arguments in [7], it is easy to show that a natural renorming of our stable bases is L_p-stable for all $p \in [1, \infty]$.

Acknowledgments. The second author was supported in part by the National Science Foundation under grant DMS-9803340 and by the Army Research Office under grant DAAD-19-99-1-0160.

References

1. Alfeld, P., B. Piper, and L. L. Schumaker, Minimally supported bases for spaces of bivariate piecewise polynomials of smoothness r and degree $d \geq 4r+1$, Comput. Aided Geom. Design **4** (1987), 105–123.
2. Carnicer, J. M. and J. M. Peña, Least supported bases and local linear independence, Numer. Math. **67** (1994), 289–301.

3. Dahmen, W., P. Oswald, and X.-Q. Shi, C^1-hierarchical bases, J. Comput. Appl. Math. **51** (1994), 37–56.
4. Davydov, O., Locally linearly independent basis for C^1 bivariate splines, in *Mathematical Methods for Curves and Surfaces II,* Morten Dæhlen, Tom Lyche, Larry L. Schumaker (eds), Vanderbilt University Press, Nashville & London, 1998, 71–78.
5. Davydov, O., G. Nürnberger, and F. Zeilfelder, Bivariate spline interpolation with optimal approximation order, manuscript, 1998.
6. Davydov, O. and L. L. Schumaker, Locally linearly independent bases for bivariate polynomial spline spaces, manuscript, 1999.
7. Davydov, O. and L. L. Schumaker, On stable local bases for bivariate polynomial spline spaces, manuscript, 2000.
8. Davydov, O., M. Sommer, and H. Strauss, Locally linearly independent systems and almost interpolation, in *Multivariate Approximation and Splines, ISNM 125,* G. Nürnberger, J. W. Schmidt and G. Walz (eds), Birkhäuser, Basel, 1997, 59–72.
9. Davydov, O., M. Sommer, and H. Strauss, On almost interpolation and locally linearly independent bases, East J. Approx. **5** (1999), 67–88.
10. Hong, D., Spaces of bivariate spline functions over triangulation, Approx. Theory Appl. **7** (1991), 56–75.
11. Ibrahim, A. and L. L. Schumaker, Super spline spaces of smoothness r and degree $d \geq 3r + 2$, Constr. Approx. **7** (1991), 401–423.
12. Lai, M. J. and L. L. Schumaker, On the approximation power of bivariate splines, Advances in Comp. Math. **9** (1998), 251–279.
13. Le Méhauté, A., Nested sequences of triangular finite element spaces, in *Multivariate Approximation: Recent Trends and Results*, W. Haussman, K. Jetter and M. Reimer (eds), Akademie-Verlag, 1997, 133–145.
14. Morgan, J. and R. Scott, A nodal basis for C^1 piecewise polynomials of degree $n \geq 5$, Math. Comp. **29(131)** (1975), 736–740.
15. Schumaker, L. L., On super splines and finite elements, SIAM J. Numer. Anal. **26** (1989), 997–1005.

Oleg Davydov
Mathematisches Institut
Justus-Liebig-Universität
D-35392 Giessen, Germany
`oleg.davydov@math.uni-giessen.de`

Larry L. Schumaker
Department of Mathematics
Nashville, TN 37240, USA
`s@mars.cas.vanderbilt.edu`

On Lacunary Multiresolution Methods of Approximation in Hilbert Spaces

Lubomir T. Dechevski and Wolfgang L. Wendland

Abstract. We study lacunary multiresolution methods from the point of view of their analogy to the use of near-degenerate elements in finite and boundary element methods. The main results are characterization of the best N-term approximation of solutions of nonlinear operator equations and best N-term approximation by near-degenerate normal approximating families in Hilbert spaces.

§1. Introduction

This communication is part of a sequence of papers exploring the use of near-degenerate elements in finite- and boundary-element methods (see also [5,6] and their wavelet-based analogues, lacunary multiresolution methods. The use of near-degenerate and lacunary methods for solving operator equations is of considerable practical significance because in many important problems arising in industry, engineering and natural sciences, the use of such methods leads to a dramatic reduction of execution time and/or computer resources. The theoretical justification for the use of such methods is, however, very challenging: it has been successfully carried out only in a number of special cases, by specific techniques which vary from case to case. The purpose of this sequence of papers is to develop a general approach to overcoming the challenges of the use of lacunary multiresolution and near-degenerate finite and boundary element methods. Because of the limited space available, we shall consider only multiresolution methods for operator equations in Hilbert spaces, with an outline of the main ideas of the proofs, which in the Hilbert-space case are simpler and relatively short. A much more technically involved and detailed discussion of both near-degenerate finite elements and lacunary multiresolution methods and the important parallel between them will be given for more general types of nonlinear operators in quasi-Banach spaces in a later paper.

Curve and Surface Fitting: Saint-Malo 1999
Albert Cohen, Christophe Rabut, and Larry L. Schumaker (eds.), pp. 181–190.

ISBN 0-8265-1357-3.

§2. Approximate Solutions of Nonlinear Operator Equations

In this section we consider a general class of nonlinear operator equations, and study the numerical solutions of these equations obtained by iterative and projection methods.

Let X, Y be real Hilbert spaces. The class of nonlinear operators to be considered is the space $LH(X,Y)$ of all Lipschitz homeomorphisms F between X and Y, that is, $\exists\, F^{-1}$ on Y and $\exists\, C(F,X,Y) < \infty : ||F(x_1) - F(x_2)||_Y \leq C||x_1 - x_2||_X$, $\forall x_1 \forall x_2 \in X$, and analogously for F^{-1}.

Let H be a Hilbert space, such that $X \cap Y \subset H$ and $X \cap Y$ is dense on H, and let Y be the dual of X pivotal to H, i.e., the dual with respect to the duality functional defined by the scalar product of H. We shall denote this dual by $Y = X^* = X^*(H)$.

Definition 1. *Let $Y = X^*(H)$. The (generally nonlinear) operator $F : X \to X^*$ is called* Lipschitz, *if*

$$\exists C(F,X,H) < \infty : ||F(x_1) - F(x_2)||_{X^*} \leq C||x_1 - x_2||_X, \tag{1}$$

$\forall x_1 \forall x_2 \in X$, and strongly monotone, *if*

$$\exists c(F,X,H) > 0 : \langle F(x_1) - F(x_2), x_1 - x_2\rangle_H \geq c||x_1 - x_2||_X^2, \tag{2}$$

$\forall x_1 \forall x_2 \in X$. The class $LSM = LSM(X,H)$ consists of exactly those $F : X \to X^*(H)$ which satisfy (1,2).

It can be shown that the constants C and c in (1,2) are related by $c \leq C$. It should be noted that the typical case here is $X \hookrightarrow H \hookrightarrow X^*$ or $X \hookleftarrow H \hookleftarrow X^*$, where, as usual, $A \hookrightarrow B$ or $B \hookleftarrow A$ denotes continuous embedding: $A \subset B$ and $||.||_B \leq C||.||_A$.

Theorem 1. *(Generalization of Theorem 18.5 in [11] and strengthening of Theorem 18.5 in [15] for the case of Lipschitz operators in Hilbert spaces.) Let X and H be Hilbert spaces with the same cardinality. Then, $LSM(X,H) \subset LH(X, X^*(H))$.*

Proof: (Outline.) By duality arguments, it can be shown that the cardinality of $X^*(H)$ is equal to that of X and H. Therefore, since all spaces are Hilbertian with the same cardinality, there exist linear invertible operators $R : H \to X$ and $S : H \to X^*(H)$ which are isometric together with their inverses. Hence, the equation $F(x) = y$, $x \in X$, $y \in X^*(H)$ is equivalent to the equation $Av = w$, $v, w \in H$, where $Av = S^{-1}FR$. Now, since $F \in LSM(X,H)$, it follows from $||S^{-1}||_{X^* \to H} = ||R||_{H \to X} = 1$, that $A \in LSM(H,H)$. Therefore, by Theorem A (see below), F is bijective from X to X^*. By a condition of the theorem, F is Lipschitz; it remains to prove the same for F^{-1}. Indeed, by the strong monotonicity of F, setting $x_1 = F^{-1}(y_1)$, $x_2 = F^{-1}(y_2)$, $\forall y_1 \forall y_2 \in X^*$, we get

$$\begin{aligned}||F^{-1}(y_1) - F^{-1}(y_2)||_X^2 &\leq \frac{1}{c}\langle y_1 - y_2, F^{-1}(y_1) - F^{-1}(y_2)\rangle_H \\ &\leq \frac{1}{c}||y_1 - y_2||_{X^*}||F^{-1}(y_1) - F^{-1}(y_2)||_X,\end{aligned}$$

which completes the proof. □

Theorem A. *(See [11], Theorem 18.5). Assume that $X = X^* = H$. Then, $LSM(H,H) \subset LH(H,H)$ and the operator $T_{\varepsilon,y}(x) = x - \varepsilon[F(x) - y]$, $x \in H$, is contractive in H for $0 < \varepsilon < \frac{2c}{C^2}$, uniformly in $y \in H$, where C and c are defined in (1,2). The best contraction factor is $1 - c^2/C^2$ and is achieved for $\varepsilon = c/C^2$.*

Following the idea of the proof of Theorem 1, Theorem A can be modified for the case when $X \neq H$. We omit the details.

In the remaining part of this section we shall consider methods for approximate solution of the equation $F(x) = y$, $x \in X$, $y \in Y$, where X, Y are Hilbert spaces.

Definition 2. *(See [12]). Let X be a Hilbert space. $G \subset X$ is called an* existence set *for X, if $\forall x \in X\ \exists g_x \in G : ||x - g_x||_X = \min_{g\in G} ||x - g||_X = E_G(x)_X$. (The best approximation g_x need not be necessarily unique.) The sequence $\{G_N\}_{N=1}^{\infty}$, $G_N \subset X$ is called a* normal approximating family *in X, if for any $N \in \mathbb{N}$, G_N is an existence set, with $G_N \subset G_{N+1}$ and $G_N - G_{N-1} \subset G_{2N}$.*

Obviously, an existence set in X is closed in X (typical example: any finite-dimensional subspace of X).

Definition 3. *Let X be a Hilbert space. The sequence $\{G_N\}_{N=1}^{\infty} : G_N \subset X$ is said to have the* strong approximation property *(SAP, for short) if $\bigcup_{N=1}^{\infty} G_N$ is dense on X in the inner-product topology of X.*

Let us consider now the Galerkin–Petrov projection methods. Let $P_N : X \to X$, $Q_N : X \to X$ be projectors with $\dim P_N(x) = \dim Q_N(X) = N$, and $P_N P_{N+1} = P_{N+1} P_N = P_N$, $Q_N Q_{N+1} = Q_{N+1} Q_N = Q_N$.

Example 1. (*Galerkin–Petrov method for monotone operators.*) For a Hilbert space X, let $Y = X^*(H)$. The equation $F(x) = y$, $x \in X$, $y \in X^*$, is replaced by $Q_N^* F(P_N x) = Q_N^* y$, where $Q_N^* : X^* \to X^*, \dim Q_N^*(X^*) = N$, is the Banach adjoint of Q_N. The $N \times N$ nonlinear system is determined by

$$\langle Q_N^* F(P_N x), Q_N h\rangle_H = \langle Q_N^* y, Q_N h\rangle_H. \tag{3}$$

By Lemma 23.1 in [15], it follows that if $F \in LMS(X,H)$, where X and H are separable, then (3) has a unique solution for N large enough.

In the case $X = X^* = H$, if N is large enough, so that $Q_N^* F(P_N H) = Q_N^* H$ holds, then, by Theorem A, (3) can be computed by quickly converging contractive iterations. For small N, the condition $Q_N^* F(P_N H) = Q_N^* H$ may fail even if F is linear and $P_N = Q_N$ (see [2], Theorem 10.1.1).

If F is twice Gateau-differentiable, then Newton's method can be used where the inverse matrix involved in each iteration is usually sparse. In general, this method needs an appropriate initial approximation x_0 to the solution of $F(x) = y$, but if F is strongly monotone and potential, that is, if there exists a real functional $f : X \to \mathbb{R}$ such that $F = \operatorname{grad} f$, then, by Theorem 5.1 in [15], f is strictly convex and the solution of (3) is equivalent to minimizing the three times Gateau-differentiable functional f. Hence, Newton's method converges to the solution of (3) for any $x_0 \in X$, the rate of convergence depending on the constant c in (2). This technique is still numerically efficient if F is only Lipschitz, and Newton's method or its various modifications be replaced by the respective variants of the more general F. Clarke's subdifferential method. In the case of potential F, the Bubnov–Galerkin method ($P_N = Q_N$) coincides with the Ritz method for minimization of f.

For projection methods (see Example 1), the strong approximation property can be written as $\lim_{N\to\infty} ||(I_X - P_N)x||_X = 0$. A typical example when $G_N = P_N X$ forms a NAF having the SAP is when P_N is obtained by multiresolution.

By Theorem 23.3 in [15], if X is separable, then $G_N = P_N X$, as defined in Example 1, has the SAP; by Lemma 23.1 in [15] and in view of $F \in LMS(X, H)$, the solution x_N of (3) exists for N large enough and $||x_N - x||_X \to 0$, where x is the solution of $F(x) = y$, $y \in X^*$.

Theorem B. *(Céa's lemma for nonlinear operators (see [13], Lemma 2.8; [11], Theorems 4.1 and 18.8.) Under the assumptions of Example 1, for $F \in LSM(X, H)$, let $x = F^{-1}(y) \in X$ be the solution of $F(x) = y \in X^*(H)$, and $x_n \in X$ be the solution of (3). Then,*

$$\exists C(F, X, H) < \infty : E_N(F^{-1}(y))_X \leq ||F^{-1}(y) - x_N||_X \leq C E_N(F^{-1}(y))_X.$$

This result shows that Galerkin-Petrov methods (of any type - finite element or wavelet) achieves the best approximation rates up to a constant factor.

§3. Best N-term Approximation

For the general paradigm of best N-term approximation (BNTAP) we refer to [12], section 3.5, and [8].

Definition 4. *Let X_j, Y_j, $j = 0, 1$ be Hilbert spaces, $X_1 \hookrightarrow X_0$, $Y_1 \hookrightarrow Y_0$, and let $F \in LH(X_0, Y_0) \cap LH(X_1, Y_1)$. The NAF $\{G_N\}_{N=1}^{\infty} : G_N \subset X_1$, is called near-degenerate of order $(\lambda; \alpha, \beta)$, $\lambda > 0$, $\alpha \geq 0$, $\beta \geq 0$, if it satisfies a direct inequality of the type*

$$\exists C < \infty : \; E_N(F^{-1}(y))_{X_0} \leq C \frac{||F^{-1}(y)||_{X_1}}{N^{\lambda}}, \qquad \forall y \in Y_1, \tag{4}$$

where $C = C(N)$, with $C \asymp N^{\alpha}$; and an inverse inequality of the type

$$\exists D < \infty : \; ||x||_{X_1} \leq D N^{\lambda} ||x||_{X_0}, \; x \in G_N, \tag{5}$$

where $D = D(N)$, with $D \asymp N^{\beta}$. The partial case $\alpha = \beta = 0$ corresponds to a non-degenerate (regular) NAF.

Consider the approximation space

$$A_q^s(X_0) := \{f \in X_0 : ||f||_{A_q^s(X_0)} = \big(||f||_{X_0}^q + \sum_{j=0}^{\infty}[2^{js}E_{2^j}(f)_{X_0}]^q\big)^{1/q} < \infty\} \tag{6}$$

and the real interpolation space

$$(Y_0, Y_1)_{\theta,q} := \{f \in X_0 : \ ||f||_{(Y_0,Y_1)_{\theta,q}} = \big(||f||_{X_0}^q + \sum_{j=0}^{\infty}[2^{j\theta}K(2^{-j}, f; Y_0, Y_1)]^q\big)^{1/q} < \infty\}, \qquad (7))$$

where $K(t, f; Y_0, Y_1)$ is Peetre's K-functional (see [2,12]), $0 < t < \infty$, $s > 0$, $0 < \theta < 1$, $0 < q \leq \infty$ (with the usual sup-modification in (6,7) for $q = \infty$). (Recall that $X_1 \hookrightarrow X_0$, which explains the presence of the saturation term $||f||_{X_0}^q$ in (6,7).)

Theorem 2. *(Characterization of the best N-term approximation of solutions of nonlinear operator equations by near-degenerate NAF in Hilbert spaces). Assume that the conditions of Definition 4 hold. Let* $0 < q \leq \infty$. *Then,*

(i) if $0 \leq \alpha < \lambda$ and $s : 0 < s < \lambda - \alpha$, then, $\exists C_1 < \infty$:

$$||F^{-1}(y)||_{A_q^s(X_0)} \leq C_1[||F^{-1}(0)||_{X_1} + ||y||_{(Y_0,Y_1)_{\frac{s}{\lambda-\alpha},q}}]; \tag{8}$$

(ii) if $\beta \geq 0$ and $0 < s < \lambda + \beta$, then $\exists C_2 < \infty$:

$$||y||_{(Y_0,Y_1)_{\frac{s}{\lambda+\beta},q}} \leq C_2[||F(0)||_{Y_1} + ||F^{-1}(y)||_{A_q^s(X_0)}]. \tag{9}$$

Proof: (Outline.) By a standard technique, typical for BNTAP (see [12], Theorem 3.16 and Corollary 3.7), we prove

$$||F^{-1}(y)||_{A_q^s(X_0)} \leq c_1||F^{-1}(y)||_{(X_0,X_1)_{\frac{s}{\lambda-\alpha},q}}, \tag{10}$$

$$||F^{-1}(y)||_{(X_0,X_1)_{\frac{s}{\lambda+\beta},q}} \leq c_2||F^{-1}(y)||_{A_q^s(X_0)}. \tag{11}$$

By obtaining appropriate upper bounds for the K-functionals in the definition of $(X_0, X_1)_{\theta,q}$ and $(Y_0, Y_1)_{\theta,q}$, $0 < \theta < 1$, and using the embeddings $X_1 \hookrightarrow X_0$, $Y_1 \hookrightarrow Y_0$, it can be shown that, for Lipschitz operators F, F^{-1},

$$||F^{-1}(y)||_{(X_0,X_1)_{\frac{s}{\lambda-\alpha},q}} \leq c_3(||F^{-1}(0)||_{X_1} + ||y||_{(Y_0,Y_1)_{\frac{s}{\lambda-\alpha},q}}), \tag{12}$$

$$||F(x)||_{(Y_0,Y_1)_{\frac{s}{\lambda+\beta},q}} \leq c_4(||F(0)||_{Y_1} + ||x||_{(X_0,X_1)_{\frac{s}{\lambda+\beta},q}}) \tag{13}$$

hold. Combining (10) with (12) and (13) with (11), we arrive at (8) and (9), respectively. □

Corollary 1. *Under the conditions of Theorem 2, let $0 < s < \lambda - \alpha$, and assume that $F(0) = 0_{Y_1}$, $F^{-1}(0) = 0_{X_1}$. Then,*

$$(Y_0, Y_1)_{\frac{s}{\lambda-\alpha},q} \hookrightarrow A_q^s(X_0) \hookrightarrow (Y_0, Y_1)_{\frac{s}{\lambda+\beta},q}. \tag{14}$$

In particular, if $\alpha = \beta = 0$, then

$$(Y_0, Y_1)_{\frac{s}{\lambda},q} = A_q^s(X_0) \tag{15}$$

(isomorphism of the spaces, equivalence of the Hilbert norms).

Note that this special case corresponds to sublinear operators.

Remark. If the dependence of C in (4) and/or D in (5) on N is weaker than polynomial, e.g., logarithmic, then the left-hand and/or right-hand embedding in (14) can be sharpened by setting $\alpha = 0$ and/or $\beta = 0$ and modifying the index q. We omit the details.

Multiresolution Galerkin-Petrov methods for monotone operators (Example 1) are included as partial cases in Theorem 2 and Corollary 1. For monotone operators, we have $Y_0 = X_0^*(H_0)$, $Y_1 = X_1^*(H_1)$, where H_0, H_1 are Hilbert spaces with $H_0 \hookleftarrow H_1$ which are *sufficiently far away from each other so that $X_0 \hookleftarrow X_1$ and $X_0^*(H_0) \hookleftarrow X_1^*(H_1)$ hold simultaneously.* Here $X_1 \cap Y_1$ is assumed to be dense in H_0 and H_1. The projectors P_N and Q_N in Example 1 are assumed generated by multiresolution, which ensures that $G_N - G_{N-1} \subset G_{2N}$.

In the rest of this section and in the next section we shall discuss how to reduce the rates α and β in Theorem 2 and Corollary 1 to zero in the presence of near-degeneracy. To this end, we shall study the analogue of the phenomenon of near-degeneracy with multiresolution methods based on biorthogonal wavelets.

One equivalent norm in the inhomogeneous potential spaces H^s (cf., e.g., [14,4] for $p = q = 2$) is given by

$$||f||_{H^s} \asymp \{ \sum_{k\in\mathbb{Z}^n} |\alpha_{0k}|^2 + \sum_{j=0}^{\infty} 2^{2js} \sum_{k\in\mathbb{Z}^n} \sum_{l=1}^{2^n-1} |\beta_{jk}^{[l]}|^2 \}^{1/2}, \tag{16}$$

with $0 < s < r$, where in [8] r is the Lipschitz regularity of the compactly supported scaling functions $\varphi \in H^r$, $\tilde{\varphi} \in H^r$ and wavelets $\psi^{[l]} \in H^r$, $\tilde{\psi}^{[l]} \in H^r$ of the biorthonormal wavelet bases, with respect to which $f \in H^s$ can be expanded as follows:

$$f(x) = \sum_{k\in\mathbb{Z}^n} \alpha_{0k}\varphi_{0k}(x) + \sum_{j=0}^{\infty} \sum_{k\in\mathbb{Z}^n} \sum_{l=1}^{2^n-1} \beta_{jk}^{[l]}\psi_{jk}^{[l]}(x), \text{ a. e. } x, \tag{17}$$

where $\alpha_{0k} = \langle f, \tilde{\varphi}_{0k}\rangle_{L_2}$, $\beta_{jk}^{[l]} = \langle f, \tilde{\psi}_{jk}^{[l]}\rangle_{L_2}$. Each hypercube in the Calderon-Zygmund decomposition of $\mathbb{R}^n$ and Stein's construction of Whitney-type extension operators corresponds to $2^n - 1$ basis functions $\tilde{\psi}_{jk}^{[l]}$, $\psi_{jk}^{[l]}$, $l = 1, \dots, 2^n - 1$, in each of the two biorthonormal bases. The convergence in (17) is in the norm topology of H^s, but also Lebesgue a.e. on the domain Ω of the functions. Ω may be $\mathbb{R}^n$, hyperrectangle, correspond to the periodic case, or even general Lipschitz-graph domain. We refer to the currently most advanced work on this topic [3], as well as to the extensive account [4] (for the case of *homogeneous* potential spaces, see [7], in the special case $p = q = 2$).

Definition 5. *Let $j_1 \in \mathbb{N}$. A* non-degenerate wavelet-based projector *(NWP) is denoted by P_{j_1} and defined by*

$$P_{j_1} f(x) = \sum_k \alpha_{0k} \varphi_{0k}(x) + \sum_{j=0}^{j_1 - 1} \sum_k \sum_{l=1}^{2^n - 1} \beta_{jk}^{[l]} \psi_{jk}^{[l]}(x), \quad x \in \Omega, \tag{18}$$

cf. (17). A near-degenerate wavelet-based projector *(NDWP) is denoted by $\tilde{P}_{j_1}$ and defined by*

$$\tilde{P}_{j_1} f(x) = \sum_k \alpha_{0k} \varphi_{0k}(x) + \sum_k \sum_{j=0}^{J(j_1,k)-1} \sum_{l=1}^{2^n - 1} \beta_{jk}^{[l]} \psi_{jk}^{[l]}(x), \quad x \in \Omega, \tag{19}$$

$$\forall k \ J(j_1, k) > J(j_1 - 1, k), \ J(j_1, k) \geq j_1; \ \exists k_{j_1} : J(j_1, k_{j_1}) = j_1. \tag{20}$$

In other words, for a NWP $J(j_1, k) = j_1 = \text{const}$, uniformly in k. Thus, NDWP's are a specific partial case of lacunary wavelet-based projectors (see the concluding remarks in [4], subsection 6.2), lacunarity being with respect to the NWP corresponding to $J_1 := \max_k J(j_1, k)$.

Example 2. One example when near-degenerate FEM or lacunary wavelet-based projectors of NDWP type are needed is in the error analysis of numerical solutions in the immediate neighbourhood of the boundary $\partial\Omega$ (see, e.g., [11], Fig. 3.14, 3.15, 6.14, 6.15, 8.12). Then it is desirable to ensure that the local approximation rates near and on $\partial\Omega$ *do not deteriorate* compared to the local approximation rates in the interior of Ω. Indeed, assume that $\partial\Omega$ is regular enough (Lipschitz or smoother). Then, by the trace theorem (see, e.g., [1,9]), if $f \in H^s(\Omega)$, $\Omega \subset \mathbb{R}^n$, then the restriction of f on $\partial\Omega$ is less regular, namely, $f_{|\partial\Omega} \in H^{s-1/2}(\partial\Omega)$ holds. Then, the local approximation rate achieved via NWP, given in (18), is $O(2^{-j_1 s})$ in the interior of Ω and only $O(2^{-j_1(s-1/2)})$ near $\partial\Omega$. To achieve the desired uniform distribution of the error in the interior and near the boundary when f is smooth enough ($s > 1/2$), NDWP given in (19,20) should be employed, with $J(j_1, k) \approx j_1$ for k corresponding to the interior of Ω, and with $J(j_1, k) \asymp C_1 j_1 + C_2$ otherwise, where

$$C_2 \geq 0, \qquad C_1 = 1 + \frac{1}{2(s - 1/2)} > 1. \tag{21}$$

In the context of Theorem 2 and Corollary 1, if $X_j = H^{s_j}$, $j = 0, 1$, with $s_0 \leq s_1$ so that $X_1 \hookrightarrow X_0$ is fulfilled, then it can be verified that for NWP both the direct inequality $||f - P_{j_1}f||_{X_0} \leq C_1 2^{-j_1\lambda}||f||_{X_1}$ and the inverse inequality $||P_{j_1}f||_{X_1} \leq C_2 2^{j_1\lambda}||P_{j_1}f||_{X_0}$ hold, with $\lambda = s_1 - s_0$ and with C_1, C_2 independent of j_1. Hence, in Definition 4 $\alpha = \beta = 0$ is attained. On the contrary, for NDWP satisfying (21) the constants C_1 and C_2 depend on j_1 and $\alpha > 0$, $\beta > 0$ holds.

In the case of NDWP, is it possible to somehow reduce α and β to zero, thereby achieving isomorphism in (14)? It turns out that *the answer is positive*, and below we shall propose a general method how to achieve this.

Our approach will be consider *more general spaces* X_0, X_1 *than* H^s, so that, for the new X_0 and X_1, $\alpha = \beta = 0$ holds. Consider the Hilbert space $H^{s,w}$ with norm

$$||f||_{H^{s,w}} \asymp (\sum_k |\alpha_{0k}|^2 + \sum_{j=0}^{\infty}\sum_k 2^{2w(j,k)s}\sum_{l=1}^{2^n-1}|\beta_{jk}^{[l]}|^2)^{1/2}.$$

The spaces from this scale still admit atomic decomposition via the same Riesz bases of biorthonormal wavelets as H^s. The weight $w(j,k)$ is positive, monotonously increasing function in j for each fixed k, and depends on the choice of $J(j_1,k)$ in (20). The definition of $w(j,k)$ is

$$w(J(j_1,k),k) = j_1, \tag{22}$$

$$w(j,k) = j_1 - 1,\ \ j = J(j_1-1,k),\ J(j_1-1,k)+1,\ \ldots,\ J(j_1,k)-1, \tag{23}$$

$\forall j_1 \in \mathbb{N}\ \forall k \in \mathbb{Z}^n$.

Now, take $X_j = H^{s_j,w}$, $j = 0,1$, with $s_0 < s_1$. It can be seen that $X_1 \hookrightarrow X_0$ holds, and we can consider this pair of spaces in the context of Theorem 2 and Corollary 1.

Corollary 2. *Under the conditions of Corollary 1, assume that* $X_j = H^{s_j,w}$, $j = 0,1$, *where* $w = w(j,k)$ *is the left inverse (see (22,23)) of* $J(j,k)$ *as defined in (20). Assume also that* $N = 2^{j_1}$ *and* $G_N = \tilde{P}_{j_1}X_0$, *where the NDWP* $\tilde{P}_{j_1}$ *is defined in (19), with the same* $J(j,k)$ *in (20). Let* $s : 0 < s < \lambda = s_1 - s_0$. *Then (15) holds.*

Proof: (Outline.) It can be verified that the bounds

$$||f - \tilde{P}_{j_1}f||_{H^{s_0,w}} \leq \Big(\sum_{j=j_1}^{\infty}\sum_{k:w(j,k)\geq j_1} 2^{2w(j,k)s_0}\sum_l |\beta_{jk}^{[l]}|^2\Big)^{1/2}, \tag{24}$$

$$||\tilde{P}_{j_1}f||_{H^{s_1,w}} \leq \Big(\sum_k |\alpha_{0k}|^2 + \sum_{j=0}^{J_1}\sum_{k:w(j,k)\leq j_1} 2^{2w(j,k)s_1}\sum_l |\beta_{jk}^{[l]}|^2\Big)^{1/2}, \tag{25}$$

hold. (Recall that $j_1 = \min_k J(j_1,k)$, $J_1 = \max_k J(j_1,k)$.) After some computations, (24) and (25) imply

$$||f - \tilde{P}_{j_1}f||_{X_0} \leq C_1 2^{-j_1\lambda}||f||_{X_1},\ \ \forall f \in X_1, \tag{26}$$

$$||\tilde{P}_{j_1} f||_{X_1} \le C_2 2^{j_1\lambda} ||\tilde{P}_{j_1} f||_{X_0}, \quad \forall f \in X_0, \tag{27}$$

with $\lambda = s_1 - s_0$, and the constants C_1 and C_2 in (26,27) *do not depend on* j_1, i.e., for this choice of the spaces X_0, X_1 in Definition 4 $\alpha = \beta = 0$ holds. The result now follows from Corollary 1. □

Thus, we have solved the problem of characterizing the best approximation spaces induced by NDWP defined in (19) and (20). In this approach we remained entirely within the classical BNTAP. There is also another approach which goes beyond the general BNTAP, by abandoning the use of the real interpolation functor. This approach leads to atomic decomposition of Wiener amalgam spaces and will be considered elsewhere.

Acknowledgments. Supported in part by the Natural Sciences and Engineering Research Council of Canada and by the Priority Programme "Boundary Element Methods" of the German Research Foundation. The first author had the chance to benefit from the proficient expertise of Michal Křížek, in a valuable discussion and from his magnificent books [10,11]. Upon request, Wolfgang Dahmen kindly sent us some of his recent papers on multiresolution methods, and they proved to be of key importance for the understanding of near-degeneracy "from the wavelet side". His moral support and understanding of the seriousness of the topic are very much appreciated. Ron DeVore who, together with Vasil Popov and Pentcho Petrushev, is the founder of the theory of best N-term approximation, has also made the most important personal contribution to the further development and applications of this theory. Several recent papers authored and co-authored by him, of which I explicitly emphasize here the work of Cohen, Dahmen and DeVore [3], have contributed in a very essential way to our knowledge about wavelet approximation, and have thus helped us to successfully complete our present work. The kind attention of Larry Schumaker, Vidar Thomée and Ian Sloan is very much appreciated.

References

1. Bergh, J., and J. Löfström, *Interpolation Spaces. An Introduction*, Springer, Berlin, 1976.
2. Chen, G., and J. Zhou, *Boundary Element Methods*, Academic Press, London, 1992.
3. Cohen, A., W. Dahmen, and R. A. DeVore, Multiscale decompositions on bounded domains, Trans. Amer. Math. Soc., to appear.
4. Dahmen, W., Wavelet and multiscale methods for operator equations, Acta Numer. (1997), 55–228.
5. Dechevski, L. T., and E. Quak, On the Bramble–Hilbert lemma, Numer. Func. Anal. Optim. **11** (1990), 485–495.
6. Dechevski, L. T., and W. L. Wendland, On the Bramble–Hilbert lemma II, in preparation.

7. Dechevsky, L. T., Atomic decomposition of function spaces and fractional integral and differential operators, Fraction. Calcul. Appl. Anal., **2** (1999), 367–381.
8. DeVore, R. A., Nonlinear approximation, Acta Numer. (1998), 51–150.
9. Jonsson, A., and H. Wallin, *Function Spaces on Subsets of* $\mathbb{R}^n$, Harwood, London, 1984.
10. Křížek, M. and P. Neittaanmäki, *Mathematical and Numerical Modelling in Electrical Engineering Theory and Applications*, Kluwer, Dordrecht, 1996.
11. Křížek, M. and P. Neittaanmäki, *Finite Element Approximation of Variational Problems and Applications*, Longman, New York, 1990.
12. Petrushev, P. P., and V. A. Popov, *Rational Approximation of Real Functions*, Cambridge University Press, Cambridge, 1987.
13. Schatz, A. H., V. Thomée, and W. L. Wendland, *Mathematical Theory of Finite and Boundary Element Methods*, Birkhäuser, Basel, 1990.
14. Sickel, W., Spline representations of functions in Besov–Triebel–Lizorkin spaces on $\mathbb{R}^n$, Forum Math. **2** (1990), 451–475.
15. Vainberg, M. M., *Variational Method and Method of Monotone Operators in the Theory of Nonlinear Equations*, Halsted Press, New York – Toronto, 1973.

Lubomir T. Dechevski
Département de mathématiques et statistique
Université de Montréal, C.P. 6128, Succursale A
Montréal (Québec) H3C 3J7, Canada
dechevsk@dms.umontreal.ca

Wolfgang L. Wendland
Mathematisches Institut A
Universität Stuttgart
Pfaffenwaldring 57
70569 Stuttgart, Germany
wendland@mathematik.uni-stuttgart.de

Interpolation with Curvature Constraints

Hafsa Deddi, Hazel Everett, and Sylvain Lazard

Abstract. We address the problem of controlling the curvature of a Bézier curve interpolating a given set of data. More precisely, given two points M and N, two directions $\vec{u}$ and $\vec{v}$, and a constant k, we would like to find two quadratic Bézier curves Γ_1 and Γ_2 joined with continuity G^1, and interpolating the two points M and N, such that the tangent vectors at M and N have directions $\vec{u}$ and $\vec{v}$ respectively, the curvature is everywhere upper bounded by k, and some evaluating function, the length of the resulting curve for example, is minimized. In order to solve this problem, we first need to determine the maximum curvature of quadratic Bézier curves. This problem was solved by Sapidis and Frey in 1992. Here we present a simpler formula that has an elegant geometric interpretation in terms of distances and areas determined by the control points. We then use this formula to solve the variant of the curvature control problem in which Γ_1 and Γ_2 are joined with continuity C^1, where the length α between the first two control points of Γ_1 is equal to the length between the last two control points of Γ_2, and where α is the evaluating function to be minimized.

§1. Introduction

An important problem in CAGD is the construction of curves interpolating given sets of data that also satisfy constraints on their curvature. Such curves are visually pleasing and are said to be "fair" [1,2]. Fair curves are also important in the design of highways, railways and trajectories of mobile robots (see [9] and [6]). In these applications, curvature continuous curves with bounded curvature are desirable. Constructing fair curves has been the subject of recent research; see, for example, [4,5,7] for results about constraining the curvature at the endpoints, and [3,8] for results about monotonicity of curvature.

In this paper we consider the problem of controlling the curvature along the whole length of a Bézier curve interpolating a given set of data. More precisely, given two points M and N, two directions $\vec{u}$ and $\vec{v}$, and a constant k, we want to find two quadratic Bézier curves Γ_1 and Γ_2 joined with continuity G^1, and interpolating the two points M and N, such that the tangent vectors at M and N have directions $\vec{u}$ and $\vec{v}$ respectively, the curvature is everywhere upper bounded by k, and some evaluating function, the length of the resulting curve for example, is minimized. We call this the curvature control problem.

Curve and Surface Fitting: Saint-Malo 1999
Albert Cohen, Christophe Rabut, and Larry L. Schumaker (eds.), pp. 191–200.

ISBN 0-8265-1357-3.

In order to solve this problem, we first need to determine the maximum curvature of quadratic Bézier curves, that is, to find an exact formula in terms of the control points. Note that, for our problem, it is not sufficient to compute the maximum curvature of a particular Bézier curve using numerical methods. Note also that a quadratic Bézier curve is a parabola and, although it presents no special difficulties to compute the maximum curvature of a parabola in terms of the coefficients of its implicit equation, what we require is a formula in terms of the control points.

In [8], Sapidis and Frey give a formula for finding the maximum curvature for quadratic Bézier curves. In Section 2, we recall these results and present a simpler formula that has an elegant geometric interpretation in terms of distances and areas determined by the control points. We then use this formula to solve variants of the curvature control problem. Definitions and motivations for these variants are presented in Section 3.1. We solve in Section 3.2 the version of the curvature control problem where Γ_1 and Γ_2 are joined with continuity C^1, where the length α between the two first control points of Γ_1 is equal to the length between the two last control points of Γ_2, and where α is the evaluating function to be minimized. In Section 3.3, we prove that if we require in the previous variant a continuity G^2 instead of C^1 at the junction point, then there exist non-degenerate data for which there is no solution to the curvature control problem. However, if a solution exists, we show how it can be computed.

Throughout the paper, curvature refers to non-signed curvature, unless otherwise indicated. We denote by $\|pq\|$ the distance between points p and q, and by " $\times$ " and " $\cdot$ " the outer and inner products, respectively, between two vectors.

§2. Maximum Curvature of Quadratic Bézier Curves

Let Γ be a quadratic Bézier curve with control points p_0, p_1 and p_2 (see Figure 1). Recall that Γ is defined for every t in $[0,1]$ by $\Gamma(t) = (1-t)^2 p_0 + 2t(1-t)p_1 + t^2 p_2$. Let $\mathcal{A}$ be the area of the control triangle $p_0p_1p_2$ and m be the midpoint of the segment p_0p_2. We assume that Γ does not degenerate into a line segment, *i.e.*, p_0, p_1 and p_2 are not collinear.

Theorem 1. *The maximum curvature of a quadratic Bézier Γ is either equal to $\|p_1m\|^3/\mathcal{A}^2$ if p_1 lies strictly outside the two disks of diameter p_0m and mp_2, or is equal to $\max\{\kappa_0, \kappa_1\}$ where $\kappa_0 = \mathcal{A}/\|p_0p_1\|^3$ and $\kappa_1 = \mathcal{A}/\|p_1p_2\|^3$ are the curvature of $\Gamma(t)$ at the endpoints $\Gamma(0)$ and $\Gamma(1)$.*

Before proving Theorem 1, we recall the result by Sapidis and Frey [8] characterizing quadratic Bézier curves with monotone curvature.

Theorem 2 [8]. *The quadratic Bézier curve Γ has monotone curvature if and only if one of the angles $\angle(p_0p_1m)$ and $\angle(mp_1p_2)$ is equal to or larger than $\frac{\pi}{2}$. In other words, Γ has monotone curvature if and only if p_1 lies on or inside one of the two circles having as diameter p_0m and mp_2 (see Figure 1).*

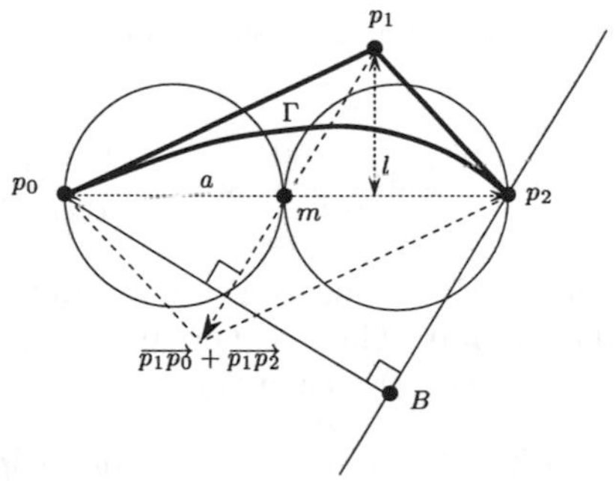

Fig. 1. The quadratic Bézier curve Γ has non-monotone curvature because p_1 lies strictly outside the two circles.

Sapidis and Frey also present in [8] the following expressions for the maximum curvature of quadratic Bézier curves. When the curvature is not monotone along Γ, then its maximum curvature is $4al/\|p_0B\|^3$, where (see Figure 1), a is the distance between p_0 and p_2, l is the distance between p_1 and the line joining p_0 and p_2, and $\|p_0B\|$ is the distance between p_0 and the line passing through p_2 and directed by $\overrightarrow{p_1p_0} + \overrightarrow{p_1p_2}$. When the curvature is monotone along Γ, its maximum is reached at one endpoint p_0 or p_2 of the curve, and is equal to $\frac{al}{2\|p_0p_1\|^3}$ or $\frac{al}{2\|p_1p_2\|^3}$ respectively.

We are now ready to prove Theorem 1. Note that the area $\mathcal{A}$ of the control triangle $p_0p_1p_2$ is equal to $al/2$. Thus, in order to prove Theorem 1, based on the results by Sapidis and Frey, it suffices to prove that $8\mathcal{A}/\|p_0B\|^3 = \|p_1m\|^3/\mathcal{A}^2$ or $2\mathcal{A} = \|p_1m\|.\|p_0B\|$. For completeness, we show how our result is derived from Theorem 2.

We assume first that p_1 lies strictly outside the two disks of diameter p_0m and mp_2. Thus, the curvature $\kappa(t)$, $t \in [0,1]$, of the quadratic Bézier curve Γ is not monotone by Theorem 2. It follows that the maximum curvature of Γ is obtained when the derivative of $\kappa(t)$ is zero.

The first and second derivatives of the Bézier curve Γ are

$$\begin{aligned}\Gamma'(t) &= 2((1-t)(p_1-p_0)+t(p_2-p_1)) \\ &= 2(p_1-p_0)+2t(p_2-2p_1+p_0), &(1)\\ \Gamma''(t) &= 2(p_2-2p_1+p_0). &(2)\end{aligned}$$

The curvature of Γ at $\Gamma(t)$ is thus, for any $t \in [0,1]$,

$$\kappa(t) = \frac{|\Gamma'(t) \times \Gamma''(t)|}{\|\Gamma'(t)\|^3} = \frac{|4(p_1-p_0)\times(p_2-p_1)|}{\|\Gamma'(t)\|^3},$$

giving

$$\kappa(t) = \frac{8\mathcal{A}}{\|\Gamma'(t)\|^3}, \tag{3}$$

where $\mathcal{A} = |(p_1 - p_0) \times (p_2 - p_1)|/2$ is the area of the control triangle $p_0p_1p_2$. The derivative of $\kappa(t)$ is

$$\kappa'(t) = \frac{-24\mathcal{A}(\|\Gamma'(t)\|)'}{\|\Gamma'(t)\|^4} = \frac{-12\mathcal{A}(\|\Gamma'(t)\|^2)'}{\|\Gamma'(t)\|^5}.$$

Since we assumed that the Bézier curve Γ is not degenerate, p_0, p_1 and p_2 are not collinear and thus $\mathcal{A} \neq 0$. Thus, $\kappa'(t) = 0$ if and only if $(\|\Gamma'(t)\|^2)' = 0$, or alternatively, $\Gamma'(t) \cdot \Gamma''(t) = 0$. Using (1) and (2), we get

$$\Gamma'(t) \cdot \Gamma''(t) = 4[(p_2 - 2p_1 + p_0)t + (p_1 - p_0)] \cdot [p_2 - 2p_1 + p_0] = 4(\alpha t - \beta),$$

where $\alpha = \|p_2 - 2p_1 + p_0\|^2$ and $\beta = -(p_1 - p_0) \cdot (p_2 - 2p_1 + p_0)$.

Thus, the derivative of the curvature $\kappa(t)$ vanishes if and only if $t = \tau = \beta/\alpha$. Note that τ is in $(0, 1)$ because the curvature of Γ is not monotone by assumption. Therefore, the maximum curvature along Γ is obtained for $t = \tau$.

Lemma 3. $\|\Gamma'(\tau)\| = \dfrac{2\mathcal{A}}{\|p_1m\|}$.

Proof: By (1), the square of the first derivative of $\Gamma(t)$ at τ is

$$\begin{aligned}\|\Gamma'(\tau)\|^2 &= 4[(p_2 - 2p_1 + p_0)\tau + (p_1 - p_0)]^2 = 4(\alpha\tau^2 - 2\tau\beta + \|p_0p_1\|^2) \\ &= 4(\alpha\frac{\beta^2}{\alpha^2} - 2\frac{\beta}{\alpha}\beta + \|p_0p_1\|^2) = \frac{4}{\alpha}(\alpha\|p_0p_1\|^2 - \beta^2),\end{aligned}$$

where, as before, $\alpha = \|p_2 - 2p_1 + p_0\|^2$ and $\beta = -(p_1 - p_0) \cdot (p_2 - 2p_1 + p_0)$. Since $p_2 - 2p_1 + p_0 = \overrightarrow{p_1p_0} + \overrightarrow{p_1p_2} = 2\overrightarrow{p_1m}$, we get $\alpha = 4\|p_1m\|^2$, $\beta = -2\overrightarrow{p_0p_1} \cdot \overrightarrow{p_1m}$, and thus

$$\|\Gamma'(\tau)\|^2 = \frac{1}{\|p_1m\|^2}(4\|p_1m\|^2\|p_0p_1\|^2 - 4(\overrightarrow{p_0p_1} \cdot \overrightarrow{p_1m})^2).$$

It follows from the canonical equation $(U \times V)^2 + (U \cdot V)^2 = U^2V^2$, for any two vectors U, V, that

$$\|\Gamma'(\tau)\|^2 = \frac{4(\overrightarrow{p_0p_1} \times \overrightarrow{p_1m})^2}{\|p_1m\|^2}.$$

Now, $|\overrightarrow{p_0p_1} \times \overrightarrow{p_1m}|$ is equal to $\mathcal{A}$, the area of the control triangle $p_0p_1p_2$. Indeed, $\overrightarrow{p_1m} = (\overrightarrow{p_1p_0} + \overrightarrow{p_1p_2})/2$ and thus $|\overrightarrow{p_0p_1} \times \overrightarrow{p_1m}| = |\overrightarrow{p_0p_1} \times \overrightarrow{p_1p_2}|/2 = \mathcal{A}$. Thus, $\|\Gamma'(\tau)\|^2 = 4\mathcal{A}^2/\|p_1m\|^2$ which yields the result. □

The expression of $\kappa_{max} = \kappa(\tau)$ now follows easily. By Lemma 3, $\|\Gamma'(\tau)\|^3$ is equal to $8\mathcal{A}^3/\|p_1m\|^3$. Thus, (3) gives

$$\kappa(\tau) = \frac{\|p_1m\|^3}{\mathcal{A}^2}.$$

That ends the proof of Theorem 1 when p_1 lies strictly outside the two disks of diameter p_0m and mp_2.

When p_1 lies inside one of these disks, Sapidis and Frey (see Theorem 2) proved that the curvature of the quadratic Bézier curve Γ is monotone. The maximum curvature is thus the curvature at one endpoint $\Gamma(0)$ or $\Gamma(1)$. Equation (1) gives $\Gamma'(0) = 2(p_1 - p_0)$ and $\Gamma'(1) = 2(p_2 - p_1)$. It then follows from (3) that

$$\kappa(0) = \frac{\mathcal{A}}{\|p_0p_1\|^3} \quad \text{and} \quad \kappa(1) = \frac{\mathcal{A}}{\|p_1p_2\|^3}.$$

§3. Controlling the Curvature of Piecewise Quadratic Bézier Curves

3.1. Preliminaries

Let Γ_1 and Γ_2 denote two quadratic Bézier curves with control points (p_0, p_1, p_2) and (q_0, q_1, q_2) respectively, and let Γ denote the concatenation of Γ_1 and Γ_2. The general curvature control problem we address is:

Given two points M and N, two unit vectors $\vec{u}$ and $\vec{v}$, and a constant k, we would like to find two quadratic Bézier curves Γ_1 and Γ_2 joined with continuity G^1 (at $p_2 = q_0$), interpolating the two points M and N (at p_0 and q_2 respectively), such that the tangent vectors at M and N have directions $\vec{u}$ and $\vec{v}$, respectively, the curvature is everywhere upper bounded by k, and some evaluating function is minimized.

We consider without loss of generality $k = 1$; for any $k \neq 0$, we can obtain an equivalent problem where $k = 1$ by scaling the plane.

The curves Γ_1 and Γ_2 are connected (at $p_2 = q_0$) with continuity G^1 if and only if there exists $\mu \in (0,1)$ such that $p_2 = q_0 = \mu p_1 + (1-\mu)q_1$. The curve Γ interpolates M and N, such that the tangent vectors at M and N have directions $\vec{u}$ and $\vec{v}$, respectively, if and only if $p_0 = M$, $q_2 = N$ and there exists α and β positive real numbers such that $p_1 - p_0 = \alpha\vec{u}$ and $q_2 - q_1 = \beta\vec{v}$ (see Figure 2). One way to solve the general curvature control problem is to

1) find the set of $(\alpha, \beta, \mu) \in (0, +\infty)^2 \times (0,1)$ on which the curvature of Γ is everywhere smaller or equal to 1, and then,
2) find a value (α, β, μ) in that set for which the evaluating function is minimized.

In general, this is a non-linear optimization problem with non-linear constraints, and thus, cannot necessarily be solved quickly and accurately. Clearly, the difficulty depends on the complexity of the set of feasible solutions and on the evaluating function that is to be minimized. Here we consider simplifying assumptions. First, we require a continuity C^1 at the junction point between the two curves Γ_1 and Γ_2. This fixes μ to $1/2$ and reduces the number of variables to two. To bring the number of variables down to one, we arbitrarily consider $\alpha = \beta$. We then choose as evaluating function the length α. By minimizing α, we ensure that all the control points p_1, $p_2 = q_0$ and q_1 remain close to the the points M and N we want to interpolate; in other words, by minimizing α, we expect that the length of the resulting curve Γ will not be too far from its minimum. With these further assumptions, we

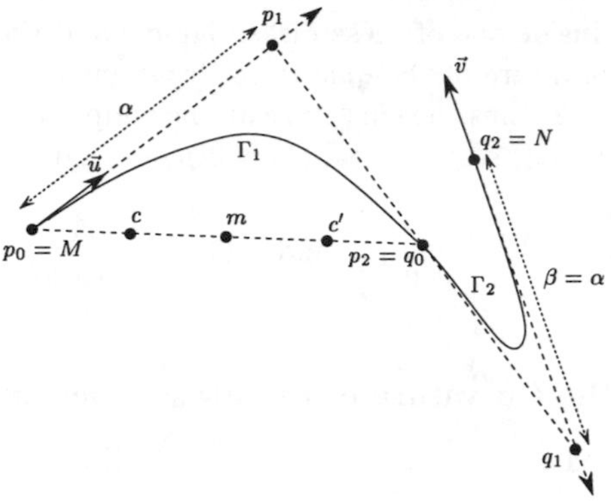

Fig. 2. Curvature control problem with continuity C^1 and $\alpha = \beta$.

solve (in Section 3.2) the given interpolation and minimization problem, except for the degenerate case when $\vec{u}$ and $\vec{v}$ are parallel, for which we prove that a solution does not necessarily exist.

In Section 3.3, we also consider $\alpha = \beta$, but we require a continuity G^2 (instead of C^1) at the junction point between the two curves Γ_1 and Γ_2. In other words, we require the signed curvature to be continuous on Γ. The variables are then reduced to (α, μ), but the constraint that the continuity is G^2 links these two variables, and thus the problem is actually one-dimensional. We prove in Section 3.3 that this set of additional constraints is too restrictive in the sense that there exists non-degenerate data $(M, N, \vec{u}, \vec{v})$ that cannot be interpolated. However, if a solution exists, we show how it can be computed.

3.2. Curvature control problem with C^1 continuity

We consider here the following variant of the curvature control problem:

Given two points M and N, and two unit vectors $\vec{u}$ and $\vec{v}$, we want to find two quadratic Bézier curves Γ_1 and Γ_2 joined with continuity C^1 (at $p_2 = q_0$), interpolating the two points M and N (at p_0 and q_2 respectively), such that the tangent vectors at M and N have directions $\vec{u}$ and $\vec{v}$ respectively, the maximum curvature of the two curves is smaller or equal to 1, the distances $\alpha = \|p_0p_1\|$ and $\beta = \|q_1q_2\|$ are equal, and such that α is minimized.

See Figure 2.

We show in this section how to solve this problem for non-degenerate data, that is when $\vec{u}$ and $\vec{v}$ are not collinear. When $\vec{u}$ and $\vec{v}$ are collinear, we show that there is not necessarily a solution.

As we said in Section 3.1, this problem is equivalent to finding the smallest $\alpha \in (0, +\infty)$ such that the curvature of Γ_1 and Γ_2 is everywhere smaller or equal to 1, where $p_0 = M$, $q_2 = N$, $p_1 = p_0 + \alpha\vec{u}$, $q_1 = q_2 - \alpha\vec{v}$ and $p_2 = q_0 = (p_1 + q_1)/2$.

We show how we compute the smallest $\alpha \in (0, +\infty)$ such that the curvature of Γ_1 is everywhere smaller or equal to 1. Computing the smallest

$\alpha \in (0, +\infty)$ for Γ_2 can be done similarly. We then return the curve Γ defined by the biggest of those two α.

First, for any value $\alpha \in (0, +\infty)$, we need to determine an expression for the maximum curvature of Γ_1. By Theorem 1, it remains to determine whether the maximum curvature of Γ_1 is given by the maximum curvature $\kappa_{max}(\Gamma_1)$ of the parabola supporting Γ_1, or by $\kappa_0(\Gamma_1)$ or $\kappa_1(\Gamma_1)$, the curvature of Γ_1 at its endpoints $\Gamma_1(0)$ or $\Gamma_1(1)$, respectively. Thus, for any value $\alpha \in (0, +\infty)$, we want to decide whether p_1 belongs to one of the disks of diameter p_0m and mp_2, where m is the midpoint of p_0p_2 (see Figure 2). Let c and c' be the respective centers of these disks and R be their radius. In order to determine whether p_1 belongs to one of these disks, we compute and compare R^2 with the distances $\|p_1c\|^2$ and $\|p_1c'\|^2$.

Since p_1 and q_1 are linear in α, and $p_2 = (p_1 + q_1)/2$, $m = (p_0 + p_2)/2$, $c = (p_0 + m)/2$, and $c' = (m + p_2)/2$, we have that $(c - p_0)^2$, $(c - p_1)^2$ and $(c' - p_1)^2$ are of degree 2 in α. Thus, $R^2 < \|p_1c\|^2$ and $R^2 < \|p_1c'\|^2$ are inequalities of degree at most 2 in α (namely $\alpha > \frac{16\vec{u}\cdot\overrightarrow{p_0q_2}}{(7\vec{u}+\vec{v})^2-(\vec{u}-\vec{v})^2}$ and $\alpha^2[(5\vec{u} + 3\vec{v})^2 - (\vec{u} - \vec{v})^2] - 2\alpha(16\vec{u} + 8\vec{v}) \cdot \overrightarrow{p_0q_2} + 8\|p_0q_2\|^2 > 0$). By solving these equations, we get a partition of $(0, +\infty)$ into two sets of intervals $\mathcal{I}$ and $\mathcal{I}'$ such that the maximum curvature of Γ_1 is given by $\kappa_{max}(\Gamma_1)$ for any $\alpha \in \mathcal{I}$, and by $\max(\kappa_0(\Gamma_1), \kappa_1(\Gamma_1))$ for any $\alpha \in \mathcal{I}'$.

With $\mathcal{A}(p_0p_1p_2)$ denoting the area of the control triangle $p_0p_1p_2$, we get by Theorem 1, when p_0, p_1 and p_2, are not collinear,

$$\kappa_{max}(\Gamma_1)^2 = \frac{\|p_1m\|^6}{\mathcal{A}(p_0p_1p_2)^4},\ \kappa_0(\Gamma_1)^2 = \frac{\mathcal{A}(p_0p_1p_2)^2}{\|p_0p_1\|^6} \text{ and } \kappa_1(\Gamma_1)^2 = \frac{\mathcal{A}(p_0p_1p_2)^2}{\|p_1p_2\|^6}.$$

A straightforward computation gives

$$\overrightarrow{p_1m} = \frac{\overrightarrow{p_0q_2} + \alpha(-3\vec{u} - \vec{v})}{4},\ \overrightarrow{p_0p_1} = \alpha\vec{u} \text{ and } \overrightarrow{p_1p_2} = \frac{\overrightarrow{p_0q_2} - \alpha(\vec{u} + \vec{v})}{2}.$$

Thus, $\mathcal{A}(p_0p_1p_2) = |\overrightarrow{p_0p_1} \times \overrightarrow{p_1p_2}|/2 = |\alpha\vec{u} \times \overrightarrow{p_0q_2} - \alpha^2\vec{u} \times \vec{v}|/4$ and

$$\kappa_{max}(\Gamma_1)^2 = \frac{(\alpha^2(3\vec{u} + \vec{v})^2 - 2\alpha(3\vec{u} + \vec{v}) \cdot \overrightarrow{p_0q_2} + \|\overrightarrow{p_0q_2}\|^2)^3}{16(\alpha^2\vec{u} \times \vec{v} - \alpha\vec{u} \times \overrightarrow{p_0q_2})^4},$$

$$\kappa_0(\Gamma_1)^2 = \frac{(\alpha^2\vec{u} \times \vec{v} - \alpha\vec{u} \times \overrightarrow{p_0q_2})^2}{16\alpha^6} \text{ and } \kappa_1(\Gamma_1)^2 = \frac{4(\alpha^2\vec{u} \times \vec{v} - \alpha\vec{u} \times \overrightarrow{p_0q_2})^2}{(\alpha(\vec{u} + \vec{v}) - \overrightarrow{p_0q_2})^6}.$$

Thus, $\kappa_{max}(\Gamma_1)^2 \leq 1$, $\kappa_0(\Gamma_1)^2 \leq 1$ and $\kappa_1(\Gamma_1)^2 \leq 1$ reduce to inequalities in α of degree at most 8, 6 and 6 respectively. Finding the intervals of $\mathcal{I}$ and $\mathcal{I}'$ on which those inequalities are satisfied can therefore simply be done by computing the roots of the corresponding equations. More precisely, the smallest of (i) the smallest root of $\kappa_{max}(\Gamma_1)^2 = 1$ in $\mathcal{I}$, and (ii) the smallest root of $\kappa_0(\Gamma_1)^2 = 1$ and $\kappa_1(\Gamma_1)^2 = 1$ in $\mathcal{I}'$, is the smallest α for which the

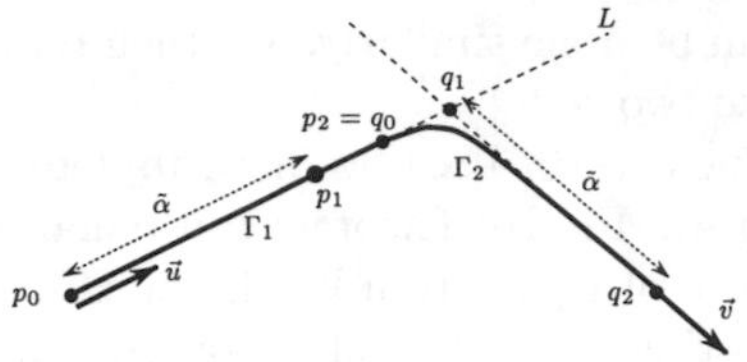

Fig. 3. Case where p_0, p_1 and p_2 are collinear and consecutive.

maximum curvature of Γ_1 is smaller or equal to 1. Such a solution exists when $\vec{u} \times \vec{v} \neq 0$ because the maximum curvature of Γ_1 goes from $+\infty$ to 0 since $\kappa_{max}(\Gamma_1)^2$, $\kappa_0(\Gamma_1)^2$ and $\kappa_1(\Gamma_1)^2$ tend to $+\infty$ when α tends to 0, and tend to 0 when α tends to $+\infty$.

We have shown that, when $\vec{u} \times \vec{v} \neq 0$, the smallest $\alpha \in (0, +\infty)$ such that the curvature of Γ_1 is everywhere smaller or equal to 1, and such that the control points p_0, p_1 and p_2 are not collinear, exists and we can compute it. Suppose now that there exists $\tilde{\alpha} \in (0, +\infty)$ such that p_0, p_1 and p_2 are collinear (see Figure 3). Assume furthermore that p_1 lies in between p_0 and p_2; otherwise, Γ_1 is not smooth and does not satisfy the constraint on the curvature. Since p_2 is the midpoint of p_1q_1, it follows that p_0, p_1, p_2 and q_1 are, in this order, on the line L passing through p_0 and directed by $\vec{u}$ (the line is necessarily directed by $\vec{u}$ because $p_1 \neq p_0$ belongs to that line). With $\vec{u} \times \vec{v} \neq 0$, q_2 does not belong to L. Thus, for $\alpha < \tilde{\alpha}$, the triangle $p_0p_1p_2$ is not flat but tends to a flat triangle, with flat vertex at p_1, as α tends to $\tilde{\alpha}$. Therefore, when α tends from below to $\tilde{\alpha}$, Γ_1 tends to a straight line segment, and the maximum curvature of Γ_1 tends to 0. Thus, there exists $\alpha < \tilde{\alpha}$ such that the maximum curvature of Γ_1 is smaller than 1. It follows that $\tilde{\alpha}$ is bigger than the smallest solution α we found previously. Therefore, when $\vec{u} \times \vec{v} \neq 0$, there is always an optimal solution with p_0, p_1 and p_2 not all collinear.

We now show that, when $\vec{u} \times \vec{v} = 0$, there may not exist a solution. Assume for example that $\overrightarrow{p_0q_2}$ is not parallel to $\vec{u}$ and $\vec{v}$, and that $\vec{u} + \vec{v} = 0$. Then, when α tends to 0, $\kappa_0(\Gamma_1)$, $\kappa_1(\Gamma_1)$ and $\kappa_{max}(\Gamma_1)$ tend respectively to $+\infty$, 0 and $+\infty$. Similarly, when α tends to $+\infty$, they tend respectively to 0, $+\infty$ and $+\infty$. It follows that $\max(\kappa_0(\Gamma_1), \kappa_1(\Gamma_1))$ and $\kappa_{max}(\Gamma_1)$ tend to $+\infty$ when α tends to 0 and $+\infty$. In addition, $\kappa_0(\Gamma_1)$, $\kappa_1(\Gamma_1)$ and $\kappa_{max}(\Gamma_1)$ are never equal to 0 because then $\|p_1m\| = 0$ or $\mathcal{A}(p_0p_1p_2) = 0$ which would imply that p_0, p_1 and p_2 are collinear, which is impossible since the two rays starting at p_0 and q_2 with direction $\vec{u}$ and $-\vec{v}$ do not intersect. Thus, $\max(\kappa_0(\Gamma_1), \kappa_1(\Gamma_1))$ and $\kappa_{max}(\Gamma_1)$ are strictly greater than a positive constant for any $\alpha \in (0, +\infty)$, and, by scaling the plane, this constant can be scaled to a value greater than 1.

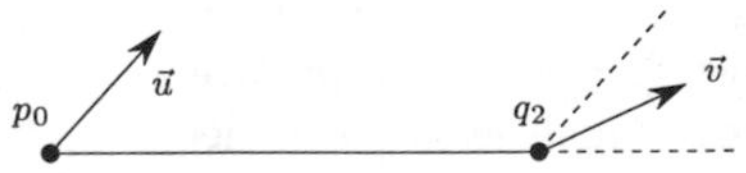

Fig. 4. Example where $\Psi > 0$ for any $\alpha > 0$ ($\overrightarrow{p_0q_2} \times \vec{u} > 0$, $\overrightarrow{p_0q_2} \times \vec{v} > 0$ and $\vec{u} \times \vec{v} < 0$).

3.3. Curvature control problem with G^2 continuity

We consider here the following variant of the curvature control problem:

> *Given two points M and N, and two unit vectors $\vec{u}$ and $\vec{v}$, we want to find two quadratic Bézier curves Γ_1 and Γ_2 joined with continuity G^2 (at $p_2 = q_0$), interpolating the two points M and N (at p_0 and q_2 respectively), such that the tangent vectors at M and N have directions $\vec{u}$ and $\vec{v}$ respectively, the maximum curvature of the two curves is smaller or equal to 1, the distances $\alpha = \|p_0p_1\|$ and $\beta = \|q_1q_2\|$ are equal, and such that α is minimized.*

As we said in Section 3.1, the problem is equivalent to finding the smallest $\alpha \in (0, +\infty)$ such that Γ_1 and Γ_2 are connected G^2 and their curvature is everywhere smaller or equal to 1, where $p_0 = M$, $q_2 = N$, $p_1 = p_0 + \alpha\vec{u}$, $q_1 = q_2 - \alpha\vec{v}$, and there exists $\mu \in (0,1)$ such that $p_2 = q_0 = \mu p_1 + (1-\mu)q_1$.

The curves Γ_1 and Γ_2 are connected G^2 if and only if the two signed curvatures of Γ_1 and Γ_2 at p_2 are equal, that is, by Theorem 1,

$$\frac{\overrightarrow{p_0p_1} \times \overrightarrow{p_1p_2}}{2\|p_1p_2\|^3} = \frac{\overrightarrow{q_0q_1} \times \overrightarrow{q_1q_2}}{2\|q_0q_1\|^3},$$

when the triplets of points (p_0, p_1, p_2) and (q_0, q_1, q_2) are not collinear. We easily get that $\overrightarrow{p_1p_2} = (1-\mu)\overrightarrow{p_1q_1}$, $\overrightarrow{q_0q_1} = \mu\overrightarrow{p_1q_1}$, $\overrightarrow{p_1q_1} = \overrightarrow{p_0q_2} - \alpha(\vec{u}+\vec{v})$, $\overrightarrow{p_0p_1} = \alpha\vec{u}$ and $\overrightarrow{q_1q_2} = \alpha\vec{v}$. Thus, we get that Γ is G^2 if and only if

$$\frac{\alpha\vec{u} \times (1-\mu)(\overrightarrow{p_0q_2} - \alpha(\vec{u}+\vec{v}))}{(1-\mu)^3\|p_1q_1\|^3} = \frac{\mu(\overrightarrow{p_0q_2} - \alpha(\vec{u}+\vec{v})) \times \alpha\vec{v}}{\mu^3\|p_1q_1\|^3} \iff$$

$$\frac{\vec{u} \times \overrightarrow{p_0q_2} - \alpha\vec{u}\times\vec{v}}{(1-\mu)^2} = \frac{\overrightarrow{p_0q_2} \times \vec{v} - \alpha\vec{u}\times\vec{v}}{\mu^2} \iff$$

$$\mu^2 - 2\mu\Psi + \Psi = 0 \quad \text{where} \quad \Psi = \frac{\overrightarrow{p_0q_2} \times \vec{v} - \alpha\vec{u}\times\vec{v}}{\overrightarrow{p_0q_2} \times (\vec{u}+\vec{v})} \quad (\text{if } \overrightarrow{p_0q_2} \times (\vec{u}+\vec{v}) \neq 0).$$

Standard calculations yield that the equation $\mu^2 - 2\mu\Psi + \Psi = 0$ admits a root in $(0,1)$ if and only if $\Psi \in (-1/3, 0)$. We can easily choose $p_0, q_2, \vec{u}$ and $\vec{v}$ such that $\Psi \notin (-1/3, 0)$. Indeed (see Figure 4), $\Psi > 0$ for any $\vec{u}$, $\vec{v}$ that are on the same side of $\overrightarrow{p_0q_2}$ (*i.e.*, $\overrightarrow{p_0q_2} \times \vec{u}$ and $\overrightarrow{p_0q_2} \times \vec{v}$ have the same sign) and such that $\vec{v}$ lies in the small wedge defined by $\overrightarrow{p_0q_2}$ and $\vec{u}$ (*i.e.*, $\vec{u}\times\vec{v}$ and $\overrightarrow{p_0q_2} \times \vec{v}$ have opposite signs). We thus proved that there is no solution to our curvature control problem for a set of non-degenerate choices of the parameters M, N, $\vec{u}$ and $\vec{v}$.

However, when a solution exists, it can be computed as in the previous section. Indeed, the curvature $\kappa_{max}(\Gamma_i)$ can be expressed as a ratio of polynomials in α and μ, and the inequality $\kappa_{max}(\Gamma_i) < 1$ reduces to a polynomial inequality of degree 28 in α. Similar remarks hold for $\kappa_0(\Gamma_i)$ and $\kappa_1(\Gamma_i)$.

§4. Concluding Remarks

It remains open to solve the curvature control problem when the length of the curve is to be minimized. Another interesting approach would be to determine how much longer than optimal our curves are. Also, we would like to consider the case when the data consist of more than two control points. Note also that, because of the high degree of the equations, it is not clear that the solutions presented in Sections 3.2 and 3.3 are usable in an interactive curve design context. This should be tested with an implementation.

References

1. Farin, G., *Curves and Surfaces for Computer Aided Geometric Design: A Practical Guide*, 4th ed. Academic Press, San Diego, 1997.
2. Farin, G. and N. S. Sapidis, Curvature and the fairness of curves and surfaces, IEEE Comp. Graph. Appl. **9** (1989), 52–57.
3. Goodman, T. N. T., Curvature of rational quadratics spline, *Curves and Surfaces in Geometric Design*, P.-J. Laurent, A. Le Méhauté, and L. L. Schumaker (eds.), A. K. Peters, Wellesley MA, 1994, 201–208.
4. Goodman, T. N. T. and K. Unsworth, Shape preserving interpolation by curvature continuous parametric curves, Comput. Aided Geom. Design **5** (1988), 323–340.
5. Juhász, I., Cubic parametric curves of given tangent and curvature, Computer-Aided Design **30** (1998), 1–9.
6. Latombe, J.-C., *Robot Motion Planning,* Kluwer Academic Publishers, Boston, 1991.
7. Paluszny, M., F. Tovar, and R. R. Patterson, G^2 composite cubic Bézier curves, J. Comput. Appl. Math. **102** (1999), 49–71.
8. Sapidis, N. S. and W. H. Frey, Controlling the curvature of a quadratic Bézier curve, Comput. Aided Geom. Design **9** (1992), 85–91.
9. Walton, D. J. and D. S. Meek, Planar G^2 curve design with spiral segments, Computer-Aided Design **30** (1998), 529–538.

Hazel Everett and Sylvain Lazard
LORIA - INRIA Lorraine, 615 rue du jardin botanique, B.P. 101
54602 Villers-les-Nancy Cedex, France
`name@loria.fr`

Hafsa Deddi
Department of mathematics and computer science
University of Lethbridge
4401 University Drive, Lethbridge, Alberta, T1K 3M4, Canada
`deddi@cs.uleth.ca`

A B-spline Tensor for Vectorial Quasi-Interpolant

Fabrice Dodu

Abstract. The aim of this paper is to introduce new techniques and new tools for vector field approximation. We do so by building the equivalent of B-splines, which are now tensor B-splines, as shown below, and by appling to the discretization based on a regular grid of a differential operator a fundamental solution of it, as done for polynomial B-splines and polyharmonic B-splines (see [5,6]). We thus obtain quasi-interpolants in the vectorial case whose properties generalize the properties of the quasi-interpolants generated by using B-splines. All this is done in the case when the data lie on a regular infinite grid.

§1. Introduction

Fluid mechanics, meteorology and more and more other applications need approximate functions from $\mathbb{R}^3$ to $\mathbb{R}^3$. Given some discrete vectorial data, we want to get a function interpolating or approximating the data. At first glance, we may think of doing this with three independent approximations of the data (one for each component of the data). Of course this can be done, but it usually gives poor results since there is no connection between the various components of the approximation function, while the applications may require, for example, a divergence-free (or a rotational-free) function. In order to take into account this kind of connection, we want to determine the function interpolating the data and minimizing a seminorm over all vectorial functions interpolating the data. The seminorm which is minimized is based on the Helmholtz decomposition of vector fields into a rotational and a gradient part ($\rho\ \|\mathrm{div}\cdot\|^2 + \|\mathrm{rot}\cdot\|^2$). This will be presented in detail in the forthcoming thesis [2]. The weight ρ is introduced to allow the rotational part of the field to dominate the gradient part and so in the so obtained function [2].

In order to determine the interpolating vector, we need to solve a linear system which is usually large (three times the number of data) and badly conditioned. This is why in this paper we do not propose to interpolate,

Curve and Surface Fitting: Saint-Malo 1999
Albert Cohen, Christophe Rabut, and Larry L. Schumaker (eds.), pp. 201–208.

ISBN 0-8265-1357-3.

but instead approximate the data by building a quasi-interpolant based on B-splines. Note that if $(z_i)_{i=1}^n$ are vectors and if we want to get a vectorial function S such that $S(x) = \sum_i B_i(x)\, z_i$ (as we do in the scalar case with B-splines B_i), we need that the functions B_i are matrix and not scalar functions as in the case of scalar approximation in $\mathbb{R}$ or $\mathbb{R}^d$. This is why the tools we build are tensors.

Now, in order to build this "B-spline Tensor", we will use the same strategy as for polynomial and polyharmonic B-splines: we will discretize the differential operator $P_{m,\rho}(D)$ defined in [2] and apply to this discretization a fundamental solution of $P_{m,\rho}(D)$, thus obtaining a kind of approximation of the Dirac tensor $\delta\, \boldsymbol{I}_3$. All this is done on a cardinal grid (i.e. a regular infinite grid).

For our three-dimensional problem [2], we choose the 3×3 differential matrix defined by

$$P_{m,\rho}(D) = (-1)^m\, \Delta^{m-1} \left[\rho \nabla \mathrm{div} \cdot - \mathrm{rot}\,(\mathrm{rot} \cdot)\right],$$

where div is the divergence operator, ∇ is the gradient operator, rot is the rotational operator and ρ is an arbitrary positive parameter.

We remark that if $\rho = 1$, then $P_{m,1}(D) = (-1)^m\, \Delta^m\, \boldsymbol{I}_3$, where $\boldsymbol{I}_3$ is the identity matrix of $\mathbb{R}^3$, so we obtain three independent operators (one for each component), each one being as in ([5]).

In the first part of this paper we give the construction of a discretization of $P_{m,\rho}(D)$. In the second part, we define polyharmonic B-spline tensors and give the main properties. In the last part, we study the associated vector quasi-interpolants, and in particular prove their $\mathcal{P}_k$-reproduction.

Notation: Let m be a integer with $m \geq 2$. $\mathbb{P}_n$ denotes the set of polynomial with variable in $\mathbb{R}^3$ of total degree at most n and $\mathcal{P}_n = \mathbb{P}_n \times \mathbb{P}_n \times \mathbb{P}_n$. D' denotes the set of distribution of $\mathbb{R}^3$ and $\mathcal{D}' = D' \times D' \times D'$. Let $f : \mathbb{R}^3 \longmapsto \mathbb{R}$ be a scalar function of a three dimensional variable. Let

$$\mathrm{rot} f = \begin{pmatrix} \partial_2 f_3 - \partial_3 f_2 \\ \partial_3 f_1 - \partial_1 f_3 \\ \partial_1 f_2 - \partial_2 f_1 \end{pmatrix}$$

be the rotational operator, and let $\mathrm{div} f = \partial_1 f_1 + \partial_2 f_2 + \partial_3 f_3$ be the divergence operator. Let $\Delta^m = \left(\sum_{i=1}^3 \partial_i^2\right)^m$. $\mathcal{F}(g)$ is the Fourier transform of the function (or distribution) g. For all $\zeta \in \mathbb{R}^3$, $\sin(\zeta)$ denotes the vector defined by $\forall 1 \leq i \leq 3$, $\sin(\zeta)_i = \sin(\zeta_i)$. δ denotes the Dirac distribution. Let $\widetilde{v}_{m+1}$ be the function such that : $\widetilde{v}_{m+1} = \frac{\Gamma(\frac{1}{2}-m)}{2^{2m+2}\pi^{3/2}} \| \cdot \|^{2m-1}$. All tensors will be denoted with bold capital letters (i.e. $\boldsymbol{X}$,...). Let $h > 0$, and $\bar{\imath}$ be such that : $\bar{\imath}^2 = -1$. $(e_i)_{i=1}^3$ denotes the canonical basis of $\mathbb{R}^3$. For $k \in \mathbb{N}$ and $i = 1, 2$, or 3, $\delta_{h,i}^k$ denotes the kth divided difference of step h defined by $\left(\delta_{h,i}^1 f\right)(x) = f\left(x + \frac{h}{2} e_i\right) - f\left(x - \frac{h}{2} e_i\right)$. $\delta_{h,i}^k = \delta_{h,i}^{(k-1)} \circ \delta_{h,i}^1$. We use standard multi-index notations. If $\alpha \in \mathbb{N}^3$ and $|\alpha| = \alpha_1 + \alpha_2 + \alpha_3$, then $D^\alpha = \frac{\partial^{\alpha_1}}{(\partial x_1)^{\alpha_1}} \frac{\partial^{\alpha_2}}{(\partial x_2)^{\alpha_2}} \frac{\partial^{\alpha_3}}{(\partial x_3)^{\alpha_3}}$, $x^\alpha = x_1^{\alpha_1} x_2^{\alpha_2} x_3^{\alpha_3}$.

§2. Discretization of $P_{m,\rho}(D)$

The goal of this section is to define an approximation of $P_{m,\rho}(D)$ which should reproduce polynomials of largest possible degree (in the sense defined below). We now introduce the $\mathbb{P}_k$-exactness of an operator, and a $\mathbb{P}_k$-exact approximation of D^α.

Definition 1. *An operator E' is said to be a $\mathbb{P}_k$-exact (resp. $\mathcal{P}_k$-exact) approximation of an operator E iff for any function f, $E'f$ is a linear combination of translates of f and for any $p \in \mathbb{P}_k$ (resp. $p \in \mathcal{P}_k$), $E'p = Ep$.*

Remark. $h^{-1}\partial^1_{h,i}$ *is a $\mathbb{P}_2$-exact approximation of $\frac{\partial}{\partial x_i}$.*

Definition 2 . *Let $D^1_{h,i,N}$ be the operator defined by*

$$D^1_{h,i,N} = \sum_{k=0}^{N} \frac{(-1)^k}{2h} \frac{(k!)^2}{(2k+1)!} \left(\delta^1_{2h,i} \circ \delta^{2k}_{h,i}\right).$$

For any $\alpha \in \mathbb{N}^3$, we define an approximation of step h and level N of D^α to be the operator defined by

$$D^\alpha_{h,N} = \left(D^1_{h,1,N}\right)^{\alpha_1} \circ \left(D^1_{h,2,N}\right)^{\alpha_2} \circ \left(D^1_{h,3,N}\right)^{\alpha_3}.$$

In the same way, $P_{m,\rho}(D_{h,N})$ denotes the approximation of $P_{m,\rho}(D)$.

Proposition 3. *Let $\alpha \in \mathbb{N}^3$ and $D^\alpha_{h,N}$ be defined as above. Then, for all mappings f from $\mathbb{R}^3$ to $\mathbb{R}$, there exists real constants (c_γ) such that*

$$\mathcal{F}\left(D^\alpha_{h,N} f\right) = \left(\frac{1}{h^{|\alpha|}} \sum_{|\gamma|=(2N+1)|\alpha|} c_\gamma \left(\sin(\pi h \cdot)\right)^\gamma\right) \mathcal{F}(f).$$

Proof: For every $1 \le i \le 3$,

$$\mathcal{F}\left(\delta^1_{h,i} f\right)(\zeta) = \left(\exp(\imath\pi h\zeta_i) - \exp(-\imath\pi h\zeta_i)\right) \cdot \mathcal{F}(f)(\zeta) = 2\imath \sin(h\zeta)_i \cdot \mathcal{F}(f)(\zeta)$$

By applying the Fourier's transform to the $(2N+1)|\alpha|$ centered differences of $D^\gamma_{h,N}$, we obtain the result. □

Proposition 4. *Let $\alpha \in \mathbb{N}^3$. Then*

i) *$D^\alpha_{h,N}$ is a $\mathbb{P}_{|\alpha|+2N+1}$-exact approximation of D^α.*

ii) *$P_{m,\rho}(D_{h,N})$ is an approximation of $P_{m,\rho}(D)$ which is $\mathcal{P}_{2m+2N+1}$-exact.*

Proof: We prove by induction on $|\alpha|$ that for all $f \in C^{2N+2+|\alpha|}$, there exist c_γ real constants such that

$$D^\alpha_{h,N} f = D^\alpha f + \sum_{|\gamma|=2N+2+|\alpha|} c_\gamma D^\gamma f(\zeta_\gamma).$$

In the following, for any $\gamma \in \mathbb{N}^3$, d_γ, e_γ are real constants. The proof for $|\alpha| = 1$ is due to Steffensen in ([7]). Suppose that it is true for $|\alpha| = m$. Let $|\alpha| = m+1$, $f \in C^{2N+2+m+1}$ and let us choose i such that $\alpha_i \neq 0$. Then

$$\begin{aligned}
D^{\alpha}_{h,N} f &= D^{1}_{h,i,N}\left(D^{\alpha-e_i}_{h,N} f\right) \\
&= D^{1}_{h,i,N}\left(D^{\alpha-e_i} f + \sum_{|\gamma|=2N+2+m} c_\gamma D^\gamma f(\zeta_\gamma)\right) \\
&= D^1_i D^{\alpha-e_i} f + \sum_{|\gamma|=2N+3} d_\gamma D^\gamma\left(D^{\alpha-e_i} f(\zeta_\gamma)\right) + \sum_{|\gamma|=2N+3+m} c_\gamma D^\gamma f(\zeta_\gamma) \\
&= D^\alpha f + \sum_{|\gamma|=2N+3+m} e_\gamma f(\zeta_\gamma).
\end{aligned}$$

Thus, we get the result if we note that $P_{m,\rho}(D)$ is a differential operator of degree $2m$. □

§3. B-spline Tensors Associated with $P_{m,\rho}(D)$

Lemma 5. *Let $\widetilde{v}_{m+1}$ be such that $(-1)^{m+1} \triangle^{m+1} \widetilde{v}_{m+1} = \delta$ in D'. Let*

$$\begin{aligned}
\widetilde{\boldsymbol{X}} &= \frac{(-1)^m}{\rho} \nabla^2 \widetilde{v}_{m+1} \\
&+ (-1)^m \begin{pmatrix} \left(\partial^2_{2,2} + \partial^2_{3,3}\right)\widetilde{v}_{m+1} & -\partial^2_{1,2}\widetilde{v}_{m+1} & -\partial^2_{1,3}\widetilde{v}_{m+1} \\ -\partial^2_{1,2}\widetilde{v}_{m+1} & \left(\partial^2_{1,1} + \partial^2_{3,3}\right)\widetilde{v}_{m+1} & -\partial^2_{2,3}\widetilde{v}_{m+1} \\ -\partial^2_{1,3}\widetilde{v}_{m+1} & -\partial^2_{2,3}\widetilde{v}_{m+1} & \left(\partial^2_{1,1} + \partial^2_{2,2}\right)\widetilde{v}_{m+1} \end{pmatrix}.
\end{aligned}$$

Then $P_{m,\rho}(D)\,\widetilde{\boldsymbol{X}} = \delta \boldsymbol{I}_3$.

Definition 6. *Let $\widetilde{\boldsymbol{X}}$ be a fundamental solution tensor of $P_{m,\rho}(D)\cdot\boldsymbol{X} = \delta\cdot\boldsymbol{I}_3$. We define the level N and step h B-spline tensor associated with the operator $P_{m,\rho}(D)$, to be the tensor $\boldsymbol{B}^m_{h,N,\rho}$ defined by $\boldsymbol{B}^{m,\rho}_{h,N} = h^3\, P_{m,\rho}(D_{h,N})\, \tilde{\boldsymbol{X}}$.*

Remarks.

a) *$\boldsymbol{B}^{m,\rho}_{h,N}$ is not a symmetrical tensor.*

b) *If m is even, we can prove the existence of a differential matrix $R_{\frac{m}{2},\rho}(D)$ of degree m such that : $P_{m,\rho}(D) = R_{\frac{m}{2},\rho}(D)\, R_{\frac{m}{2},\rho}(D)$ and we can construct a symmetrical tensor $\boldsymbol{C}^m_{h,N,\rho}$ defined by*

$$\boldsymbol{C}^{m,\rho}_{h,N} = h^3\, R_{\frac{m}{2},\rho}(D_{h,N})\, \tilde{\boldsymbol{X}}\, R_{\frac{m}{2},\rho}(D_{h,N}).$$

Lemma 7. *The elements of $\left(\boldsymbol{B}^{m,\rho}_{h,N}\right)_{1\leq i,j,\leq 3}$ are in the set of tempered distributions on $\mathbb{R}^3$.*

Denoting by $\mathcal{F}\left(\boldsymbol{B}_{h,N}^{m,\rho}\right)$ the Fourier transform applied to each element of the tensor $\boldsymbol{B}_{h,N}^{m,\rho}$, we obtain the following theorem:

Theorem 8. *With the above notation,*

i) $\mathcal{F}\left(\boldsymbol{B}_{h,N}^{m,\rho}\right)(0) = h^3 \cdot \boldsymbol{I}_3$,

ii) $\forall \gamma \in \mathbb{N}_*^3,\ |\gamma| \le 2N+1\ ;\ D^\gamma \mathcal{F}\left(\boldsymbol{B}_{h,N}^{m,\rho}(0)\right) = 0$,

iii) $\forall 1 \le j,l \le 3,\ \left(\boldsymbol{B}_{h,N}^{m,\rho}\right)_{j,l}(t) \underset{|t|\to+\infty}{=} \mathcal{O}\left(|t|^{-2N-5}\right)$,

iv) $\forall k \in \mathbb{Z}_*^3,\ \forall \gamma \in \mathbb{N}^3,\ |\gamma| \le 2m-1\ ;\ D^\gamma \mathcal{F}\left(\boldsymbol{B}_{h,N}^{m,\rho}\right)\left(\frac{k}{h}\right) = 0$,

v) $\boldsymbol{B}_{h,N}^{m,\rho} = \boldsymbol{B}_{1,N}^{m,\rho}\left(\frac{\cdot}{h}\right)$.

Proof: For i)–ii), we use Definition 2 and Proposition 3. By the Taylor expansion of sin, near 0 we obtain $\mathcal{F}\left(\boldsymbol{B}_{h,N}^{m,\rho}\right)(\zeta) = h^3 \cdot \boldsymbol{I}_3 + \boldsymbol{Q}(\zeta)$, where $\boldsymbol{Q}(\zeta)$ is such that for all i and j such that $1 \le i,j \le 3$, there exist real coefficients $c_\alpha^{i,j}$ such that $\boldsymbol{Q}_{i,j}(\zeta) = \sum_{|\alpha| \ge 2N+2m+4} c_\alpha^{i,j} \frac{\zeta^\alpha}{\|2\pi\zeta\|^{2m+2}}$. Thus, we obtain the results.

We prove iii). By using the above expression of $\boldsymbol{Q}$, $D^\gamma \mathcal{F}\left(\boldsymbol{B}_{h,N}^{m,\rho}\right)_{j,l}$ are integrable near 0 if and only if $2N+2-|\gamma| > -3$. Using Proposition 3 and Lemma 5, there exist real coefficients $d_\alpha^{i,j,k}$ and for all i and j such that $1 \le i,j \le 3$

$$\mathcal{F}\left(\boldsymbol{B}_{h,N}^{m,\rho}\right)_{i,j}(\zeta) = \sum_{k=1}^{3} \frac{1}{h^{2m}} \sum_{|\alpha|=2m(2N+1)} d_\alpha^{i,j,k} \frac{(\sin(\pi h \zeta))^\alpha}{\|2\pi\zeta\|^{2m+2}} \zeta_k \zeta_j.$$

As a consequence, the elements of $\mathcal{F}\left(\boldsymbol{B}_{h,N}^{m,\rho}\right)$ are bounded at infinity by rational fractions of degree $-2m$. $D^\gamma \mathcal{F}\left(\boldsymbol{B}_{h,N}^{m,\rho}\right)_{j,l}$ are integrable near infinity if and only if $-2m-|\gamma| < -3$. Then, $D^\gamma \mathcal{F}\left(\boldsymbol{B}_{h,N}^{m,\rho}\right)_{j,l}$ are integrable in $\mathbb{R}^3$ iff $2N+2-|\gamma| > -3$ and $-2m-|\gamma| < -3$. Thus, $D^\gamma \mathcal{F}\left(\boldsymbol{B}_{h,N}^{m,\rho}\right)_{j,l} \underset{|t|\to+\infty}{=} o\left(|t|^{-2N-4}\right)$, for all $1 \le j,l \le 3$. Using [4], the last expression may be strengthened to

$$D^\gamma \mathcal{F}\left(\boldsymbol{B}_{h,N}^{m,\rho}\right)_{j,l} \underset{|t|\to+\infty}{=} \mathcal{O}\left(|t|^{-2N-5}\right), \qquad 1 \le j,l \le 3.$$

We now establish iv). For all j in $\mathbb{Z}^3$, γ in $\mathbb{N}^3$ and positive h, we have $D^\gamma \mathcal{F}\left(\boldsymbol{B}_{h,N}^{m,\rho}\right)\left(\frac{j}{h}\right) = 0$ iff $D^\gamma \mathcal{F}\left(P_{m,\rho}(D_{h,N})\right)(0) = 0$. Now by using Proposition 3, we have, for all tensors $\boldsymbol{Y}$ and all γ in $\mathbb{N}^3$ such that $|\gamma| \le 2N+2m+1$:

$$D^\gamma\Big(\mathcal{F}(P_{m,\rho}(D_{h,N})\,\boldsymbol{Y})\Big)(0) = D^\gamma\Big(\mathcal{F}(P_{m,\rho}(D))(0) \cdot \mathcal{F}(\boldsymbol{Y})(0)\Big).$$

Furthermore, $\mathcal{F}(P_{m,\rho}(D))$ is a polynomial matrix of degree $2m$ so $\forall|\gamma| \leq 2m-1$, $D^\gamma \mathcal{F}(P_{m,\rho}(D))(0) = 0$. This gives iv).

Finally, we prove v). According to the Fourier transform's properties, v) is equivalent to

$$\mathcal{F}\left(\boldsymbol{B}_{h,N}^{m,\rho}\right)(\zeta) = \mathcal{F}\left(\boldsymbol{B}_{1,N}^{m,\rho}\left(\frac{\cdot}{h}\right)\right)(\zeta) = h^3 \cdot \mathcal{F}\left(\boldsymbol{B}_{1,N}^{m,\rho}\right)(h\cdot\zeta), \qquad \forall\zeta \in \mathbb{R}^3.$$

Now, from the definition of $\boldsymbol{B}_{h,N}^{m,\rho}$, we derive for any ζ in $\mathbb{R}^3$,

$$\begin{aligned}\mathcal{F}\left(\boldsymbol{B}_{1,N}^{m,\rho}\right)(h\cdot\zeta) &= \mathcal{F}\left(P_{m,\rho}(D_{1,N})\tilde{\boldsymbol{X}}\right)(h\cdot\zeta) \\ &= \mathcal{F}\Big(P_{m,\rho}(D_{1,N})\Big)(h\cdot\zeta)\cdot\mathcal{F}(P_{m,\rho})(h\cdot\zeta)^{-1}.\end{aligned}$$

Using Proposition 3, we have

$$\mathcal{F}(P_{m,\rho}(D_{1,N}))(h\cdot\zeta) = h^{2m}\mathcal{F}(P_{m,\rho}(D_{h,N}))(\zeta),$$

$$\mathcal{F}(P_{m,\rho}(D))^{-1}(h\cdot\zeta) = h^{-2m}\mathcal{F}(P_{m,\rho}(D))^{-1}(\zeta),$$

and thus $\forall\zeta \in \mathbb{R}^3$,

$$\begin{aligned}\mathcal{F}\left(\boldsymbol{B}_{1,N}^{m,\rho}\right)(h\cdot\zeta) &= \mathcal{F}(P_{m,\rho}(D_{h,N}))(\zeta)\cdot\mathcal{F}(P_{m,\rho}(D))^{-1}(\zeta) \\ &= h^{-3}\cdot\mathcal{F}\left(B_{h,N}^{m,\rho}\right)(\zeta). \quad \square\end{aligned}$$

Remarks.

a) *These polyharmonic B-spline tensors may be considered as a regularisation of the Dirac distribution tensor* $\delta\,\boldsymbol{I}_3$.

b) We obtain the same properties with $\boldsymbol{C}_{h,N}^{m,\rho}$.

§4. Associated Vector Quasi-Interpolant

Given vectorial data $(z_j)_{j\in\mathbb{Z}^3}$, in this section we define a vector field S approximating the data $(jh, z_j)_{j\in\mathbb{Z}^3}$ (i.e. such that $S(jh) \simeq z_j$ for all j in $\mathbb{Z}^3$), by using the above defined tensor B-splines. This vector generalizes the polyharmonic B-spline quasi-interpolant (see [5,6]).

Definition 9. *Let* $\boldsymbol{B}_{h,N}^{m,\rho}$ *be the level* N *and step* h *B-spline tensor associated with* $P_{m,\rho}(D)$. *For all* $j \in \mathbb{Z}^3$, *let* $z_j \in \mathbb{R}^3$, *and let* $z = (z_j)_{j\in\mathbb{Z}^3}$. *Then the* vector quasi-interpolant of step h and level N associated with the operator $P_{m,\rho}(D)$ and the $(jh, z_j)_{j\in\mathbb{Z}^3}$ data, *is the vector function defined by*

$$S_{h,N;z}^{m,\rho} = \sum_{j\in\mathbb{Z}^3} \boldsymbol{B}_{h,N}^{m,\rho}(\cdot - jh)\ z_j.$$

Theorem 10. *Let $l = \inf\{2N+1, 2m-1\}$, and suppose there exists $p \in \mathcal{P}_l$ such that for all j in $\mathbb{Z}^3$, $z_j = p(jh)$. Then, $S^{m,\rho}_{h,N;z} = p$. We say that the vector quasi-interpolant of step h and level N reproduces $\mathcal{P}_l$. As a particular case, $S^{m,\rho}_{h,m-1}$ reproduces $\mathcal{P}_{2m-1}$.*

Proof: The proof is based mainly on Poisson's egality and Theorem 8. It follows along the same lines as the proof in [3] in the scalar case. □

Remarks.

a) *$2m-1$ is the maximal order of reproduction.*

b) *We obtain the same properties if we define the vector quasi-interpolant using by the symmetrical tensor $C^{m,\rho}_{h,N}$.*

c) *A similar problem is studied in $\mathbb{R}^2$ in ([1]), where using another discretization of $\Gamma_{2,\rho}(D)$, the authors obtain a vector quasi-interpolant which is $\left(\mathbb{P}_1\left(\mathbb{R}^2\right)\right)^2$-reproducing.*

Theorem 11. *Let f be a vector function of $C\left(\mathbb{R}^3\right)^k$-class and all partial derivatives of order k being bounded over $\mathbb{R}^3$. Let $S^{m,\rho}_{h,N;f}$ be the above defined vector quasi-interpolant associated to the $\Big(jh, f(jh)\Big)_{j\in\mathbb{Z}^3}$ data. Then*

$$\sup_{t\in\mathbb{R}^3} \|S^{m,\rho}_{h,N;f}(t) - f(t)\| \underset{h\to 0}{=} \begin{cases} \mathcal{O}\left(h^k\right) & \text{if } k \le 2N+1, \\ \mathcal{O}\left(h^{2N+2}|\ln(h)|\right) & \text{if } k \ge 2N+2. \end{cases}$$

Proof: The proof follows that of Theorems 4.11, 5.1 and 5.6 in [3]. □

References

1. Amodei, L., and M. N. Benbourhim, A vector spline quasi-Interpolation, in *Wavelets, Images, and Surface Fitting*, P.-J. Laurent, A. Le Méhauté, and L. L. Schumaker (eds.), A. K. Peters, Wellesley MA, 1994, 1–10.
2. Dodu, F., Thesis to appear.
3. Jackson, I. R. H., Radial basis function methods for multivariate approximation, Thesis University of Cambridge, 1988.
4. Powell, M. J. D., The theory of radial basis function approximation in 1990, DAMTP report NA 11, University of Cambridge, 1990.
5. Rabut, C., How to build quasi-interpolant: Application to polyharmonic B-splines, in *Curves and Surfaces*, P.-J. Laurent, A. Le Méhauté, and L. L. Schumaker (eds.), Academic Press, New York, 1991, 391–402.
6. Rabut, C., B-splines polyharmoniques cardinales : interpolation, quasi-interpolation, filtrage, Thèse d'Etat, Université de Toulouse, 1990.
7. Steffensen, J. F., *Interpolation*, Chelsea publishing company, New-York, 1950.

Fabrice Dodu
Institut National des Sciences Appliquées de Toulouse
U.M.R. 9974 C.N.R.S./I.N.S.A./M.I.P.
Génie Mathématique et Modélisation
135 Avenue de Rangueil
31077 Toulouse Cedex 4
France
dodu@gmm.insa-tlse.fr

Analysis of Scalar Datasets on Multi-Resolution Geometric Models

Alexandre Gerussi and Georges–Pierre Bonneau

Abstract. Recently, multi-resolution methods based on non-nested spaces were introduced to allow the visualization and approximation of functions defined on irregular triangulations [3,4,5]. This paper comes back to these methods and shows more precisely how the subdivision/prediction /correction scheme of ordinary wavelet-based multi-resolution analysis (MRA)is also present in that framework. As an illustration, it is demonstrated how it can be applied in two of the classical issues of MRA: compression and level-of-detail editing. We also show that the framework can be used for the analysis and approximation of scalar data defined on meshes with arbitrary topology, thus extending our previous results in the plane and the sphere. Here again, the link with the corresponding classical multi-resolution scheme of [6] as well as decimation methods is made.

§1. Introduction

In the last few years, the problem of simplifying huge 3D triangular meshes, for the purpose of *e.g.*, visualization, transmission or storage, has received considerable attention. Among those works, two major approaches can be found. In the case of regular meshes, the use of a wavelet-based framework has proven to be a powerful solution [6,9,14]. On the other hand, when meshes are not regular, the approach has been to simplify the mesh by applying a sequence of elementary geometric simplification operations, such as vertex removals, edge collapses or triangle collapses, the order of removal being driven by a greedy algorithm [1,8,11]. We refer to this latter approach as a decimation approach.

This paper is concerned with the simplification of data that is defined on a surface by means of a triangulation. This topic is closely related to surface simplification, since a triangular mesh can be seen, at least locally, as the graph of a piecewise linear function supported by a triangulation. Here again, when the surface is well-known, for example a plane square, a sphere, a cylinder, *etc.*, several wavelet based approaches have been employed [9,14], with regular underlying meshes.

Curve and Surface Fitting: Saint-Malo 1999
Albert Cohen, Christophe Rabut, and Larry L. Schumaker (eds.), pp. 209–218.

ISBN 0-8265-1357-3.

In [3,4,5] the concept of non-nested MRA was introduced and applied to the approximation and progressive visualization of piecewise constant or linear functions defined on arbitrary planar or spherical meshes. In Section 2, we investigate the relationship between the non-nested framework and irregular subdivision. Examples of compression and level-of-detail editing in that framework are given in Section 3.

Section 4 focuses on the approximation and visualization of functions defined on triangular meshes with arbitrary topology. Here, an additional difficulty is that the surface which supports the function is also altered by the approximation process. Like in the case of surface simplification, a wavelet approach can successfully be applied when the original mesh has subdivision connectivity [6,12]. Otherwise, in an irregular setting, a decimation approach is usually employed, and the function is approximated during the simplification process [1,11]. We show how to apply our framework to functions defined on such general meshes. We will see that it makes the link between the wavelet-based approach available for subdivision surfaces and the decimation model. As in our previous papers, the function in its multi-resolution form is described by a coarse approximation defined on the simplified mesh, and a sequence of detail coefficients that are used for the reconstruction of the function on every LOD up to the original mesh. Our scheme is fully bijective: The function multi-resolution representation has the same size as the original one. The approximation process performs L_2 approximation of the data, but other types of approximation are also possible.

§2. Non-Nested Framework and Irregular Subdivision

In Section 2.1 we briefly review the non-nested decomposition scheme described in [3,5]. Section 2.2 makes the link with the notion of irregular subdivision.

2.1. Decomposition scheme

For simplicity, every space is supposed to have finite dimension. Let Ω be a measurable domain and V^i, $i = 0, \ldots, N$, a sequence of subspaces of $L_2(\Omega)$. These spaces do not have to be nested but will in general be"growing" in the sense that $\dim(V^i) \leq \dim(V^{i+1})$. Now let $f = f_N$ be a function in the finest subspace V^N. In classical MRA, the spaces are nested and the link between V^{N-1} and V^N is made by taking a complementary space W^{N-1} of V^{N-1} in V^N, that is

$$V^N = V^{N-1} \oplus W^{N-1}.$$

Now if we write $f_N = f_{N-1} + g_{N-1}$ according to the space decomposition, f_{N-1} can be seen as an approximation of f_N in V^{N-1}, and g_{N-1} as the detail needed to recover the original function from its approximation. By repeating the decomposition, one obtains

$$f_N = f_0 + g_0 + \cdots + g_{N-1}$$

which corresponds to the space decomposition $V^N = V^0 \oplus \bigoplus_0^{N-1} W^i$.

Notice that in this case $f_i = P_{V^i}(f_{i+1})$, where P_{V^i} is the projector on V^i with direction W^i. We return to the general case and suppose that a linear "projector" $P^i : V^{i+1} \to V^i$ is given. To avoid technical details, these projectors are required to be *surjective*, but the results in this paragraph also hold if this is not the case. Let W^i be the kernel of P^i, and $\tilde{V}^i$ be a complementary space of W^i in V^{i+1}. We now observe that the restriction of P^i to $\tilde{V}^i$ is a bijective operator, having the same range as P^i. Thus if $f_i = P^i(f_{i+1})$ and $g_i = Q^i(f_{i+1})$, where Q^i is the projector on W^i (defined by the choice of $\tilde{V}^i$), the following reconstruction formula holds:

$$f_{i+1} = \mathrm{Inv}(P^i_{|\tilde{V}^i})(f_i) + g_i. \tag{1}$$

Again, by iterating this decomposition, we obtain a coarse approximation $f_0 = P^0 \circ \cdots \circ P^{N-1}(f_N)$ and "detail" functions $g_0, \ldots, g_{N-1}$.

We now take a look at the reconstruction process. Denote by S^i the inverse of $P^i_{|\tilde{V}^i}$, $S^i : V^i \to \tilde{V}^i \subset V_{i+1}$. The complete coarse-to-fine reconstruction formula is obtained by iterating the reconstruction formula (1):

$$f_N = S^{N-1} \circ \cdots \circ S^0(f_0) + S^{N-1} \circ \cdots \circ S^1(g_0) + \cdots + S^{N-1}(g_{N-2}) + g_{N-1}. \tag{2}$$

2.2. Approximating spaces, subdivision spaces and scaling spaces

We are going to see that the previous scheme can actually be considered in two different ways. Until now, it was implicitly assumed that the spaces V^i were playing the role of the scaling spaces in classical MRA. Under this assumption, we conceptually have a really *non-nested* framework; if the spaces were nested, the operators S^i would be the identity (formally injecting V^i in V^{i+1}). However, we will not call them scaling spaces but approximation spaces, and keep the term "scaling" for other spaces that are going to be defined below.

In the non-nested framework, one loses the notion of subdivision (or cascade algorithm). However, looking at things slightly differently allows subdivision to fit into the non-nested scheme. To show this, consider the operators S^i as *subdivision* operators, and call the spaces V^i subdivision spaces accordingly. This means that we start from a function $f_0 \in V^0$ and iteratively subdivide it into f_1, $f_2, \ldots f_N$ using the formula $f_{i+1} = S^i(f_i)$. The notion of subdivision used here is very general. In that context, classical regular or semi-regular subdivision schemes would give rise to a nested sequence of subdivision spaces. But completely irregular schemes would require the non-nested framework to be fitted in. The use of non-nested MRA was introduced in [2], and was later applied to triangular schemes in [3,4,5]. Recently, another approach on general irregular schemes was proposed in the work of Sweldens and Guskov [10,13].

We now define the scaling spaces. Like in classical MRA, they are the spaces containing the limit functions resulting from the subdivision process. Since very little is known about the convergence of such schemes, we won't push the subdivision to infinity, but restrict ourselves to an integer N. This makes sense since, when using such a framework, one usually starts from an initial triangulation $\mathcal{T}_N$ which is *coarsened* to $\mathcal{T}_0$. This contrasts with the traditional approach in subdivision where one starts from a base mesh and subdivides it according to a systematic rule.

The scaling spaces, for $i = 0, \ldots, N$, are defined as

$$V^{i,N} = \{f_k^{i,N} \mid f_k \in V^i\},$$

where $f_k^{i,N} = S^{i,N}(f_k)$, and $S^{i,N} = S^{N-1} \circ \cdots \circ S^i$. This operator carries functions in V^i through $N - i$ subdivision steps to functions in $V^{i,N}$.

Notice that, for every i, $V^{i,N} \subset V^N$ and moreover, that $V^{i,N} \subset V^{i+1,N}$. Now fix a basis $(\varphi_\cdot^i)$ for V^i and $(\psi_\cdot^i)$ for W^i. If $(a_\cdot^i)$ and $(b_\cdot^i)$ denote the coordinates of $f_i \in V^i$ and $g_i \in W^i$ with respect to these basis, the reconstruction formula can be re-written as

$$f_N = \sum_k a_k^j \varphi_k^{j,N} + \sum_{i=j}^{N-1} \sum_k b_k^i \psi_k^{i,N},$$

for any $j = 0, \ldots, N$, which corresponds to the decomposition of f_N in $V^{N,N} = V^N$ (using the same notations as above: $\varphi_k^{j,N} = S^{j,N}(\varphi_k^j)$, and $\psi_k^{i,N} = S^{i+1,N}(\psi_k^i)$).

Although formally identical, considering the V^i's as approximation or subdivision spaces changes the aspect of several questions. For example, in the problem of error measure in the context of approximation spaces, we are interested in $\|f_N - f_i\|$, whereas in the other context we are looking for $\|f_N - f_i^{i,N}\|$.

§3. Application to Data Compression and LOD Editing

In this section we show how the framework can be used on functions defined over irregular triangulations to achieve data compression and LOD editing, which are both standard applications of MRA. The context here is the planar or spherical setting of [4,5], only the filters need to be changed. Indeed, the analysis operator used in those papers was the orthogonal projector (the goal being progressive visualization). However, for compression it is often useful to know in advance the error between the original function and its approximation, in terms of the wavelet coefficients that were used in the reconstruction. This will be achieved by designing new filters.

3.1. Isometric subdivision

In order for the detail coefficients to have the error measure property, the synthesis operator is required to be an isometry. Indeed, suppose that

- $S^i : V^i \to V^{i+1}$ is an isometric operator, $\forall i = 0, \ldots, N-1$,
- the complementary spaces $\tilde{V}^i$ are chosen orthogonal to the W^i, $\forall i = 0, \ldots, N-1$.

When the latter condition is fulfilled, we say that we are in a semi-orthogonal framework. If f_N denotes the original function, then, according to the global reconstruction formula (2), the quantity $\|f_N - f_0^{0,N}\|^2$ which measures the contribution of the correction steps in the reconstruction process is

$$\|S^{N-1} \circ \cdots \circ S^1(g_0) + \cdots + S^{N-1}(g_{N-2}) + g_{N-1}\|^2.$$

Because of semi-orthogonality, it is equal to

$$\|S^{N-1} \circ \cdots \circ S^1(g_0) + \cdots + S^{N-1}(g_{N-2})\|^2 + \|g_{N-1}\|^2,$$

and because S^{N-1} is an isometric operator, we can factorize and remove it from the first term above, and then iterate the operation to get

$$\|f_N - f_0^{0,N}\|^2 = \sum_{i=0}^{N-1} \|g_i\|^2.$$

Notice that even if the subdivision operators are not isometric,

$$\|f_N - f_0^{0,N}\|^2 \leq \|S^{N-1}\|^2 \cdots \|S^1\|^2 \|g_0\|^2 + \cdots + \|S^{N-1}\|^2 \|g_{N-2}\|^2 + \|g_{N-1}\|^2$$

still holds in the semi-orthogonal setting. Let $\widehat{f}_N$ denote the partially reconstructed function, and suppose in addition that we are in an orthonormal framework, that is, the functions $\psi^i_\cdot$ form an orthonormal basis of W^i. Proceeding as above leads to

$$\|f_N - \widehat{f}_N\|^2 = \sum_{i=0}^{N-1} \sum_k \varepsilon_k^i (b_k^i)^2,$$

where the b_k^i's are the wavelet coefficients and ε_k^i equals 1 whenever b_k^i is taken in the reconstruction and 0 otherwise. Consequently, in this setting we have an error measure in terms of wavelet coefficients.

3.2. An isometric subdivision operator

The idea behind this construction is that the corresponding analysis operator should have reasonable approximation quality, which seems intuitively required to achieve compression. Accordingly, the projection operator used in [4,5] is taken as a starting point, the problem being to approximate it by means of an isometry. Let $P^i : V^{i+1} \to V^i$ be that operator. The first step is to find the matrix of P^i with respect to some orthonormal basis $(e^{i+1}_\cdot)$ and $(e^i_\cdot)$ of V^{i+1} and V^i. Now let UDV be the singular value decomposition of that matrix. This decomposition admits the following interpretation:

- V is the matrix of an isometric operator of V^{i+1} in the basis $(e^{i+1}_\cdot)$ since it is unitary and the basis is orthonormal.

- D is a $\ell \times c$ matrix ($\ell < c$) whose diagonal coefficients are all non-negative and ≤ 1. The diagonal matrix formed by the positive entries of D is the matrix of a bijective operator mapping $\tilde{V}^i$ onto $\mathrm{Ran}(P^i)$.
- Like V, the matrix U is an isometric operator of V^i.

Let $\tilde{D}$ be the matrix obtained by replacing every positive diagonal element of D by 1. This amounts to turning the bijective operator above into an isometric one, and thus $U\tilde{D}V$ is the matrix of an operator $\tilde{P}^i$ whose corresponding subdivision operator S^i is isometric. A few remarks can be made about this construction:

- $\tilde{P}^i$ does not depend on a particular choice of orthonormal basis.
- $\tilde{P}^i$ is not the best isometric approximation of P^i in terms of L_2 norm of operators, but it can be shown that it is the best with respect to the Frobenius norm.
- The diagonal coefficient of D are by definition the cosines of the angles between the spaces V^i and V^{i+1} (see, *e.g.*, [7] Chapter 1). In the nested case, they would all be equal to 1, and the corresponding subdivision operator would be the identity.
- This method could be used to approximate operators by means of similarities, by replacing the entries of D by an appropriate scalar α instead of 1. Although better in terms of approximation quality, this leads to bad visual results since the resulting subdivision operator doesn't reproduce constants if $\alpha \neq 1$.

Notice that in the context of [4,5], this approximation is always computed locally, leading to a global algorithm in linear time.

3.3. Examples

As it is mentioned in the beginning of this section, these examples were created using the setting described in [3,4,5]; the reader is invited to look there for details. The initial triangulation is completely irregular, generated by random vertex insertion. In Figure 1, a piecewise linear setting is used for LOD editing. The function is edited at a coarse resolution, by pulling values up ($\rightarrow$ white) at some vertices, and then adding detail coefficients back. Figure 2 shows an example of data compression when the approximation spaces are spaces of piecewise constant functions over triangulations generated from the original one by homogeneous decimation. This last setting is the full generalization of Haar wavelets to irregular grids, as it would lead to them in the regular case.

§4. Scalar Datasets on Irregular Meshes with Arbitrary Topology

In this section we describe how the non-nested framework can be used to handle scalar attributes defined on meshes with any topology. As a starting point, we take a multi-resolution decimation model, based on the vertex–removal (VR) operation to simplify the geometry. This means we assume that an initial fine mesh is given along with its associated sequence of VR's

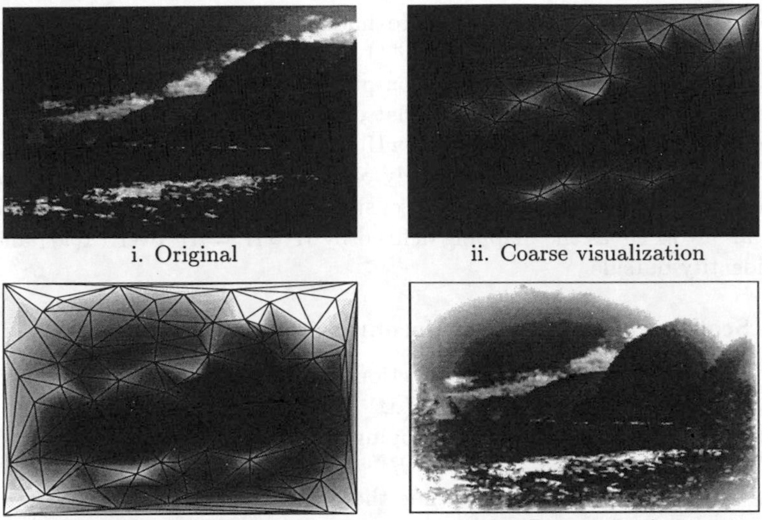

i. Original ii. Coarse visualization

iii. 50 points edited iv. Reconstruction

Fig. 1. Picture design using level-of-detail editing.

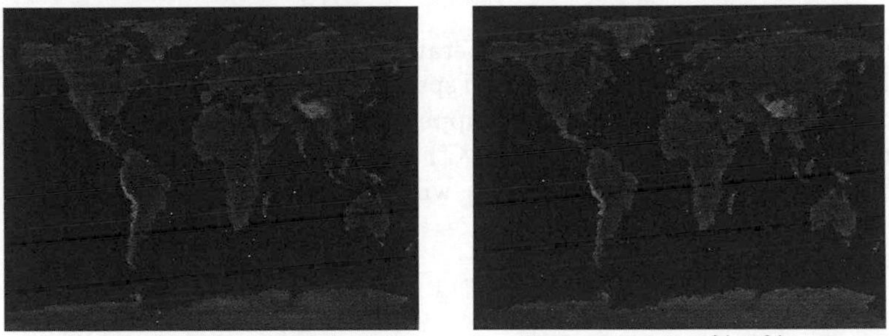

Original: 50000 triangles Compressed to 5%, 6% error

Fig. 2. Compression of a piece-wise constant function.

that can be progressively applied to decimate the mesh down to a base mesh. In addition, we suppose that the scalar attributes are defined by means of piecewise constant and/or linear functions parameterized on the initial mesh. In what follows, PI($\mathcal{V}$) denotes the polygon of influence of a vertex $\mathcal{V}$; it is the polygonal area delimited by the 1–neighbours of $\mathcal{V}$. In order to apply the non-nested framework, approximation spaces and approximation operators need to be defined.

4.1. Local mapping

Let M^i, $i = N_0, \ldots, N$, denote the triangular mesh consisting of i vertices (the original mesh after $N - i$ VR operations). Let $\mathcal{F}(M^i)$ be the space of real-valued functions defined on M^i, and $\mathcal{C}^i$ (resp. $\mathcal{L}^i$) be the subspace of functions of $\mathcal{F}(M^i)$ that are piecewise constant (resp. linear) on each triangle of M^i. We refer to these spaces as the upper approximation spaces. Each VR alters locally the surface, thus functions of $\mathcal{F}(M^{i+1})$ and $\mathcal{F}(M^i)$ are defined on different domains. To define an approximation problem, a common parameterization for these functions is required. To that end, we assume that for every VR of vertex $\mathcal{V}_i$, a *local one to one projection* Π^i of $\mathrm{PI}(\mathcal{V}_i)$ onto a plane is also known. The reader can refer to [8] for a study of the existence and determination of such a projection. We use Π^i to consider the *change of parameterization* $H^i : M^i \to M^{i+1}$ as the mapping defined by $\Pi^i \circ H^i = \Pi^i$ over $\mathrm{PI}(\mathcal{V}_i)$ and by the identity outside.

4.2. Scaling spaces and data decomposition

Let $\mathcal{K}$ stand for $\mathcal{C}$ or $\mathcal{L}$. To a function $f \in \mathcal{K}^i$ we associate the function $f^{i,i-1} = f \circ H^{i-1}$, and let $\mathcal{K}^{i,i-1} \subset \mathcal{F}(M^{i-1})$ be the space of all these functions. This can be iterated: The local mappings also define a *global* mapping from the base mesh to the mesh M^i by $H^{N_0,i} = H^{i-1} \circ \cdots \circ H^{N_0}$, for each $i = N_0 + 1, \ldots, N$. This allows to define the approximation spaces $\bar{\mathcal{K}}^i$ from the upper approximation spaces as

$$\bar{\mathcal{K}}^i = \{f \circ H^{N_0,i} \mid f \in \mathcal{K}^i\} \subset \mathcal{F}(M^{N_0}).$$

The second step is to define the operators $\bar{P}^i : \bar{\mathcal{K}}^{i+1} \to \bar{\mathcal{K}}^i$. Fortunately, working directly in the approximation spaces is not required: Because they are isomorphic (by construction) to the upper approximation spaces, it suffices to define some operators $P^i : \mathcal{K}^{i+1,i} \to \mathcal{K}^i$, and for the purpose of visualization, the least-square projection operators will be used. The operator $\bar{P}^i$ is thus defined by

$$\bar{P}^i : \bar{f}_{i+1} = f_{i+1} \circ H^{N_0,i+1} \longmapsto P^i(f_{i+1}^{i+1,i}) \circ H^{N_0,i}.$$

$f_{i+1} \in \mathcal{K}^{i+1}$ defines $f_{i+1}^{i+1,i}$ to which we apply one decomposition step to get an approximation $f_i \in \mathcal{K}^i$ and detail coefficients (1 in the linear case, and 2 in the constant case). The entire operation is then repeatedly applied to f_i, f_{i-1}, *etc.* Consequently, in practice, everything happens in the upper approximation spaces which have a much simpler structure than the corresponding approximation spaces.

4.3. Results and remarks

The output of this algorithm is a coarse function f_{N_0} defined on the base mesh and a list of detail coefficients that allow the exact reconstruction of the original function through the hierarchy of LODs. Figure 3 shows some results in the linear setting. In these examples, the geometric criterion guiding the decimation priority-queue is simply the distance from the 1–neighbours of a candidate vertex to their least-square approximation plane. On the upper right snapshot, we see the drawback of a geometric-only driven priority queue: Some quasi-planar areas on the object have been severely decimated, leading

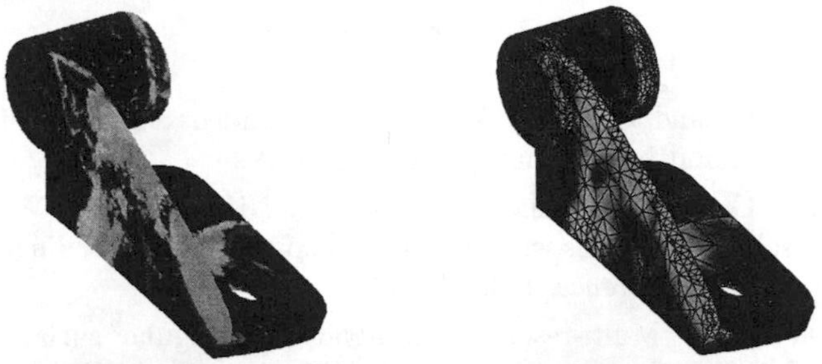

Original: 32000 vertices Geometric criterion: 3000 v.

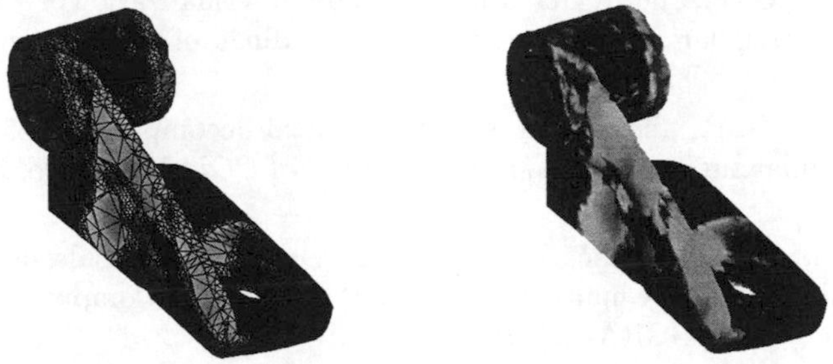

Geom. & data criterion: 3000 v. Same without vertices

Fig. 3. Bracket: linear approximation examples.

to a quite coarse approximation of the function. The lower snapshots show the result using the geometric($\frac{2}{3}$) and data($\frac{1}{3}$) based criterion. The resulting approximations are better, but the geometry presents some visible deformations (this is also partially due to our simple geometric criterion). Finding a compromise in an automated way seems to be a difficult task. Moreover, if the simplification is just a process prior to other computations, such as can be the case, *e.g.* in mechanics, then a high accuracy in the approximated function might be the primary interest. Thus, it seems better to let the weights depend on the application, under user control.

4.4. Comparison to classical MRA

In [6], a MRA for subdivision surfaces is used to handle both the geometry and the scalar attributes of a mesh. The presentation given above makes the link between these methods and the decimation approach. Indeed, from the "upper" point of view — the decomposition using the operators P^i — it compares to decimation in many respects, whereas it is also a decimation step corresponding to the approximation spaces, which is exactly what is done is [6] in the nested case. However, parameterizations are then obtained without a local projection hypothesis, thanks to the particular 1-to-4 splitting strategy that is performed on the base mesh.

Acknowledgments. We would like to thank Jean-Claude Leon from the laboratory of Mechanic 3S, Grenoble, for providing us with the 3D meshes.

References

1. Bajaj, C. L., and D. R. Schikore, Error bounded reduction of triangle meshes with multivariate data, SPIE, 2656, 1996.
2. Bonneau, G.-P., S. Hahmann, and G. M. Nielson, BLaC-Wavelets: a multi-resolution analysis with non-nested spaces, in *Proceedings of IEEE Visualization Conference*, 1996, 43–48.
3. Bonneau, G.-P., Multi-resolution analysis on irregular surface meshes, IEEE TVCG **4**, Number 4 (1998), 365–378.
4. Bonneau, G.-P., and A. Gerussi, Level of detail visualization of scalar data sets on irregular surface meshes, in Proceedings of IEEE Visualization Conference, 1998, 73–77.
5. Bonneau, G.-P., and A. Gerussi, Hierarchical decomposition of datasets on irregular surface meshes, in *Proceedings of CGI Conference*, 1998, 59–63.
6. Certain, A., J. Popović, T. DeRose, T. Duchamp, D. Salesin, and W. Stuetzle, Interactive multi-resolution surface viewing, Computer Graphics Proceedings (SIGGRAPH), 1996, 91–98.
7. Chatelin, F., *Valeurs Propres de Matrices*, col. Mathématiques Appliquées pour la Maîtrise, MASSON, 1989.
8. Cohen, J., D. Manosha, and M. Olano, Simplifying polygonal models using successive mappings, in *Proceedings of IEEE Visualization Conference*, 1997.
9. Gross, M. H., R. Gatti, and O. Staadt, Fast multi-resolution surface meshing, Comp. Sc. Dept. internal report no. 230, 1995.
10. Guskov, I., Multivariate subdivision schemes and divided differences, preprint, Princeton University, 1998.
11. Hoppe, H., Progressive meshes, Computer Graphics Proceedings, (SIGGRAPH), 1996, 99–108.
12. Kolarov, K., and W. Lynch, Compression of functions defined on surfaces of 3D objects, in *Proceedings of IEEE Data Compression Conference*, 1997.
13. Sweldens, W., P. Schröder, and I. Guskov, Multi-resolution signal processing for meshes, Computer Graphics Proceedings (SIGGRAPH), 1999.
14. Sweldens, W., and P. Schröder, Spherical wavelets: efficiently representing functions on the sphere, Computer Graphics Proceedings (SIGGRAPH), 1995, 161–172.

Alexandre Gerussi, Georges–Pierre Bonneau
LMC–IMAG, Université Joseph Fourier
BP 53, 38041 Grenoble cedex 9, France
`[Alexandre.Gerussi, Georges-Pierre.Bonneau]@imag.fr`

Biorthogonal Refinable Spline Functions

Tim N. T. Goodman

Abstract. We give a construction for refinable spline functions of degree n with compact support and simple knots in $\frac{1}{4}\mathbb{Z}$ which are biorthogonal to uniform B-splines of degree n with simple knots at $\frac{1}{3}\mathbb{Z}$.

§1. Introduction

A function is refinable if it is a linear combination of dilates of integer translates of itself. Such functions are central to multiresolution methods, in particular in the construction of wavelets. In general, refinable functions can be defined implicitly from the refinement equation which they satisfy, but explicit constructions of refinable functions are restricted mainly to spline functions, i.e. piecewise polynomials. If we require the natural condition that the integer translates of the univariate refinable spline function ϕ with compact support form a Riesz basis, then ϕ can only be a uniform B-spline with simple knots [4]. However there is more flexibility if we replace the single function ϕ by a refinable vector of spline functions $(\phi_1, \ldots, \phi_r)$. For a survey on refinable spline functions, see [2].

In multiresolution methods, orthogonality plays an important role. In [1], constructions are given for refinable functions whose integer translates are biorthogonal to a given refinable function ϕ, in particular when ϕ is a uniform B-spline with simple knots. However these dual functions are not defined explicitly. We give, in Section 3, constructions for refinable spline functions of compact support which are biorthogonal to uniform B-splines with simple knots. This requires refinable vectors of three functions: the uniform B-splines have knots in $\frac{1}{3}\mathbb{Z}$, while the dual functions have the same degree and simple knots in $\frac{1}{4}\mathbb{Z}$. The construction is based on a general result in Section 2 giving necessary and sufficient conditions for biorthogonality of certain vectors of (not necessarily refinable) functions in terms of a Grammian matrix.

Curve and Surface Fitting: Saint-Malo 1999
Albert Cohen, Christophe Rabut, and Larry L. Schumaker (eds.), pp. 219–226.

ISBN 0-8265-1357-3.

§2. Biorthogonal Basic Sets

Let $\phi_1, \ldots, \phi_r$ be compactly supported real-valued functions in $L^2(\mathbb{R})$. We say $\{\phi_1, \ldots, \phi_r\}$ is a basic set for a space V if V comprises all real, finite, linear combinations of integer translates of $\phi_1, \ldots, \phi_r$.

We say basic sets $\{\phi_1, \ldots, \phi_r\}$ and $\{\psi_1, \ldots, \psi_r\}$ are biorthogonal (or the basic set $\{\psi_1, \ldots, \psi_r\}$ is dual to $\{\phi_1, \ldots, \phi_r\}$) if

$$\int_{-\infty}^{\infty} \phi_i \psi_j(. - k) = \delta_{ij}\delta_{0k}, \quad i, j = 1, \ldots, r, \qquad k \in \mathbb{Z}.$$

A basic set $\{\phi_1, \ldots, \phi_r\}$ is said to be stable if $\{\phi_i(. - j) : i = 1, \ldots, r, j \in \mathbb{Z}\}$ forms a Riesz basis, i.e. for some A, $B > 0$,

$$A \sum_{i=1}^{r} \sum_{j=-\infty}^{\infty} a_{ij}^2 \leq \int_{-\infty}^{\infty} [\sum_{i=1}^{r} \sum_{j=-\infty}^{\infty} a_{ij}\phi_i(. - j)]^2 \leq B \sum_{i=1}^{r} \sum_{j=-\infty}^{\infty} a_{ij}^2$$

for any $a_{ij} \in \mathbb{R}$, $i = 1, \ldots, r$, $j \in \mathbb{Z}$.

It is shown in [3] that $\{\phi_1, \ldots, \phi_r\}$ is stable if and only if for each u in $\mathbb{R}$, there are integers $k_1, \ldots, k_r$ with

$$\det \left[\hat{\phi}_i(u + 2\pi k_j)\right]_{i,j=1}^{r} \neq 0. \tag{1}$$

We shall say a matrix $M(z) = [M(z)_{ij}]_{i,j=1}^{r}$ of Laurent polynomials is invertible if it has an inverse which is a matrix of Laurent polynomials, i.e.

$$\det M(z) = az^l, \quad \text{some } a \neq 0, \quad l \in \mathbb{Z}.$$

Lemma 1. *If $\{\phi_1, \ldots, \phi_r\}$ is a stable basic set for V, then $\{\psi_1, \ldots, \psi_s\}$ in V also forms a stable basic set for V if and only if $s = r$ and*

$$\psi_i = \sum_{j=1}^{r} \sum_{k=-\infty}^{\infty} A_{ij}(k)\phi_j(. - k), \quad i = 1, \ldots, r, \tag{2}$$

where the matrix of Laurent polynomials

$$A(z) := \left[\sum_{k=-\infty}^{\infty} A_{ij}(k)z^k\right]_{i,j=1}^{r}$$

is invertible.

Before proving this lemma, it will be useful to introduce the following vector notation. For a basic set $\{\phi_1, \ldots, \phi_r\}$, we let ϕ denote the column vector $(\phi_1, \ldots, \phi_r)^T$. Then we can write (2) as

$$\psi = \sum_{k=-\infty}^{\infty} A_k \phi(. - k), \tag{3}$$

where A_k is the matrix $[A_{ij}(k)]_{i,j=1}^r$. Taking Fourier transforms then gives

$$\hat{\psi}(u) = A(z)\hat{\phi}(u), \tag{4}$$

where $z = e^{-iu}$.

Proof of Lemma 1: Suppose that (2) holds, where $A(z)$ is invertible. From (4) we have

$$\hat{\phi}(u) = A(z)^{-1}\hat{\psi}(u).$$

Since $A(z)^{-1}$ is a matrix of Laurent polynomials, it follows that $\phi_1, \ldots, \phi_r$ are finite linear combinations of integer translates of $\psi_1, \ldots, \psi_r$. Since V comprises all finite linear combinations of integer translates of $\phi_1, \ldots, \phi_r$, it follows that $\{\psi_1, \ldots, \psi_r\}$ is a basic set for V.

Also for any u in $\mathbb{R}$, we may choose integers $k_1, \ldots, k_r$ so that (1) holds, and thus from (4),

$$\det[\hat{\psi}_i(u + 2\pi k_j)]_{i,j=1}^r = \det A(z) \det[\hat{\phi}_i(u + 2\pi k_j)]_{i,j=1}^r \neq 0.$$

So $\{\psi_1, \ldots, \psi_r\}$ is stable.

Conversely suppose that $\{\psi_1, \ldots, \psi_s\}$ is a stable basic set for V. Then there exist an $s \times r$ matrix $A(z)$ and an $r \times s$ matrix $B(z)$ of Laurent polynomials such that for $z = e^{-iu}$,

$$\hat{\psi}(u) = A(z)\hat{\phi}(u)$$

and

$$\hat{\phi}(u) = B(z)\hat{\psi}(u) = B(z)A(z)\hat{\phi}(u).$$

For any $u \in \mathbb{R}$ we may choose integers $k_1, \ldots, k_r$ so that (1) holds. Since

$$[\hat{\phi}_i(u + 2\pi k_j)]_{i,j=1}^r = B(z)A(z)[\hat{\phi}_i(u + 2\pi k_j)]_{i,j=1}^r,$$

it follows that $B(z)A(z) = I_r$, the $r \times r$ identity matrix. Similarly $\hat{\psi}(u) = A(z)B(z)\hat{\psi}(u)$ and since $\{\psi_1, \ldots, \psi_s\}$ is stable, we can deduce as above that $A(z)B(z) = I_s$. Thus $s = r$ and $A(z)$ is invertible. □

Theorem 2. *Suppose that $\{\phi_1, \ldots, \phi_r\}$ and $\{\psi_1, \ldots, \psi_r\}$ are stable basic sets for V and W respectively. For $k \in \mathbb{Z}$, $i, j = 1, \ldots, r$, we define*

$$M_{ij}(k) := \int_{-\infty}^{\infty} \phi_i \psi_j(\cdot - k), \tag{5}$$

and let M denote the $r \times r$ matrix of Laurent polynomials given by

$$M(z) := \left[\sum_{k=-\infty}^{\infty} M_{ij}(k) z^k \right]_{i,j=1}^r . \tag{6}$$

Then there exist biorthogonal basic sets for V and W if and only if M is invertible. Moreover in this case, for any stable basic set for V there is a unique dual stable basic set for W.

Proof: Let $\{\tilde{\phi}_1, \dots, \tilde{\phi}_r\}$ and $\{\tilde{\psi}_1, \dots, \tilde{\psi}_r\}$ be any stable basic sets for V and W respectively. Then by Lemma 2.1 we have

$$\tilde{\phi} = \sum_{k=-\infty}^{\infty} A_k \phi(. - k), \quad \tilde{\psi} = \sum_{k=-\infty}^{\infty} B_k \psi(. - k), \tag{7}$$

where A_k, B_k are $r \times r$ matrices such that

$$A(z) := \sum_{k=-\infty}^{\infty} A_k z^k, \quad B(z) := \sum_{k=-\infty}^{\infty} B_k z^k, \tag{8}$$

are invertible.

For $k \in \mathbb{Z}$, $i, j = 1, \dots, r$, define

$$\tilde{M}_{ij}(k) = \int_{-\infty}^{\infty} \tilde{\phi}_i \tilde{\psi}_j(. - k), \tag{9}$$

and let $\tilde{M}$ denote the $r \times r$ matrix of Laurent polynomials given by

$$\tilde{M}(z) := \left[\sum_{k=-\infty}^{\infty} \tilde{M}_{ij}(k) z^k \right]_{i,j=1}^{r}. \tag{10}$$

Then from (2.5)-(2.10) we have for $z = e^{-iu}$,

$$\tilde{M}(z) = A(z) M(z) B(z)^*, \tag{11}$$

where $B(z)^* = \overline{B(z)}^T = B(z^{-1})^T$.

Now by (2.9) and (2.10), $\tilde{\phi}$ and $\tilde{\psi}$ are biorthogonal if and only if $\tilde{M}(z) = I$. If this holds, then by (11), M must be invertible. Conversely, if M is invertible, then for any choice of $A(z)$ we can define $B(z)$ uniquely by

$$B(z) = (M(z)^{-1} A(z)^{-1})^*$$

so that (11) holds with $\tilde{M}(z) = I$. Thus if M is invertible, then for any stable basic set $\tilde{\phi}$ for V, there is a unique dual basic set $\tilde{\psi}$ for W. □

We remark that from the definition, any biorthogonal basic sets must have linearly independent integer translates. Thus if M as in Theorem 2.1 is invertible, any basic set for V or W which is stable must in fact have linearly independent integer translates.

Now for $m \in \mathbb{Z}$, $m \geq 2$, we say a space V of functions on $\mathbb{R}$ is m-refinable if

$$f \in V \Rightarrow f\left(\frac{\cdot}{m}\right) \in V.$$

If ϕ is a basic set for an m-refinable space V, then $\phi(\frac{\cdot}{m})$ is in V, and so

$$\phi(\frac{\cdot}{m}) = \sum_{k=-\infty}^{\infty} C_k \phi(. - k)$$

for some finite set of matrices C_k. Thus, ϕ satisfies the refinement equation

$$\phi = \sum_{k=-\infty}^{\infty} C_k \phi(m. - k).$$

In this paper we shall, for simplicity, consider only the case $m = 2$.

§3. Biorthogonal Refinable Splines

For any integer $n \geq 1$ we let N_n denote the uniform B-spline of degree n with simple knots at $0, 1, \ldots, n+1$. We now fix n, and define

$$\phi_1(x) = N_n(3x), \quad \phi_2(x) = N_n(3x - 1), \quad \phi_3(x) = N_n(3x - 2).$$

Then $\{\phi_1, \phi_2, \phi_3\}$ is a stable basic set for the space V of all spline functions of degree n with compact support and simple knots at $\frac{1}{3}\mathbb{Z}$. Clearly V is refinable. We wish to find a refinable space W of spline functions of compact support which has a basic set which is dual to $\{\phi_1, \phi_2, \phi_3\}$. From Theorem 2 we see that this is equivalent to finding a basic set $\{\psi_1, \psi_2, \psi_3\}$ for W such that the matrix M in (6) is invertible.

We shall choose

$$\psi_1(x) = N_n(2x), \quad \psi_2(x) = N_n(2x - 1),$$

and ψ_3 to be a spline function of degree n with knots in $\frac{1}{4}\mathbb{Z}$ and support in $[\frac{1}{4}, 2n - \frac{1}{4}]$. Thus W is a space of spline functions of degree n and simple knots in $\frac{1}{4}\mathbb{Z}$, which contains all spline functions of degree n with compact support and simple knots in $\frac{1}{2}\mathbb{Z}$. For any function f in W, $f(\frac{\cdot}{2})$ has knots in $\frac{1}{2}\mathbb{Z}$ and so lies in W. Thus W is refinable.

Theorem 3. *We can choose ψ_3 as above so that there are biorthogonal basic sets for V and W, or equivalently that there is a unique basic set for W dual to the basic set $\{\phi_1, \phi_2, \phi_3\}$.*

Proof: The space W_0 of spline functions of degree n with simple knots in $\frac{1}{2}\mathbb{Z}$ and support in $[0, 2n]$ has dimension $3n$, while the space W_1 of spline functions of degree n with simple knots in $\frac{1}{4}\mathbb{Z}$ and support in $[0, 2n]$ has dimension $7n$. Thus we may choose linearly independent functions $f_1, \ldots, f_{4n}$ in W_1 with support in $[\frac{1}{4}, 2n - \frac{1}{4}]$ which together with W_0 span W_1. We write

$$\psi_3 = \sum_{k=1}^{4n} a_k f_k.$$

It remains to choose $a_1,\ldots,a_{4n}$ so that M in (6) is invertible. Now $M_{ij}(k)=0$ except in the following cases: $M_{11}(k)$, $-\frac{n}{2}\le k\le\frac{n}{3}$; $M_{12}(k)$, $-\frac{n+1}{2}\le k\le\frac{n-1}{3}$; $M_{13}(k)$, $-2n+1\le k\le\frac{n}{3}$; $M_{21}(k)$, $-\frac{n}{2}\le k\le\frac{n+1}{3}$; $M_{22}(k)$, $-\frac{n+1}{2}\le k\le\frac{n}{3}$; $M_{23}(k)$, $-2n+1\le k\le\frac{n+1}{3}$; $M_{31}(k)$, $-\frac{n-1}{2}\le k\le\frac{n+2}{3}$; $M_{32}(k)$, $-\frac{n}{2}\le k\le\frac{n+1}{3}$; $M_{33}(k)$, $-2n+1\le k\le\frac{n+2}{3}$. Then

$$\det M(z)=\sum_{k=-3n+1}^{n} b_k z^k,$$

for some numbers b_k, $-3n+1\le k\le n$, which are linear functions of $a_1,\ldots,a_{4n}$. The condition $\det M(z)\equiv 1$ then gives $4n$ linear equations in $4n$ unknowns $a_1,\ldots,a_{4n}$ and we shall show that this system is non-singular.

Suppose, to the contrary, that the system is singular. Then we may choose $a_1,\ldots,a_{4n}$, not all zero, so that $\det M(z)\equiv 0$. Then the columns of M are linearly dependent in the sense that there are Laurent polynomials p_1, p_2, p_3, not all zero, so that

$$\sum_{j=1}^{3} M(z)_{ij}p_j(z)\equiv 0,\quad i=1,2,3.$$

Writing

$$p_j(z)=\sum_{k=-\infty}^{\infty} c_j(k)z^k,\quad j=1,2,3,$$

this becomes

$$\sum_{j=1}^{3}\sum_{k=-\infty}^{\infty} M_{ij}(k)z^k\sum_{l=-\infty}^{\infty} c_j(l)z^l\equiv 0,\quad i=1,2,3,$$

i.e.

$$\sum_{j=1}^{3}\sum_{l=-\infty}^{\infty} M_{ij}(l)c_j(k-l)=0,\quad i=1,2,3,\quad k\in\mathbb{Z},$$

which on recalling (5) gives

$$\int_{-\infty}^{\infty}\phi_i(.-k)f=0,\quad i=1,2,3,\quad k\in\mathbb{Z},$$

where

$$f=\sum_{j=1}^{3}\sum_{l=-\infty}^{\infty} c_j(-l)\psi_j(.-l).$$

Thus f is orthogonal to V. Note that since the integer translates of ψ_1, ψ_2, ψ_3 are linearly independent, f is not identically zero. Since only a finite number of coefficients $c_j(-l)$, $j=1,2,3$, $l\in\mathbb{Z}$, are non-zero, f has compact support. Suppose that the support is $[\alpha, \beta + 2n]$, α, $\beta \in \mathbb{Z}$, but not in $[\alpha + 1, \beta + 2n]$ or $[\alpha, \beta + 2n - 1]$. It is easily seen that $\alpha \leq \beta$ and

$$f = \sum_{j=\alpha}^{\beta} c_j \psi_3(. - j) + g,$$

where $c_\alpha \neq 0 \neq c_\beta$ and g has support in $[\alpha, \beta + 2n]$ and lies in the space of splines of degree n with knots in $\frac{1}{2}\mathbb{Z}$, which we shall denote by Z.

We first note that for some h_1 in Z with support in $[\beta + \frac{3n}{2} - 1, \beta + 2n]$,

$$\eta_1 := c_\beta \psi_3(. - \beta) + h_1$$

is orthogonal to those elements of V with support in $[\beta + 2n - 1, \infty)$. Now

$$f - \frac{1}{c_\beta} \sum_{j=\alpha}^{\beta-1} c_j \eta_1(. - j + \beta)$$

is orthogonal to those elements of V with support in $[\beta + 2n - 2, \infty)$ and on this interval coincides with a function

$$\eta_2 := c_\beta \psi_3(. - \beta) + h_2,$$

where h_2 is in Z with support in $[\beta + \frac{3n}{2} - 2, \beta + 2n]$. Continuing in this way we recursively construct

$$\eta_j := c_\beta \psi_3(. - \beta) + h_j, \quad j = 1, \ldots, 4n,$$

which is orthogonal to those elements of V with support in $[\beta + 2n - j, \infty)$, where h_j is in Z with support in $[\beta + \frac{3n}{2} - j, \beta + 2n]$. In particular, η_{4n} has support in $[\beta - \frac{5n}{2}, \beta + 2n]$ and is orthogonal to those elements of V with support in $[\beta - 2n, \infty)$.

Choose F with $F^{(n+1)} = \eta_{4n}$ and with support in $(-\infty, \beta + 2n]$. Now for $j = 0, \ldots, 12n - 1$, let B_j be the B-spline of degree n with knots $\beta - 2n + \frac{j}{3}$, $\beta - 2n + \frac{j+1}{3}, \ldots, \beta - 2n + \frac{j+n+1}{3}$. Then since B_j is in V,

$$\begin{aligned} 0 = \int_{-\infty}^{\infty} B_j \eta_{4n} &= \int_{-\infty}^{\infty} B_j F^{n+1} \\ &= \left[\beta - 2n + \frac{j}{3}, \beta - 2n + \frac{j+1}{3}, \ldots, \beta - 2n + \frac{j+n+1}{3}\right] F. \end{aligned}$$

Thus F vanishes at $\beta - 2n + \frac{j}{3}$, $j = 0, \ldots, 12n - 1$. Now F coincides on $[\beta - 2n, \beta + 2n]$ with a spline G of degree $2n+1$ with support $[\beta - 3n - \frac{1}{2}, \beta + 2n]$ with knots at

$$\beta - 3n - \frac{1}{2}, \beta - 3n, \ldots, \beta - \frac{1}{2}, \beta, \beta + \frac{1}{4}, \beta + \frac{1}{2}, \ldots, \beta + 2n - \frac{1}{4}, \beta + 2n.$$

It then follows from the Schoenberg-Whitney Theorem [5] that G vanishes identically on $[\beta - 2n, \beta + 2n]$ and hence so does η_{4n}. So η_{4n} has support in $[\beta - \frac{5n}{2}, \beta - 2n]$ with knots in $\frac{1}{2}\mathbb{Z}$, and so η_{4n} vanishes identically, which is a contradiction. Thus the linear system is non-singular, which completes the proof. □

Finally we note that if ψ_3 is as in Theorem 3 and M is given by (6), then the basic set for W dual to $\{\phi_1, \phi_2, \phi_3\}$ is $\{\tilde{\psi}_1, \tilde{\psi}_2, \tilde{\psi}_3\}$ given by

$$\tilde{\psi} := \sum_{k=-\infty}^{\infty} C_k \psi(. - k),$$

where

$$\sum_{k=-\infty}^{\infty} C_k z^k = (M(z)^{-1})^* = (\text{adj } M(z))^* = (\text{adj } M(z^{-1}))^T.$$

Acknowledgments. This work was undertaken during a visit to the National University of Singapore funded by the Wavelets Strategic Research Programme. Our thanks go to Prof. S. L. Lee both for arranging the visit and for helpful discussions.

References

1. Cohen, A., I. Daubechies, and J.-C. Feauveau, Biorthogonal bases of compactly supported wavelets, Comm. Pure Appl. Math. **45** (1992), 485–560.
2. Goodman, T. N. T., Refinable spline functions, in *Approximation Theory IX, Vol. 2: Computational Aspects*, Charles K. Chui and Larry L. Schumaker (eds.), Vanderbilt University Press, Nashville, 1998, 71–96.
3. Jia, R. Q., and C. A. Micchelli, On linear independence of integer translates of a finite number of functions, Proc. Edinburgh Math. Soc. **36** (1992), 69–85.
4. Lawton, W., S. L. Lee, and Z. Shen, Complete characterization of refinable splines, Advances in Comp. Math. **3** (1995), 137–145.
5. Schoenberg, I. J., and A. Whitney, On Pólya frequency functions III: The positivity of translation determinants with an application to the interpolation problem by spline curves, Trans. Amer. Math. Soc. **74** (1953), 246–259.

Tim N. T. Goodman
Dept. of Mathematics
The University
Dundee DD1 4HN, U.K.
tgoodman@mcs.dundee.ac.uk

Fitting Parametric Curves to Dense and Noisy Points

A. Ardeshir Goshtasby

Abstract. Given a large set of irregularly spaced points in the plane, an algorithm for partitioning the points into subsets and fitting a parametric curve to each subset is described. The points could be measurements from a physical phenomenon, and the objective in this process could be to find patterns among the points and describe the phenomenon analytically. The points could be measurements from a geometric model, and the objective could be to reconstruct the model by a combination of parametric curves. The algorithm proposed here can be used in various applications, especially where given points are dense and noisy.

§1. Introduction

In many science and engineering problems there is a need to fit a curve or curves to an irregularly spaced set of points. Curve fitting has been studied extensively in Approximation Theory and Geometric Modeling, and there are numerous books on the subject [1,5,6,12,23]. Existing techniques typically find a single curve segment that approximates or interpolates the given points. Many techniques assume that the points are ordered and fit a curve to them by minimizing an error criterion [3,7,8,14,16,22,27,29,31,34]. If the points are ordered, piecewise polynomial curves can also be fitted to them [19,30]. Difficulties arise when the points are not ordered.

To fit curves to an irregularly spaced set of points, 1) the set should be partitioned into subsets, 2) the points in each subset should be ordered, and 3) a curve should be fitted to points in each subset. This paper will provide solutions to the first two problems; that is, partitioning a point set into subsets and ordering the points in each subset. Once the points in each subset are ordered, existing techniques can be used to find the curves.

Given a large set of irregularly spaced points in the plane, $\{\mathbf{p}_i = (x_i, y_i) : i = 1, \ldots, N\}$, we would like to fit one or more parametric curves to the points, with the number of the curves to depend on the organization of the

Curve and Surface Fitting: Saint-Malo 1999
Albert Cohen, Christophe Rabut, and Larry L. Schumaker (eds.), pp. 227–236.

ISBN 0-8265-1357-3.

points and the resolution of the representation. When fitting a parametric curve to an irregularly spaced set of points, the main problem is to find the nodes of the curve. The nodes of a parametric curve determine the adjacency relation between the points and order them. The curve will then approximate the points in the order specified. Methods to order sparse points [11,17,24] as well as dense points [25,26,32] have been developed. Existing methods, however, fit a single curve segment to an entire data set. Sometimes it is not desirable to fit a single curve segment to a large and complex point set, and it is necessary to represent the geometric structure present in the point set by many curve segments. In this paper it will be shown how to partition a point set into subsets and how to fit a parametric curve to each subset. A new method to order a set of dense and noisy points for curve fitting will also be presented.

In the proposed model, a radial field is centered at each point such that the strength of the field monotonically decreases as one moves away from the point. The sum of the fields has the averaging effect and reduces the effect of noise, and local maxima of the sum of the fields has the effect of tracing the spine of the points. Therefore, we will use the local maxima of the sum of the fields (the ridges of the obtained field surface) as an approximation to the curves to be determined. Based on the organization of the points, disjoint ridges may be obtained, each suggesting a curve. The ridges will be used to partition the points into subsets and fit a curve to each subset. In the following, the steps of this process are described in detail.

§2. Approach

A desirable property of an approximating curve is for it to pass as close as possible to the given points while providing a certain smoothness appearance. For a dense point set, the curve cannot pass close to all the points, so it is desired that the curve trace the spine of the points. In the model proposed here, an initial estimation to a curve is obtained by taking points in the xy plane whose sum of inverse distances to the given points is locally maximum. That is, if the sum of inverse distances of point (x, y) to given points $\{(x_i, y_i) : i = 1, \ldots, N\}$ is larger than the sum of inverse distances of points in the neighborhood of (x, y) to the given points, then point (x, y) is considered an initial estimation to a point on the curve. Therefore, by tracing points in the xy plane that locally maximize

$$f(x,y) = \sum_{i=1}^{N} \left[(x - x_i)^2 + (y - y_i)^2 + 1\right]^{-\frac{1}{2}}, \tag{1}$$

we find an approximation to the curves we want to find.

The function f can also be interpreted as follows: Suppose a radial field of strength 1 is centered at point (x_i, y_i), $i = 1, \ldots, N$, such that the strength of the field decreases with inverse distance as one moves away from the point. Then, the strength of the field at point (x, y) will be $\left[(x - x_i)^2 + (y - y_i)^2\right]^{-\frac{1}{2}}$, and the curves to be found can be considered points in the xy plane whose sum of field values are locally maximum.

Once a set of points is given, the function f becomes fixed, and the obtained ridges will have a fixed shape. In order to have control over the shape or smoothness of obtained ridges, we revise formula (1) as follows. If instead of inverse distances defined by $[(x-x_i)^2+(y-y_i)^2+1]^{-\frac{1}{2}}$, we use

$$[(x-x_i)^2+(y-y_i)^2+r^2]^{-\frac{1}{2}} \qquad (2)$$

in equation (1), we obtain

$$g(x,y)=\sum_{i=1}^{N}[(x-x_i)^2+(y-y_i)^2+r^2]^{-\frac{1}{2}}. \qquad (3)$$

The basis functions defined by (2) are known as inverse multiquadrics [13]. The parameter r of the basis functions can be varied to generate different surfaces [21]. Figure 1b shows the field surface obtained when using the points of Fig. 1a and inverse multiquadric basis functions with $r=5$.

Instead of inverse multiquadric basis functions, other radial basis functions [2,4,10,28,33,35] also can be used to define function g. The choice of the basis functions influences the shape of the obtained field surface, the shape of the obtained ridges, and, consequently, the shape of the obtained curves.

By tracing the local maxima of the field surface g in the xy plane, we will obtain an approximation to the curves. Parameter r changes the shape of the basis functions and affects the shape of the field surface.

Local maxima of surface g can result in structures that contain branches and loops. The proposed model, therefore, can recover very complex patterns in dense and noisy point sets. Note also that the proposed method does not require any knowledge about the adjacency relation between the points. This method, in fact, provides the means to determine the adjacency relation between the points.

§3. Implementation

Derivation of an analytic formula that represents the local maxima of the surface g may not be possible. Digital approximation to the local maxima, however, is possible. This approximation is found in the form of digital contours and is used to partition the points into subsets. To digitally trace surface ridges, the surface is digitized into a digital image. The digitization process involves starting from $x=x_{min}$ and $y=y_{min}$ and incrementing x and y by some small increment δ until reaching $x=x_{max}$ and $y=y_{max}$. For each discrete (x,y), the value for $g(x,y)$ is then found from formula (3). x_{min} and x_{max} could be the smallest and largest x coordinates, and y_{min} and y_{max} could be the smallest and largest y coordinates of the given points. The parameter δ is used as the increment for both x and y because radially symmetric basis functions are used to define g. This parameter determines the resolution of the obtained image. For a finer resolution, this parameter should be reduced, while for a coarser resolution this parameter should be increased. If this parameter is to be chosen automatically, it should be selected such that most given points map to unique pixels in the obtained image.

Digitizing the surface g in this manner will result in a digital image whose pixel values show uniform samples from surface g. Figure 1b shows digitization of a field surface into an image of 256×256 pixels. To find the image ridges, pixels with locally maximum intensities are located. To find locally maximum image intensities, the gradient magnitude and the gradient direction [20] of the image at each pixel are determined. The gradient direction at a pixel is the direction at which change in intensity at the pixel is maximum, and gradient magnitude is the magnitude of the intensity change in the gradient direction at the pixel.

To find the ridges, we find each pixel A in the image where two pixels B and C that are adjacent to it and are at its opposite sides have intensities that are smaller than that at A. Assuming that the image obtained after digitizing surface g is represented by I, we mark the pixel at (i, j) as A if one of the following is true:

$$I(i-1,j) < I(i,j) \quad \& \quad I(i+1,j) < I(i,j); \tag{4}$$
$$I(i,j-1) < I(i,j) \quad \& \quad I(i,j+1) < I(i,j); \tag{5}$$
$$I(i-1,j-1) < I(i,j) \quad \& \quad I(i+1,j+1) < I(i,j); \tag{6}$$
$$I(i-1,j+1) < I(i,j) \quad \& \quad I(i+1,j-1) < I(i,j). \tag{7}$$

Using the image of Fig. 1b, we find that pixels in the contours shown in Fig. 1c are marked as A. We will call the contours obtained in this manner the minor ridges of the image. Next, we find each pixel D whose value is not only larger than those of B and C adjacent to it and at its opposite sides, but which also has a gradient direction that is the same as the direction obtained by connecting pixels B and C. The gradient direction at a pixel is quantized with 45-degree steps to ensure that only directions that are possible to obtain when connecting pixels B and C in an image are obtained. The pixels marked as D are shown in Fig. 1d. We will call these contours the major ridges of the image. As can be observed, major ridges are a subset of minor ridges. We also see that major ridge points do not fall on small and noisy branches of the minor ridges but rather fall on contours that represent the spines of the points. If the minor ridges are cut at the branch points, and branches that do not contain a major ridge point are removed, and the remaining contours are thinned, we obtain Fig. 1e. The obtained contours will be called the local-maxima contours, or simply the contours. These contours will be taken as approximations to the curves to be found. We will use them not only to partition the points into subsets but also to order the points in the subsets.

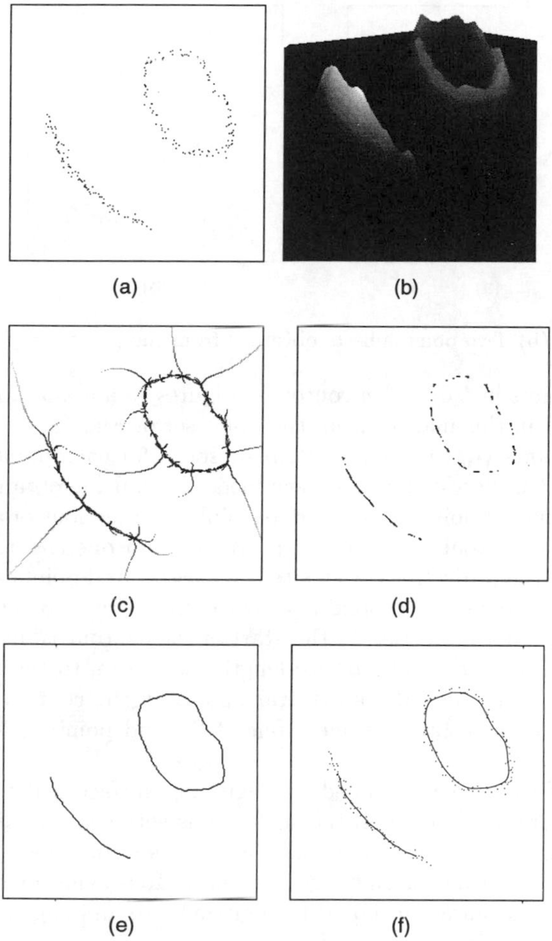

Fig. 1. (a) An irregularly spaced set of points. (b) A digitized field surface. (c) Contours representing the minor ridges. (d) Contours representing the major ridges. (e) Local-maxima contours. (f) RaG curves with $\sigma = 0.04$ approximating points shown in (a).

§4. Node Estimation

The method outlined in the preceding section determines contours that are approximations to the curves to be found. These contours will be used to partition a point set into subsets and order the points in each subset.

Suppose a point set has produced m contours; then, a point is assigned to contour j ($1 \le j \le m$) if it is closest to a pixel in contour j than to a pixel in any other contour. In this manner, a point is assigned to one of m contours. This process, when completed, will partition a point set into m subsets by

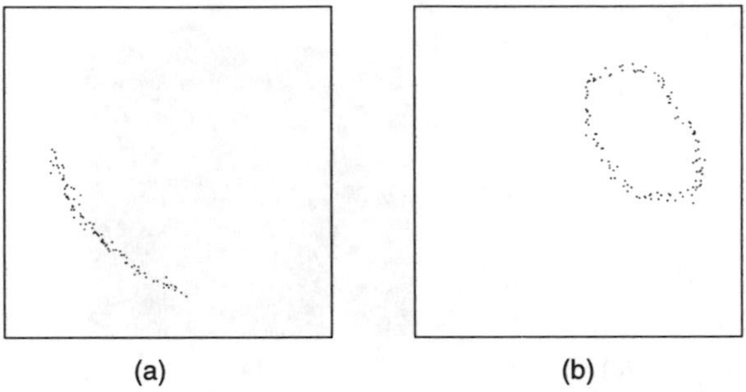

Fig. 2. (a), (b) Two point subsets obtained from the point set of Fig. 1a.

assigning the points into one of m contours. Figures 2a and 2b show the point subsets obtained in this manner from the point set of Fig. 1a.

To order points $\{\mathbf{q}_i : i = 1, \ldots, n\}$ in subset j, for each point $\mathbf{q}_i$ a point in contour j that is closest to it is determined. We call the obtained contour point the projection of point $\mathbf{q}_i$. After determining projections of all points in the subset to the contour, the contour is traced from one end to the other, and in the order the projections are visited, the associated points are ordered.

Since the contours are approximations to the curves to be found, the contour length from a projection to the start of the contour is divided by the length of the contour to obtain an arc-length estimation to the node of the point. If the contour is closed, an arbitrary point on the contour is taken as the start point. If the contour is open, one of the end points is taken as the start point.

The size of the image obtained by digitizing surface g determines the accuracy of the obtained nodes. If the surface g is very coarsely digitized, the obtained contours will be very short, and numerous points may produce the same node, especially when given points are dense. To provide a more accurate node estimation, the surface g should be digitized into an image large enough to produce unique nodes.

Once the coordinates of given points and the associated nodes are known, a parametric curve can be fitted to the points by one of the existing methods [9,11,16,18,30]. Fitting rational Gaussian (RaG) curves [9] to the points shown in Fig. 1a with nodes as determined above, we obtain the curves shown in Fig. 1f. The curves are overlaid with the original points to show the quality of the curve fitting. Note that these curves were obtained using the points in Fig. 1a and not the contour points in Fig. 1e. The contour points were used only to partition a point set into subsets and to determine the nodes of the points.

§5. Observations

To observe the behavior of the proposed curve-fitting method, results on three additional point sets are shown in Fig. 3. Figure 3a shows noisy points along

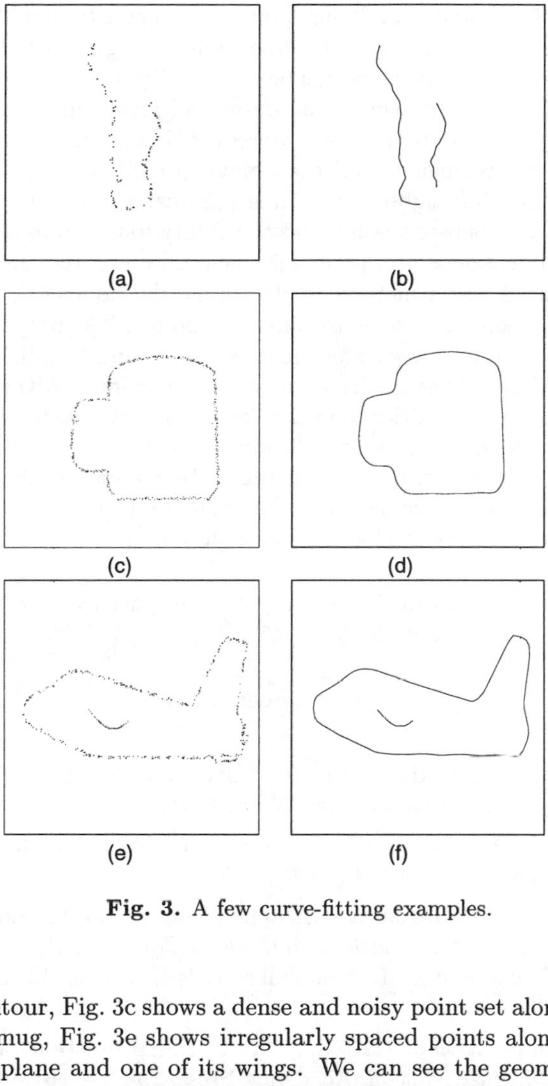

Fig. 3. A few curve-fitting examples.

an open contour, Fig. 3c shows a dense and noisy point set along the silhouette of a coffee mug, Fig. 3e shows irregularly spaced points along the silhouette of a model plane and one of its wings. We can see the geometric structures in these point sets and, if asked, can trace the structures manually without any difficulty. The algorithm proposed here is intended to do the same. The curves obtained are shown in Figs. 3b, 3d, and 3f.

The point sets shown in Fig. 3 did not contain geometric structures with branches and loops. If a point set contains branches and loops, the local-maxima contours will also contain branches and loops. A single curve segment, however, cannot represent branching structures. The solution we propose is to segment a complex contour into simple ones by cutting it at the branch points and fitting a curve to each branch.

§6. Summary and Conclusions

A large number of techniques for fitting parametric curves to irregularly spaced points have been developed. These techniques fit a single curve to the given points and often require that the points be ordered. In science and engineering problems that deal with measurement data, the given points may not be ordered and they may contain noise. Moreover, it may not be appropriate to fit a single curve segment to all the points. In this paper, a method to partition a point set into subsets and fit a parametric curve to each subset was described. The proposed method has the ability to take into consideration the noisiness and denseness of a point set when obtaining the curves.

Also introduced was a method to determine the nodes of a parametric curve that approximates a set of dense and noisy points. The proposed method provides the means to fit any parametric curve, including B-Splines and Non-Uniform Rational B-Splines, to irregularly spaced points. Although in this paper only inverse multiquadrics were used as basis functions to obtain a field surface, from which the curve segments were determined, other radial basis functions [33] can be used in the same manner. Depending on the parametric curve formulation and the radial basis functions used, the number and the shapes of the curves fitting to a set of points may vary.

Acknowledgments. This work was supported in part by a grant from the National Science Foundation: IRI-9529045.

References

1. Beach, R. C., *An Introduction to the Curves and Surfaces of Computer-Aided Design,* Van Nostrand, New York, 1991.
2. Buhmann, M. D., Multivariate cardinal interpolation with radial basis functions., Constr. Approx. **6** (1990), 225–255.
3. Cohen, E. and C. L. O'Dell, A data dependent parametrization for spline approximation, in *Mathematical Methods in Computer Aided Geometric Design,* T. Lyche and L. L. Schumaker (eds.), Vanderbilt Univ. Press, 1989, 155–166.
4. Dyn, N., Interpolation of scattered data by radial functions, in *Topics in Multivariate Approximation,* Academic Press, 1987, 47–61.
5. Farin, G., *Curves and Surfaces for Computer Aided Geometric Design,* Academic Press, New York, 1988.
6. Faux, I. D. and M. J. Pratt, *Computational Geometry for Design and Manufacture,* Ellis Horwood, Chichester, 1979.
7. Foley, T. A. and G. M. Nielson, Knot selection for parametric spline interpolation, in *Mathematical Methods in Computer Aided Geometric Design,* T. Lyche and L. L. Schumaker (eds.), Academic Press, New York, 1989, 261–271.

8. Fritsch, F. N. and R. E. Carlson, Monotone piecewise cubic interpolation, SIAM J. Numer. Anal. **17**, 2 (1980), 238–246.
9. Goshtasby, A., Geometric modeling using rational Gaussian curves and surfaces, Computer-Aided Design **27**, 5 (1995), 363–375.
10. Goshtasby, A. and W. D. O'Neill, Surface fitting to scattered data by a sum of Gaussians, Comput. Aided Geom. Design **10** (1993), 143–156.
11. Grossman, M., Parametric curve fitting, Computer J. **14**, 2 (1970), 169–172.
12. Hagen, H., *Curve and Surface Design,* SIAM, 1992.
13. Hardy, R. L., Theory and applications of the multiquadrics–biharmonic method, Comput. Math. Appl. **19**, 8/9 (1990), 163–208.
14. Hartley, P. J. and C. J. Judd, Parametrization and shape of B-spline curves for CAD, Computer-Aided Design **12**, 5 (1980), 235–238.
15. Hölzle, G. E., Knot placement for piecewise polynomial approximation of curves, Computer-Aided Design **15**, 5 (1983), 295–296.
16. Hoschek, J., Approximate conversion of spline curves, Comput. Aided Geom. Design **4** (1987), 59–66.
17. Hoschek, J., Intrinsic parametrization for approximation, Comput. Aided Geom. Design **5** (1988), 27–31.
18. Hoschek, J., Spline approximation of offset curves, Comput. Aided Geom. Design **5** (1988), 33–40.
19. Ichida, K. and T. Kiyono, Curve fitting by a one-pass method with a piecewise cubic polynomial, ACM Trans. Math. Software **3**, 2 (1977), 164–174.
20. Jain, A. K., *Fundamentals of Digital Image Processing,* Prentice Hall, 1989, 347–349.
21. Kansa, E. J., Multiquadrics–A scattered data approximation scheme with applications to computational fluid dynamics–I, Comput. Math. Appl. **19**, 8/9 (1990), 127–145.
22. Kosters, M., Curvature-dependent parametrization of curves and surfaces, Computer-Aided Design **23**, 8 (1991), 569–578.
23. Lancaster, P. and K. Salkauskas, *Curve and Surface Fitting,* Academic Press, New York, 1986.
24. Lee, E. T. Y., Choosing nodes in parametric curve interpolation, Computer-Aided Design **6** (1989), 363–370.
25. Lee, I.-K., Curve reconstruction from unorganized points, Comput. Aided Geom. Design **17** (2000), 161–177.
26. Levin, D., The approximation power of moving least-squares, Math. Comp. **67** (1998), 1517–1531.
27. Marin, S. P., An approach to data parametrization in parametric cubic spline interpolation problem, J. Approx. Theory **41** (1984), 64–86.

28. Meinguet, J., An intrinsic approach to multivariate spline interpolation at arbitrary points, *Polynomial and Spline Approximation,* B. N. Sahney (ed.), D. Reidel Publishing, 1979, 163–190.
29. Mullineux, M., Approximating shapes using parametrized curves, IMA J. Applied Mathematics **29** (1982), 203–220.
30. Piegl, L., A technique for smoothing scattered data with conic sections, Computers in Industry **9** (1987), 223–237.
31. Plass, M. and M. Stone, Curve-fitting with piecewise parametric cubics, Computer Graphics **17**, 3 (1983), 229–239.
32. Pottmann, H. and T. Randrup, Rotational and helical surface approximation for reverse engineering, Computing **60** (1998), 307–322.
33. Powell, M. J. D., Radial basis functions for multivariable interpolation: A review, *Algorithms for Approximation,* J. C. Mason and M. G. Cox (eds.), Clarendon Press, Oxford, 1987, 143–167.
34. Sarkar, B. and C-H Menq, Parameter optimization in approximating curves and surfaces in measurement data, Comput. Aided Geom. Design **8** (1991), 267–290.
35. Schagen, I. P., The use of stochastic processes in interpolation and approximation, Int. J. Computer Math., Section B, **8** (1980), 63–76.

A. Ardeshir Goshtasby
Department of Computer Science and Engineering
Wright State University
Dayton, OH 45435, USA
agoshtas@cs.wright.edu
http://www.cs.wright.edu/~agoshtas/

Smooth Irregular Mesh Interpolation

Stefanie Hahmann, Georges-Pierre Bonneau,
and Riadh Taleb

Abstract. The construction of a smooth surface from an irregular mesh in space is considered. The mesh vertices can either be interpolated or approximated as a control net. A collection of triangular Bézier patches results from a local, affine invariant and visually smooth interpolation scheme that can represent surfaces of arbitrary topological type. It is based on a domain 4-split. Beside the surface construction scheme, the optimal employment of the numerous degrees of freedom is crucial for an overall pleasing shape. Different local minimum norm criteria are tested to see if they produce satisfactory shapes.

§1. Introduction

The general problem of constructing a parametric triangular G^1 continuous surface interpolating an irregular mesh in space has been considered by many authors. In [5] a survey of such schemes is given, and it is concluded that local polynomial interpolants have similar shape defects due to the absence of an optimization strategy for using the free parameters (a special solution has been proposed in [6] for one of these schemes).

A different method has been developed by Loop [4] producing a collection of patches that meet each other with G^1 continuity. The vertex enclosure problem, which occurs when joining with G^1 continuity an even number of polynomial patches around a vertex, is solved by first constructing C^2-consistent boundary curves and cross-boundary tangents and then filling in the patches. In one-to-one correspondence to the mesh faces, sextic triangular Bézier patches are constructed, which lead to a very small number of degrees of freedom. One scalar value per vertex controls the length of the tangents of the boundary curves at the end points and one control point per patch is free. This is not enough for sufficient control of the shape of a sextic patch. In [5], it was stated that well-shaped boundary curves are a necessary condition. Loop's scheme doesn't provide any influence on the second derivatives of the boundary curves, which can lead to undulations. It was therefore proposed to

Curve and Surface Fitting: Saint-Malo 1999
Albert Cohen, Christophe Rabut, and Larry L. Schumaker (eds.), pp. 237–246.

ISBN 0-8265-1357-3.

relax the interpolation condition, which leads to an extra free parameter per vertex controlling the distance from the patch vertices to their corresponding mesh vertices. This is clearly improving the shape.

Recently, another triangular interpolation scheme has been developed [3]. A regular domain 4-split leads to the construction of four quintic Bézier patches which form a macro-patch in one-to-one correspondence to a mesh face. They have one polynomial degree less than Loop's scheme, but one degree more than Piper's or Shirman-Séquin's method [8,9]. The domain 4-split is a new approach in triangular mesh interpolation, and has several obvious advantages: the boundary curves and cross-boundary tangents are piecewise polynomial. They can therefore be of low degree and simultaneously separate first and second derivatives of the boundary curves of the macro-patch corners from the neighbours. The scheme is completely local. Furthermore, the 4-split leads to four patches per macro-patch which leaves enough control points free for inner shape control. Finally this scheme offers two parameters per vertex for controlling first and second derivatives of the boundary curves, and six free control points inside the macro-patch. Additionally, the interpolation condition can also be relaxed to gain one more free parameter per vertex.

The present paper investigates the problem of how the free parameters and control points of the 4-split domain method can be employed optimally. The challenge is to get an overall satisfactory shape, which is a global requirement, while maintaining the locality property of the scheme. Various geometric and variational criteria are proposed and compared.

§2. Triangular G^1 Interpolation by 4-splitting Domain Triangles

2.1 Notations

The surface mesh $\mathcal{M}$ is input, and consists of a list of vertices and edges describing a 2-manifold triangulated mesh in $\mathbb{R}^3$. The surface $\mathcal{S}$ which interpolates the vertices of $\mathcal{M}$ is composed of triangular macro-patches M^i which are in one-to-one correspondence with the mesh facets. It is therefore convenient for the construction of $\mathcal{S}$ to choose a parameterization of the macro-patches M^i around a common vertex, sharing pairwise a common boundary as illustrated in Fig. 1. All subscripts $i = 1, \ldots, n$ are taken modulo n, where n is the order of the mesh vertex corresponding to $M^i(0,0)$. The parameter u_i lies in the interval $[0,1]$.

The fundamental idea of the present triangular interpolation scheme is to subdivide the domain triangle into four subtriangles by joining the edge midpoints together, see Fig. 1. Each macro-patch M^i will therefore be a piecewise polynomial image of the unit triangle in $\mathbb{R}^2$, composed of four quintic Bézier triangles [2] each. The macro-patches will join together with G^1 continuity. The resulting surface $\mathcal{S}$ will also be G^1.

The G^1 conditions which are used in this paper are subject to some simplifying assumption in order to keep the interpolation scheme of low degree. Two adjacent patches $M^{i-1}(u_{i-1}, u_i)$ and $M^i(u_i, u_{i+1})$ join at a common boundary

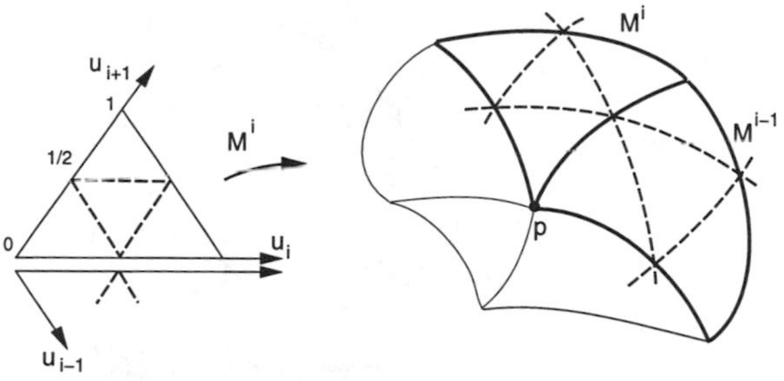

Fig. 1. Parameterization and domain 4-split.

with G^1 continuity if there exists a scalar function Φ_i such that

$$\Phi_i(u_i)\, M^i_{u_i}(u_i, 0) = \frac{1}{2}\, M^i_{u_{i+1}}(u_i, 0) + \frac{1}{2}\, M^{i-1}_{u_{i-1}}(0, u_i)\,, \qquad (1)$$

where n is the order of the vertex corresponding to $u_i = 0$. $M^i_{u_i}$ denotes the partial derivative of M^i with respect to u_i.

The algorithm for constructing the spline surface consists of three steps

- constructing boundary curves,
- constructing cross-boundary tangents,
- filling in the patches,

which will be briefly presented in the following three subsections. For more details and complete explanations of this method, the reader is referred to [3].

It is important to keep all these functions of the lowest degree possible. The main contribution to this comes from the domain 4-split. It allows for piecewise polynomial functions of low degree while simultaneously fulfilling all other requirements, such as continuity and localness.

2.2. Boundary curves and vertex consistency

First the boundary curves of the macro-patches are constructed in correspondence to the edges of $\mathcal{M}$ by interpolating the mesh vertices at the end points by satisfying the G^1 conditions at the vertices and by keeping the surface scheme local. They are called C^2-consistent.

Each boundary curve between two adjacent patches is a piecewise (2 pieces) cubic Bézier curve parameterized on $\{0, \frac{1}{2}, 1\}$. Around each vertex of $\mathcal{M}$, the control points $\boldsymbol{b}^i_0, \boldsymbol{b}^i_1, \boldsymbol{b}^i_2$, $i = 1, \ldots, n$, of all incident boundary curves are constructed independently from the joining curve piece of the opposite vertices. The "midpoints" $\boldsymbol{b}^i_3$ are then constructed in order to have C^1 boundary curves. See Fig. 2 for the notation.

At a vertex $\boldsymbol{v}$ the Φ_i-functions which are defined on the incident edges to $\boldsymbol{v}$ are first determined by calculating $\Phi_i(0)$ and $\Phi_i(1)$ from system (1) by solving it for $u_i = 0$ and $u_i = 1$ resp., which gives $\Phi^0 = \Phi_i(0) = \cos\left(\frac{2\pi}{n}\right)$

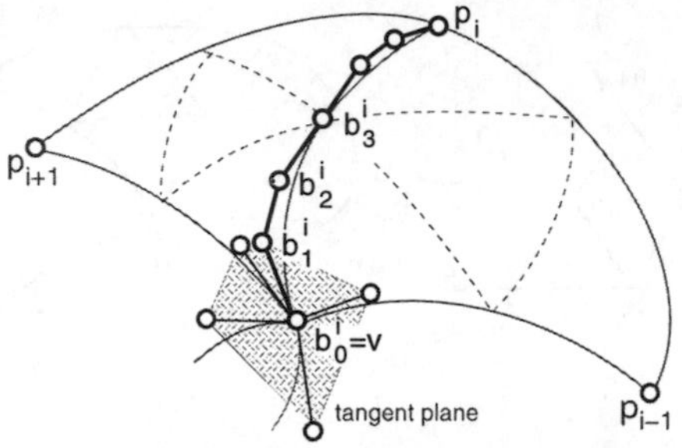

Fig. 2. Control points of the boundary curves at vertex $\boldsymbol{v}$.

and $\Phi_i(1) = 1 - \cos(\frac{2\pi}{n_i})$. The domain 4-split now enables to seperate vertex derivatives, and to take the Φ_i-function to be piecewise linear:

$$\Phi_i(u_i) = \begin{cases} \cos\frac{2\pi}{n}(1-2u_i) + u_i, & \text{for } u_i \in [0, \frac{1}{2}], \\ (1-u_i) + (1-\cos\frac{2\pi}{n_i})(2u_i - 1), & \text{for } u_i \in [\frac{1}{2}, 1]. \end{cases} \tag{2}$$

Let us now adopt a matrix notation for the boundary curve control points between $\boldsymbol{v}$ and $\boldsymbol{p}_i$, $i = 1, \ldots, n$:

$$\bar{\boldsymbol{b}}_0 := \begin{bmatrix} \boldsymbol{b}_0^1 \\ \vdots \\ \boldsymbol{b}_0^n \end{bmatrix}, \quad \bar{\boldsymbol{b}}_1 := \begin{bmatrix} \boldsymbol{b}_1^1 \\ \vdots \\ \boldsymbol{b}_1^n \end{bmatrix}, \quad \bar{\boldsymbol{b}}_2 := \begin{bmatrix} \boldsymbol{b}_2^1 \\ \vdots \\ \boldsymbol{b}_2^n \end{bmatrix}, \quad \bar{\boldsymbol{p}} := \begin{bmatrix} \boldsymbol{p}_1 \\ \vdots \\ \boldsymbol{p}_n \end{bmatrix}, \quad \bar{\boldsymbol{v}} := \begin{bmatrix} \boldsymbol{v} \\ \vdots \\ \boldsymbol{v} \end{bmatrix},$$

where $\bar{\boldsymbol{p}}$ is referred to as the vertex neighborhood of $\boldsymbol{v}$.

The following choice for the boundary curve Bézier points near the vertex $\boldsymbol{v}$ enables us to find a solution to system (1) which at the same time solves the vertex consistency problem [3,4]:

$$\begin{aligned} \bar{\boldsymbol{b}}_0 &= \alpha\bar{\boldsymbol{v}} + B^0\bar{\boldsymbol{p}}, \\ \bar{\boldsymbol{b}}_1 &= \alpha\bar{\boldsymbol{v}} + B^1\bar{\boldsymbol{p}}, \\ \bar{\boldsymbol{b}}_2 &= \left[(\gamma_0+\gamma_1)\alpha + \frac{\gamma_2}{3}\right]\bar{\boldsymbol{v}} + B^2\bar{\boldsymbol{p}}, \end{aligned} \tag{3}$$

where B^0, B^1, B^2 are $n \times n$ matrices defined as

$$\begin{aligned} B_{ij}^0 &= \frac{1-\alpha}{n}, \\ B_{ij}^1 &= \frac{1-\alpha+\beta\cos\left(\frac{2\pi(j-i)}{n}\right)}{n}, \\ B_{ij}^2 &= \frac{(\gamma_0+\gamma_1)(1-\alpha)+\gamma_1\beta\cos\left(\frac{2\pi(j-i)}{n}\right)}{n} + \gamma_2\begin{cases} 1/6 & \text{if } j = i-1, i+1, \\ 1/3 & \text{if } j = i, \\ 0 & \text{otherwise.} \end{cases} \end{aligned} \tag{4}$$

The free parameters $\alpha, \beta, \gamma_1, \gamma_2$ control the interpolation and the first and second derivatives. In Section 3 it will be shown how they can be set optimally. The control points of the joining curves pieces $\boldsymbol{b}_0^k, \boldsymbol{b}_1^k, \boldsymbol{b}_2^k$ and $\boldsymbol{b}_3^k = \boldsymbol{b}_3^i$ are found by applying the formulas (3) and (4) to the neighbouring mesh points $\boldsymbol{p}_i$ of $\boldsymbol{v}$. k is the index of $\boldsymbol{v}$ relative to the neighborhood of $\boldsymbol{v}$.

This curve network construction is local in that changes of one mesh vertex only affect the boundary curve pieces relative to the neighbourhood of that vertex.

2.3. Cross-boundary tangents

The cross-boundary tangents are subject to the G^1 conditions (1), the vertex consistency constraints, and the curve network of Sect. 2.2, and are set to be equal

$$\begin{aligned} M^i_{u_{i+1}}(u_i, 0) &= \Phi_i(u_i) M^i_{u_i}(u_i, 0) + \Psi_i(u_i) \boldsymbol{V}_i(u_i), \\ M^{i-1}_{u_{i-1}}(0, u_i) &= \Phi_i(u_i) M^i_{u_i}(u_i, 0) - \Psi_i(u_i) \boldsymbol{V}_i(u_i). \end{aligned} \tag{5}$$

The scalar function Ψ_i and the vector function $\boldsymbol{V}_i$ are built of minimal degree so as to interpolate the values of the cross-derivatives and the twists at the vertices $\boldsymbol{p}$ and $\boldsymbol{p}_i$:

$$\begin{aligned} \Psi_i(u_i) &= \sin \frac{2\pi}{n}(1 - u_i) + \sin \frac{2\pi}{n_i} u_i, && \text{(linear)} \\ \boldsymbol{V}_i(u_i) &= \sum_{k=1}^{n} \boldsymbol{v}_i^k B_k^2(2u_i) \quad u_i \in [0, \frac{1}{2}], && \text{(piecewise quadratic)} \end{aligned} \tag{6}$$

where

$$\bar{\boldsymbol{v}}_0 = V^0 \bar{\boldsymbol{p}}, \quad \bar{\boldsymbol{v}}_1 = V^1 \bar{\boldsymbol{p}}, \quad \bar{\boldsymbol{v}}_2 = \frac{1}{2}\bar{\boldsymbol{v}}_1 + \frac{1}{2}\bar{\boldsymbol{v}}_1^{\,opp}. \tag{7}$$

The $n \times n$ matrices V^0 and V^1 are given by

$$V^0_{ij} = \frac{6\beta}{n} \sin\left(\frac{2\pi(j-i)}{n}\right), \quad i, j = 1, \ldots, n,$$

$$\begin{aligned} V^1_{ij} = \frac{1}{\psi_i^0} & \left[(6\phi^1 - 48\phi^0 + 24\phi^0) \tan\left(\frac{\pi}{n}\right) - 6\psi_i^1 \right] \frac{\beta}{n} \sin\left(\frac{2\pi(j-i)}{n}\right) \\ & + \frac{4}{\psi_i^0} \gamma_2 \phi^0 \begin{cases} 1 & \text{if } j = i+1, \\ -1 & \text{if } j = i-1, \end{cases} \end{aligned} \tag{8}$$

where $\Phi^0 = \Phi_i(0), \Phi^1 = \Phi_i'(0)$ and $\Psi_i^1 = \Psi_i'(0)$ are known from (2) and (6).

2.4. Filling-in the macro-patches

Each macro-patch is composed of four C^1 quintic triangular Bézier patches. The boundary curves of the macro-patch are the twice degree elevated curves of Section 2.2. The cross-boundary tangents of Sect. 2.3 determine the first

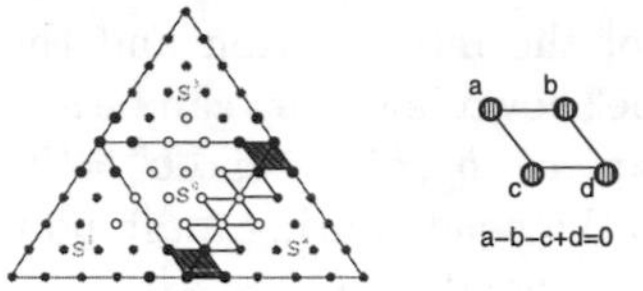

Fig 3. C^1-conditions between two adjacent quintic Bézier patches.

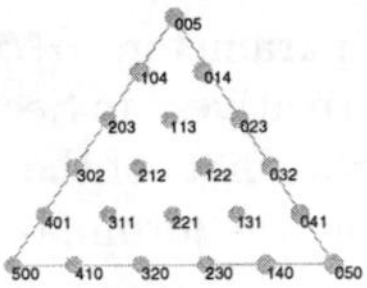

Fig 4. Labelling the control points of a quintic Bézier patch.

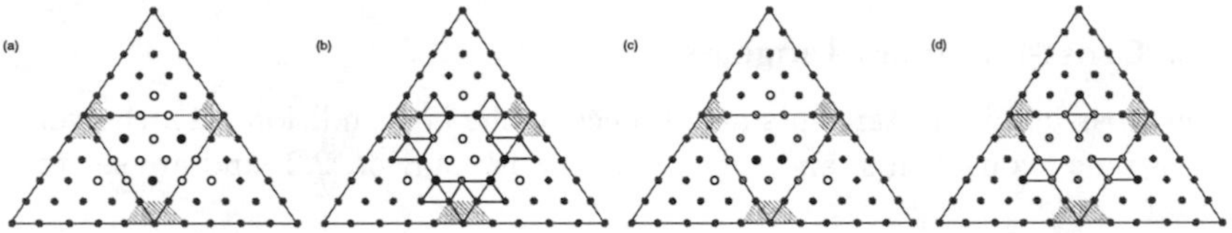

Fig. 5. Four steps for filling in the macro-patch M with C^1-continuity.

inner row of control points after one degree elevation [3]. The remaining 15 inner control points, which are highlighted in Fig. 5a, are now computed by joining the four inner patches with C^1 continuity. The necessary and sufficient C^1-continuity conditions between two internal Bézier patches inside one macro-patch are shown in Fig. 3: all pairs of adjacent triangles must form a parallelogram. In [3] it was shown that the first and last pairs of adjacent triangles in Fig. 3 already form parallelograms.

It remains to compute the free Bézier points such that the other three pairs of triangles along each edge inside the macro-patch also form parallelograms. This is done in four steps:

- choose the three twists points of the internal Bézier patch arbitrarily; these are free shape parameters (see Fig. 5a),
- compute the third and fourth Bézier points along each internal curve joining two Bézier patches using the second and fourth parallelogram conditions (see Fig. 5b),
- choose the remaining three unknown Bézier points of the central patch arbitrarily; these are free shape parameters (see Fig. 5c),
- compute the three remaining unknown Bézier points of the outer patches using the third parallelogram condition along each edge (see Fig. 5d).

§3. Local Optimization of the Boundary Curves

The present triangular interpolation method offers several degrees of freedom for shape control. They can be set manually or by using simple heuristics. An interactive design system can allow for manually adjusting these parameters in order to improve the shape. This procedure seems not to be sufficient if the given triangulated point set is very large, or if the data points are irregularly distributed.

Our goal is to investigate some optimization techniques. Two groups of degrees of freedom have to be distinguished. First there are 4 scalar parameters per vertex controlling the curve network, then 6 free inner Bézier control points are available for each macro-patch. Let us first concentrate on the curve network. It was stated in [5] that triangular interpolants often suffer from undulating curve networks. It can be confirmed here that a "well shaped" curve network is not sufficient, but is necessary for the construction of a pleasing shape. As pointed out in Sect. 2, the 4-split method is local. This property should not be altered by an optimization procedure. Local optimization criteria are therefore needed. This localness requirement conflicts in some sense with the global requirement of a "well shaped" surface. Every local scheme has to accept this conflict, otherwise it loses its localness property. Nevertheless, it will be shown that good results can be obtained.

In detail, four curve parameters per vertex are available, see (3) and (4). α is not really free. It allows us to switch between interpolation and approximation of the surface mesh. At first, we only consider the interpolation problem and set $\alpha = 1.0$. β affects the length of the tangent vectors of the boundary curves at the vertex $\boldsymbol{v} = \boldsymbol{b}_0^i$. The control points $\bar{\boldsymbol{b}}_1$ are obtained by a first order Fourier approximation of the neighbourhood $\boldsymbol{p}_i$, $i = 1, \ldots, n$ of $\boldsymbol{v}$. In other words, the $\boldsymbol{b}_1^i$ are an affine image of a regular n-gon whose centroid is $\boldsymbol{b}_0^i$. Too short tangent vectors lead to sharp corners at the patch vertices, while too long tangent vectors can lead to unwanted undulations. γ_1, γ_2 control the second derivatives of the vertices. The control points $\boldsymbol{b}_2^i$ depend linearly on them. But they don't depend linearly on $\beta, \gamma_1, \gamma_2$.

Due to the previous observations, the optimization of $\beta, \gamma_1, \gamma_2$ should mainly avoid undulations and allow for more or less bent or stretched curves. If for computation-time reasons one wants to perform only linear optimization in a least-squares sense, as we do, it should be done in two steps separating β from γ_1, γ_2.

In the following, the computation of optimal $\beta, \gamma_1, \gamma_2$ is separated into two steps. Otherwise, the problem would become non-linear. Each boundary curve consists of two cubic pieces joining with C^1 continuity. For locality reasons of the whole scheme, the pieces have to be constructed independently one another. In a *first step* the points $\boldsymbol{b}_1^i$ of the boundary curves incident in $\boldsymbol{v}$, Fig. 2, are determined. Each boundary curve corresponds to an edge of the surface mesh. In this case we are looking at the edge connecting $\boldsymbol{v}$ and $\boldsymbol{p}_i$. In order to reproduce the shape inherent to the underlying surface mesh, geometrical considerations imply that $\boldsymbol{b}_1^i$ would optimally lie in the plane spanned by this edge and a vector between $\boldsymbol{v}_0$ and the orthogonal projection of $\boldsymbol{p}_i$ on the tangent plane in $\boldsymbol{b}_0^i$, as Piper does in [8]. We call these points $\bar{\boldsymbol{b}}_1^* = [\boldsymbol{b}_1^{1*}, \cdots, \boldsymbol{b}_1^{n*}]^T$. The tangent planes should be estimated first. The constraints on the boundary curves in the present method don't allow for setting $\bar{\boldsymbol{b}}_1$ equal $\bar{\boldsymbol{b}}_1^*$, this is why these points are approximated in a least-squares sense. The key point here is that the locality of the equations (3) and (7) enables us to replace the true neighbourhood points $\boldsymbol{p}_i$ of $\boldsymbol{v}$ in these equations by new "virtual" neighbourhood points $\boldsymbol{p}_i^*$ that are only used in these equations, i.e. to compute the boundary curves, and the cross-boundary tangents. Therefore, we are able to solve the following linear least-square problem in order to compute the "virtual" points $\boldsymbol{p}_i^*$:

$$\sum_{i=1}^{n} \|\boldsymbol{b}_i^{1^*} - \boldsymbol{b}_i^1\|^2 \longrightarrow \min. \tag{9}$$

The new, optimal, control points are now given by

$$\bar{\boldsymbol{b}}_1^{optimal} = \alpha\bar{\boldsymbol{v}} + B^1\bar{\boldsymbol{p}}^*. \tag{10}$$

In a *second step*, $\bar{\boldsymbol{b}}_2$ has to be determined. $\bar{\boldsymbol{p}} := \bar{\boldsymbol{p}}^*$ and β are already fixed. Two parameters per vertex, γ_1, γ_2 are left free for optimization. $\boldsymbol{b}_3$ is then fixed as the midpoint between $\boldsymbol{b}_2^i$ of the two curve pieces. The requirements on $\bar{\boldsymbol{b}}_2$ are twofold: avoid undulations and bend the curve on request. The second requirement concerns the whole boundary curve between $\boldsymbol{v}$ and $\boldsymbol{p}_i$, and depends on the choice of $\bar{\boldsymbol{b}}_2$. The idea is to cope with that problem by introducing a target point $\boldsymbol{t}$ for each boundary curve. The control points $\boldsymbol{b}_2^i$ of each curve piece are then determined so that $\boldsymbol{b}_3 = \frac{1}{2}(\boldsymbol{b}_2^i + \tilde{\boldsymbol{b}}_2^i)$ approaches the target point by minimizing an appropriate energy functional on each curve piece locally. The introduction of the target point allows for global control of the boundary curves, while still keeping the scheme local.

The target point is fixed by a subdivision rule in terms of $\boldsymbol{b}_0^i, \boldsymbol{b}_1^i$ and $\tilde{\boldsymbol{b}}_0^i, \tilde{\boldsymbol{b}}_1^i$ of the joining curve piece, such as

$$\boldsymbol{t} = \frac{1}{16}(-\boldsymbol{b}_0 + 9\boldsymbol{b}_1^i + 9\tilde{\boldsymbol{b}}_1^i - \tilde{\boldsymbol{b}}_0). \tag{11}$$

$\bar{\boldsymbol{b}}_2$, which depends linearly on the free parameters γ_1, γ_2, is now determined by minimizing the linearized version of the bending energy combined with a curve length component [1]

$$E_\omega = \int_0^1 \|X''(t)\|^2 dt + \omega \int_0^1 \|X'(t)\|^2 dt\,, \quad \omega \geq 0. \tag{12}$$

The solution of a linear 2×2 system gives the optimal values for γ_1, γ_2 for $\bar{\boldsymbol{b}}_2$.

Different ways for finding an optimal, i.e. well shaped, curve network have been studied. Within the local schemes, the concept of target points can be replaced by target tangent vectors. This leads to a $2n \times 2n$ linear system of equations per vertex. $\beta, \gamma_1, \gamma_2$ can also be determined by a non-linear optimization method in only one step. When relaxing the localness requirement of the scheme, plenty of curve network schemes are possible, like a variant of Nielson's MNN [7] or the integration of given curvature values or second fundamental forms at the mesh vertices in the optimization process. This is not a subject of the present paper.

§4. Minimum Norm Criteria for Macro-patches

Once the curve network and the cross-boundary tangents are constructed, 15 inner Bézier control points remain for each macro-patch. They are related to each other by the C^1 continuity conditions which are imposed between the four quintic sub-patches. Six of them are completely free. They are drawn as full black dots in Figs. 5a and 5b. The remaining 9 points depend linearly on

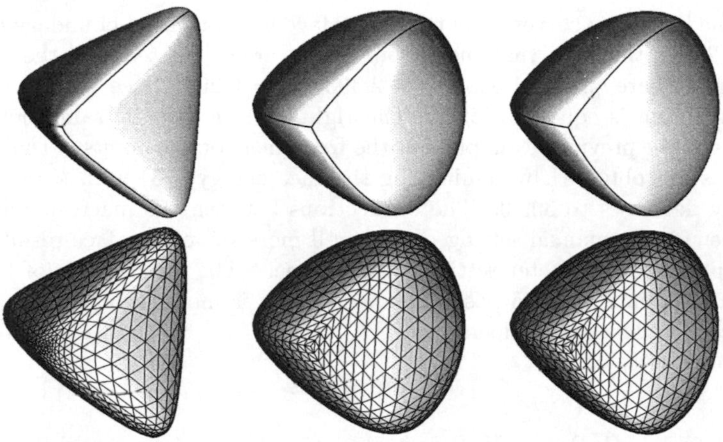

Fig. 6. *Left*: manually fixed free parameters, *Middle*: optimized boundary curves, *Right*: optimized boundary curves and patches.

them. It is therefore possible to use one of the two quadratic functionals $E_k(X, X)$, where $k = 2, 3$ and

$$E_k(X,Y) = \int_\triangle \left(\frac{\partial^k X}{\partial u^k} \cdot \frac{\partial^k Y}{\partial u^k} + \frac{\partial^k X}{\partial v^k} \cdot \frac{\partial^k Y}{\partial v^k} + \frac{\partial^k X}{\partial w^k} \cdot \frac{\partial^k Y}{\partial w^k} \right) dudvdw.$$

The free points $\bar{\mathbf{s}}$ are determined by minimizing

$$\sum_{j=0}^{3} E_k(\boldsymbol{S}^j, \boldsymbol{S}^j) = \sum_{j=0}^{3} (M_j \bar{\mathbf{s}} + \boldsymbol{a}_j)^T A (M_j \bar{\mathbf{s}} + \boldsymbol{a}_j) \,, \quad k = 2, 3, \tag{13}$$

where $\boldsymbol{S}^j = \sum_{|\mathsf{i}|=5} \boldsymbol{b}_\mathsf{i} B_\mathsf{i}^5(u, v, w)$ denote the four quintic Bézier sub-patches, and $\bar{\mathbf{s}} = [\boldsymbol{s}_{113}^0, \boldsymbol{s}_{131}^0, \boldsymbol{s}_{311}^0, \boldsymbol{s}_{122}^0, \boldsymbol{s}_{212}^0, \boldsymbol{s}_{221}^0]^t$ denotes the vector of the 6 unknown control points of the middle patch (see Figs. 4 and 3). The (21×6) matrices M_j and the vectors $\boldsymbol{a}_j$ contain the linear relations between control points of the 4 sub-patches and the 6 unknown points. The (21×21) matrix A is given by $A_{\mathsf{ij}} = E_k(B_\mathsf{i}, B_\mathsf{j})$ for $|\mathsf{i}| = 5$, $|\mathsf{j}| = 5$.

§5. Results

Fig. 6 shows the interpolation of a tetrahedron by our method. This very simple example was chosen because it illustrates clearly the influence of the free parameters and control points. The upper row shows three surfaces with the boundary curves of the macro-patches, while the lower row shows their iso-parametric lines. The left surface is obtained by manually setting $\alpha = 1.0$, $\beta = 0.15$, $\gamma_0 = -1.0$, $\gamma_1 = 2.0$, $\gamma_2 = 0.0$. These values are identical for all vertices due to the regularity of the surface mesh. The free inner control points are set by a rule combining the mesh face normal, cross-boundary tangents and the boundary curves. The middle surface has optimized boundary curves, (9)-(12). Optimal "virtual" neighbour points are calculated, and the optimal parameters are $\gamma_0 = -1.598$, $\gamma_1 = 2.393$, $\gamma_2 = 0.205$. Face energy without minimization is equal to 732.2. The right surface has the same boundary curves as the previous example, but the free inner control points of the macro-patches are obtained by minimizing the face energy (13) with $k = 3$. The energy decreases to 309.0. The connections between the macro-patches are sharper for the manual setting. An overall more smooth surface results from the optimized parameter setting of this paper. The distribution of the iso-parametric lines shows a positive side-effect: it is more regular at the patch vertices for the optimized surfaces.

References

1. Bonneau, G.-P., and H. Hagen, Variational design of rational Bézier curves and surfaces, in *Curves and Surfaces in Geometric Design*, P.-J. Laurent, A. Le Méhauté, and L. L. Schumaker (eds.), A. K. Peters, Wellesley MA, (1994), 51–58.
2. Farin, G., *Curves and Surfaces for Computer Aided Geometric Design*, Academic Press, New York, 4th edition, 1997.
3. Hahmann, S., and G.-P. Bonneau, Triangular G^1 interpolation by 4-splitting domain triangles, to appear in Computer Aided Geometric Design.
4. Loop, C., A G^1 triangular spline surface of arbitrary topological type, Computer Aided Geometric Design **11** (1994), 303–330.
5. Mann, S., C. Loop, M. Lounsbery, D. Meyers, J. Painter, T. DeRose, and K. Sloan, A survey of parametric scattered data fitting using triangular interpolants, in *Curve and Surface Design*, H. Hagen (ed.), SIAM (1992), 145–172.
6. Mann, S., Using Local optimization in surface fitting, in *Mathematical Methods for Curves and Surfaces* Morten Dæhlen, Tom Lyche, Larry L. Schumaker (eds.), Vanderbilt University Press, Nashville & London, 1995, 323–332.
7. Nielson, G., A method for interpolating scattered data based upon a minimum norm network, Mathematics of Computation **40** (1983), 253–271.
8. Piper, B. R., Visually smooth interpolation with triangular Bézier patches, in *Geometric Modeling: Algorithms and New Trends*, G. Farin (ed.), SIAM (1987), 221–233.
9. Shirman, L. A., and C. H. Séquin, Local surface interpolation with Bézier patches, Computer Aided Geometric Design **4** (1987), 279–295.

S. Hahmann, G.-P. Bonneau, and R. Taleb
Laboratoire de Modélisation et Calcul, CNRS-IMAG, B.P. 53
F-38041 Grenoble Cedex 9, France
`hahmann@imag.fr`

Multi–Level Approximation to Scattered Data Using Inverse Multiquadrics

S. J. Hales and J. Levesley

Abstract. A method of finding local approximations is used to thin data before a hierarchical iterative refinement scheme is employed in conjunction with domain decomposition. The interpolation problem on each sub–domain is solved by using the same stored inverse. The approximation power of the inverse multiquadric is exploited whilst overcoming the computational difficulties associated with globally supported basis functions.

§1. Introduction

Radial basis functions have been widely used for multivariate interpolation of scattered data, see [4] for a summary. An interpolant is generated by a linear combination of basis functions ϕ at distinct centres x_i, $i = 1, \ldots, N$;

$$s(x) = \sum_{i=1}^{N} \lambda_i \phi(\|x - x_i\|), \tag{1}$$

constrained by $s(x_i) = f_i$, $i = 1, \ldots, N$, where $\mathcal{F} : \mathbb{R}^d \mapsto \mathbb{R}$ and $f_i = \mathcal{F}(x_i)$. The interpolation matrix $A \in \mathbb{R}^N \times \mathbb{R}^N$ is given by $A_{1 \leq i,j \leq N} = \phi(\|x_i - x_j\|)$, and λ satisfies

$$A\lambda = f, \tag{2}$$

where $\lambda = [\lambda_1 \cdots \lambda_N]^T$ and $f = [f_1 \cdots f_N]^T$.

Common choices for ϕ in this setting are given in [6],

$$\phi(\|x - x_i\|) = \exp(-c^2 \|x - x_i\|^2), \qquad \text{Gaussian,}$$
$$\phi(\|x - x_i\|) = (c^2 + \|x - x_i\|^2)^{-1/2}, \qquad \text{Inverse multiquadric,}$$

where c is a constant shape parameter. With a small modification to the scheme, the thin plate spline and multiquadric are also used.

Curve and Surface Fitting: Saint-Malo 1999
Albert Cohen, Christophe Rabut, and Larry L. Schumaker (eds.), pp. 247–254.

ISBN 0-8265-1357-3.

The above parameter–dependent functions are good at approximating data for certain values of c, but these cause inherent ill–conditioning in A. Schaback [7] explains this phenomenon by means of an "Uncertainty Relation" between upper bounds on errors for interpolants of the form (1), and lower bounds on the smallest eigenvalue of A. Iterative techniques for solving such badly conditioned systems often suffer from poor rates of convergence, and therefore computationally expensive direct methods have to be employed.

In the inverse multiquadric case, large values of c achieve good initial approximations to smooth data, whilst smaller values produce functions capable of resolving fine detail. Ideally, such properties could be exploited without having to solve (2) directly.

Since s is evaluated at *points* $y \neq x_i$ where an approximation is required, a global solution incorporating all N centres may be inappropriate. Further, it is unnecessary to find $\hat{\lambda}$ such that $\|f - A\hat{\lambda}\|_\infty \ll |\mathcal{F}(y) - s(y)|$, since the accuracy of $s(y)$ is limited by the approximating power of ϕ. Rather than searching for a complete global solution, this suggests that attention may be focussed on small regions around evaluation points. Moreover, the aim is to obtain a solution such that the residual and approximation accuracy are comparable, for little is to be gained by having a small residual, while the approximation power of the basis functions limits the final accuracy.

In Section 2, local approximations are used to convert irregular data to a regular mesh of approximate function values. Whilst the method can be generalised to $\mathbb{R}^d$, the description and examples are given in $\mathbb{R}^2$. The system of equations associated with the gridded data is inverted and used to solve subsequent systems.

Floater & Iske [1] demonstrate the benefits of a multi–level approach to approximation, and the theoretical foundation is provided by Narcowich, Schaback & Ward [5]. Section 3 describes the present hierarchical iterative refinement algorithm, and explains the computational advantages of domain decomposition and the use of a stored inverse.

§2. Local Solutions and Gridding Data

If the function $\mathcal{F}$ is not arbitrary, but arises from a physical system, then some degree of smoothness can be assumed. A smooth data set can be significantly thinned whilst retaining general information about its behaviour. Floater & Iske [1] demonstrate that Delaunay triangulation can be used to optimise the uniformity of data, and provide a good thinning algorithm. Such triangulation and assembling of data is computationally expensive for excessively large N. An $\mathcal{O}(N)$ method of finding uniform approximate data is presented.

An approximation to $\mathcal{F}$ at a point $y \in \mathbb{R}^d$ is achieved by solving a small interpolation problem centred on y. The closest q points in X to y are interpolated by inverse multiquadrics with shape parameter c_{local}, and evaluated at y. Since q can be as low as $20 \sim 30$, c_{local} can be relatively large before the matrix ill-conditioning becomes unacceptable, thus yielding a good approximation.

This method is highly parallelizable, and large data-sets can be dealt with without the need of assembling or storing the matrix A.

Finding the optimum shape parameter on regular or scattered data remains an open problem, as shown in [3]. There is no obvious correlation between point spacing and a good choice of c_{local}. The best shape parameter is generally found by increasing the value of c_{local} until just prior to machine precision breakdown.

Let $Y = \{y_1, \ldots, y_{n^2}\}$ be the set of points on the $n \times n$ regular unit grid. If the previous local approximation technique is applied to each y_i, then the irregular data can be transformed to a regular grid with approximate function values $\hat{f}_i$. The aim is to find a global approximation using the new data at the grid points.

After converting scattered data to a regular grid, certain approximation techniques become available which would otherwise have been difficult to implement. Polynomial tensor product splines can be efficiently employed to approximate a solution from the given gridded data. To find such an approximation at a point z, z must lie inside a $(d+1) \times (d+1)$ subgrid of the regular points, where d is the degree of the Lagrange polynomials to be used. Let the points of such a subgrid be labelled ξ_{ij} and have function values $\hat{f}_{ij}$ for $i, j = 1, \ldots, d+1$. The univariate Lagrange polynomials $L_i(x)$ and $L^j(y)$ are constructed such that

$$L_i(\xi_{k\bullet}) = \delta_i^k \quad \text{and} \quad L^j(\xi_{\bullet k}) = \delta_j^k.$$

The polynomial tensor product spline ϕ_{ij} is defined to be

$$\phi_{ij}(z) = L^i(z) \cdot L_j(z).$$

The approximation at z is given by

$$\sum_{i=1}^{d+1} \sum_{j=1}^{d+1} \phi_{ij}(z) \hat{f}_{ij}.$$

Alternatively, a thinned global interpolant of the form (1) can be achieved by solving

$$B\mu = \hat{f}, \tag{3}$$

where $B_{1 \le i,j \le n} = \phi(\|y_i - y_j\|)$, $\mu = [\mu_1 \cdots \mu_n]^T$ and $\hat{f} = [\hat{f}_1 \cdots \hat{f}_n]^T$.

This amounts to finding an interpolant $\hat{s}$ to a thinned approximation of the initial data. The local approximation errors $|f(y_i) - \hat{f}(y_i)|$ limit the final accuracy of $\hat{s}$.

The inverse of B need only be computed once, and then stored for future use. All scattered data problems can then be scaled and transformed to the regular grid Y, whereupon μ is given by the matrix–vector product $\mu = B^{-1}\hat{f}$. Only half of the entries of B^{-1} need to be stored since $B^{-1} = (B^{-1})^T$.

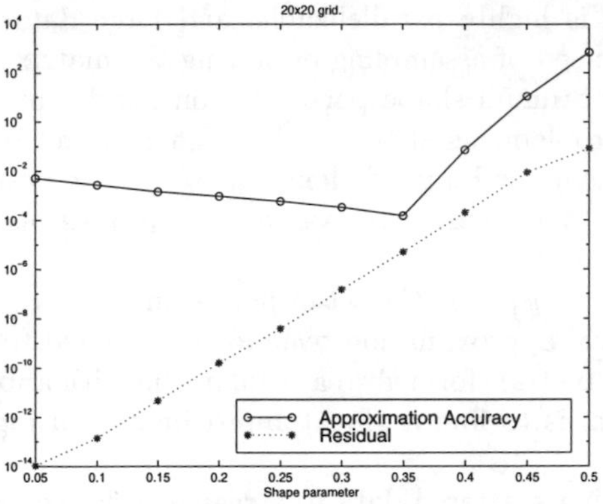

Fig. 1. An example of approximation vs. residual.

Too large a matrix B causes storage problems, and difficulties in calculating the inverse. As c increases, the approximation improves, but the residual $\|\hat{f} - B\mu\|_\infty$ grows. A value for c is chosen before the approximation begins to deteriorate due to the rise in the residual. As an example, the function $\mathcal{F} \equiv 1$ is approximated on the unit square using inverse multiquadrics by interpolating $\hat{f}_i = \mathcal{F}(y_i)$, $i = 1, \ldots, 400$ using (3). The approximation is evaluated at 1000 random points. The results in Figure 1 are typical for smooth functions, but the consequent choice of c is only a guide, and does not guarantee success for all $\mathcal{F}$.

§3. Hierarchical Iterative Refinement

The hierarchical method uses increasingly dense subsets of X to refine the current approximation; see [5]. Let $\chi_k = \{\hat{x}_1, \ldots, \hat{x}_{N_k}\} \subseteq X$, such that $N_{k+1} > N_k$. Let s_k be the current approximation, and r_k be the full global residual at the kth level,

$$r_k(x_i) = f(x_i) - s_k(x_i).$$

Let $\hat{r}_k$ be the kth residual over the points in χ_k. This thinned global residual is interpolated by

$$t_k(x) = \sum_{i=1}^{N_k} \gamma_i \phi_k(\|x - \hat{x}_i\|), \tag{4}$$

where $\phi_k(\|x - \hat{x}_i\|) = (c_k^2 + \|x - \hat{x}_i\|^2)^{-1/2}$, and $t_k(\hat{x}_i) = \hat{r}_k(\hat{x}_i)$, $i = 1, \ldots, N_k$.

The initial interpolant $s_1 \equiv 0$ is updated by

$$s_{k+1} = s_k + t_k. \tag{5}$$

The technique of gridding data in Section 2 is used to find an approximate function value for every point in Y_p. Therefore, (5) is replaced by $\hat{s}_{k+1} = \hat{s}_k + t_k$, where $\hat{s}_k$ is the current approximation to regular approximate data, $\hat{s}_1 \equiv 0$.

The value c_1 can be relatively large to give a good initial approximation. As N_k increases, c_k has to be reduced to ensure computational solvability. The decrease in c_k introduces tighter basis functions which improve the resolution of the approximation.

A method of data thinning is required to determine the points in χ_k. The dense systems arising from (4) have to be solved directly, but this is impractical for large N_k. To overcome such complications, domain decomposition is applied to each χ_k.

The levels of the hierarchy have to be computed sequentially, but by using domain decomposition each sub–domain can be dealt with in parallel. Moreover, each such solution only requires a single matrix–vector product.

To put this in the current context, each χ_k is constructed from overlapping square grids Y_p , where $p = 1, \ldots, m_k$. These square grids need not be the same size or of similar orientation, but must contain an equal number of points. Each sub–domain Y_p consists of an inner region, where the approximation is finally evaluated, and an overlap. Special attention has to be given to sub–domains whose edges coincide with the boundary of X.

At the kth level, m_k sub–domain interpolation problems need to be solved. Since B is invariant under shifts and rotations of the centres y_i, the stored B^{-1} can be invoked. If the centres are scaled $y_i \mapsto \alpha y_i$, this amounts to a change in the shape parameter.

Recall that $B_{1\le i,j\le n} = \phi(\|y_i - y_j\|)$, where $y_i \in [0,1] \times [0,1]$. Now,

$$\begin{aligned}\phi(\|y - y_i\|) &= (c^2 + \|y - y_i\|^2)^{-1/2} \\ &= \alpha(\alpha^2 c^2 + \|\alpha y - \alpha y_i\|^2)^{-1/2}.\end{aligned}$$

Let $w_i = \alpha y_i$ and define $\psi(\|w - w_i\|) = \alpha(\alpha^2 c^2 + \|w - w_i\|^2)^{-1/2}$. Then $B_{1\le i,j\le n} = \phi(\|y_i - y_j\|) = \psi(\|w_i - w_j\|)$ where $w_i \in [0,\alpha] \times [0,\alpha]$. Therefore by using the matrix B, a new inverse multiquadric is created at scaled points with shape parameter αc.

Each of the thinned global interpolation problems (4) can be decomposed and solved by multiple applications of the stored inverse B^{-1}. Continued use of the same inverse naturally introduces tighter basis functions suitable for approximating typical residuals.

§4. Numerical Results

We give an example where the above scheme is used to approximate Franke's function [2] over 10000 scattered points in the unit square in $\mathbb{R}^2$:

$$\begin{aligned}\mathcal{F}(u,v) = \;& 0.75e^{-0.25(9u-2)^2 - 0.25(9v-2)^2} + 0.75e^{-(9u-2)^2/49 - (9v-2)^2/10} \\ &+ 0.5e^{-0.25(9u-7)^2 - 0.25(9v-3)^2} - 0.2e^{-(9u-4)^2 - (9v-7)^2}.\end{aligned}$$

Level	No. of domains m_k	Shape parameter c_k	Overlap	Max. error in gridded data $\|f_i - \hat{f}_i\|_\infty$	Max. error in solution $\|\mathcal{F} - \hat{s}_k\|_\infty$
1	1	0.25	0	8.241×10^{-5}	1.371×10^{-4}
2	4	0.138	1/36	2.424×10^{-6}	1.838×10^{-5}
3	16	0.0688	1/72	3.189×10^{-6}	3.189×10^{-6}

Tab. 1. Error for Franke's function.

The localised interpolation problems are solved directly using Gauss Elimination, with $q = 20$ and $c_{local} = 0.2$. It is the error function at each level which is approximated locally, and not the original function $\mathcal{F}$. The square sub–domain grids Y_p are comprised of 21×21 equally spaced points. For ease of implementation, the sub–domains used for a particular level are of equal size. The overlaps between sub–domains therefore consist of one or two mesh points, depending on position. The key interpolation matrix B is constructed from inverse multiquadrics with $y_i \in [0,1]^2$, $c = 0.25$, and B^{-1} is generated using Matlab. The domain decomposition is straightforward on the unit square with $m_k = 4^{k-1}$. The thinned global interpolants $\hat{s}_k$ are evaluated at points $t_i \in [0,1]^2$. Table 1 shows the error in the approximated data at the regular grid points $|f(y_i) - \hat{f}(y_i)|$, and the error in the approximation $|\mathcal{F}(t_i) - \hat{s}_k(t_i)|$. Figures 2 and 3 show the approximation error for each level.

The error function from Level 1 clearly demonstrates the ability of the inverse multiquadric to approximate smooth data. The error near the boundary is scaled by an order of magnitude at each level, but has the same general behaviour. The final iteration leaves error near the boundary, aggravated by test points being outside the original scattered data set. Such evaluation points ought to be included since, although they require the extrapolation of $\hat{s}_k$ to evaluate, the experiment was specified to be conducted on the unit square. The original aim of finding a solution where the residual is comparable to the approximation accuracy is fulfilled at Level 3.

Example I is repeated as far as the regularization of data, and then polynomial tensor product splines are used to find the final approximation, as described in Section 2. Such splines cannot replace the inverse multiquadric approximation on the regular grid without an increase in error. Such an error is then propagated to the next level where the discrepancy is amplified. However, if the hierarchical refinement procedure is abandoned, then these basis functions efficiently yield a good approximation. Table 2 shows the approximation accuracy for such splines of different polynomial degree without iterative refinement. The grid sizes are comparable to those used in the original example.

Grid Size	Linear	Quadratic	Cubic
21×21	2.2×10^{-2}	6.4×10^{-3}	3.5×10^{-3}
41×41	5.3×10^{-3}	9×10^{-4}	2.6×10^{-4}
81×81	1.4×10^{-3}	1×10^{-4}	5.7×10^{-5}

Tab. 2. Approximation error for the various splines.

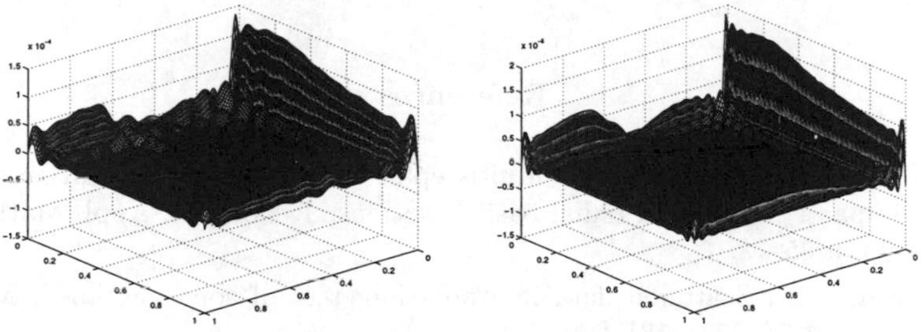

Fig. 2. Approximation Error for Levels 1 and 2.

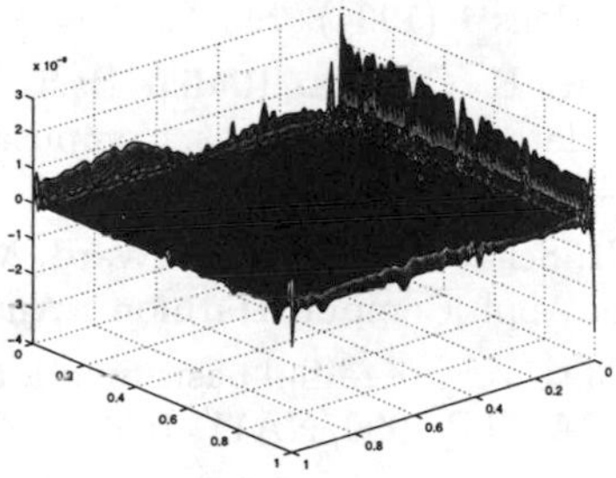

Fig. 3. Approximation Error for Level 3.

§5. Conclusion

A global solution to an interpolation problem involving a large number of data points is too expensive to compute directly if inverse multiquadrics are to be used effectively. However, if the aim is to generate approximations to a function, then such a solution is unnecessary, and an alternative method has been presented.

The underlying idea is to transform given scattered data f_i at points x_i to regular approximate data $\hat{f}_i$ at y_i, which is easier to solve for. The aim is then no longer to interpolate the initial data, but to find a good approximation to it. The final solution $\hat{s}$ is an approximation to $\hat{f}$, which is close to f. Success relies on minimising the local approximation errors $|f(y_i) - \hat{f}(y_i)|$.

The algorithm is $\mathcal{O}(N)$ since the only work related to the number of initial points is the search for the q closest points to each y_i. Such a search can be improved by making assumptions as to which x_i are unlikely to qualify.

The time required to solve each sub–domain problem is reduced due to the use of the stored $n \times n$ key inverse matrix. Solving directly would be $\mathcal{O}(n^3)$, but the required matrix–vector product is $\mathcal{O}(n^2)$.

The hierarchical iterative refinement strategy produces good approximations, and is the only sequential aspect of the method. The search for approximate regular data, and the solutions for each decomposed sub–domain are parallelizable operations, although this has yet to be implemented. These features mean that large data sets can be dealt with in acceptable computing time.

References

1. Floater, M. S. and A. Iske, Multistep scattered data interpolation using compactly supported radial basis functions, J. Comput. Appl. Math. **73** (1996), 65–78.
2. Franke, R., Scattered data interpolation: test of some methods, Math. Comp. **38** (1982), 181–200.
3. Hagan, R. E. and E. J. Kansa, Studies of the R parameter in the multiquadric function applied to groundwater pumping, Journal of Applied Science & Computations **1** (1994), 266–282.
4. Kansa, E. J. and R. E. Carlson, Radial Basis Functions: A class of grid–free, scattered data approximations, Computational Fluid Dymanics Journal **3** (1995), 479–496.
5. Narcowich, F. J., R. Schaback, and J. D. Ward, Multilevel interpolation and approximation, Appl. Comput. Harmonic Anal. to appear.
6. Powell, M. J. D., The theory of radial basis function approximation, Cambridge University, DAMTP 1990/NA11, 1990.
7. Schaback, R., Multivariate Interpolation and Approximation by Translates of a Basis Function, in *Approximation Theory VIII,* C. K. Chui & L. L. Schumaker (eds), Approximation and Interpolation, Texas, 1995, 491–514.

Stephen Hales
Mathematics & Computer Science Dept.
University of Leicester
University Road
Leicester, LE1 7RH, UK
`sjh16@mcs.le.ac.uk`

Jeremy Levesley
Mathematics & Computer Science Dept.
University of Leicester
University Road
Leicester, LE1 7RH, UK
`jl1@mcs.le.ac.uk`

Best Approximation Algorithms: A Unified Approach

V. V. Kovtunets

Abstract. A generalization of the Remez algorithm is proposed. The new approach uses differential properties of the best approximation operator. The method was developed for polynomial approximation of complex-value functions. In this paper the convergence of algorithm is proved for Banach spaces.

§1. Introduction

Let us consider the best approximation operator

$$\mathbf{P} : \mathbf{B} \to \mathbf{P}_n,$$

where $\mathbf{B}$ is a Banach space (complex in general), $\mathbf{P}_n$ an n-dimensional subspace. Suppose that $\mathbf{P}_n$ is univalent and one-side differentiable in any direction [1]. This assumption is valid when:

i) $\mathbf{B} = C(Q, \mathbf{R})$ Q-compact, and $\mathbf{P}_n$ is a Chebyshev subspace (in particular, when P_n is the subspace of algebraic polynomials of degree less or equal to $n-1$ [2,9]);

ii) $\mathbf{B} = C(Q, \mathbf{C}), Q$ is finite set, and $\mathbf{P}_n$ is an n-dimensional Chebyshev subspace [3];

iii) $\mathbf{B} = L_p,\ p > 1$, and $\mathbf{P}_n$ is an n-dimensional subspace (R. Holmes and B. Kripke).

Originally the differential properties of the best approximation operator were applied to the development of algorithms in [4,5]. The goal was to generalize the Remez algorithm for complex-valued functions. In [6] the new best approximation algorithm was applied to the approximation of conformal mappings by polynomials. In [7] it was shown that for real polynomial approximation,

Curve and Surface Fitting: Saint-Malo 1999
Albert Cohen, Christophe Rabut, and Larry L. Schumaker (eds.), pp. 255–262.

ISBN 0-8265-1357-3.

such an approach generates exactly the Remez algorithm, and a stronger convergence theorem was proven using differentiation technique. In [8,1], the new approach was applied successfully to nonlinear approximation, including rational and generalized rational uniform approximations.

Here we show that the method is applicable to the best approximation from a finite-dimensional subspace in the arbitrary Banach space if the best approximation operator is one-sided differentiable.

§2. Description of the Algorithm

Suppose that the i-th step of the algorithm is performed to find the best approximation of the element $f \in \mathbf{B}$, and that the element $P_i \in \mathbf{P}_n$ is found. If $\|f - P_i\| = E(f)$, then the process is finished. Otherwise the inequality

$$\|f - P_i\| > E(f)$$

holds, and the next step should be performed.

In order to construct the next approximation P_{i+1}, we construct an auxiliary element $g_i \in \mathbf{B}$ such that the equality

$$\|f + g_i - P_i\| = \|f - P_i\| \tag{1}$$

holds. Suppose the following assumptions are true:

Assumption 1. *The mapping $G = G(f) : \mathbf{P}_n \to \mathbf{B}$, which defines the auxiliary element $g_i = G(f, P_i)$ is continuous.*

Assumption 2. *For all functionals $x \in \mathbf{B}^*$ with properties $|x(f - P_i)| = \|f - P_i\|, \|x\|_{B^*} = 1$, equality $x(g_i) = 0$ holds. Moreover, for every such extremal functional x, a weak neighbourhood $V(x) \subset \mathbf{B}^*$ exists such that*

$$Re\overline{y(f - P_i)}y(g_i) \geq 0, \quad \forall y \in V(x) \cap \{z \in \mathbf{B}^*, \|z\| = 1\}.$$

Assumption 3. *For given fixed f the mapping $D = D(f) : \mathbf{B} \to \mathbf{P}_n$, which defines the derivative $D_i = D(f, g_i)$, is continuous.*

These assumptions may be satisfied easily for real and complex uniform approximations [4,5,7]. When the derivative

$$D_i = \left.\frac{d\mathbf{P}(f + (1-t)g_i)}{dt}\right|_{t=+0}$$

is calculated (usually as the solution of system of linear algebraic equations), the next element P_{i+1} is computed as

$$P_{i+1} = P_i + t_i D_i, \tag{2}$$

where $c\tau_i < t_i \leq \tau_i, 0 < c = const \leq 1$, and τ_i is the minimal value of t, for which

$$\hat{E}_i(\tau_i) = \min\{\hat{E}_i(t) = \|f - P_i - tD_i\|, 0 \leq t \leq 1\}.$$

§3. Main Theorem

Theorem 1. *For given $f \in \mathbf{B} \backslash \mathbf{P_n}$, the sequence $\{P_i\}_{i=0}^{\infty}$ constructed according to the general scheme (see the previous section) converges to the element $\mathbf{P(f)}$ of best approximation of $f \in \mathbf{B}$.*

Proof: Let us write

$$\begin{aligned} E_i &= \|f - P_i\|, \\ E_i(t) &= \|f + (1-t)g_i - \mathbf{P}(f + (1-t)g_i)\|, \\ \tilde{E}_i(t) &= \|f + (1-t)g_i - P_i - tD_i\|, \\ \hat{E}_i(t) &= \|f - P_i - tD_i\|, \\ \alpha_i &= \hat{E}'_i(+0), \\ E_i^* &= \min\{\hat{E}_i(t), 0 \le t \le 1\} = \hat{E}_i(\tau_i), \qquad i \ge 0. \end{aligned} \tag{3}$$

The convexity of the function $E_i(t)$ implies

$$E'_i(+0) \le E_i(1) - E_i(0) = E(f) - E(f + g_i) = E(f) - \|f - P_i\| < 0.$$

Since

$$|E_i(t) - \tilde{E}_i(t)| \le \|\mathbf{P}(f + (1-t)g_i) - P_i - tD_i\| = o(t), \ t \to +0,$$

the equality

$$\tilde{E}'_i(+0) = E'_i(+0)$$

holds. Therefore,

$$\tilde{E}'_i(+0) \le E(f) - E_i < 0. \tag{4}$$

Now we show that there exists $\epsilon > 0$ such that

$$\hat{E}_i(t) \le \tilde{E}_i(t) + \frac{2t^2\|D_i\|^2}{\|f - P_i\|} \tag{5}$$

for all $0 \le t < \epsilon$. Suppose that the last statement is invalid. Then there is a sequence $\{t_l\}, \quad t_l \to +0, \quad l \to \infty$, such that

$$\|f - P_i - t_l D_i\| > \|f + (1-t_l)g_i - P_i - t_l D_i\| + \frac{2t_l^2\|D_i\|^2}{\|f - P_i\|}, \qquad l \ge 1.$$

Choosing a subsequence if necessary, we may consider the weakly convergent sequence of functionals $\{x_l\} \subset \mathbf{B}^*, \quad \forall l \ \|x_l\| = 1$, such that

$$|x_l(f - P_i - t_l D_i)| > \|f + (1-t_l)g_i - P_i - t_l D_i)\| + \frac{2t_l^2\|D_i\|^2}{\|f - P_i\|}, \qquad l \ge 1. \tag{6}$$

Let $x_0 = \lim_{l \to \infty} x_l$. Then inequality (6) implies

$$|x_0(f - P_i)| \ge \|f + g_i - P_i\| = \|(f - P_i)\|,$$

and since $\|x_0\| = 1$, we finally have

$$|x_0(f - P_i)| = \|(f - P_i)\|.$$

Hence in accordance with Assumption 2

$$x_0(g_i) = 0. \tag{7}$$

Since $\|x_l\|_{\mathbf{B}^*} = 1$, it follows from (6) that

$$|x_l(f - P_i - t_l D_i)| > |x_l(f + (1 - t_l)g_i - P_i - t_l D_i))| + \frac{2t_l^2 |x_l(D_i)|^2}{\|f - P_i\|}, \tag{6$'$}$$

for $l \geq 1$.

Now we temporarily write

$$\begin{aligned} a &= x_l(f - P_i - t_l D_i), \\ b &= (1 - t_l)x_l(g_i), \\ s &= \frac{2t_l^2 |x_l(D_i)|^2}{\|f - P_i\|} > 0. \end{aligned}$$

Using this notation, the inequality (6$'$) may be rewritten in the form

$$|a| > |a + b| + s,$$

which implies $|a| > |a + b|$. Consequently,

$$|a|^2 > |a + b|^2 + 2|a + b|s > |a + b|^2 + 2|a|s.$$

Thus,

$$|a|^2 + |b|^2 + 2\mathrm{Re}\bar{a}b < |a|^2 - 2|a|s,$$

and

$$2\mathrm{Re}\bar{a}b < -|b|^2 - 2|a|s. \tag{8}$$

Now we substitute the values of a, b, s in (8) and obtain

$$\begin{aligned} &2\mathrm{Re}\overline{x_l(f - P_i - t_l D_i)}x_l(g_i)(1 - t_l) \\ &\qquad < -(1 - t_l)^2 |x_l(g_i)|^2 - 4|x_l(f - P_i - t_l D_i)\frac{t_l^2 |x_l(D_i)|^2}{\|f - P_i\|}. \end{aligned}$$

Since $t_l \to +0$, $x_l(f - P_i - t_l D_i) \to \|f - P_i\|$, when $l \to \infty$, there exists number l_0 such that inequalities

$$
\begin{aligned}
&2\mathrm{Re}\,\overline{x_l(f-P_i)}x_l(g_i)\\
&\quad < -(1-t_l)|x_l(g_i)|^2 + 2t_l\mathrm{Re}\overline{x_l(D_i)}x_l(g_i) - \frac{4t_l^2|x_l(f-P_i-t_lD_i)||x_l(D_i)|^2}{(1-t_l)\|f-P_i\|}\\
&\quad < -\frac{|x_l(g_i)|^2}{2} + 2t_l\mathrm{Re}\overline{x_l(D_i)}x_l(g_i) - 2t_l^2|x_l(D_i)|^2\\
&\quad = -\frac{1}{2}|x_l(g_i) + 2x_l(D_i)|^2 < 0
\end{aligned}
$$

are valid for all $l \geq l_0$. Therefore,

$$\mathrm{Re}\overline{x_l(f-P_i)}x_l(g_i) < 0, \qquad \forall l \geq l_0.$$

But taking into consideration (7), we see that this inequality contradicts Assumption 2, so (5) is proven.

Since $\tilde{E}_i(0) = \hat{E}_i(0) = E_i(0)$, (5) implies

$$\hat{E}_i'(+0)| \leq E_i'(+0) < E(f) - E_i. \tag{4'}$$

Therefore, in accordance with (2),

$$\hat{E}_i(\tau_i) = min\{\|f - P_i - tD_i\|, 0 \leq t \leq 1\} < E_i \text{ and } \tau_i > 0.$$

So

$$E_{i+1} < E_i.$$

Hence the sequence $\{E_i\}_{i=0}^{\infty}$ converges to some value $E_* \geq E(f)$, i.e.,

$$\lim E_i = E_*. \tag{9}$$

From (2), (4′) and convexity of $E_i(t)$, it follows that

$$E_{i+1} < E_i - \frac{E_i - \hat{E}_i(\tau_i)}{\tau_i}t_i \leq E_i - c(E_i - \hat{E}_i(\tau_i)).$$

Consequently,

$$\Delta E_i = E_{i+1} - E_i \geq c(E_i - \hat{E}_i(\tau_i)).$$

Since $\Delta E_i \to 0$, also $E_i - \hat{E}_i(\tau_i) \to 0$, and

$$\lim \hat{E}_i(\tau_i) = \lim E_i = E_*. \tag{9'}$$

To complete the proof we must show that

$$E_* = E(f).$$

Suppose that this statement isn't valid, i.e.,

$$E_* > E(f). \tag{10}$$

Since the subspace $\mathbf{P}_n$ is finite-dimensional and Assumptions 1 and 3 hold, the subsequence $\{P_{i_k}\} = \{\tilde{P}_k\}$ may be chosen so that the following limits exist:

i) $\lim \tilde{P}_k = P_*$;

ii) $\lim g_{i_k} = g_*$;

iii) $\lim \alpha_{i_k} = \alpha_*$;

iv) $\lim \tau_{i_k} = \tau_*$.

From Assumption 3 it follows that also

$$\lim D_{i_k} = D_*$$

exists, and

$$\lim \|f - P_{i_k} - tD_{i_k}\| = \|f - P_* - tD_*\| \tag{11}$$

uniformly for $t \in [0,1]$. Equalities (9) and (9') imply that at least one of following statements

$$\alpha_* = \lim \alpha_{k_i} = 0; \tag{12}$$

or

$$\tau_* = \lim \tau_{k_i} = 0 \tag{13}$$

is valid.

Using the assumption (10) and the scheme of the algorithm, we construct the auxiliary element $\tilde{g} \neq 0$ for the approximation P_*. Due to Assumption 1, we have

$$g_* = \tilde{g}.$$

For the following two convex functions

$$E_*(t) = \|f + (1-t)g_* - \mathbf{P}(f + (1-t)g_*)\|,$$

$$\hat{E}_*(t) = \|f - P_* - tD_*\|,$$

where

$$D_* = \frac{d\mathbf{P}(f + (1-t)g_*)}{dt}\,|_{t=+0},$$

analogously to (4′), we obtain inequalities

$$E'_*(+0) = \hat{E}'_*(+0) < E(f) - E_* < 0 \tag{14}$$

and $\tilde{\tau} > 0$, where τ is a minimal value of t, for which

$$\hat{E}_*(\tilde{\tau}) = \min\{\hat{E}_*(t),\ 0 \le t \le 1\}.$$

From (11), it follows that $\tau_* = \tilde{\tau}$ and therefore (13) is impossible. So (12) is true. But (14) implies that there is an integer k_0 such, that for every $k, k > k_0$, the inequality

$$\alpha_{i_k} < \frac{E(f) - E_*}{2} < 0$$

is valid, and therefore

$$\alpha_* < 0.$$

This inequality contradicts (12). Hence assumption (10) is invalid. The theorem is proven. □

§3. Applications

As was mentioned above, the proposed algorithm may be considered as a wide generalization of the classical Remez algorithm. Applied to polynomial approximations of complex-valued functions, the method generates an algorithm which possesses in general a linear convergence as numerical experiments show (see also [6]). For finite sets, the convergence of the algorithm is quadratical.

When applied to real polynomial approximations, the method generates exactly the Remez algorithm [7]. But even in this case an approach which uses differential properties of the best approximation operator allows better estimations of convergence.

Theorem 2. ([7]). *Let* $\mathbf{P_n}$ *be a* n*-dimensional Chebyshev subspace in* $C[a,b]$, *and let* $\Delta_h^2(u,x)$ *be the second difference of the function* u *at the point* x *with step* h. *If the function* $f \in C[a,b]\backslash\mathbf{P_n}$ *has the best approximation* $\mathbf{P}(f) \in \mathbf{P_n}$ *such that the difference* $f - \mathbf{P}(f)$ *possesses exactly* $n+1$ *extremal points* $x_0, x_1, \ldots, x_n$ *and the inequality*

$$|\Delta_h^2(f - \mathbf{P}(f), x_j)| \geq \gamma h^2, \ \gamma = const, \quad j = 0, 1, \ldots, n$$

holds in points x_j, *then the Remez algorithm for* f *converges quadratically.*

In [7] a modified Remez algorithm for twice continuously differentiable functions is proposed. A procedure for extremal points calculation, using differential properties, is developed to reduce the complexity of the most complicated part of the Remez algorithm.

This method may be applied to the best polynomial L_p- approximation.

References

1. Ivashchuk, Ya. G., Algorithms of best approximation of functions from unisolvent families, in *Approximation Theory and its Applications, Proceedings of the Int. Conf., dedicated to the memory of V. K. Dziadyk, May, Kyiv*. Kyiv, Inst. of Mathematics Publ., 1999, 26–31, to appear.

2. Kolushov, A., About differentiability of the operator of best approximation, Mat. Zametki, **29** (1981), No.4, 577–596 (Russian); Engl. transl. in Math. Notes **29** (1981).
3. Kovtunets, V. V., Differentiability of the operator of best uniform approximation of complex-valued functions, Soviet Math. Dokl. **42** (1991), No. 1, 41–44.
4. Kovtunets, V. V., Algorithm for computing the best approximation polynomial of the complex-valued function, *Issledovaniya Po Teorii Approximacii Funkcij (Researchs on Function Approximation Theory)*, Kiev, Inst. of Mathematics Publ., 1987, 35–42 (Russian).
5. Kovtunets, V. V., Algorithm for computing the best approximation polynomial of the complex-valued function on the compact set, *Nekotorye Voprosy Teorii Priblizheniya Funkcij I Ih Prilozheniya (Some Problems of Approximation theory and its Applications)*, Kiev, Inst. of Mathematics Publ., 1988, 71–78 (Russian).
6. Kovtunets, V. V., About application of polynomials with the least deviation from zero to the problem of conformal mapping, Ukr. Mathem. Journal, **41** (1989), No. 4, 566–567 (Russian).
7. Kovtunets, V. V., Quasi-newtonian approach to the development of algorithms of the best uniform approximation, Volynskiy Mathem. Visnyk, L'viv Math. Society Publ., **1** (1994), 14–29. (Ukrainian)
8. Kovtunets, V. V., The algorithm for the best uniform nonlinear approximation, *Approximation Theory and its Applications, Proceedings of the Int. Conf., dedicated to the memory of V. K. Dziadyk, May, Kyiv*. Kyiv, Inst. of Mathematics Publ., 1999, 26–31, to appear.
9. Kroò, A., Differential properties of the operator of best approximation, Acta math. Acad. sci. hung. **30** No. 3-4 (1977), 185–203.

blvrd. Lesi Ukrainky 21a, ap. 17
Kyiv,
Ukraine 01133
`ktp@carrier.kiev.ua`

On Curve Interpolation in $\mathbb{R}^{\mathrm{d}}$

Jernej Kozak and Emil Žagar

Abstract. In this paper the interpolation by G^2 continuous spline curves of degree n in $\mathbb{R}^d$ is studied. There are r interior and two boundary data points interpolated on each segment of the spline curve. The general form of the spline curve, as well as the defining system of nonlinear equations are derived. The asymptotic existence of the solution, and the approximation order are studied for the polynomial case only. It is shown that the optimal approximation order is achieved, and asymptotic existence is established provided the relation $r = n - 2$ is satisfied. These conclusions hold independently of d. It is also pointed out that the underlying analysis could not be carried over to the case $r = n - 1$.

§1. Introduction

The interpolation problem considered is the following. Let the points

$$\boldsymbol{T}_0, \boldsymbol{T}_1, \ldots, \boldsymbol{T}_N \in \mathbb{R}^d, \quad \boldsymbol{T}_j \neq \boldsymbol{T}_{j+1}, \text{ all } j, \ d \geq 2, \tag{1}$$

and the tangent directions

$$\boldsymbol{d}_0, \boldsymbol{d}_N \tag{2}$$

at the boundary points be given. Find a G^2 continuous spline curve $\boldsymbol{B}_n$ of degree n which interpolates the prescribed data.

The problem appeared first as a particular limit case in [2], and was further generalized in several papers, among them in [3–5,6,9–10]. A general approach to the approximation order achieved can be found in [8].

Here, the general setup is tackled. The interpolating spline curve in the Lagrange form is established and the defining system of nonlinear equations is derived in general. However, the asymptotic existence of the solution (i.e. the existence of the solution when given points are sampled densely enough) and the approximation order turned out too comprehensive to be studied here in a general framework. The positive conclusions for the single segment case when $r = n - 2$ are established. It is possible to extend these results to

Curve and Surface Fitting: Saint-Malo 1999
Albert Cohen, Christophe Rabut, and Larry L. Schumaker (eds.), pp. 263–272.

ISBN 0-8265-1357-3.

the m-segment spline curve, but the proofs are not short, and will appear elsewhere. On the contrary, as one could guess from [8], the case $r = n - 1$ is not encouraging.

Why would one use the G^2-continuous splines as interpolating curves? Quite clearly, the derivative continuity at the breakpoints becomes in this way independent of the local parametrisation. Also, these curves could be seen as a generalization of the odd order spline function interpolation at knots, applied so successfully in many cases. The order of G-continuity 2 is pinned down by the human eye, sometimes quite important in CAGD: it can detect the continuity, the continuity of the tangent direction and the curvature, but hardly higher order geometric quantities.

Throughout the paper bold faced letters will stand for vectors, and ordinary ones for scalars. The dot product on $\mathbb{R}^d$ will be denoted by $\cdot$ and its implied norm by $||\cdot||$. Derivatives with respect to the global (or local) parameter will be denoted by $\dot{}$ (or $d/d\zeta$), and those with respect to the natural parameter by $'$.

Now let $\boldsymbol{B}_n$ be a continuous spline curve of degree n with m segments

$$\boldsymbol{B} := \boldsymbol{B}_n : [\zeta_0, \zeta_m] \to \mathbb{R}^d$$

corresponding to the breakpoints

$$\zeta_0 < \zeta_1 < \ldots < \zeta_m.$$

We suppose that the pieces are locally parametrized on $[0,1]$ as

$$\boldsymbol{B}(\zeta) = \boldsymbol{B}^\ell\left(\frac{\zeta - \zeta_{\ell-1}}{\Delta\zeta_{\ell-1}}\right), \quad \zeta \in [\zeta_{\ell-1}, \zeta_\ell].$$

Suppose $\boldsymbol{B}$ interpolates the data (1), and (2). If r interior and two boundary points are to be met on each segment, then $N = m(r+1)$. Further, on the ℓ-th segment the interpolation conditions read

$$\boldsymbol{B}^\ell(t_{\ell,j}) = \boldsymbol{T}_{\ell,j} := \boldsymbol{T}_{(\ell-1)(r+1)+j}, \quad j = 0, 1, \ldots, r+1, \quad \ell = 1, 2, \ldots, m, \tag{3}$$

where

$$0 =: t_{\ell,0} < t_{\ell,1} < \cdots < t_{\ell,r+1} := 1,$$

and $(t_{\ell,j})_{j=1}^r$ are the unknown parameters to be determined. Let $\boldsymbol{x} \wedge \boldsymbol{y} \mapsto (x_i y_j - x_j y_i)_{i<j}$ denote the 2-wedge product. The geometric continuity of $\boldsymbol{B}$ requires the tangent direction

$$\frac{1}{||\dot{\boldsymbol{B}}||}\dot{\boldsymbol{B}} \tag{4}$$

as well as the curvature

$$\frac{1}{||\dot{\boldsymbol{B}}||^3}\dot{\boldsymbol{B}} \wedge \ddot{\boldsymbol{B}} \tag{5}$$

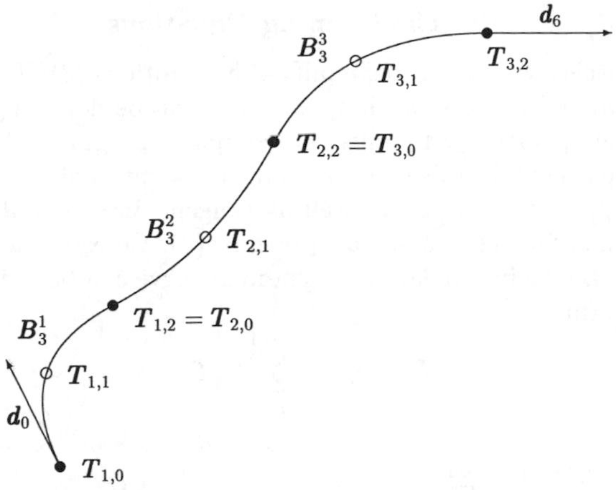

Fig. 1. An interpolating spline curve with three segments .

to be continuous at the breakpoints. Additionally, at the boundary points the tangent directions $\boldsymbol{d}_0$ and $\boldsymbol{d}_N$ have to be interpolated too, i.e.,

$$\boldsymbol{d}_0 \wedge \dot{\boldsymbol{B}}(\zeta_0) = \dot{\boldsymbol{B}}(\zeta_m) \wedge \boldsymbol{d}_N = \boldsymbol{0}. \tag{6}$$

Fig. 1 gives an example of such an interpolating spline curve for $r = 1$, $n = 3$, and $d = 3$. A brief look at the conditions (3)–(6) reveals that the number of independent equations would be equal to the number of independent unknowns if

$$d\,n - (d-1)\,r = 3\,d - 2. \tag{7}$$

As already observed in [5], for fixed d this Diophantine equation always has an infinite number of nonnegative solutions. The following lemma gives its general solution.

Lemma 1. *The possible choices of pairs r and n that satisfy (7) for fixed d are given by*

$$r = d - 2 + dk, \quad n = d + (d-1)k, \quad k = 0, 1, \ldots . \tag{8}$$

Proof: The relation (7) can be rewritten as

$$d(n-d) - (d-1)(r-d+2) = 0.$$

Since $d \geq 2$, the numbers d and $d-1$ are relatively prime. So d must divide $r - d + 2$, and $d - 1$ must divide $n - d$, i.e.,

$$\frac{r-d+2}{d} = \frac{n-d}{d-1} = k$$

for an integer k. But $r = d - 2 + dk \geq 0$ implies $k \geq \frac{2}{d} - 1 > -1$, and the conclusion follows. □

§2. The Defining Equations

Several approaches were used to simplify the conditions (3)–(6) for particular choices of d, n, and r. Here we show that this can be done in general, which will provide an opportunity to unify the computer programs. Let us consider a single segment first. In this case, the data to be interpolated are the points $\boldsymbol{T}_0, \boldsymbol{T}_1, \ldots, \boldsymbol{T}_{r+1}$, $\boldsymbol{T}_j \neq \boldsymbol{T}_{j+1}$, as well as tangent directions $\boldsymbol{d}_0$, $\boldsymbol{d}_{r+1}$ at the boundary points. Suppose r and n are given by (8). Consider the case $n = r+2$ first, i.e., $k = 0$. The interpolating polynomial curve can be written explicitly in Lagrange form as

$$\boldsymbol{B} := \boldsymbol{b}\omega + \sum_{j=0}^{r+1} \boldsymbol{T}_j \mathcal{L}_j$$

with

$$\omega(t) := \prod_{j=0}^{r+1}(t - t_j), \quad \mathcal{L}_j(t) := \frac{\omega(t)}{(t - t_j)\omega'(t_j)}, \tag{9}$$

$t_j := t_{1,j}$, and the values $(t_j)_{j=1}^{r}$, to be determined. Here $\boldsymbol{b} \in \mathbb{R}^d$ denotes the unknown leading coefficient vector. If $k \geq 1$, one has

$$r + 2 = d(k+1) > d(k+1) - k = n,$$

and $\boldsymbol{B}$ is of degree at most $r+1$, i.e.,

$$\boldsymbol{B} = \sum_{j=0}^{r+1} \boldsymbol{T}_j \mathcal{L}_j.$$

In particular, this imposes additional conditions

$$\text{degree} \sum_{j=0}^{r+1} \boldsymbol{T}_j \mathcal{L}_j \leq n \tag{10}$$

for $k > 1$. An easy way to meet the tangent direction conditions (6) is to introduce two additional (strictly positive) real unknowns, α_0 and α_{r+1}, and require

$$\dot{\boldsymbol{B}}(t_0) = \alpha_0 \boldsymbol{d}_0, \quad \dot{\boldsymbol{B}}(t_{r+1}) = \alpha_{r+1} \boldsymbol{d}_{r+1}. \tag{11}$$

Let

$$\tau_{-1} = \tau_0 := t_0, \quad \tau_j := t_j,\ j = 1, 2, \ldots, r, \quad \tau_{r+2} = \tau_{r+1} := t_{r+1}. \tag{12}$$

Since $\boldsymbol{B}$ is a polynomial of degree $\leq n$, the divided difference, based upon

$$n + 2 = r + 4 - k$$

points maps it to zero. So the conditions (11) and (10) can be written in a compact form as

$$[\tau_{j-1}, \tau_j, \ldots, \tau_{j+r+2-k}]\boldsymbol{B} = 0,\ j = 0, 1, \ldots, k, \tag{13}$$

which is a system of $d(k+1)$ nonlinear equations for $r+2 = d(k+1)$ scalar unknowns

$$\alpha_0, t_1, t_2, \ldots, t_r, \alpha_{r+1}. \tag{14}$$

In the case $n = r+2$, one has to determine additionally the coefficient vector $\boldsymbol{b}$, for example as

$$\boldsymbol{b} = [t_0, t_0, t_1, \ldots, t_r, t_{r+1}]\boldsymbol{B} = [t_0, t_1, \ldots, t_r, t_{r+1}, t_{r+1}]\boldsymbol{B}. \tag{15}$$

Now, for an m-segment spline curve, the directions $\boldsymbol{d}_\ell$, $\ell = 1, 2, \ldots, m-1$, are unknown, as well as

$$\alpha_{\ell,0}, t_{\ell,1}, t_{\ell,2}, \ldots, t_{\ell,r}, \alpha_{\ell,r+1}, \quad \ell = 1, 2, \ldots, m. \tag{16}$$

But one can still write the interpolation conditions on the ℓ-th segment as

$$[\tau_{\ell,j-1}, \tau_{\ell,j}, \ldots, \tau_{\ell,j+r+2-k}]\boldsymbol{B}^\ell = 0, \quad j = 0, 1, \ldots, k, \tag{17}$$

where $\tau_{\ell,j}$ are defined as in (12), but this time for the composite case. In addition, the missing $(d-1)(m-1)$ equations are supplied by the continuity conditions of the curvature (5).

§3. Asymptotic Existencc and Approximation Order

The system of equations based on (17) and continuity of curvature (5) is nonlinear, and one of the approaches to study it is to assume that the data (1) and (2) are based upon a smooth underlying regular parametric curve $\boldsymbol{f} : I \to \mathbb{R}^d$, parametrized by the arclength s. The local expansion of the curve $\boldsymbol{f}$, and the data $\boldsymbol{T}_{\ell,j}$ (sampled densely enough), give rise to an asymptotic analysis of the nonlinear system. The simplest way to obtain the local expansion is to use the Frenet frame as the local coordinate system, and the Frenet-Serret formulae to obtain this expansion. Let $(\boldsymbol{e}_i(s))_{i=1}^d$ denote the Frenet frame, with

$$\boldsymbol{f}' = \boldsymbol{e}_1. \tag{18}$$

The Frenet-Serret formulae read

$$\begin{aligned} \boldsymbol{e}_1'(s) &= \kappa_1(s)\boldsymbol{e}_2(s), \\ \boldsymbol{e}_i'(s) &= -\kappa_{i-1}(s)\boldsymbol{e}_{i-1}(s) + \kappa_i(s)\boldsymbol{e}_{i+1}(s), \quad i = 2, 3, \ldots, d-1, \\ \boldsymbol{e}_d'(s) &= -\kappa_{d-1}(s)\boldsymbol{e}_{d-1}(s), \end{aligned} \tag{19}$$

where κ_i are first $d-1$ principal curvatures of $\boldsymbol{f}$, expanded as

$$\kappa_i(s) = \kappa_{i0} + \frac{1}{1!}\kappa_{i1}s + \frac{1}{2!}\kappa_{i2}s^2 + \cdots \tag{20}$$

Since $\boldsymbol{f}$ is a regular curve, $\kappa_{i0} \geq \text{const} > 0$, $i = 1, 2, \ldots, d-2$. We will additionally assume that $\kappa_{d0} \geq \text{const} > 0$. Beginning with (18), the higher derivatives of $\boldsymbol{f}$ can be computed by (19) and (20). This produces the required expansion

$$\begin{aligned}\boldsymbol{f}(s) &= \boldsymbol{f}(0) + \boldsymbol{f}'(0)s + \frac{1}{2!}\boldsymbol{f}''(0)s^2 + \cdots \\ &= \boldsymbol{f}(0) + (s - \frac{1}{6}\kappa_{1,0}^2 s^3 + \cdots)\boldsymbol{e}_1(0) \\ &+ (\frac{1}{2}\kappa_{1,0}s^2 + \frac{1}{6}\kappa_{1,1}s^3 + \cdots)\boldsymbol{e}_2(0) + (\frac{1}{6}\kappa_{1,0}\kappa_{2,0}s^3 + \cdots)\boldsymbol{e}_3(0) + \cdots\end{aligned} \tag{21}$$

Let us now consider the single segment case of the interpolation problem with data based on a smooth $\boldsymbol{f} : [0,h] \to \mathbb{R}^d$,

$$\boldsymbol{d}_0 = \boldsymbol{f}'(\eta_0 h), \quad \boldsymbol{T}_j = \boldsymbol{f}(\eta_j h), \ j = 0, 1, \ldots, r+1, \quad \boldsymbol{d}_{r+1} = \boldsymbol{f}'(\eta_{r+1} h),$$

with points separated independently of h, i.e.,

$$0 =: \eta_0 < \eta_1 < \cdots < \eta_r < \eta_{r+1} := 1.$$

Since translation and rotation do not influence the asymptotic analysis, we may assume $\boldsymbol{f}(0) = \boldsymbol{0}$, and

$$\boldsymbol{e}_i(0) = (\delta_{i,j})_{j=1}^d, \quad i = 1, 2, \ldots, d. \tag{22}$$

Then, with the help of (21), one obtains

$$\boldsymbol{f}(\eta_j h) = \left(\frac{1}{i!}\eta_j^i h^i \prod_{q=0}^{i-1} \kappa_{q,0} + \mathcal{O}(h^{i+1}) \right)_{i=1}^d, \tag{23}$$

and a similar expression for $\boldsymbol{f}'(h)$. Since the divided difference is a linear functional, we can normalize the system (13) by multiplying the data values by D^{-1}, with

$$D := \operatorname{diag}\left(\frac{1}{i!} h^i \prod_{q=1}^{i-1} \kappa_{q,0} \right)_{i=1}^d.$$

Let $\widetilde{\boldsymbol{f}}(s) := (s^i)_{i=1}^d$ denote the leading part of the normalized $\boldsymbol{f}$. Then

$$[t_0, t_0, t_1, \ldots, t_r, t_{r+1}, t_{r+1}] D^{-1} \boldsymbol{B} = [t_0, t_0, t_1, \ldots, t_r, t_{r+1}, t_{r+1}] \widetilde{\boldsymbol{B}} + \mathcal{O}(h), \tag{24}$$

and $\widetilde{\boldsymbol{B}}$ is a polynomial of degree $\le n = r+2$ that satisfies the interpolation conditions

$$\begin{aligned}\widetilde{\boldsymbol{B}}'(t_j) &= \widetilde{\alpha}_j \widetilde{\boldsymbol{f}}'(\eta_j), \quad j = 0, r+1, \\ \widetilde{\boldsymbol{B}}(t_j) &= \widetilde{\boldsymbol{f}}(\eta_j), \quad j = 0, 1, \ldots, r, r+1,\end{aligned}$$

where

$$\widetilde{\alpha}_0 := \frac{\alpha_0}{h}, \quad \widetilde{\alpha}_{r+1} := \frac{\alpha_{r+1}}{h}.$$

Note that all the components of $\widetilde{\boldsymbol{f}}$ are polynomials of degree $\le d = r+2$. This implies that

$$[\eta_0, \eta_0, \eta_1, \ldots, \eta_r, \eta_{r+1}, \eta_{r+1}]\widetilde{\boldsymbol{f}} = 0, \tag{25}$$

and the solution of (24) in the limit $h \to 0$ now reads as

$$\boldsymbol{t}^* := (\widetilde{\alpha}_0^*, t_1^*, t_2^*, \ldots, t_r^*, \widetilde{\alpha}_{r+1}^*) = (1, \eta_1, \eta_2, \ldots, \eta_r, 1). \tag{26}$$

To prove the existence of the solution for h small enough, it is sufficient to show that the Jacobian of the system (24) is nonsingular at the limit (26). The Jacobian will be determined with the help of the following fact: if x_j is different from all the other points x_i, and if a function g is smooth enough, one has

$$\begin{aligned}\left(\frac{\partial}{\partial x_j}[\ldots, x_j, \ldots]\right)g &= \frac{d}{dx_j}\Big([\ldots, x_j, \ldots]g\Big) - \frac{g'(x_j)}{\prod_{i\neq j}(x_j - x_i)} \\ &= [\ldots, x_j, x_j, \ldots]g - \frac{g'(x_j)}{\prod_{i\neq j}(x_j - x_i)}.\end{aligned} \tag{27}$$

Consider now $\widetilde{\boldsymbol{B}} = (\widetilde{\boldsymbol{B}} - \widetilde{\boldsymbol{f}}) + \widetilde{\boldsymbol{f}}$. Since $\widetilde{\boldsymbol{B}} - \widetilde{\boldsymbol{f}} = 0$ at $\boldsymbol{t}^*$, all its partial derivatives with respect to t_j vanish, and this difference contributes to the Jacobian at the limit point $\boldsymbol{t}^*$ only in the first and last column, i.e.,

$$\begin{aligned}\frac{\partial}{\partial \alpha_0}[t_0, t_0, t_1, \ldots, t_r, t_{r+1}, t_{r+1}](\widetilde{\boldsymbol{B}} - \widetilde{\boldsymbol{f}})\Big|_{\boldsymbol{t}^*} &= \frac{1}{(\eta_0 - \eta_{r+1})\widetilde{\omega}'(\eta_0)}\widetilde{\boldsymbol{f}}'(\eta_0), \\ \frac{\partial}{\partial \alpha_{r+1}}[t_0, t_0, t_1, \ldots, t_r, t_{r+1}, t_{r+1}](\widetilde{\boldsymbol{B}} - \widetilde{\boldsymbol{f}})\Big|_{\boldsymbol{t}^*} &= \frac{1}{(\eta_{r+1} - \eta_0)\widetilde{\omega}'(\eta_{r+1})}\widetilde{\boldsymbol{f}}'(\eta_{r+1}),\end{aligned} \tag{28}$$

where ω is given by (9), and

$$\widetilde{\omega} := \omega\Big|_{\boldsymbol{t}^*}.$$

The polynomial curve $\widetilde{\boldsymbol{f}}$ does not depend on $\widetilde{\alpha}_0, \widetilde{\alpha}_{r+1}$, and from (27) and (25) one obtains the columns $2, 3, \ldots, r+1$ with $j = 1, 2, \ldots, r$ as

$$\left(\frac{\partial}{\partial t_j}[t_0, t_0, t_1, \ldots, t_r, t_{r+1}, t_{r+1}]\right)\widetilde{\boldsymbol{f}}\Big|_{\boldsymbol{t}^*} = -\frac{1}{(\eta_j - \eta_0)(\eta_j - \eta_{r+1})\widetilde{\omega}'(t_j)}\widetilde{\boldsymbol{f}}'(\eta_j).$$

It is now straightforward to see that the Jacobian at $\boldsymbol{t}^*$ is the Vandermonde matrix $V(\eta_0, \eta_1, \ldots, \eta_{r+1})$, multiplied by $D_1 := \operatorname{diag}(i)_{i=1}^d$ from the left, and by

$$D_2 := \operatorname{diag}\left(-\frac{1}{\widetilde{\omega}'(\eta_0)}, \frac{1}{\eta_1(1-\eta_1)\widetilde{\omega}'(\eta_1)}, \ldots, \frac{1}{\eta_r(1-\eta_r)\widetilde{\omega}'(\eta_r)}, \frac{1}{\widetilde{\omega}'(\eta_{r+1})}\right)$$

from the right. This prepares the proof of the following theorem.

Theorem 2. *The system (13) has a unique solution for h small enough. The approximation order of the resulting interpolating polynomial curve $\boldsymbol{B}_n$ is optimal, i.e., $r+4=n+2$.*

Proof: Since the matrices $V(\eta_0,\eta_1,\ldots,\eta_{r+1})$, D_1 and D_2 are nonsingular, the Jacobian at the limit point $\boldsymbol{t}^*$ is nonsingular, too, and the existence of a unique solution for h small enough is established. Furthermore, the unknown parameters are of the form

$$\alpha_0=\alpha_{r+1}=h+\mathcal{O}(h^2), \quad t_j=\eta_j+\mathcal{O}(h), \quad j=1,2,\ldots,r. \tag{29}$$

Since there are $r+2$ points, as well as two directions interpolated, the optimal approximation order is quite clearly $\leq r+4$. The proof will now follow the approach applied in [2], and extended in [5]. It is based on a reparametrisation that transforms the direction interpolation to the derivative interpolation, and gives an estimate of the parametric approximation order as defined in [7]. Recall (22), and the fact that interpolation conditions are satisfied. By [2] and [5], it is now enough to confirm that all the components of $\boldsymbol{f}$ and $\boldsymbol{B}$ can be reparametrized by the ordinate of the first component of both curves. As to $\boldsymbol{f}$, for h small enough this fact is obvious. The first component behaves by (21) as $s+\mathcal{O}(s^3)$, and the others at least as $\mathcal{O}(s^2)$. To establish the same conclusion for $\boldsymbol{B}$, it is enough to show that

$$\dot{\boldsymbol{B}}=c\,h(\delta_{1i})_{i=1}^d+\mathcal{O}(h^2), \ c\neq 0. \tag{30}$$

Further, the optimal approximation order proof depends on the additional relations

$$\boldsymbol{B}^{(q)}=\mathcal{O}(h^q), \quad q=2,3,\ldots,r+2. \tag{31}$$

The result required then follows from the standard error estimate of interpolation, and the fact that the $(r+4)$-th derivative of $\boldsymbol{B}$ with respect to the new parameter is bounded independently of h. Let us verify the relations (30) and (31). Recall first

$$t^q=\sum_{j=0}^{r+1}t_j^q\mathcal{L}_j(t), \quad q=0,1,\ldots,r+1, \quad t^{r+2}=\omega(t)+\sum_{j=0}^{r+1}t_j^{r+2}\mathcal{L}_j(t). \tag{32}$$

The divided difference $[t_0,t_0,t_1,\ldots,t_r,t_{r+1}]$ maps polynomials of degree $\leq r+1=d-1$ to zero, and depends continuously on its arguments if applied to a smooth function. Thus $\boldsymbol{b}$ by (15) and (23) near the limit point $\boldsymbol{t}^*$ behaves like

$$\boldsymbol{b}=\left(\mathcal{O}(h^d),\mathcal{O}(h^d),\ldots,\mathcal{O}(h^d),\chi_d\,h^d+\mathcal{O}(h^{d+1})\right)^T,$$

where $\chi_i=\prod_{q=0}^{i-1}\kappa_{q,0}>0$. On the other hand, (29) and (32) imply that

$$\begin{aligned}\sum_{j=0}^{r+1}\boldsymbol{T}_j\mathcal{L}_j(t)&=\sum_{j=0}^{r+1}\boldsymbol{f}(\eta_jh)\mathcal{L}_j(t)\\&=\left(\chi_1\,h\,t,\ldots,\chi_{d-1}\,h^{d-1}\,t^{d-1},\chi_d\,h^d\,(t^d-\omega(t))\right)^T(1+\mathcal{O}(h)).\end{aligned}$$

But

$$\boldsymbol{B}^{(q)}(t) = \boldsymbol{b}\,\omega(t)^{(q)} + \sum_{j=0}^{r+1} \boldsymbol{T}_j \mathcal{L}_j(t)^{(q)}, \quad q = 1, 2, \ldots, r+2,$$

and (31) follows. The proof is complete. □

There is no hope that this approach could be used for all k. In fact, it fails already for $k = 1$, as we will show now. By (13), the equation (24) is replaced by

$$[t_0, t_0, t_1, \ldots, t_r, t_{r+1}] D^{-1} \boldsymbol{B} = [t_0, t_0, t_1, \ldots, t_r, t_{r+1}] \widetilde{\boldsymbol{B}} + \mathcal{O}(h),$$
$$[t_0, t_1, \ldots, t_r, t_{r+1}, t_{r+1}] D^{-1} \boldsymbol{B} = [t_0, t_1, \ldots, t_r, t_{r+1}, t_{r+1}] \widetilde{\boldsymbol{B}} + \mathcal{O}(h).$$

Further, as in the proof of Theorem 2, the first column of the Jacobian is determined from

$$\frac{\partial}{\partial \alpha_0} [t_0, t_0, t_1, \ldots, t_r, t_{r+1}] \big(\widetilde{\boldsymbol{B}} - \widetilde{\boldsymbol{f}}\big)\Big|_{\boldsymbol{t}^*} = \frac{1}{\widetilde{\omega}'(\eta_0)} \widetilde{\boldsymbol{f}}'(\eta_0),$$
$$\frac{\partial}{\partial \alpha_0} [t_0, t_1, \ldots, t_r, t_{r+1}, t_{r+1}] \big(\widetilde{\boldsymbol{B}} - \widetilde{\boldsymbol{f}}\big)\Big|_{\boldsymbol{t}^*} = \boldsymbol{0},$$

the last column from

$$\frac{\partial}{\partial \alpha_{r+1}} [t_0, t_0, t_1, \ldots, t_r, t_{r+1}] \big(\widetilde{\boldsymbol{B}} - \widetilde{\boldsymbol{f}}\big)\Big|_{\boldsymbol{t}^*} = \boldsymbol{0},$$
$$\frac{\partial}{\partial \alpha_{r+1}} [t_0, t_1, \ldots, t_r, t_{r+1}, t_{r+1}] \big(\widetilde{\boldsymbol{B}} - \widetilde{\boldsymbol{f}}\big)\Big|_{\boldsymbol{t}^*} = \frac{1}{\widetilde{\omega}'(\eta_{r+1})} \widetilde{\boldsymbol{f}}'(\eta_{r+1}),$$

and the other columns from

$$\left(\frac{\partial}{\partial t_j} [t_0, t_0, t_1, \ldots, t_r, t_{r+1}]\right) \widetilde{\boldsymbol{f}}\big|_{\boldsymbol{t}^*} = -\frac{1}{(\eta_j - \eta_0)\widetilde{\omega}'(t_j)} \widetilde{\boldsymbol{f}}'(\eta_j),$$
$$\left(\frac{\partial}{\partial t_j} [t_0, t_1, \ldots, t_r, t_{r+1}, t_{r+1}]\right) \widetilde{\boldsymbol{f}}\big|_{\boldsymbol{t}^*} = -\frac{1}{(\eta_j - \eta_{r+1})\widetilde{\omega}'(t_j)} \widetilde{\boldsymbol{f}}'(\eta_j).$$

After normalizing the Jacobian from the left by D_1^{-1}, and by D_2^{-1} from the right one obtains the matrix $A := (a_{ij})_{i,j=1}^{2d}$ with

$$a_{i,1} = \delta_{i,1}, \; i = 1, 2, \ldots, 2d,$$
$$a_{i,2d} = 0, \; a_{i+d,2d} = 1, \; i = 1, 2, \ldots, d,$$

and

$$a_{i,j} = \eta_{j-1}^{i} - \eta_{j-1}^{i-1}, \; a_{i+d,j} = \eta_{j-1}^{i}, \; i = 1, 2, \ldots, d, \; j = 2, 3, \ldots, 2d-1.$$

A simple rank preserving transformation

$$a_{i,j} \mapsto a_{i,j} - a_{i-1,j}, \; i = 2d, 2d-1, \ldots, d+1, \; j = 1, 2, \ldots, 2d,$$

transforms A to a matrix with row i equal to row $i+d$ for $i = 2, 3, \ldots, d$. It is now easy to see that the rank of the matrix A is $d+1$, and consequently $\dim \ker A = d - 1$. Thus, since the Jacobian is singular, some other approach such as [1], pp. 154–155, should be applied to carry out the asymptotic analysis.

Acknowledgments. Jernej Kozak was supported by the Ministry of Science and Technology of Slovenija.

References

1. Berger, M. S., *Nonlinearity and Functional Analysis*, Academic Press, 1977.
2. de Boor, C., K. Höllig, and M. Sabin, High accuracy geometric Hermite interpolation, Comput. Aided Geom. Design **4** (1987), 269–278.
3. Feng, Y. Y. and J. Kozak, On G^2 continuous interpolatory composite quadratic Bézier curve, J. Comput. Appl. Math. **72** (1996), 141–159.
4. Feng, Y. Y. and J. Kozak, On G^2 continuous cubic spline interpolation, BIT **37** (1997), 312–332.
5. Feng, Y. Y. and J. Kozak, On spline interpolation of space data, in *Mathematical Methods for Curves and Surfaces II*, M. Dæhlen, T. Lyche, and L. L. Schumaker (eds.), Vanderbilt University Press, Nashville, 1998, 167–174.
6. Kozak, J. and M. Lokar, On piecewise quadratic G^2 approximation and interpolation, in *Mathematical Methods in Computer Aided Geometric Design II*, T. Lyche and L. L. Schumaker (eds.), Academic Press, New York, 1992, 359–366.
7. Lyche, T. and K. Mørken, A metric for parametric approximation, in *Curves and Surfaces in Geometric Design*, P.-J. Laurent, A. Le Méhauté, and L. L. Schumaker (eds.), A. K. Peters, Wellesley MA, 1994, 311–318.
8. Mørken, K. and K. Scherer, A general framework for high-accuracy parametric interpolation, Math. Comp. **66**, 217 (1997), 237–260.
9. Schaback, R., Interpolation in $\mathbb{R}^2$ by piecewise quadratic visually C^2 Bézier polynomials, Comput. Aided Geom. Design **6** (1989), 219–233.
10. Schaback, R., Planar curve interpolation by piecewise conics of arbitrary type, Constr. Approx. **9** (1993), 373–389.

Jernej Kozak
FMF and IMFM
University of Ljubljana
Jadranska 19, 1000 Ljubljana, Slovenia
`Jernej.Kozak@FMF.Uni-Lj.Si`

Emil Žagar
FRI and IMFM
University of Ljubljana
Tržaška 25, 1000 Ljubljana, Slovenia
`Emil@Gollum.Fri.Uni-Lj.Si`

Interpolating Involute Curves

Mitsuru Kuroda and Shinji Mukai

Abstract. We propose a straightforward method for designing an interpolating involute curve whose radius of curvature is piecewise linear or quadratic with respect to winding angle. Designers can specify and control the curvature radius profile to a certain extent. End radii of a circle involute are solved in terms of end tangent angles, and a G^1 involute curve is derived by the Hermite interpolation. For G^2 and G^3 involute curves, relevant nonlinear equations are solved by the Newton-Raphson method. NC machines with an involute generator can draw the resulting curves with "reduced data".

§1. Introduction

We present a new method for designing two kinds of smooth interpolating curves, smooth in the sense of consisting of less segments with continuous monotone curvature radius plot. In the method, we derive a G^2 (curvature continuous) involute of circular arcs or a G^3 involute of circle involute arcs through describing its radius of curvature that is piecewise linear or quadratic with respect to winding angle.

"The most important curve in engineering is arguably the circle involute ... it has played key historical roles in a variety of scientific and technological applications" [3]. This curve has excellent shape properties which make it interesting for CAGD (Computer Aided Geometric Design). One can draw the involute curve manually with simple equipment if necessary. NC (Numerical Control) machines with an involute generator are available [1,5].

In our straightforward design method, designers can specify tangents and curvatures at junction points, and control the curvature profile directly to a certain extent. End radii of a circle involute are solved in terms of the end tangent angles, and so an interpolating G^1 involute curve is derived span by span by the two-point Hermite interpolation. For G^2 and G^3 involute curves, continuity conditions and other requirements lead to a system of nonlinear equations. We solve this equation system by the Newton-Raphson method, using initial values from the conventional C^2 cubic spline curve. We obtain examples of the curves satisfying additional requirements, and illustrate the properties of the newly developed curves.

Curve and Surface Fitting: Saint-Malo 1999
Albert Cohen, Christophe Rabut, and Larry L. Schumaker (eds.), pp. 273–280.

ISBN 0-8265-1357-3.

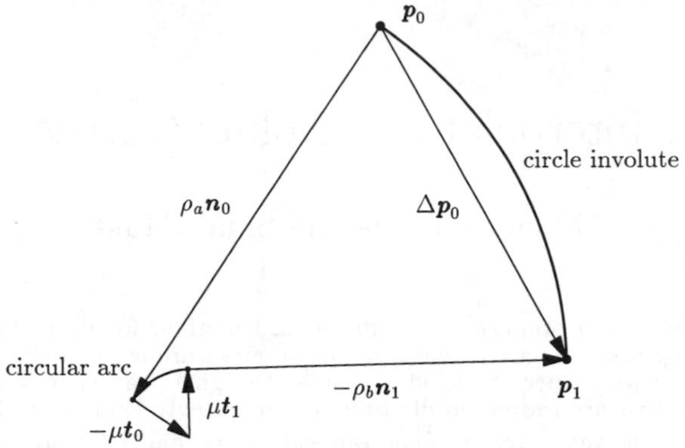

Fig. 1. Circle involute and its evolute.

§2. Circle Involute Arc

A planar curve $\boldsymbol{r}(s)$ is expressed as

$$\boldsymbol{r}(s) = \boldsymbol{p}_0 + \int_0^s \begin{pmatrix} \cos\theta \\ \sin\theta \end{pmatrix} ds, \quad -\infty < \theta < \infty, \tag{1}$$

where s is arclength from the starting point $\boldsymbol{p}_0$ and θ is a winding angle (the angle between tangent vector and the direction of the x axis). The following relations hold among the curve $\boldsymbol{r}$, unit tangent vector $\boldsymbol{t}$, unit normal vector $\boldsymbol{n}$ and radius of curvature ρ:

$$\frac{d^2\boldsymbol{r}}{ds^2} = \frac{d\boldsymbol{t}}{ds} = \frac{d\theta}{ds}\boldsymbol{n} = \frac{1}{\rho}\boldsymbol{n}. \tag{2}$$

The radius of curvature of circle involute is proportional with respect to θ:

$$\rho = \frac{ds}{d\theta} = \rho_a + \mu(\theta - \theta_0), \tag{3}$$

$$\mu \equiv \frac{\rho_b - \rho_a}{\Delta\theta_0} = const.,$$

$$\rho(\theta_0) = \rho_a, \quad \rho(\theta_1) = \rho_b,$$

where μ is the radius of circular arc that is the evolute of $\boldsymbol{r}$, and Δ is the forward difference operator defined by $\Delta z_i \equiv z_{i+1} - z_i$. We change the variable of the expression (1) from s to θ by the relation (3) and integrate it:

$$\Delta\boldsymbol{p}_0 = \int_{\theta_0}^{\theta_1} \rho\boldsymbol{t}d\theta = \rho_a\boldsymbol{n}_0 + \mu(\boldsymbol{t}_1 - \boldsymbol{t}_0) - \rho_b\boldsymbol{n}_1, \tag{4}$$

$$\boldsymbol{r}(\theta_0) = \boldsymbol{p}_0, \quad \boldsymbol{t}(\theta_0) = \boldsymbol{t}_0, \quad \boldsymbol{n}(\theta_0) = \boldsymbol{n}_0,$$
$$\boldsymbol{r}(\theta_1) = \boldsymbol{p}_1, \quad \boldsymbol{t}(\theta_1) = \boldsymbol{t}_1, \quad \boldsymbol{n}(\theta_1) = \boldsymbol{n}_1.$$

The vector equation (4) is understood easily as in Fig. 1.

The expression (4) is a system of equations with respect to unknowns ρ_a and ρ_b. We can solve this as follows:

$$\Delta \boldsymbol{p}_0 \equiv L_0 \begin{pmatrix} \cos\phi_0 \\ \sin\phi_0 \end{pmatrix},$$

$$\begin{aligned} \rho_a &= \frac{\Delta\theta_0 \cos(\theta_1 - \phi_0) + \sin(\theta_0 - \phi_0) - \sin(\theta_1 - \phi_0)}{-2 + 2\cos\Delta\theta_0 + \Delta\theta_0 \sin\Delta\theta_0} L_0, \\ \rho_b &= \frac{\Delta\theta_0 \cos(\theta_0 - \phi_0) + \sin(\theta_0 - \phi_0) - \sin(\theta_1 - \phi_0)}{-2 + 2\cos\Delta\theta_0 + \Delta\theta_0 \sin\Delta\theta_0} L_0. \end{aligned} \tag{5}$$

Since the involute arc is given in terms of start and end points as well as corresponding tangent vectors by the expression (5), we can obtain a G^1 involute curve by the Hermite interpolation which satisfies the equations

$$\boldsymbol{r}(\theta_i) = \boldsymbol{p}_i, \quad i = 0, 1, \ldots, n. \tag{6}$$

Radii of segments of the evolute of $\boldsymbol{r}$ are rewritten as

$$\mu_i = \frac{\cos(\theta_i - \phi_i) - \cos(\theta_{i+1} - \phi_i)}{-2 + 2\cos\Delta\theta_i + \Delta\theta_i \sin\Delta\theta_i} L_i, \quad i = 0,\ 1, \ldots, n-1.$$

§3. Interpolating G^2 Involute Curve

We can also derive an interpolating G^2 involute curve of circular arcs. Using the equation (5), we can solve the following nonlinear equation system (7) with respect to unknowns $\theta_0,\ \theta_1, \ldots, \theta_n$ by the Newton-Raphson method:

$$\rho_i \equiv \rho(-\theta_i) = \rho(+\theta_i), \quad i = 1,\ 2,\ \ldots, n-1. \tag{7}$$

However, the Jacobian matrix necssary for the method makes a programming code long and convergence relatively slow. Therefore, from the practical point of view, we prefer to solve the following equations (8) based on the equation (4) directly. Adding unknowns $\rho_0,\ \rho_1, \ldots, \rho_n$ to the previous ones $\theta_0,\ \theta_1, \ldots, \theta_n$, we get

$$\rho_i \boldsymbol{n}_i + \mu_i \Delta \boldsymbol{t}_i - \rho_{i+1} \boldsymbol{n}_{i+1} = L_i \begin{pmatrix} \cos\phi_i \\ \sin\phi_i \end{pmatrix}, \quad i = 0,\ 1, \ldots, n-1, \tag{8}$$

$$\mu_i = \frac{\Delta\rho_i}{\Delta\theta_i}, \quad i = 0,\ 1, \ldots, n-1.$$

Fig. 2 shows an example of interpolating G^2 involute curves and its profile of curvature and radius of curvature. Initial values were from the conventional C^2 cubic splines. The Newton-Raphson method converged after three iterations. In spite of the unpleasant configuration of data points, the curve derived is quite smooth. Its evolute curve (circular arcs) is G^1 continuous except for two cusp points that correspond to extremal points of the radius of curvature.

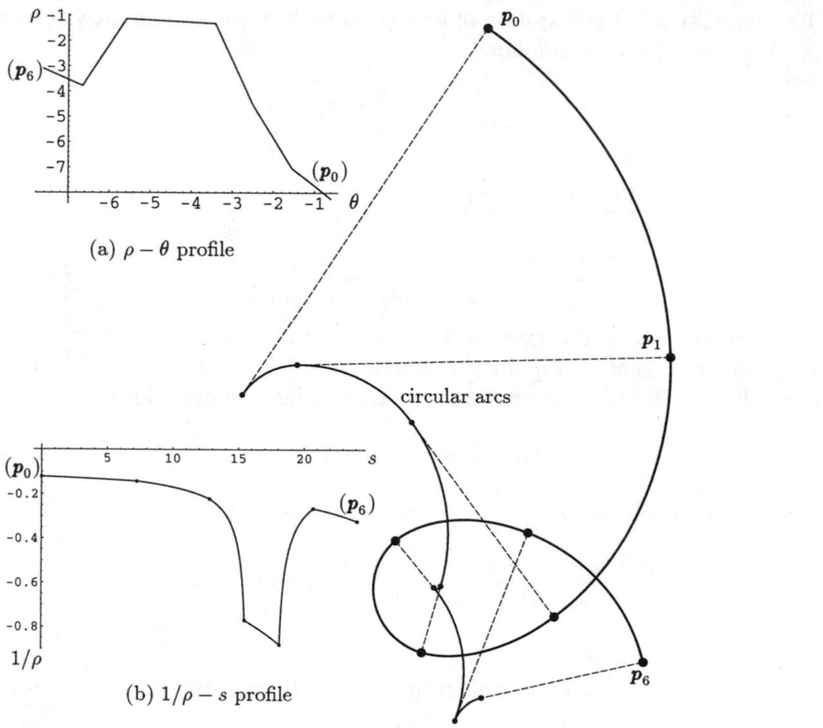

Fig. 2. Interpolating G^2 involute curve.

§4. Interpolating G^3 Involute Curve

Expressing ρ as a quadratic B-spline function of variable θ, we extend the interpolating G^2 involute curve in the previous section. The radius $\rho(\theta)$ is C^1 continuous. The arclength s is cubic with respect to θ, since $\rho = ds/d\theta$ and $\rho(\theta)$ is quadratic. In this case, we get the indefinite integral

$$\boldsymbol{r} = \int \rho \boldsymbol{t} d\theta = -\rho \boldsymbol{n} + \mu \boldsymbol{t} + \nu \boldsymbol{n}, \tag{9}$$

$$\mu \equiv \frac{d\rho}{d\theta}, \quad \nu \equiv \frac{d^2\rho}{d\theta^2} = const.$$

Using (9) span by span, we derive the continuity conditions

$$\rho_i \boldsymbol{n}_i - \rho_{i+1} \boldsymbol{n}_{i+1} - \mu_i \boldsymbol{t}_i + \mu_{i+1} \boldsymbol{t}_{i+1} - \nu_i \boldsymbol{n}_i + \nu_i \boldsymbol{n}_{i+1} = \Delta \boldsymbol{p}_i, \ i = 0,\ 1, \ldots, n-1, \tag{10}$$

$$\rho_i = \rho(\theta_i), \quad \mu_i = \left.\frac{d\rho}{d\theta}\right|_{\theta=\theta_i}, \quad i = 0,\ 1, \ldots, n,$$

$$\nu_i = \left.\frac{d^2\rho}{d\theta^2}\right|_{\theta=\theta_i}, \quad i = 0,\ 1, \ldots, n-1.$$

Based on these conditions, we are going to derive a set of equations with unknown parameters of $\rho(\theta)$ and solve. We use the following notation in $\rho-\theta$ space:

1) Knots: $\theta_{-1},\ \theta_0, \ldots,\ \theta_{n+1}$, where the end knots are of multiplicity 2.

$$\theta_{-1} = \theta_0, \quad \theta_{n+1} = \theta_n.$$

2) de Boor ordinates [2]: $d_0,\ d_1, \ldots,\ d_{n+1}$, where d_i corresponds to the Greville abscissa $(\theta_{i-1} + \theta_i)/2$.
3) Bézier ordinates [2]: $b_0,\ b_1, \ldots,\ b_{2n}$.

For easy manipulation, we break down the non-uniform B-spline function $\rho(\theta)$ into the following quadratic Bézier functions with local parameter t:

$$\rho(\theta) = (1-t)^2 b_{2i} + 2(1-t)t b_{2i+1} + t^2 b_{2i+2}, \tag{11}$$

$$0 \le t = \frac{\theta - \theta_i}{\Delta\theta_i} \le 1, \quad i = 0, 1, \ldots, n-1.$$

$$b_{2i} = \frac{d_{i+1}\Delta\theta_{i-1} + d_i \Delta\theta_i}{\Delta\theta_{i-1} + \Delta\theta_i}, \qquad i = 0,\ 1, \ldots, n,$$

$$b_{2i+1} = d_{i+1}, \qquad i = 0,\ 1, \ldots, n-1.$$

From (11), we obtain

$$\rho_i = b_{2i}, \quad \mu_i = \frac{2\Delta b_{2i}}{\Delta\theta_i}, \quad i = 0,\ 1, \ldots, n,$$

$$\nu_i = \frac{2\Delta^2 b_{2i}}{(\Delta\theta_i)^2}, \quad i = 0,\ 1, \ldots, n-1.$$

We solve the equations (10) with unknowns $\theta_0, \theta_1, \ldots, \theta_n, d_0, d_1, \ldots, d_{n+1}$ by the Newton-Raphson method, using initial values from the conventional C^2 cubic splines. The number of equations is $2n$, while the number of unknowns is $2n+3$. Accordingly we can give 3 more additional requirements. Since the radius of curvature $\rho(\theta)$ determines a unique curve shape, we can specify and control an interpolating curve by the control polygon (Greville abscissae and de Boor ordinates) of $\rho(\theta)$. Therefore the curve includes circular arc, circle involute and involute of the circle involute because $\rho(\theta)$ is a quadratic B-spline function.

Fig. 3 shows an example of an interpolating G^3 involute curve with the same data points and the same end tangents as in Fig. 2. The computation converged after four iterations. The evolute of this curve has cusp points within segments, since the radius of curvature is quadratic, while the evolute in Fig. 2 has cusp points only at junction points. The evolute of the evolute (circular arcs) has three cusp points. The curvature profile shows the smoothness of this G^3 involute curve.

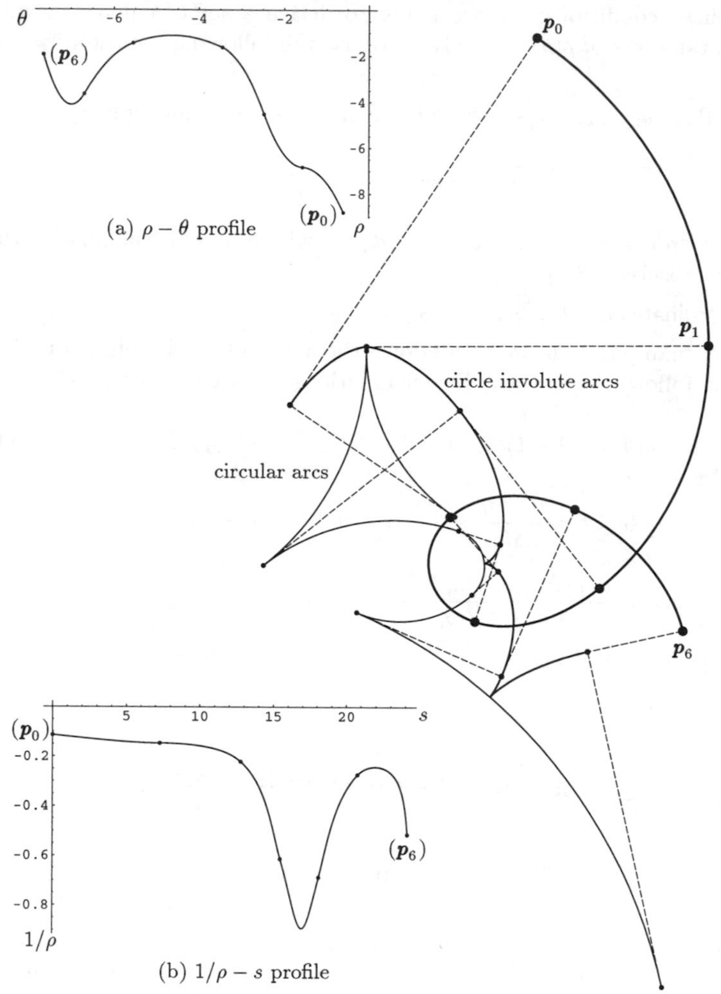

Fig. 3. Interpolating G^3 involute curve.

§5. Some More Numerical Results

To illustrate the properties of the newly developed curve, we show some more examples of the curves. The same data points are used in Figs. 1 to 5.

The G^2 involute curve is practically more important than the G^1 and G^3 ones. Accordingly, in Fig. 4 the G^2 involute in Fig. 2 is compared with other curves: (a) the G^3 involute curve with the same end tangents in Fig. 3, (b) the conventional C^2 cubic spline curve which is used as an initial curve by the Newton-Raphson method, and (c) a G^1 biarc curve derived by minimum difference between curvatures of two arcs [4]. The labels "G^2","G^3" or "C^2" in Fig. 4 point out which side the corresponding curve passes through. Small circles in Fig. 4(c) are centers of circular arcs. It is understood from

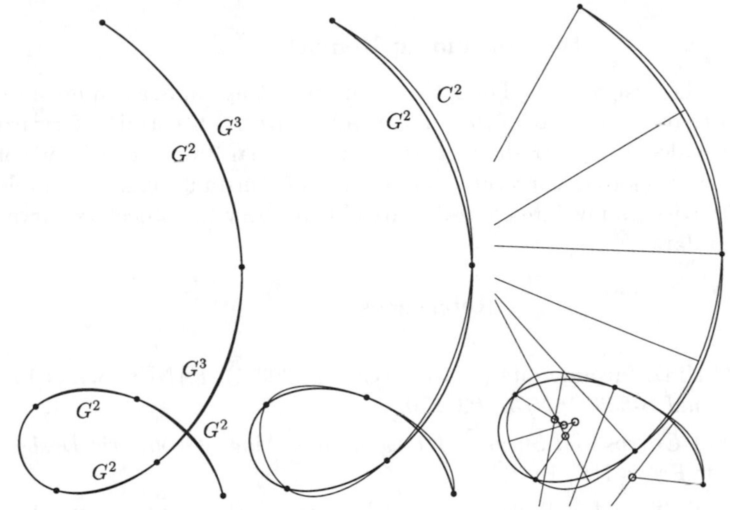

(a) with a G^3 involute (b) with a C^2 cubic curve (c) with a G^1biarc curve

Fig. 4. G^2 involute and comparison with other curves.

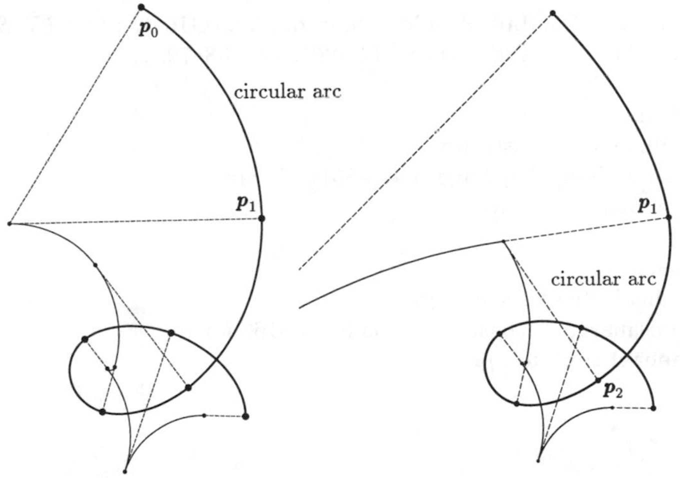

(a) a circular arc $\boldsymbol{p}_0\boldsymbol{p}_1$ (b) a circular arc $\boldsymbol{p}_1\boldsymbol{p}_2$

Fig. 5. G^2 involute including a circular arc.

observation that the involute curves are quite smooth.

Fig. 5 illustrates a G^2 involute curve with additional requirements, which includes (a) a circular arc $\boldsymbol{p}_0\boldsymbol{p}_1$ and (b) a circular arc $\boldsymbol{p}_1\boldsymbol{p}_2$.

§6. Concluding Remarks

We proposed the design method of up to G^3 interpolating curve as an involute of circular arcs or an involute of the circle involute arcs. This straightforward approach provides a tool for the construction of planar curves consisting of segments with monotone curvature radius plots of constant sign. Available NC machines with an involute generator are able to draw the objective curves with reduced data.

References

1. FANUC LTD, Involute interpolation (G02.2, G03.2), *FANUC Series 15i-MA Manual B-63324EN/01*, 92–100.
2. Farin, G., *Curves and Surfaces for Computer Aided Geometric Design*, Academic Press, NY, 1988.
3. Farouki, R. T. and J. Rampersad, Cycles upon cycles: an anecdotal history of higher curves in science and engineering, in *Mathematical Methods for Curves and Surfaces II*, Morten Dæhlen, Tom Lyche, Larry L. Schumaker (eds), Vanderbilt University Press, Nashville & London, 1998, 95–116.
4. Su, B.-Q. and D.-Y. Liu, *Computational Geometry: Curve and Surface Modeling*, G.-Z. Chang (Trans.), Academic Press, San Diego, 1989.
5. Toshiba Machine Co., Ltd., Involute interpolation (G105), *TOSNUC 888 Programing Manual (Additional) STE 42864-11*, 68–79.

Mitsuru Kuroda
Toyota Technological Institute
2-12 Hisakata, Tempaku, Nagoya 468-8511, Japan
`kuroda@toyota-ti.ac.jp`

Shinji Mukai
Maebashi Institute of Technology
460-1 Kamisanaru, Maebashi, Gunma 371-0816, Japan
`mukai@maebashi-it.ac.jp`

Interpolation from Lagrange to Holberg

Michel Léger

Abstract. As the order $2n$ tends to infinity, Lagrange interpolators of periodically sampled 1D functions converge to the *sinc* function modulated by two exponentials. One is related to instabilities and the other to Gaussian apodizing. The Hermite interpolation of Lagrange interpolators gives convolutive $\mathcal{C}^{k+1}$-differentiable Lagrange-Hermite interpolators. Whereas their support has width of order $2n+2$, the active part of their impulse response is width of order $\sqrt{2n}$, instead of $2n$ for Holberg interpolators, which are optimal combinations of Lagrange-Hermite interpolators, and therefore much more efficient. Efficient filters can be derived from these differentiable interpolators, as well as numerical schemes of derivatives at any abscissa.

§1. Introduction

Some applications require very large 2D or 3D regular grids, such as finite-difference modeling of seismic waves, for instance. The processing of these grids involves the computing of numerical schemes of first or second derivatives, and also interpolators and filters. These quantities need to be evaluated very efficiently because the requirements in terms of computation time and memory use are critical.

Numerical schemes, interpolators and filters are interrelated issues, and I choose to study them from the viewpoint of interpolation, which is the most general. For sake of simplicity, I assume 1D interpolation of periodically sampled functions, with unit-sampling rate and even orders.

§2. Lagrange Interpolators

Definition 1. *For some function f with known values f_i at abscissae $x_i = i$, $i \in [-n, n]$, the Lagrange interpolator* ([9,6,2,3]) *of order $2n$ is*

$$\mathcal{L}_{2n}^{0}(\{f_k\}_{k=-n}^{n}, x) = \sum_{i=-n}^{n} f_i \prod_{\substack{j=-n \\ j\neq i}}^{n} \frac{x-j}{i-j} = \sum_{i=-n}^{n} f_i \, p_{2n,\,i}^{L}(x). \qquad (1)$$

Curve and Surface Fitting: Saint-Malo 1999
Albert Cohen, Christophe Rabut, and Larry L. Schumaker (eds.), pp. 281–290.

ISBN 0-8265-1357-3.

It can be shifted and centered at any integral abscissa i:

$$\mathcal{L}^i_{2n}(\{f_k\}^{i+n}_{k=i-n}, x) = \sum_{j=-n}^{n} f_{i+j}\, p^L_{2n,\,j}(x-i).$$

Proposition 2. *As $2n \to \infty$, Lagrange polynomials converge to the sinc function modulated by two exponentials:*

$$p^L_{2n,\,i}(x) = \frac{\sin\pi(x-i)}{\pi(x-i)}\ \exp(\frac{x^2}{n+\frac{1}{2}})\ \exp(\frac{-i^2}{n+\frac{1}{2}})\,(1+O(n^{-3})). \qquad (2)$$

Proof: Simple changes in (1) give

$$p^L_{2n,\,i}(x) = \frac{(-1)^{2n-i}(n!)^2 x}{(n+i)!\,(n-i)!\,(x-i)} \prod_{j=1}^{n}(1-\frac{x^2}{j^2}). \qquad (3)$$

Since $\prod_{j=1}^{\infty}(1-\frac{x^2}{j^2}) = \frac{\sin(\pi x)}{\pi x}$, we have

$$\prod_{j=1}^{n}(1-\frac{x^2}{j^2}) = \frac{\sin(\pi x)}{\pi x} / \prod_{j=n+1}^{\infty}(1-\frac{x^2}{j^2}). \qquad (4)$$

Since $\ln\prod_{j=n+1}^{\infty}(1-\frac{x^2}{j^2}) = -x^2\sum_{n+1}^{\infty}(\frac{1}{j^2}+O(j^{-4}))$, and since $\frac{1}{j^2} = O(j^{-4}) + \int_{j-1/2}^{j+1/2}\frac{dt}{t^2}$, we obtain

$$\prod_{j=n+1}^{\infty}(1-\frac{x^2}{j^2}) = \exp(-\frac{x^2}{n+\frac{1}{2}} + O(n^{-3})). \qquad (5)$$

Moreover, by using Stirling formula $n! = n^n\exp(-n)\sqrt{2\pi n}\,(1+\frac{1}{12n}+\frac{1}{288n^2}+O(n^{-3}))$, and noting that $\ln(1+\frac{1}{12n}+\frac{1}{288n^2}+O(n^{-3})) = \frac{1}{12n}+O(n^{-3})$ and that $\frac{1}{12n}(2-\frac{n}{n+i}-\frac{n}{n-i}) = O(n^{-3})$, we obtain $\ln\frac{(n!)^2}{(n+i)!\,(n-i)!} = -(n+i+\frac{1}{2})\ln(1+\frac{i}{n}) - (n-i+\frac{1}{2})\ln(1-\frac{i}{n}) + O(n^{-3})$. Since $\ln(1+x) = x - \frac{x^2}{2} + \frac{x^3}{3} + O(x^4)$, we have $\ln\frac{(n!)^2}{(n+i)!\,(n-i)!} = \frac{-i^2}{n+\frac{1}{2}} + O(n^{-3})$ and hence

$$\frac{(n!)^2}{(n+i)!\,(n-i)!} = \exp(\frac{-i^2}{n+\frac{1}{2}} + O(n^{-3})). \qquad (6)$$

Noting that $(-1)^i\sin\pi x = \sin\pi(x-i)$ and inserting (4), (5) and (6) in (3) concludes the proof. □

According to the first term of (2), Lagrange polynomials converge to the perfect interpolator *sinc* function (Fig. 1). Away from the center of the interval of the data points, the first exponential explains the well-known instabilities (a small change in the data results in a large change in the interpolation), and the second one corresponds to Gaussian apodizing (see *apodization* in [10]), that is, the vanishing effect of the non-centered data points. Note that these two exponentials compensate one another as $x \to i$.

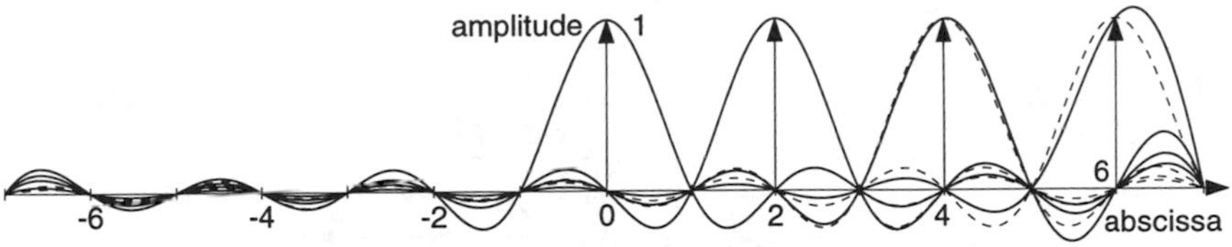

Fig. 1. Lagrange polynomials of order 14 divided by the two exponentials of (2), for $i = 0, 2, 4, 6$. Related *sinc* functions in dotted lines.

§3. Stationary Lagrange Interpolators

Since a Lagrange interpolator is stable only near the middle of the interval of the contributing data points, a natural idea is to change it for each interval between two successive points, in such a way that the interpolator is always used at its best.

Definition 3. *The stationary Lagrange interpolator of order* $2n$ *uses at abscissa* x *the Lagrange interpolator of order* $2n$ *centered at* i_x, *with* $x = i_x + d_x$, $i_x \in \mathbb{N}$, $d_x \in [0,\ 1[$:

$$\mathcal{L}_{2n}(\{f_k\}_{k\in\mathbb{Z}}, x) = \sum_{j=-n}^{n} f_{i_x+j}\ p^L_{2n,j}(d_x) = \mathcal{L}^{i_x}_{2n}(x). \tag{7}$$

Remark 4. *Any interpolator such that* $\mathcal{I}(\{f_k\}_{k\in\mathbb{Z}}, x) = \sum_{j=a}^{b} f_{i_x+j}\, p_j(d_x)$, *with* a, b *and* p_j *independent of* i_x, *is convolutive, that is, there exists a continuous function* $\lambda(x)$ *such that* $\mathcal{I}(\{f_k\}_{k\in\mathbb{Z}}, x) = \Delta(f)*\lambda(x)$, *with* $\Delta(f) = \sum_{i\in\mathbb{Z}} f_i\ \delta(x-i)$ *being the "Dirac comb" modulated by function* f.

This is obvious by considering the impulse response $\lambda(x) = \mathcal{I}(\{\delta_{k0}\}_{k\in\mathbb{Z}}, x) = p_{-i_x}(d_x)$, with δ_{k0} the Kronecker's symbol.

As a particular consequence, stationary Lagrange interpolators are convolutive with impulse responses $\lambda_{2n}(x) = p^L_{2n,\,-i_x}(d_x)$. Moreover, they are also subject to Gaussian apodizing since, as $2n \to \infty$,

$$\lambda_{2n}(x) \sim \frac{\sin \pi x}{\pi x} \exp\left(\frac{-x^2}{n+\frac{1}{2}}\right). \tag{8}$$

It is clear from Fig. 2 that these impulse responses vanish very rapidly as compared to the *sinc* function.

Remark 5. *Since* $\forall k$, $k \in [1,\ 2n]$, $\int_{-\infty}^{+\infty} x^k \lambda_{2n}(x)dx = 0$, *then, with* $\mathcal{F}$ *denoting the Fourier transform,* $\forall k$, $k \in [1,\ 2n]$, $(\mathcal{F}(\lambda_{2n})(u))^{(k)}_{u=0} = 0$ ([1]).

Hence, stationary Lagrange interpolators are good at low frequencies.

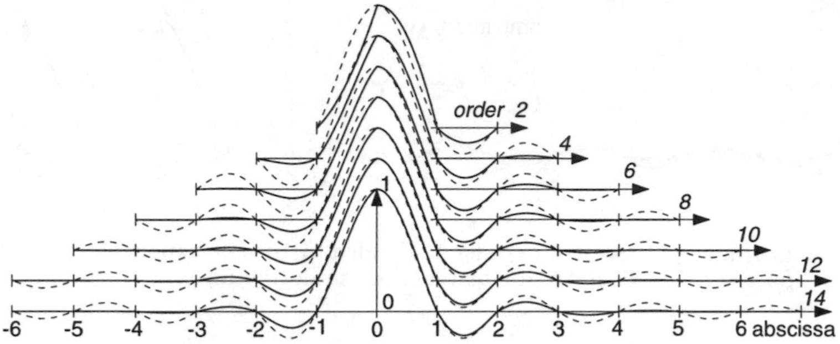

Fig. 2. Impulse responses of stationary Lagrange interpolators for orders 2 to 14 (the last one is truncated). *sinc* function in dotted lines.

§4. Hermite Interpolators

In a general way, Hermite interpolators are consistent with the values of a function and its k (or k_i, $1 \leq i \leq l$) first derivatives at l distinct points [4,2,3]. Here, I consider convolutive interpolators for any k and $l = 2$.

Definition 6. *The $\mathcal{C}^k$ piecewise polynomial Hermite interpolator is*

$$\mathcal{H}_k(f) = \Delta(f) * \eta_{k0} + \ldots + \Delta(f^{(j)}) * \eta_{kj} + \ldots + \Delta(f^{(k)}) * \eta_{kk},$$

where $\eta_{kj}(x)$ are $[-1,\ +1]$-supported basis functions such that $\forall m,\ m \in [0,\ k]$, $\eta_{kj}^{(m)}(-1) = 0 = \eta_{kj}^{(m)}(+1)$, and $\eta_{kj}^{(m)}(0) = \delta_{mj}$.

From now on, I consider $\eta_{k0}(x)$ functions only.

Proposition 7. *The polynomial expression of Hermite interpolators η_{k0} is*

$$1 - \eta_{k0}(x) = (2k+1)\, C_{2k}^{k} \sum_{j=k}^{j=2k} \frac{(-1)^{j-k} C_k^{j-k}}{j+1}\, x^{j+1} \tag{9}$$

for $x \in [0,\ 1]$, and $\eta_{k0}(x) = \eta_{k0}(-x)$ for $x \in [-1,\ 0]$. Moreover, asymptotically,

$$\lim \eta_{k0}(x) \sim \sqcap_0^1(x) * 2\sqrt{\frac{k}{\pi}} \exp(-4kx^2), \quad when \quad k \to \infty, \tag{10}$$

with $\sqcap_a^b(x) = 1$ if $x \in [a - \frac{b}{2},\ a + \frac{b}{2}]$ and $\sqcap_a^b(x) = 0$ otherwise.

Proof: For $x \in [0,\ 1]$, it is clear from Def. 6 that $(1-\eta_{k0}(x))^{(1)} = \alpha x^k (1-x)^k$. Since $\eta_{k0}(0) = 1$ and $\eta_{k0}(1) = 0$, we have

$$\frac{1}{\alpha} = \int_0^1 (1 - \eta_{k0}(x))^{(1)} dx. \tag{11}$$

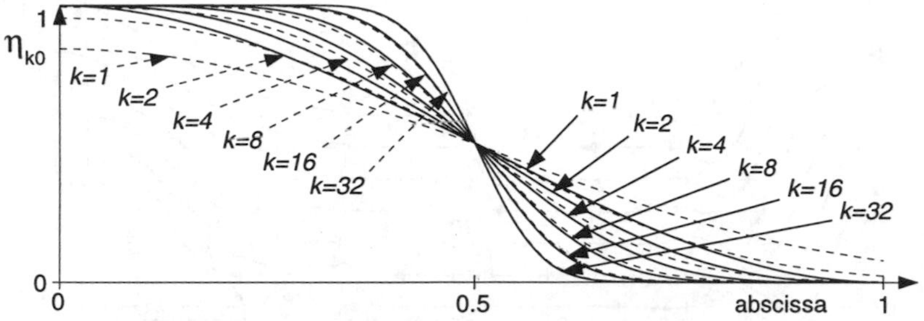

Fig. 3. Right part of impulse responses of Hermite interpolators according to (9), their asymptotic form according (10) in dotted lines.

Let us define $I_{k,l} = \int_0^1 x^k(1-x)^l dx$. Integrating by parts gives $I_{k,l} = \frac{l}{k+1} I_{k+1,l-1}$, that is, $I_{k,k} = \frac{(k!)^2}{(2k)!} I_{2k,0} = C_{2k}^k I_{2k,0}$ by recurrence. Since $I_{2k,0} = \frac{1}{2k+1}$, we obtain $\alpha = (2k+1)C_{2k}^k$, and therefore, from (11),

$$(1 - \eta_{k0}(x))^{(1)} = (2k+1)\, C_{2k}^k\, x^k (1-x)^k. \tag{12}$$

The binomial theorem and a simple integration concludes the proof of (9). Moreover, by the change of variable $x = \frac{1}{2} + u$ in (12), we have $-\eta_{k0}^{(1)}(\frac{1}{2}+u) = (2k+1)\, C_{2k}^k\, 4^{-k}\,(1-4u^2)^k$. For any abscissa $u_1 > 0$ we may define $u_k = \frac{u_1}{\sqrt{k}}$ and we have $\exp(-4ku_k^2) = \exp(-4u_1^2)$ and $(1-4u_k^2)^k = (1-4\frac{u_1^2}{k})^k$. Since $\lim_{k\to\infty}(1-4\frac{u_1^2}{k})^k = \exp(-4u_1^2)$, then $(1-4u^2)^k \sim \exp(-4ku^2)$ as $k \to \infty$. Moreover, by using $C_{2k}^k = \frac{(2k)!}{(k!)^2}$ and Stirling formula, we obtain, as $k \to \infty$, $(2k+1)C_{2k}^k 4^{-k} \sim 2\sqrt{\frac{k}{\pi}}$, and then $-\eta_{k0}^{(1)}(x) \sim 2\sqrt{\frac{k}{\pi}}\exp(-4k(x-\frac{1}{2})^2)$. Symmetrically, we have $\eta_{k0}^{(1)}(x) \sim 2\sqrt{\frac{k}{\pi}}\exp(-4k(x+\frac{1}{2})^2)$ for $x \in [-1,0]$. Therefore, $\eta_{k0}^{(1)}(x) \sim 2\sqrt{\frac{k}{\pi}}\exp(-4kx^2) * (\delta(x+\frac{1}{2}) - \delta(x-\frac{1}{2}))$, which concludes the proof of (10) since $\sqcap_0^{1(1)}(x) = \delta(x+\frac{1}{2}) - \delta(x-\frac{1}{2})$. □

Note that Lagrange and Hermite interpolators could be considered as Fourier pairs since the right members of (8) and (10) are mutual Fourier transforms if $\pi^2 n = 4k$.

§5. Lagrange-Hermite Interpolators

Since stationary Lagrange interpolators are good at low frequencies and since Hermite interpolators are differentiable, hence good at high frequencies, combining them is a natural idea.

Definition 8. *For* $k \geq 0$, *the* Lagrange-Hermite interpolator of order $2n$ *is the smooth Hermite interpolation of two successive Lagrange interpolators, that is,*

$$\mathcal{M}_{2n}^{k+1}(x) = \eta_{k0}(x-i_x)\,\mathcal{L}_{2n}^{i_x}(x) + (1-\eta_{k0}(x-i_x))\,\mathcal{L}_{2n}^{i_x+1}(x).$$

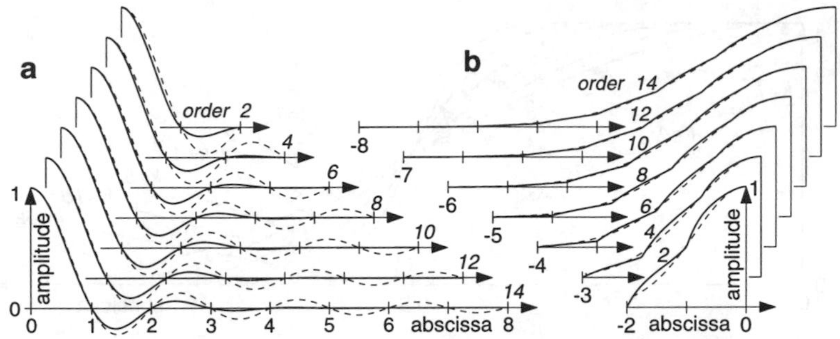

Fig. 4. Right part of the impulse responses of $\mathcal{C}^1$-differentiable Lagrange-Hermite interpolators (**a**), *sinc* in dotted lines. Their left part divided by the *sinc* (**b**), as compared to Gaussian curves $\exp(\frac{-x^2}{n+1/2})$, in dotted lines.

Proposition 9. *For $k \geq 0$, Lagrange-Hermite interpolators $\mathcal{M}_{2n}^{k+1}$ are $\mathcal{C}^{k+1}$-differentiable.*

Proof: Since it is piecewise polynomial, $\mathcal{M}_{2n}^{k+1}$ is $\mathcal{C}^\infty$-differentiable for all non-integral abscissas x. Moreover, $\mathcal{M}_{2n}^{k+1}$ is convolutive, and thus it is sufficient to examine it around $x = 0$. From Def. 8, for $x > 0$, we have $\mathcal{M}_{2n}^{k+1} = \eta_{k0}\,\mathcal{L}_{2n}^0 + (1-\eta_{k0})\,\mathcal{L}_{2n}^1$. From (12), we have $\eta_{k0}(x) = 1 - O(x^{k+1})$, and then $\mathcal{M}_{2n}^{k+1}(x) = \mathcal{L}_{2n}^0(x) + O(x^{k+1})\,(\mathcal{L}_{2n}^1 - \mathcal{L}_{2n}^0)$. Since the interpolators $\mathcal{L}_{2n}^0$ and $\mathcal{L}_{2n}^1$ are continuous, $\mathcal{L}_{2n}^1 - \mathcal{L}_{2n}^0 = O(x)$ and thus $\mathcal{M}_{2n}^{k+1} = \mathcal{L}_{2n}^0 + O(x^{k+2})$. Therefore, for any $j \in [1,\, k+1]$, $\mathcal{M}_{2n}^{k+1\,(j)}(0^+) = \mathcal{L}_{2n}^{0\,(j)}(0^+)$. The same argument with $\mathcal{M}_{2n}^{k+1} = \eta_{k0}\,\mathcal{L}_{2n}^{-1} + (1-\eta_{k0})\,\mathcal{L}_{2n}^0$ for $x < 0$ leads to $\mathcal{M}_{2n}^{k+1\,(j)}(0^-) = \mathcal{L}_{2n}^{0\,(j)}(0^-)$, which concludes the proof since $\mathcal{L}_{2n}^0(x)$ is a polynomial. □

The Fourier transform of Lagrange-Hermite impulse responses (which I call μ_{2n}^k for interpolator $\mathcal{M}_{2n}^k$) are very similar to those of stationary Lagrange interpolators below the sampling frequency. Beyond the sampling frequency, for the orders 2 to 14, the rejection in dB of the greatest secondary lobe is 27, 30, 32, 33, 34, 35 and 35 in the $\mathcal{C}^0$ case, 42, 47, 51, 53, 55, 57 and 58 in the $\mathcal{C}^1$ case, 33, 36, 37, 38, 39, 40 and 41 in the $\mathcal{C}^2$ case, and slowly decreases for higher differentiabilities. From this viewpoint, the $\mathcal{C}^1$ Lagrange-Hermite interpolators are the best choice.

§6. Holberg Interpolators

Lagrange-Hermite interpolators are good at low and high frequencies, but unsatisfactory inbetween. Indeed, Gaussian apodizing makes λ_{2n} as well as μ_{2n}^k gradually ineffective, since their cost, which is proportional to the length of their support, increases like n, whereas their active part widens like $\sqrt{n}$. Faced with the same problem in terms of numerical schemes, Holberg ([5]) had the idea of combining several orders and optimizing the passband for given tolerance and maximal order. Holberg interpolators just proceed from the same idea.

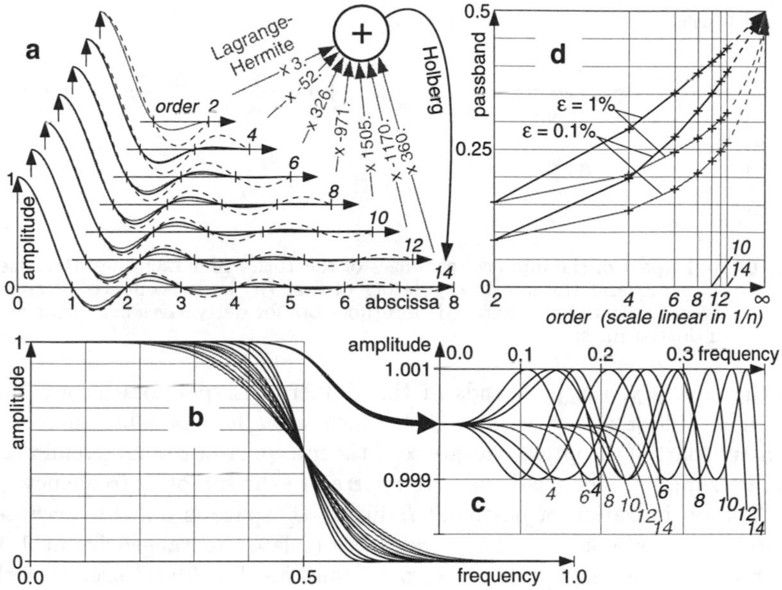

Fig. 5. Impulse responses of $\mathcal{C}^1$-differentiable and 0.1%-precise Holberg interpolators (**a**) (Lagrange-Hermite in grey, *sinc* in dotted lines). Their Fourier transform (**b**) and an enlargement of them (**c**). The passband-*vs*-order diagram (**d**) showing the linear behaviour of Holberg interpolators and the parabolic behaviour of the Lagrange-Hermite interpolators.

Definition 10. *The $2n$-hybrid order, $\mathcal{C}^k$-differentiable and ε-precise Holberg interpolator is the linear combination $\tilde{\mathcal{M}}_{2n}^{k,\varepsilon} = \sum_{i=1}^{i=n} \beta_i^\varepsilon \mathcal{M}_{2i}^k$ such that $\sum_1^n \beta_i^\varepsilon = 1$ and such that the following passband is maximized,*

$$B(\beta_1^\varepsilon, \dots, \beta_{n-1}^\varepsilon) = \sup(\{\nu;\ \forall \xi \le \nu;\ |\sum_i \beta_i^\varepsilon \mathcal{F}(\mu_{2i}^k)(\xi) - 1| \ \le\ \varepsilon\}).$$

Clearly, these Holberg interpolators are convolutive, with impulse responses $\tilde{\mu}_{2n}^{k,\varepsilon} = \sum_1^n \beta_i^\varepsilon \mu_{2n}^k$.

Fig. 5 illustrates the case of $\varepsilon = 0.1\%$ and $k = 1$ ($\mathcal{C}^1$-differentiability). Fig. 5a shows that the impulse responses of Holberg interpolators decrease much slower than the Lagrange-Hermite interpolators. The sum of the β_i^ε is one, but their absolute sum may be far from one, for instance about 4300 in the case of the 14th order. In the Fourier domain, Fig. 5b, together with its enlargement Fig. 5c, shows that the passband is increased a lot from Lagrange-Hermite to Holberg.

For tolerances 1% and 0.1%, Fig. 5d shows the *parabolic* behaviour around infinite order of the passbands of the Lagrange-Hermite interpolators (grey arrows are parabolas with vertical tangent at the corner). This is due to Gaussian apodizing, since the Fourier transform of (8) gives the convolution of a box function with a Gaussian function that narrows like $\frac{1}{\sqrt{n}}$ as $n \to \infty$.

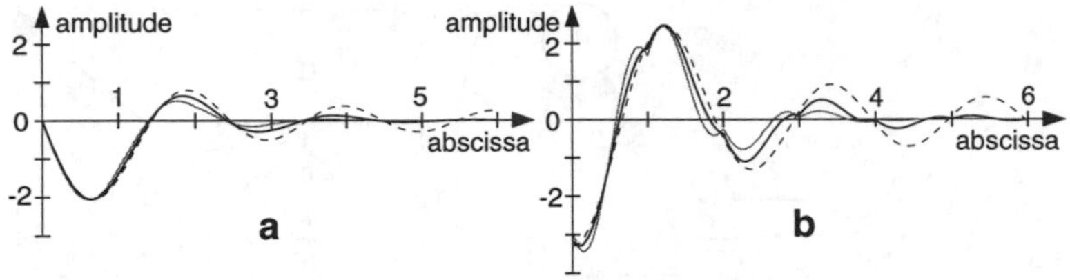

Fig. 6. Right part of the impulse responses of the 10th-order Lagrange-Hermite (in grey) and Holberg interpolators of first (**a**) and second (**b**) derivatives. Perfect (but truncated) interpolators for derivatives are displayed in dotted lines.

On the contrary, the passbands of the Holberg interpolators have a *linear* behaviour (black arrows are straight), which is the best possible one because the active part of the impulse response of the interpolator cannot expand faster than its support. The passbands are referred to the sampling frequency.

The optimization of passband B in the β_i^ε space is difficult because B is discontinuous along $n-1$ hypersurfaces (related to tangencies at $1 \pm \varepsilon$) which intersect at the optimum. To overcome this difficulty, I used a method consisting of the following steps:

1) choose a strictly increasing sequence of frequencies ν_i in $]0, 0.5[$, with $1 \leq i \leq n-1$,
2) set at $1+\varepsilon$ the value of the combination at frequency ν_{n-1}, $1-\varepsilon$ at ν_{n-2}, $1+\varepsilon$ at ν_{n-3}, and so on until ν_1, and finally 1 at frequency $\nu_0 = 0$,
3) solve the linear system for the β_i^ε,
4) detect the frequencies at which the combination is extremal,
5) if these frequencies are close to the ν_i, then stop, else update the ν_i and come back to step 2.

A priori, the feasability of steps 3 and 4 and the convergence are not guaranteed. In practice however, this method works simply well, with less than ten iterations. See also [7,8] on optimal filtering.

§7. Applications

The main application of Holberg interpolators are Holberg numerical schemes ([5]) because they are cost-effective in the field of numerical simulation of acoustic wave propagation. Especially in 3D, this effectiveness is of considerable importance because of the huge amount of computing time needed.

From a $2n$-order $\mathcal{C}^k$-differentiable Lagrange-Hermite or Holberg interpolator, $2n$-order $\mathcal{C}^{k-m}$-differentiable interpolators of the mth derivative ($m \leq k$) can be easily derived, as well as numerical schemes of these derivatives at any abscissa. Fig. 6 shows the responses of the 10th-order Lagrange-Hermite and Holberg ($\varepsilon = 0.1\%$) interpolators of first (a) and second (b) derivatives. The values at integral abscissas give standard numerical schemes. In a similar way, the integration of these impulse responses could result in Newton-Cotes-like Holberg formulas (see *Newton-Cotes Formulas* in [10]).

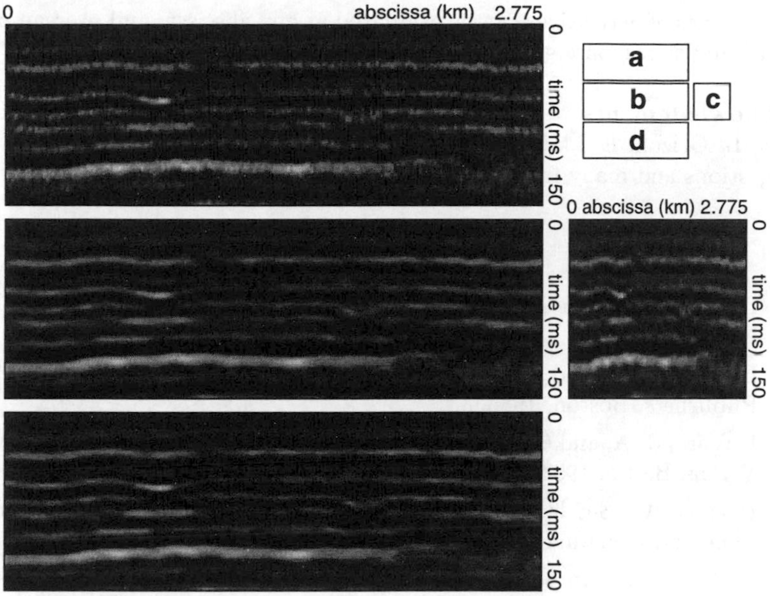

Fig. 7. A seismic section (**a**) has been horizontally filtered (**b**) for antialiasing, threefold undersampled (**c**), and finally interpolated (**d**).

Holberg interpolators, and numerical filters that can be generated from them, are also interesting for their efficiency. For instance, the response of a 1%-precise, 6th-order, $\mathcal{C}^1$ Holberg interpolator has been fourfold oversampled and used to filter horizontally Fig. 7a into Fig. 7b for antialiasing. The threefold undersampling of Fig. 7b gives Fig. 7c. The 1%-precise, 6th-order, $\mathcal{C}^1$ threefold Holberg interpolation of Fig. 7c gives Fig. 7d. The similarity of (**b**) and (**d**) measures the quality of the filtering and of the interpolation.

§8. Conclusions

In the case of periodic data points, Lagrange interpolators converge to a *sinc* function multiplied by two exponentials. The first one explains the well-known instabilities of Lagrange polynomials, which only vanish at the center of the interval of the data points. The second exponential explains the vanishing influence of non-centered data points (Gaussian apodizing). Stationary Lagrange interpolators are stable and convolutive. Hermite interpolators are as differentiable as desired and convolutive and their impulse response converges to the convolution of a box function with a Gaussian function. Lagrange-Hermite interpolators combine the advantages of unlimited order and differentiability. Because of Gaussian apodizing, these interpolators become ineffective at high orders. On the other hand, Holberg interpolators have a much better quality/cost ratio since they are optimal combinations of Lagrange-Hermite interpolators. From Holberg interpolators, efficient numerical schemes of derivatives can be evaluated at any abscissa, and oversampling their impulse response gives short but efficient filters.

Acknowledgments. I would like to thank my colleagues J. Brac, F. Coppens, L. Grizon, E. Maffiolo, L. Nicolétis, J. Pirot and T. Tonellot for their suggestions and many fruitful discussions.

References

1. Bracewell, R., *The Fourier Transform and its Applications*, McGraw-Hill, New-York, 1965.
2. Burden, R. L. and J. D. Faires, *Numerical Analysis*, 3rd edition, PWS Publishers, Boston, 1985.
3. DeVore, R. A. and G. G. Lorentz, *Constructive Approximation*, Springer-Verlag, Berlin, 1993.
4. Hermite, C., Sur la formule d'interpolation de Lagrange. Journal für die reine und angewandte Mathematik **84** (1878), 70–79.
5. Holberg, O., Computational aspects of the choice of operator and sampling interval for numerical differentiation in large-scale simulation of wave phenomena. Geophysical Prospecting **35** (1987), 629–655.
6. Lancaster, P. and K. Šalkauskas, *Curve and Surface Fitting, an Introduction*, Academic Press, London, 1986.
7. McClellan, J. H., T. W. Parks and L. R. Rabiner, A computer program for designing optimum FIR linear phase digital filters. IEEE Transactions on Audio and Electroacoustics **AU-21** (1973) 506–526.
8. Oppenheim, A. V. and R. W. Shafer, *Digital Signal Processing*, Prentice-hall, Englewood Cliffs, New Jersey, 1975.
9. Waring, E., Problems concerning interpolations, Philosophical Transactions of the Royal Society of London **69** (1779), 59–67.
10. Weisstein, E. W., *CRC Concise Encyclopedia of Mathematics*, CRC Press, Boca Raton, 1999.

Michel Léger
1 & 4 avenue de Bois-Préau
92852 Rueil-Malmaison, France
`michel.leger@ifp.fr`

Local Approximation on Manifolds Using Radial Functions and Polynomials

Jeremy Levesley and David L. Ragozin

Abstract. The main focus of this paper is to give error estimates for interpolation on compact homogeneous manifolds, the sphere being an example of such a manifold. The notion of a radial function on the sphere is generalised to that of a spherical kernel on a compact homogeneous manifold. Reproducing kernel Hilbert space techniques are used to generate a pointwise error estimate for spherical kernel interpolation using a positive definite kernel. By exploiting the nice scaling properties of Lagrange polynomials in the tangent space, the error estimate is bounded above by a power of the point separation, recovering, in particular, the convergence rates for radial approximation on spheres.

§1. Introduction

There is currently significant interest in approximation on spheres, related to many interesting geophysical problems. There are a number of different approximation methods currently available on spheres, including wavelets [3], splines [1], and the subject of this paper, radial functions (sometimes called spherical splines) [3,6]. Error estimates and convergence rates for radial approximation on spheres, of an optimal nature, are recent in vintage [5,4], and rely on some technically demanding mathematics. In this paper we build on an idca of Bos and de Marchi [2] in order to provide convergence rates for radial interpolation on a much wider class of manifolds: the reflexive, compact homogeneous spaces. We will conclude the paper by proving a local spherical harmonic polynomial approximation result on spheres.

Let M^d be a d-dimensional compact manifold with a metric $d(\cdot,\cdot)$ which possesses a transitive group G of isometries. The group acts transitively in that for every $x, y \in M^d$, there exists $g \in G$ such that $gx = y$. If, furthermore, there exists $g \in G$ such that $gx = y$ and $gy = x$, then M^d is termed reflexive (for more details see [11]). Such a manifold is a reflexive, compact, metric, homogeneous space. We comment that we can always embed M^d in

Curve and Surface Fitting: Saint-Malo 1999
Albert Cohen, Christophe Rabut, and Larry L. Schumaker (eds.), pp. 291–300.

ISBN 0-8265-1357-3.

some higher dimensional Euclidean space $\mathbb{R}^{d+r}$, the group G being a compact subgroup of the isometries of $\mathbb{R}^{d+r}$. We assume that the metric $d(\cdot,\cdot)$ on M^d is inherited from some Euclidean embedding.

We will be interested in interpolation on M^d using continuous zonal kernels $k(\cdot,\cdot)$ which have the property that $k(gx,gy) = k(x,y)$ for all $x,y \in M^d$ and $g \in G$. Such kernels are natural generalisations of radial functions, which are functions only of distance, which is itself G-invariant. Given a set $\{x_1,\dots,x_N\} \subset M^d$ and data $f_1,\dots,f_N \in \mathbb{R}$, we seek a function of the form

$$s_k(x) = \sum_{i=1}^{N} \alpha_i k(x,x_i)$$

such that $s_k(x_i) = f_i$, $i = 1,\dots,N$.

Given the data $f_i = f(x_i)$, $i = 1,\dots,N$, we wish to bound the pointwise error between s_k and f at $x \in M^d$. We make no assumption on the data points except that they satisfy a point separation criteria in some subset of M^d (see Section 3).

In Section 2 we will introduce some necessary harmonic analysis on M^d, discuss the notion of positive definiteness on M^d in brief, and give a standard error estimate, which we will use in Section 3 to obtain convergence rates. In Section 4 we prove a Whitney type error estimate for local spherical harmonic approximation on the sphere.

§2. Harmonic Analysis and Error Estimates

For a more complete version of the brief description we give here, see [10,11]. Let Π_n^{d+r} be the degree n polynomials in $\mathbb{R}^{d+r}$, the space in which M^d is homogeneously embedded. Then, let $\mathcal{P}_n$ (the spherical polynomials of degree n) be the restriction of these polynomials to M^d. Furthermore, let $\mathcal{H}_n := \mathcal{P}_n \cap \mathcal{P}_{n-1}^{\perp}$, where the orthogonality is with respect to $d\mu$, the unique normalised G-invariant measure on M^d:

$$[f,g] := \int_{M^d} fg\,d\mu.$$

Then, we can uniquely decompose $\mathcal{H}_n$ into irreducible G-invariant subspaces Ξ_{nj}, each of dimension d_{nj}, $j = 1,\dots,h_n$, resulting in the G-invariant decomposition

$$L_2(M^d) = \oplus_{n=0}^{\infty} \oplus_{j=1}^{h_n} \Xi_{nj}.$$

Let $\mathcal{X}_{nj}$ be the orthogonal projection onto Ξ_{nj}, $n = 0,1,\dots$, and $j = 1,\dots,h_n$, and $\mathcal{R}_{nj}(\cdot,\cdot)$ be the kernel of this projection. We will consider interpolation using strictly positive definite kernels of the form

$$k(x,y) = \sum_{n=0}^{\infty} \sum_{j=1}^{d_n} a_{nj} \mathcal{R}_{nj}(x,y),$$

where $a_{nj} > 0$, $n = 0, 1, \ldots$, $j = 1, \ldots, h_n$, and

$$\sum_{n=0}^{\infty} \sum_{j=1}^{d_n} d_{nj} a_{nj} < \infty. \tag{1}$$

We will approximate functions from the Hilbert space

$$\mathcal{W} = \{f \in L_2(M^d) : \|f\|^2 := \sum_{n=0}^{\infty} \sum_{j=1}^{d_n} \|\mathcal{X}_{nj} f\|_2^2 / a_{nj} < \infty\},$$

where $\|\cdot\|_2$ denotes the $L_2(M^d)$ norm. The associated inner product in $\mathcal{W}$ is

$$(f, g) := \sum_{n=0}^{\infty} \sum_{j=1}^{d_n} [\mathcal{X}_{nj} f, \mathcal{X}_{nj} g] / a_{nj}.$$

The condition (1) ensures that point evaluation is a continuous linear functional in $\mathcal{W}$. It is straightforward to show that k is the reproducing kernel for the $\mathcal{W}$: $f(x) = (f, k(x, \cdot)), f \in \mathcal{W}, x \in M^d$. An immediate consequence of the reproducing kernel property is that s_k is the interpolant of minimum $\mathcal{W}$ norm. For, if g is another interpolant,

$$(s_k - g, s_k) = \sum_{i=1}^{N} \alpha_i (f - s_k, k(x_i, \cdot)) = \sum_{i=1}^{N} \alpha_i (f(x_i) - s_k(x_i)) = 0.$$

Therefore,

$$\begin{aligned}(g, g) = (g - s_k + s_k, g - s_k + s_k) &= (g - s_k, g - s_k) + 2(g - s_k, s_k) + (s_k, s_k) \\ &= (g - s_k, g - s_k) + (s_k, s_k),\end{aligned} \tag{2}$$

and the norm minimisation property is established. Now, following the standard arguments, see e.g. [8,9], we have, using the fact that s_k interpolates f at $x_1, \ldots, x_N$,

$$\begin{aligned}|f(x) - s_k(x)| &= |(f - s_k, k(x, \cdot)| \\ &= |(f - s_k, k(x, \cdot) + \sum_{i=1}^{N} \beta_i k(x_i, \cdot)| \\ &\leq \|f - s_k\| \|k(x, \cdot) + \sum_{i=1}^{N} \beta_i k(x_i, \cdot)\| \\ &\leq \|f\| \|k(x, \cdot) + \sum_{i=1}^{N} \beta_i k(x_i, \cdot)\|\end{aligned}$$

for arbitrary β_i, $i = 1, \dots, N$, where we have used (2) in the final step. Our final error estimate follows from the fact that

$$\|k(x,\cdot) + \sum_{i=1}^{N} \beta_i k(x_i,\cdot)\| = (k(x,\cdot) + \sum_{i=1}^{N} \beta_i k(x_i,\cdot), k(x,\cdot) + \sum_{i=1}^{N} \beta_i k(x_i,\cdot))^{\frac{1}{2}}$$
$$= (k(x,x) - 2\sum_{i=1}^{N} \beta_i k(x,x_i) + \sum_{i,j=1}^{N} \beta_i\beta_j k(x_i,x_j))^{\frac{1}{2}}.$$

Defining

$$P(x,x_1,\dots,x_N) := \inf_{\beta_1,\dots,\beta_N \in \mathbf{R}} (k(x,x) - 2\sum_{i=1}^{N} \beta_i k(x,x_i) + \sum_{i,j=1}^{N} \beta_i\beta_j k(x_i,x_j))^{\frac{1}{2}},$$

we sum these results up in

Theorem 1. *Let s_k be the k-spline interpolant at $x_1, \dots, x_N \in M^d$ to $f \in \mathcal{W}$. Then, for every $x \in M^d$,*

$$|f(x) - s_k(x)| \leq \|f\| P(x, x_1, \dots, x_N).$$

§3. Convergence Rates for Radial Kernels

In this section we shall give pointwise error estimates in terms of the point separation

$$\rho := \max_{y \in V} \min_{i=1,\dots,N} d(y, x_i),$$

where $V \subset M^d$ contains x, the point at which we are measuring the error. As we shall see later in this section, producing a pointwise convergence rate from the error estimate of Theorem 1 requires us to bound Lagrange polynomials related to a subset of the interpolation points. Efforts to produce convergence rates on the sphere foundered because it is difficult to bound the Lagrange polynomials for spherical harmonic interpolation as the interpolation set, with a fixed number of points, scale towards x. The early error estimates of [3], of $\mathcal{O}(\rho)$, were the best known until recently, and only required bounding of the constant Lagrange polynomial for a single point. Light and v. Golitschek [4] proved boundedness for all polynomials on S^d, $d \geq 2$, and consequently achieved $\mathcal{O}(\rho^r)$ approximation for radial kernels with $2r$ continuous derivatives on the sphere.

A very simple proof of the result of Light and v. Golitschek was given by Bos and de Marchi in [2]. What we will do is introduce an analytic coordinate transformation, and construct Lagrange polynomials in the tangent space, which is a d-dimensional Euclidean space. We will quote a result which uses scaling arguments in Euclidean space which are easy to perform, observing that distance on the manifold and in the tangent are comparable.

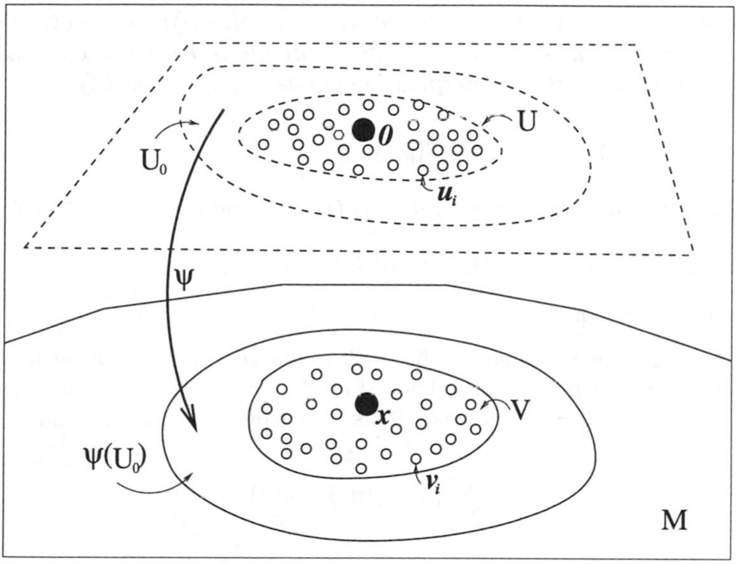

Fig. 1. Coordinate chart at x.

Let $x \in V \subset M$. We shall assume the existence of a C^∞-chart (U_0, ψ) with an open subset $U \subset U_0$ satisfying the following (see Figure 1):

1) $\psi(U) = V$ with $\psi(0) = x$,
2) $U_0 = \{y - z : y, z \in U\}$.

These conditions ensure the validity of Taylor series arguments which follow. Also, since U is precompact, ψ is bi-differentiable and the metric d is assumed boundedly equivalent to the Euclidean distance on $\mathbb{R}^{d+r}$,

$$c_1\|y - z\| \leq d(\psi(y), \psi(z)) \leq c_2\|y - z\|, \quad y, z \in U.$$

Let $v_1, \ldots, v_Q = V \cap X$ be the interpolation points in V, and redefine $\rho := \sup_{v \in V} \min_{i=1,\ldots,Q} d(v, v_Q)$. Let $u_i = \psi^{-1}(v_i)$, $i = 1, \ldots, Q$. Then, from the previous equation we have

$$\eta := \sup_{u \in U} \min_{i=1,\ldots,Q} \|u - u_i\| \leq \rho/c_1. \tag{3}$$

It is shown in [7] that provided ρ and hence η are sufficiently small to guarantee that $Q > t$, we can make a selection of interpolation points $v_1, \ldots, v_t$ (assuming a convenient ordering of the points), where $t = \dim(\Pi_{2r-1}^d)$, such that the Lagrange polynomials $p_1, \ldots, p_t$ for $u_1, \ldots, u_t$ are bounded at the origin:

$$p_i(0) \leq C_L, \qquad i = 1, \ldots, t, \tag{4}$$

where C_L is independent of i and ρ. Furthermore, $u_1, \ldots, u_t$ are all contained in $\eta B_b := \{\eta w : \|w\| \leq b\}$. We are now ready to prove the main theorem of this paper:

Theorem 2. *Let $\phi : \mathbb{R} \to \mathbb{R}$ and $k(\cdot,\cdot) = \phi(d(\cdot,\cdot))$ be strictly positive definite, and $2r$-times continuously differentiable in each variable. Let $x \in S \subset M$. Suppose that the interpolation points $x_1, \ldots, x_n$ satisfy*

$$\sup_{y \in S} \min_{i=1,\ldots,n} d(y, x_i) = \rho.$$

Then for all sufficiently small ρ, if s_k is the k-spline interpolant to $f \in \mathcal{W}$,

$$|f(x) - s_k(x)| \leq C\|f\|\rho^r,$$

where C is independent of ρ.

Proof: First we now choose the coefficients $\beta_1, \ldots, \beta_n$ (appearing in the statement of Theorem 1) as follows. Let $\beta_i = p_j(0)$ if $x_i = \psi(u_j)$ for some $j = 1, \ldots, t$. Set $\beta_i = 0$ otherwise. This choice of coefficients is made since

$$\sum_{j=1}^{t} p_j(0) q(u_j) = q(0), \tag{5}$$

for all $q \in \Pi_{2r-1}^d$. Then,

$$\begin{aligned} P(x, x_1, \ldots, x_N) \leq \Big[&\phi(d(\psi(0), \psi(0))) - 2\sum_{j=1}^{t} p_j(0)\phi(d(\psi(0), \psi(u_j))) \\ &+ \sum_{i,j=1}^{t} p_i(0)p_j(0)\phi(d(\psi(u_i), \psi(u_j)))\Big]^{1/2}. \end{aligned} \tag{6}$$

Since $\phi(d(\cdot,\cdot))$ is $2r$ times continuously differentiable in each variable, for fixed $w \in U$ we may expand

$$\phi(d(\psi(w), \psi(z))) = \sum_{|\alpha|<2r} c_\alpha^w (z - w)^\alpha + R_{2r}^w(z), \quad z \in U, \tag{7}$$

where R_{2r}^w is a Taylor series remainder satisfying

$$R_{2r}^w(z) \leq C_R \|z - w\|^{2r}, \tag{8}$$

for some constant C_R independent of z and w. Putting $z = w$ in the above expansion we see that

$$c_0^w = \phi(0). \tag{9}$$

Putting (7) into (6) gives

$$\begin{aligned} P(x, x_1, \ldots, x_N) \leq \Big[&\phi(0) - 2\sum_{j=1}^{t} p_j(0)\Big(\sum_{|\alpha|<2r} c_\alpha^0 (u_j)^\alpha + R_{2r}^0(u_j)\Big) \\ &+ \sum_{i,j=1}^{t} p_i(0)p_j(0)\Big(\sum_{|\alpha|<2r} c_\alpha^{u_i}(u_j - u_i)^\alpha + R_{2r}^{u_i}(u_j)\Big)\Big]^{1/2}. \end{aligned}$$

Using the polynomial reproduction (5), we get

$$P(x, x_1, \ldots, x_N) \le \Big[-\phi(0) - 2\sum_{j=1}^{t} p_j(0) R_{2r}^0(u_j) + \sum_{i=1}^{t} p_i(0) \Big(\sum_{|\alpha|<2r} c_\alpha^{u_i} (-u_i)^\alpha \Big) + \sum_{i,j=1}^{t} p_i(0) p_j(0) R_{2r}^{u_i}(u_j) \Big]^{1/2} \tag{10}$$

where we have used (6) and (9) in the above argument.

Now, since the distance function is symmetric, for any $u, w \in \mathbb{R}^d$,

$$\begin{aligned} \phi(d(\psi(u), \psi(w))) &= \sum_{|\alpha|<2r} c_\alpha^w (u-w)^\alpha + R_{2r}^w(u) \\ &= \sum_{|\alpha|<2r} c_\alpha^u (w-u)^\alpha + R_{2r}^u(w). \end{aligned}$$

In particular, with $w = 0$ we get

$$\sum_{|\alpha|<2r} c_\alpha^u (-u)^\alpha = \sum_{|\alpha|<2r} c_\alpha^0 (u)^\alpha + R_{2r}^0(u) - R_{2r}^u(0).$$

Substituting the last equation into (10) and again using (5) gives

$$\begin{aligned} P(x, x_1, \ldots, x_N) &\le \Big[-\phi(0) - 2\sum_{j=1}^{t} p_j(0) R_{2r}^0(u_j) \\ &\quad + \sum_{i=1}^{t} p_i(0) \Big(\sum_{|\alpha|<2r} c_\alpha^0 (u_i)^\alpha + R_{2r}^0(u) - R_{2r}^u(0) \Big) \\ &\quad + \sum_{i,j=1}^{t} p_i(0) p_j(0) R_{2r}^{u_i}(u_j) \Big]^{1/2} \\ &\le \Big[-\sum_{j=1}^{t} p_j(0) (R_{2r}^0(u_j) + R_{2r}^{u_j}(0)) \\ &\quad + \sum_{i,j=1}^{t} p_i(0) p_j(0) R_{2r}^{u_i}(u_j) \Big]^{1/2} \\ &\le C_R \big(\max_{i,j=1,\ldots,t} \{ \|u_i - u_j\|, \|u_i\| \} \big)^r \sum_{i,j=1}^{t} |p_i(0)| |2 + p_j(0)| \\ &\le C_R \eta^r \sum_{i,j=1}^{t} C_L (2 + C_L) \\ &\le C \rho^r, \end{aligned}$$

using (3), (4) and (8), recalling that $u_1, \ldots, u_t$ are contained in an $\eta \le \rho/c_1$ scaled ball of radius b. Substituting this result in Theorem 1 concludes the proof. □

§4. Whitney-type Estimates

We will now use the coordinate system suggested by Bos and de Marchi [2] to prove Whitney type estimates for approximation using spherical harmonics on $\mathbb{R}^d$. This result answers a question posed by L. L. Schumaker during the conference for which these are the proceedings. For $x = (x_2, \ldots, x_{d+1}) \in \mathbb{R}^d$, define $\Theta : \mathbb{R}^d \to S^d$ by $\Theta(x_2, \ldots, x_{d+1}) = ((1 - \sum_{i=2}^{d+1} x_i^2)^{1/2}, x_2, \ldots, x_{d+1})$. This is a smooth parametrisation of a neighbourhood of $e_1 = (1, 0, \ldots, 0) \in S^d$. As long as $\sum_{i=2}^{d+1} x_i^2 < \sin^2 \rho$, then $d(e_1, \Theta(x)) < \rho$.

Without loss of generality, we shall consider the approximation of a function $f \in C^k(S^d)$, using spherical polynomials, in any spherical neighbourhood X_ρ of e_1, where $\max_{y \in X_\rho} d(e_1, y) = \rho < \pi/2$. In fact, we will prove

Theorem 3. *Let $f \in C^k(S^d)$. Then, there exists a degree k harmonic polynomial p_{k-1} such that, for every $\pi/2 > \rho > 0$,*

$$\max_{x \in X_\rho} |f(x) - p_{k-1}(x)| \le C(f)\rho^k,$$

where the constant $C(f)$ does not depend on ρ.

Proof: The crucial element of this proof is that the coordinate mapping Θ maps polynomials of degree k in $\mathbb{R}^d$, the tangent plane at e_1 coordinatised by $x_2, \ldots, x_{n+1}$, to polynomials of degree k on the sphere. Since $f \circ \Theta \in C^k(\Theta^{-1} X_\rho)$, we can perform the multivariate Taylor series expansion

$$f \circ \Theta(x) = \sum_{|\alpha|<k} c_\alpha x^\alpha + R_k(f, x), \tag{11}$$

where $R_k(f, x)$ is the remainder satisfying

$$R_k(f, x) \le C(f)(\max_{y \in X_\rho} |\Theta^{-1}(y)|)^k. \tag{12}$$

Letting $\theta = \Theta(x)$, and defining the degree $k-1$ spherical polynomial

$$p_{k-1}(\theta) := \sum_{|\alpha|<k} c_\alpha (\Theta^{-1}(\theta))^\alpha = \sum_{|\alpha|<r} c_\alpha x^\alpha,$$

equations (11) and (12) tell us that

$$\max_{\theta \in X_\rho} |f(\theta) - p_{k-1}(\theta)| \le C(f)(\max_{y \in X_\rho} |\Theta^{-1}(y)|)^k.$$

The result follows because $\max_{y \in X_\rho} |\Theta^{-1}(y)| \le \rho$. □

Acknowledgments. David Ragozin's work was partially supported by NSF Grant DMS-9972004.

References

1. Alfeld, P., M. Neamtu, and L. L. Schumaker, Fitting scattered data on sphere-like surfaces using spherical splines, J. Comput. Appl. Math. **73** (1996), 5–43.

2. Bos, L. and S. De Marchi, Limiting values under scaling of the Lebesgue function for polynomial interpolation on spheres, J. Approx. Theory **96** (1999), 366–377.

3. Freeden W., On spherical spline approximation and interpolation, Math. Meth. in Appl. Sci. **3** (1981), 551–575.

4. von Golitschek, M. and W. A. Light, Interpolation by polynomials and radial basis functions on spheres, Constr. Approx., to appear.

5. Jetter, K., J. Stöckler, and J. D. Ward, Error estimates for scattered data interpolation on spheres, Math. Comp. **68** (1999), 733–747.

6. Levesley, J., W. Light, D. L. Ragozin, and X. Sun, A simple approach to the variational theory for interpolation on spheres, in *New Developments in Approximation Theory*, M. W. Müller, M. D. Buhmann, D. H. Mache, and M. Felten (eds.), ISNM **132**, Birkhäuser (1999), 117–143.

7. Levesley J. and D. L. Ragozin, Radial basis interpolation on compact homogeneous manifolds: Convergence rates, preprint.

8. Light, W. A. and H. Wayne, Error estimates for approximation by radial basis functions, in *Wavelet Analysis and Approximation Theory*, S.P. Singh and A. Carbone (eds.), Kluwer Academic, Dordrecht, 1995, 215–246.

9. Madych, W. R. and S. A. Nelson, Multivariate interpolation and conditionally positive definite functions II, Math. Comp. **54** (1990), 211–230.

10. Ragozin, D. L. and J. Levesley, Zonal kernels, approximations, and positive definiteness on spheres and compact homogeneous spaces, *Surface Fitting and Multiresolution Methods*, A. Le Méhauté, C. Rabut, and L. L. Schumaker (eds.), Vanderbilt University Press, Nashville, 1997, 143–150.

11. Ragozin, D. L. and J. Levesley, The density of translates of zonal functions on compact homogeneous spaces, J. Approx. Theory **103** (2000), 252–268.

Jeremy Levesley
Department of Mathematics
University of Leicester
Leicester LE1 7RH, UK
`jl1@mcs.le.ac.uk`

David L. Ragozin
University of Washington
Department of Mathematics
Seattle, WA 98195-4350, USA
rag@math.washington.edu

Characterizations of Native Spaces

Lin-Tian Luh

Abstract. In the theory of radial basis functions, linear combinations of the translates of a single function Φ are used as interpolants. The space spanned by all of these linear combinations carries an inner product defined via Φ itself. It can be completed and becomes a Hilbert space, called the native space for Φ, which is of great importance for further investigation of radial basis functions. The native space will contain abstract elements which are not linear combinations of radial basis functions, and require some work to be recognized as functions. This paper provides some characterizations of native spaces and relates some of the different approaches used to define them. Finally, embedding results for native spaces into Sobolev spaces are proven.

§1. Introduction

Our goal is to describe properties of the set of functions

$$\sum_{j=1}^{N} c_j \Phi(x, x_j), \quad x \in \Omega, \quad c_j \in \mathbb{C}, \tag{1}$$

where Ω is a subset of $\mathbb{R}^d$ and Φ is a real-valued symmetric function on $\Omega \times \Omega$. These functions depend on sets $X = \{x_1, \dots, x_N\} \subset \Omega$ of N pairwise distinct points called "centers", while the number N of centers and their placement within Ω are arbitrary. Functions of the form (1) arise naturally as tools for multivariate approximation, especially if Φ is a radial basis function $\Phi(x, y) := \phi(\|x - y\|_2)$ with a real-valued function ϕ on $[0, \infty)$. We shall study the closure of the linear span of functions (1) under a natural topology that comes from Φ itself, provided that Φ has a crucial property:

Definition 1. *A function $\Phi \in C(\Omega \times \Omega)$ is called* conditionally positive definite *(abbreviated as* **c.p.d.***) of order m on Ω if the quadratic form*

$$\sum_{j,k=1}^{N} c_j \bar{c}_k \Phi(x_j, x_k)$$

Curve and Surface Fitting: Saint-Malo 1999
Albert Cohen, Christophe Rabut, and Larry L. Schumaker (eds.), pp. 301–308.

ISBN 0-8265-1357-3.

is positive for all sets $X = \{x_1, \ldots, x_N\} \subset \Omega$ of N pairwise distinct points and all vectors $c = (c_1, \ldots, c_N)^T \in \mathbb{C}^N \setminus \{0\}$ satisfying

$$\sum_{i=1}^{N} c_i p(x_i) = 0 \text{ for all } p \in P_m^d, \tag{2}$$

where P_m^d is the space of d-variate complex-valued polynomials of order not exceeding m.

There are various possibilities to proceed from here. Already in their early pioneering papers, Madych and Nelson already took two different approaches, via finitely supported functionals [3] and via a specific version of generalized Fourier transforms [4] in the spirit of Gelfand–Shilov. The latter requires measure–theoretic arguments at certain places, and is rather complicated to deal with. The dissertation of Iske [1] used variational inequalities, while Weinrich [6] proceeded via regularized distributions in the sense of Schwartz. Our goal here is to show, as far as possible, the equivalence of the cited approaches. Since the access via generalized Fourier transforms has problems in dealing with arbitrary domains $\Omega \subseteq \mathbb{R}^d$, we proceed as in [5] in order to start with the most general approach known so far.

§2. Construction via Finitely Supported Functionals

Consider the space

$$(P_m^d)_\Omega^\perp := \left\{ \sum_{i=1}^{N} c_i \delta_{x_i} \mid c_i \in \mathbb{C},\ x_i \in \Omega \text{ for } 1 \le i \le N \text{ with (2)} \right\}$$

of all functionals that are finitely supported in Ω and vanish on the polynomials in P_m^d. Starting with a c.p.d. function Φ of order m in Ω, we define

$$(\lambda, \mu)_\Phi := \sum_{i=1}^{N} \sum_{j=1}^{M} \lambda_i \bar{\mu}_j \Phi(x_i, y_j)$$

for $\lambda, \mu \in (P_m^d)_\Omega^\perp$ with $\lambda = \sum_{i=1}^N \lambda_i \delta_{x_i}$, $\mu = \sum_{j=1}^M \mu_j \delta_{y_j}$ to get an inner product $(\cdot,\cdot)_\Phi$ which induces a Φ–dependent norm in the Φ–independent space $(P_m^d)_\Omega^\perp$. To relate functionals with functions, we use the map

$$R_\Phi \ : \ (P_m^d)_\Omega^\perp \to C(\Omega),\ R_\Phi(\lambda) := \lambda^x \Phi(x, \cdot) =: \lambda * \Phi,$$

where λ^x stands for the action of λ with respect to the variable x. By standard Hilbert space arguments, the fundamental identity

$$\mu(R_\Phi(\lambda)) = (\lambda, \mu)_\Phi \text{ for all } \lambda, \mu \in (P_m^d)_\Omega^\perp \tag{3}$$

proves that $(P_m^d)_\Omega^\perp$ and its image under R_Φ form a dual pair. Furthermore, this equation carries over to the Hilbert space closures $P_{\Phi,\Omega}$ of $(P_m^d)_\Omega^\perp$ and $F_{\Phi,\Omega}$ of $R_\Phi((P_m^d)_\Omega^\perp)$, respectively.

This construction is simple, but it leads to rather abstract elements instead of classical functionals and functions. To overcome this problem, one assumes that P_m^d and Ω allow a Lagrange–type basis $l_1, \ldots, l_{m'}$ with $m' = \dim P_m^d$ and points $x_1, \ldots, x_{m'} \in \Omega$ such that $l_i(x_j) = \delta_{ij}$ for $1 \le i, j \le m'$. Then the functional $\delta_{(x)} := \delta_x - \sum_{i=1}^{m'} l_i(x)\delta_{x_i}$ lies in $(P_m^d)_\Omega^\perp$, and the map S_Φ with

$$S_\Phi(\mu)(x) := (\mu, \delta_{(x)})_\Phi = \delta_{(x)} R_\Phi(\mu) = \mu(R_\Phi(\delta_{(x)})) \text{ for all } \mu \in P_{\Phi,\Omega}, x \in \Omega$$

uses (3) to define a classical function $S_\Phi(\mu)$ for each abstract element $\mu \in P_{\Phi,\Omega}$. The space $G_{\Phi,\Omega} := S_\Phi((P_m^d)_\Omega^\perp)$ now is a much more concrete space. The first of our results can be found in [2] with full proofs.

Theorem 2. *The spaces $F_{\Phi,\Omega} := R_\Phi(P_{\Phi,\Omega})$ and $G_{\Phi,\Omega} := S_\Phi(P_{\Phi,\Omega})$ are isometrically isomorphic via the mapping $S_\Phi \circ R_\Phi^{-1}$ and the inner product it introduces on $G_{\Phi,\Omega}$. Furthermore,*

$$\mu(S_\Phi(\lambda)) = (\lambda, \mu)_\Phi = \mu(R_\Phi(\lambda)) \tag{4}$$

holds for all functionals in $(P_m^d)_\Omega^\perp$ and its closure $P_{\Phi,\Omega}$.

It is not straightforwardly possible to associate classical function values to the elements $R_\Phi(\lambda)$. But (4) indicates that $R_\Phi(\lambda)$ and $S_\Phi(\lambda)$ should agree up to a polynomial from P_m^d on Ω. The function $S_\Phi(\lambda)$, however, vanishes on the points we used for the Lagrange interpolation in P_m^d, and thus realizes a very special assignment of function values modulo P_m^d. Thus, we can interpret $R_\Phi(\lambda)$ as an equivalence class of functions mod P_m^d on Ω, one representer of which is $S_\Phi(\lambda)$. Thus we should add P_m^d to the spaces we dealt with so far.

Definition 3. *Let Φ be c.p.d. of order $m \ge 0$ in Ω. Then the direct sum*

$$N_\Phi(\Omega) := P_m^d(\Omega) \oplus G_{\Phi,\Omega}$$

is called the native space *of Φ.*

The above construction allows us to define a semi–inner product $(.,.)_\Phi$ on this space such that the nullspace is P_m^d. Theorem 2 now implies the isometric isomorphisms $N_\Phi(\Omega) \cong P_m^d(\Omega) \oplus P_{\Phi,\Omega}$ and $N_\Phi(\Omega) \cong P_m^d(\Omega) \oplus F_{\Phi,\Omega}$ as two characterizations of the native space. We add two others, with proofs in [2] dating partially back to [3]:

Theorem 4. *Assume Ω is a subset of $\mathbb{R}^d$ and $m \ge 0$. Then $N_\Phi(\Omega)$ is the unique subspace of $C(\Omega)$ with a semi-inner product $(\cdot,\cdot)_\Phi$ satisfying*

(a) the null-space of the semi-norm is $P_m^d(\Omega)$,

(b) $N_\Phi(\Omega)/P_m^d(\Omega)$ is a Hilbert space,

(c) if $\mu \in (P_m^d)_\Omega^\perp$, then $\mu * \Phi \in N_\Phi(\Omega)$ and $(\mu * \Phi, f)_\Phi = \mu(\bar{f})$ for all $f \in N_\Phi(\Omega)$.

Theorem 5. *Fix $m \geq 0$ and a c.p.d. function Φ of order m in Ω. Then a complex-valued function f on Ω is in $N_\Phi(\Omega)$ iff there is a constant $c(f)$ such that*

$$|\mu(\bar{f})| \leq c(f)\|\mu\|_\Phi$$

for all μ in $(P_m^d)_\Omega^\perp$. The smallest possible constant for such f is the seminorm $\|f\|_\Phi$.

The following sections will proceed gradually from here to other characterizations of native spaces. The main guideline is the various forms that functionals can take, starting from finitely supported functionals used in this section. We proceed via measures (finitely or compactly supported) to distributions, and we refer the reader to [2] for full proofs.

§3. Construction via Measures

Definition 6. *The family of all finitely supported measures on Ω is denoted by $M(\Omega)$.*

Theorem 7. *Let m be a nonnegative integer. Assume Φ is positive definite in Ω with the following property: for all $\lambda \in M(\Omega)$ and $\varepsilon > 0$, there exists μ_ε in $(P_m^d)_\Omega^\perp$ satisfying*

$$\|\mu_\varepsilon - \lambda\|_\Phi < \varepsilon.$$

Then $(P_m^d)_\Omega^\perp$ is contained in $M(\Omega)$, and $M(\Omega)$ is isometrically isomorphic to a dense subset of $\overline{(P_m^d)_\Omega^\perp}$. Furthermore, we have

$$N_\Phi(\Omega) \cong P_m^d(\Omega) \oplus \overline{M(\Omega)},$$

where the closure is induced by Φ. The inner product on $\overline{M(\Omega)}$ is defined as $(\lambda, \mu)_\Phi := \lambda(\overline{\mu * \overline{\Phi}})$.

Now we introduce a new space $\langle(P_m^d)_\Omega^\perp\rangle$ consisting of all compactly supported measures μ on Ω with vanishing moments for P_m^d, i.e., all integrals of polynomials from P_m^d with respect to μ are zero. If we assume

$$\int_\Omega \int_\Omega \Phi(x,y) d\nu(y) d\overline{\mu(x)} = \int_\Omega \int_\Omega \Phi(x,y) d\overline{\mu(x)} d\nu(y)$$

for all μ, ν in $\langle(P_m^d)_\Omega^\perp\rangle$, and

$$\nu(\overline{\nu * \overline{\Phi}}) > 0$$

for all nonzero ν, it is easily checked that

$$(\nu, \mu) := \nu(\overline{\mu * \overline{\Phi}})$$

forms an inner product on $\langle(P_m^d)_\Omega^\perp\rangle$. Then we have the following theorem:

Theorem 8. *Under the above assumptions, $\overline{\langle (P_m^d)_\Omega^\perp \rangle}$ is isometrically isomorphic to $\overline{(P_m^d)_\Omega^\perp}$. Furthermore, the native space $N_\Phi(\Omega)$ is equivalent to $P_m^d(\Omega) \oplus \overline{\langle (P_m^d)_\Omega^\perp \rangle}$.*

The proof of Theorem 8 in [2] is quite hard. It involves *weak** topology and the Krein-Milman theorem. So far, Theorem 8 is the best result concerning interpretation of the dual of the native space as a space of measures.

§4. Construction via Tempered Test Functions

Starting from [4] there is an approach to native spaces via generalized Fourier transforms in the sense of Gelfand and Shilov. Here, we want to avoid distributions and generalized Fourier transforms as far as possible. The key point is to use variational equations on spaces of tempered test functions as a convenient substitute for generalized Fourier transforms.

Let $\mathcal{S}(\Omega)$ denote the space of tempered test functions in the sense of Laurent Schwartz with supports contained in Ω, and define $\mathcal{S}_m^\perp(\Omega)$ as the space of tempered test functions with support in Ω and vanishing moments up to order m. For all $v, w \in \mathcal{S}_m^\perp(\Omega)$

$$\langle v, w \rangle_\Phi := \int_\Omega \int_\Omega \Phi(x,y) v(x) \overline{w(y)} dx dy$$

is a bilinear form, and we would like to base a second construction of the native space on it. To this end, it would be a reasonable possibility to define a property like "tempered conditional positive definiteness" to require that this form is positive definite on $\mathcal{S}_m^\perp(\Omega)$. The result would be a different theory, but we want to blend this approach into our previous setting. Thus we look at conditions that allow to relate this bilinear form to the earlier one.

Following [1], we assume a continuous positive function $\varphi : \mathbb{R}^d \backslash \{0\} \longrightarrow \mathbb{R}$ exists such that

$$\langle v, w \rangle_\Phi = (2\pi)^{-d} \int_{R^d} \varphi(x) \hat{v}(x) \overline{\hat{w}(x)} dx \qquad (5)$$

for all $v, w \in \mathcal{S}_m^\perp(\Omega)$. Here

$$\hat{v}(w) := \int_\Omega e^{-ix^T w} v(x) dx$$

denotes the classical Fourier transform of v. By approximation of functionals from $(P_m^d)_\Omega^\perp$ by regular distributions generated by functions from $\mathcal{S}_m^\perp(\Omega)$, Iske [1] proved that this assumption is slightly stronger than c.p.d. of Φ on $\mathbb{R}^d$, and that $(v, w)_\Phi = \langle v, w \rangle_\Phi$ holds for all $v, w \in \mathcal{S}_m^\perp(\Omega)$.

Definition 9. *If a function Φ of difference form $\Phi(x,y) = \phi(x-y)$ with a continuous and even function ϕ on $\mathbb{R}^d$ satisfies (5), we call Φ* variationally positive definite (v.p.d) *of order $m \geq 0$ on $\mathbb{R}^d$.*

Definition 10. *Let Φ be v.p.d. of order $m \geq 0$ on $\mathbb{R}^d$. A complex-valued function f is in the space $\mathcal{C}_{\Phi,m}(\Omega)$ if and only if $f \in C(\Omega)$ and there exists a constant $c(f)$ such that*

$$|\int_\Omega f(x)v(x)dx| \leq c(f)\{\int_\Omega\int_\Omega \Phi(x,y)v(x)\overline{v(y)}dxdy\}^{1/2} \text{ for all } v \in \mathcal{S}_m^\perp(\Omega).$$

Theorem 11. *Let Ω be open and Φ be v.p.d. of order $m \geq 0$ in $\mathbb{R}^d$. Then $N_\Phi(\Omega) \cong \mathcal{C}_{\Phi,m}(\Omega)$. Furthermore, $\overline{\mathcal{S}_m^\perp(\Omega)}$ is isometrically isomorphic to $\overline{(P_m^d)_\Omega^\perp}$.*

Theorem 11 provides a nice unification of the theories of Weinrich [6] and Iske [1]. Their work is based on $(P_m^d)_\Omega^\perp$ and $\mathcal{S}_m^\perp(\Omega)$, respectively. The proof of Theorem 11 is rather involved [2].

§5. Embedding Theorems

We now construct continuous embeddings of native spaces into well-known spaces. Madych and Nelson's discovery that $N_\Phi(\mathbb{R}^d) \subset C(\mathbb{R}^d)$ can be regarded as the first step towards embedding theorems, but it was just an inclusion result. In this paper, all the embedding theorems concern continuous embeddings with respect to the topologies of the spaces. Even the embeddings of native spaces into L_2 spaces can be nontrivial, provided that the underlying domains are unbounded (see [5] for the bounded case).

In this section we first assume Φ to be v.p.d. of order 0 on $\mathbb{R}^d$ with a positive classical Fourier transform $\varphi \in L_1(\mathbb{R}^d)$ of ϕ with $\Phi(x,y) = \phi(x-y)$. All functions f of the form (1) have a classical Fourier transform

$$\widehat{f}(\omega) = \varphi(\omega)\sum_{j=1}^N c_j e^{-i\omega\cdot x_j},$$

and there is an isometry $B : R_\Phi((P_0^d)_\Omega^\perp) \to L_2(\mathbb{R}^d),\quad f \mapsto \widehat{f}/\sqrt{\varphi}$ mapping these functions into $L_2(\mathbb{R}^d)$. It is now easy to see that the equation $\widehat{f} = \sqrt{\varphi}\cdot B(f)$ holds for all functions in $R_\Phi((P_0^d)_\Omega^\perp)$ and its closure $F_{\Phi,\Omega}$ which can be identified with the native space of Φ.

Theorem 12. *For variationally positive definite functions on $\mathbb{R}^d$ of order zero with a positive L_1 Fourier transform φ, the functions in the native space of Φ have Fourier transforms of the form $\sqrt{\varphi}\cdot g$ with an L_2 function g. The native space for Φ can be continuously embedded [2] into $L_2(\mathbb{R}^d)$.*

The last statement was generalized in [2] to

Theorem 13. *Let Φ be symmetric and translation–invariant on $\Omega \times \Omega$ and c.p.d. of order $m \geq 0$ on a domain $\Omega \subseteq \mathbb{R}^d$ containing points $\xi_1, \ldots, \xi_N$ which uniquely determine polynomials of $P_m^d(\Omega)$. If there exists a positive continuous $g \in L^1(\Omega)$ which decays exponentially at infinity and satisfies*

$$\int_\Omega |p(x)\phi(x)|g(x)dx < \infty$$

for all $p(x) \in P_m^d(\Omega)$, then $F_{\Phi,\Omega}$ can be continuously embedded in $L^2(\Omega)$.

Theorem 12 characterizes native spaces as spaces of functions whose Fourier transforms lie in a weighted L_2 space. The same holds for Sobolev spaces on $\mathbb{R}^d$, and this similarity can be used to derive theorems for embedding of native spaces into global Sobolev spaces on $\mathbb{R}^d$. For embeddings of local native spaces on domains $\Omega \subseteq \mathbb{R}^d$, we refer the reader to the fact (proven in [2] and [5]) that functions in native spaces always have an extension to the largest domain where Φ has the c.p.d. property. This yields embeddings of local native spaces into spaces of restrictions of global Sobolev spaces for globally defined functions Φ, but the case of purely locally defined Φ is unsolved.

If Φ is v.p.d.of positive order m on $\mathbb{R}^d$, the function φ of (5) will have a singularity at zero, and thus the notion of Fourier transforms needs generalization. We simply view (5) as a variational property satisfied by the generalized Fourier transform φ of Φ, and we want to prove

Theorem 14. *For v.p.d. functions Φ on $\mathbb{R}^d$ of order $m > 0$ the functions f in the native space of Φ have generalized Fourier transforms $\widehat{f} = \sqrt{\varphi} \cdot g$ with an L_2 function g, where the generalized Fourier transform of f is defined via the variational property*

$$\int f \cdot w = (2\pi)^{-d} \int \widehat{f} \cdot \widehat{w} \text{ for all } w \in \mathcal{S}_m^\perp(\Omega).$$

Proof: We take two functions $v, w \in \mathcal{S}_m^\perp(\Omega)$ and form the function $f_v := \Phi * \bar{v}$. Then (5) yields

$$\begin{aligned}\int_\Omega w \cdot f_v &= (2\pi)^{-d} \int \varphi \widehat{w}\overline{\widehat{\bar{v}}} \\ &= (2\pi)^{-d} \int \widehat{w}\sqrt{\varphi}\sqrt{\varphi}\overline{\widehat{\bar{v}}} \qquad (6) \\ &= (2\pi)^{-d} \int \widehat{w}\sqrt{\varphi}B(f_v),\end{aligned}$$

if we define $B(f_v) := \sqrt{\varphi}\overline{\widehat{\bar{v}}} \in L_2(\mathbb{R}^d)$. This maps isometrically into $L_2(\mathbb{R}^d)$, because the canonical inner product of such functions is

$$(f_u, f_v)_\Phi = (2\pi)^{-d} \int \varphi \widehat{\bar{u}}\overline{\widehat{\bar{v}}}.$$

Now (6) carries over to the closure, i.e. the native space, and it yields the desired result. □

Acknowledgments. The author thanks R. Schaback for help with the final formulation of this paper.

References

1. Iske, A., Characterization of function spaces associated to conditionally positive definite functions, in *Mathematical Methods in Computer Aided Geometric Design III*, T. Lyche and L. L. Schumaker(eds.), Academic Press, New York, 1995.
2. Luh, L.-T., Characterizations of native spaces, PhD. Dissertation, Göttingen, 1998.
3. Madych, W. R. and S. A. Nelson, Multivariate interpolation and conditionally positive definite functions, Approx. Theory Appl. **4** (1988), 77–89.
4. Madych, W. R. and S. A. Nelson, Multivariate interpolation and conditionally positive definite functions II, Math. Comp. **54** (1990), 211–230.
5. Schaback, R., Native Spaces for Radial Basis Functions I, International Series of Numerical Mathematics **132**, Birkhäuser 1999, 255–282
6. Weinrich, M., Charakterisierung von Funktionenräumen bei der Interpolation mit radialen Basisfunktionen, Ph.D. thesis, Göttingen, 1994.

Lin-Tian Luh
Dept. of Applied Mathematics
Providence University
200 Chungchi Road, Shalu 43301
Taichung Hsien, Taiwan, Republic of China

ltluh@pu.edu.tw

Review of Some Approximation Operators for the Numerical Analysis of Spectral Methods

Yvon Maday

Abstract. This paper reviews some operators that are used in the numerical analysis of spectral and spectral element methods. We motivate the introduction of these different operators and sketch their approximation properties. Finally, we apply them to derive optimal error estimates for spectral type approximations of the solution of elliptic partial differential equations.

§1. Introduction

Spectral type methods are high order discretizations that allow to compute approximate solutions of partial differential equations. The recent version of spectral approximations is based on the Galerkin approach where the variational statement (equivalent to the strong formulation of the PDE) is set on discrete spaces of test and trial functions. For instance, let us consider the problem: *find $u \in X$ such that*

$$a(u,v) = \langle f, v \rangle, \quad \forall v \in X, \tag{1}$$

where X is some Hilbert space, and a is a continuous bilinear form over X. The general Galerkin approximation of this problem first requires the choice of a family of discrete spaces $X_N \subset X$, where N is a parameter that tends to infinity and is related to the dimension of the discrete space X_N. The discrete problem is then stated as follows: *find $u_N \in X_N$ such that*

$$a(u_N, v_N) = \langle f, v_N \rangle, \quad \forall v_N \in X_N. \tag{2}$$

The basic general hypothesis that makes problem (1) well-posed is that a is continuous and α-elliptic over X (i.e. $\exists\ \alpha > 0$ such that $a(u,u) \geq \alpha \|u\|_X^2$ for

Curve and Surface Fitting: Saint-Malo 1999
Albert Cohen, Christophe Rabut, and Larry L. Schumaker (eds.), pp. 309–324.

ISBN 0-8265-1357-3.

all $u \in X$. These properties remain true over each X_N (since $X_N \subset X$); thus (2) is also well-posed for each N. In addition, the solution u_N satisfies

$$\|u - u_N\|_X \leq c \inf_{v_N \in X_N} \|u - v_N\|_X. \tag{3}$$

The constant c that appears in (3) is the quotient of the continuity constant of a with the ellipticipty constant α and is thus independent of X_N.

Going back to spectral methods, the definition of X_N involves polynomials, and in the most simple cases (we shall see more general examples in Section 5) we have $X_N = X \cap \mathbb{P}_N$, where $\mathbb{P}_N$ represents the set of all polynomials of (partial) degree less than or equal to N. Here N is the parameter responsible for the convergence of the method. Due to (3), one ingredient in the numerical analysis of the spectral method is the approximation properties of the space of polynomials for given functions. The classical analysis of the approximation properties of polynomials is done in terms of L^∞–norms. This is not completely appropriate for our purpose since most often X is a Hilbert space (generally L^2 or H^1 spaces), and the approximation properties have to be measured with these norms. If a rate of convergence (with respect to N) on the best fit $\inf_{v_N \in X_N} \|u - v_N\|_X$ is sought after, some regularity has to be assumed over u. In Section 2, we give a survey of these best approximation results depending on the regularity of the function we want to approximate. We first analyze the L^2–best fit and then the H^1–best fit. The main ingredient in this analysis relies on the Legendre basis that is composed of the orthogonal polynomials for the standard Lebesgue mesure over the interval $(-1, +1)$. These polynomials, denoted as $(L_n)_n$, are defined by: degree$(L_n) = n$,

$$L_n(1) = 1, \tag{4}$$

$$\int_{-1}^{1} L_n(\zeta) L_m(\zeta) d\zeta = \frac{2\delta_{m,n}}{2n+1}. \tag{5}$$

They satisfy some standard properties (actually valid for most families of orthogonal polynomials)

$$\mathcal{A}(L_n) \equiv -\frac{d}{d\zeta}((1-\zeta^2)\frac{dL_n}{d\zeta} = n(n+1)L_n, \tag{6}$$

that one can translate by saying that the Legendre polynomials are the eigenvectors of the (Sturm-Liouville) operator $\mathcal{A}$. Since this is a possible basis set for the implementation of problem (2), this gives the name of spectral to the methods we shall consider hereafter, and that have been first analyzed in [10]. We refer also to [6] and [3] for more recent surveys on the numerical analysis of these methods. In Section 3, we introduce the notion of numerical integration and the interpolation operator, two notions that are naturally quite close and that allow to transform the "theoretical" approximation method into a "applicable" one. In Section 4, motivated by the analysis of the Stokes problem, we introduce a new operator, that, in opposition to the previous ones, is uniformly stable (in N) both in the L^2–norm and the H^1–norm and possess optimal approximation properties. It has to be said, beforehand, that in the precise analysis of these spectral (or polynomial) approximation, the Bernstein inequality runs counter to most standard tools that generally allow for deriving approximation results for a new operator from an already analyzed one. This Bernstein inequality tells about the equivalence of norms on the finite dimensional linear space of polynomials. It is well known that, for any function in H^1, the L^2 norm is smaller than the H^1–norm; of course this is true in particular on polynomials:

$$\forall \phi_N \in \mathbb{P}_N, \qquad \|\phi_N\|_{L^2} \le \|\phi_N\|_{H^1}.$$

Since all norms are equivalent on $\mathbb{P}_N$, there exists a constant (obviously depending on N) such that

$$\forall \phi_N \in \mathbb{P}_N, \qquad \|\phi_N\|_{H^1} \le c(N)\|\phi_N\|_{L^2}.$$

The behaviour of this constant is made precise by the Bernstein inequality

$$\forall \phi_N \in \mathbb{P}_N, \qquad \|\phi_N\|_{H^1} \le cN^2\|\phi_N\|_{L^2},$$

where c no longer depends on N. This estimate is optimal (in the sense that there exists a sequence of polynomials such that the ratio of the H^1–norm over the L^2–norm scales like $\mathcal{O}(N^2)$), but is bad as regards the ratio of convergence rate between the H^1–best fit and the L^2–best fit that scales like $\mathcal{O}(N^{-1})$, as we shall see below.

In the first three sections, the domains where the functions live will be very simple, actually too simple to tackle real life problems; indeed these are bricks equal to $(-1,1)^d$ where $d = 1, 2$ or 3. The generalization of spectral methods to more complex geometries is done by combining two key ingredients: the mapping of bricks onto curved bricks through regular mappings, and domain decomposition. We give some hints about this generalization in §5.

§2. Hilbert Type Projection Operators

Let us start with the one-dimensional case. In $L^2(-1,1)$, we consider the set $\mathbb{P}_N(-1,1)$ of all polynomials of degree $\le N$. From the Weierstrass density theorem, we know that any element ϕ in $L^2(-1,1)$ can be written as

$$\phi(\zeta) = \sum_{n=0}^{\infty} \widehat{\phi}^n L_n(\zeta), \tag{7}$$

where the convergence of the series holds in L^2. The coefficients $\widehat{\phi}^n$ can be derived from ϕ thanks to the orthogonality of the Legendre basis as follows:

$$\widehat{\phi}^n = \frac{2n+1}{2}\int_{-1}^{1} \phi(\zeta)L_n(\zeta)d\zeta.$$

Next, from (6) we derive that

$$\widehat{\phi}^n = \frac{2n+1}{2}\int_{-1}^{1}\phi(\zeta)\frac{\mathcal{A}L_n(\zeta)}{n(n+1)}d\zeta,$$

noticing that $\mathcal{A}$ is symmetric, and assuming ϕ regular enough, we derive that

$$\widehat{\phi}^n = \frac{2n+1}{2}\int_{-1}^{1}\mathcal{A}(\phi)(\zeta)\frac{L_n(\zeta)}{n(n+1)}d\zeta.$$

If we iterate this argument p times, we obtain

$$\widehat{\phi}^n = \frac{2n+1}{2}\int_{-1}^{1}\mathcal{A}^p(\phi)(\zeta)\frac{L_n(\zeta)}{n^p(n+1)^p}d\zeta,$$

so that, the following simple relation holds between the Legendre coefficients of ϕ and of $\mathcal{A}^p(\phi)$:

$$\widehat{\phi}^n = \frac{1}{n^p(n+1)^p}\widehat{\mathcal{A}^p(\phi)}^n.$$

Next, let π_N denote the $L^2(-1,1)$–projection over $\mathbb{P}_N(-1,1)$. Going back to (6), we deduce from (7) and (5) that

$$\pi_N(\phi) = \sum_{n=0}^{N}\widehat{\phi}^n L_n(\zeta), \tag{8}$$

so that

$$\phi - \pi_N(\phi) = \sum_{n=N+1}^{\infty}\widehat{\phi}^n L_n(\zeta) = \sum_{n=N+1}^{\infty}\frac{1}{n^p(n+1)^p}\widehat{\mathcal{A}^p(\phi)}^n L_n(\zeta)$$

and, by Parseval

$$\begin{aligned}\|\phi - \pi_N(\phi)\|^2_{L^2(-1,1)} &= \sum_{n=N+1}^{\infty}[\frac{1}{n^p(n+1)^p}]^2[\widehat{\mathcal{A}^p(\phi)}^n]^2\frac{2}{2n+1}\\ &\le [\frac{1}{N}]^{4p}\sum_{n=N+1}^{\infty}[\widehat{\mathcal{A}^p(\phi)}^n]^2\frac{2}{2n+1}\\ &\le [\frac{1}{N}]^{4p}\sum_{n=0}^{\infty}[\widehat{\mathcal{A}^p(\phi)}^n]^2\frac{2}{2n+1} = [\frac{1}{N}]^{4p}\|\mathcal{A}^p(\phi)\|^2_{L^2(-1,1)}.\end{aligned}$$

We have thus proven that, for any ϕ in the domain $\mathcal{D}[\mathcal{A}^p]$ of $\mathcal{A}^p$,

$$\|\phi - \pi_N(\phi)\|_{L^2(-1,1)} \le c(p)N^{-2p}\|\mathcal{A}^p(\phi)\|_{L^2(-1,1)}.$$

It is easy to check that $H^{2p}(-1,1) \subset \mathcal{D}[\mathcal{A}^p]$; hence the following theorem (due to Canuto and Quarteroni [7]), proven here for even values of r, holds for any r thanks to an argument of interpolation between Sobolev spaces:

Theorem 1. *For any real number $r \geq 0$, there exists a constant $c > 0$, depending only on r such that, for any function $\phi \in H^r(-1,1)$,*

$$\|\phi - \pi_N(\phi)\|_{L^2(-1,1)} \leq cN^{-r}\|\phi\|_{H^r(-1,1)}. \tag{9}$$

Let us denote now by Π_N the $L^2((-1,1)^d)$ orthogonal projection operator over the set $\mathbb{P}_N((-1,1)^d)$ of all polynomials of degree $\leq N$ with respect to each variable. By Fubini's theorem, $\Pi_N = \pi_N \otimes \pi_N$, in 2D and $\Pi_N = \pi_N \otimes o\pi_N \otimes o\pi_N$, in 3D. By tensorizing (9) we derive

Theorem 2. *For any real number $r \geq 0$, there exists a constant $c > 0$, depending only on r such that, for any function $\phi \in H^r((-1,1)^d)$,*

$$\|\phi - \Pi_N(\phi)\|_{L^2((-1,1)^d)} \leq cN^{-r}\|\phi\|_{H^r((-1,1)^d)}.$$

We are now in a position to tackle the approximation in the H^1 norms. First, we consider a function $\phi \in H_0^1(-1,1) \cap H^r(-1,1)$, with $r \geq 1$. It is quite immediate to check that the polynomial $\phi_N(\zeta) = \int_{-1}^{\zeta} \pi_{N-1}\frac{d\phi}{d\xi}(\xi)d\xi$ belongs to $\mathbb{P}_N(-1,1)$, vanishes at $\zeta = -1$, and satisfies

$$\begin{aligned}
\phi_N(1) &= \int_{-1}^{1} \pi_{N-1}\frac{d\phi}{d\xi}(\xi)d\xi = \int_{-1}^{1} \pi_{N-1}\frac{d\phi}{d\xi}(\xi)L_0(\xi)d\xi \\
&= \int_{-1}^{1} \frac{d\phi}{d\xi}(\xi)L_0(\xi)d\xi = \int_{-1}^{1} \frac{d\phi}{d\xi}(\xi)d\xi \\
&= \phi(1) - \phi(-1) = 0,
\end{aligned}$$

and hence is an element of $\mathbb{P}_N(-1,1) \cap H_0^1(-1,1)$. Finally it is a good approximation of ϕ, since from Poincarré's inequality and Theorem 1,

$$\begin{aligned}
\|\phi - \phi_N\|_{H^1(-1,1)} &\leq c\|\frac{d\phi}{d\xi} - \frac{d\phi_N}{d\xi}\|_{L^2(-1,1)} \\
&\leq c\|\frac{d\phi}{d\xi} - \pi_{N-1}(\frac{d\phi}{d\xi})\|_{L^2(-1,1)} \\
&\leq cN^{1-r}\|\frac{d\phi}{d\xi}\|_{H^{r-1}(-1,1)} \leq cN^{1-r}\|\phi\|_{H^r(-1,1)}.
\end{aligned}$$

Let us introduce now the orthogonal projection operator $\pi_N^{1,0}$ from $H_0^1(-1,1)$ onto $\mathbb{P}_N(-1,1) \cap H_0^1(-1,1)$, we can state the following result (due to Maday and Quarteroni [15]):

Theorem 3. *For any real number $r \geq 1$ and any real number $0 \leq s \leq 1$, there exists a constant $c > 0$, depending only on r and s such that for any function $\phi \in H_0^1(-1,1) \cap H^r(-1,1)$,*

$$\|\phi - \pi_N^{1,0}(\phi)\|_{H^s(-1,1)} \leq cN^{s-r}\|\phi\|_{H^r(-1,1)}. \tag{10}$$

Proof: The theorem has been obtained for $s = 1$. For $s = 0$ it is obtained through a standard Aubin-Nitsche duality argument, and then for any s by interpolation between Sobolev spaces. □

Remark. At this point it has to be said that the L^2 projection operator π_N does not have optimal approximation properties in the H^1–norm, the only (non-improvable) property that can be obtained is

$$\|\phi - \pi_N(\phi)\|_{H^1(-1,1)} \le cN^{\frac{3}{2}-r}\|\phi\|_{H^r(-1,1)}.$$

We refer to [3] for details and counter-examples.

Remark. It may also be interesting to note that, despite their definition, the previous operators have stability properties in various norms. First for the L^2-operator, we have

$$\|\pi_N\phi\|_{H^1(-1,1)} \le cN^{\frac{1}{2}}\|\phi\|_{H^1(-1,1)},$$

which is related to what we have indicated in the previous remark, but also

$$\|\pi_N^{1,0}\phi\|_{L^2(-1,1)} \le cN^{\frac{1}{2}}\|\phi\|_{L^2(-1,1)},$$

which is rather suprising since, from this (non-uniform) stability property, the H_0^1–projection operator can be extended to (irregular) functions of L^2!!

Again by tensorization of the results of the one dimensional Theorem 3, we exhibit a polynomial that approximates regular functions in $H_0^1((-1,1)^d)$ well, from which we derive approximation properties on the multidimensional projection operator $\Pi_N^{1,0}$ from $H_0^1((-1,1)^d)$ over $\mathbb{P}_N((-1,1)^d) \cap H_0^1((-1,1)^d)$ both in H^1–norm and in L^2–norm (derived by duality):

Theorem 4. *For any real number $r \ge 1$ and any real number $0 \le s \le 1$, there exists a constant $c > 0$, depending only on r and s such that, for any function $\phi \in H_0^1((-1,1)^d) \cap H^r((-1,1)^d)$,*

$$\|\phi - \Pi_N^{1,0}(\phi)\|_{H^s((-1,1)^d)} \le cN^{s-r}\|\phi\|_{H^r((-1,1)^d)}. \tag{11}$$

These results can be completed in order to derive a whole scale of approximation projectors in higher order norms. These are required, e.g. for the analysis of the approximation of fourth-order problems. The general result, concerning the orthogonal projection operator $\Pi_{N,0}^{\rho,\sigma}$ from $H^\rho((-1,1)^d \cap H_0^\sigma((-1,1)^d$ onto $\mathbb{P}_N((-1,1)^d) \cap H_0^\sigma((-1,1)^d)$ is given in the following theorem (due to Maday [11] in 1D, see also [3] for the extension to 2 and 3D):

Theorem 5. *For any real number $0 \le \sigma \le \rho$ and any $0 \le s \le \rho \le r$, there exists a constant $c > 0$, depending only on r, s, ρ, σ such that, for any function $\phi \in H_0^\sigma((-1,1)^d) \cap H^r((-1,1)^d)$,*

$$\|\phi - \Pi_{N,0}^{\rho,\sigma}(\phi)\|_{H^s((-1,1)^d)} \le cN^{s-r}\|\phi\|_{H^r((-1,1)^d)}.$$

Remark. A final remark on these operators is that improved-approximation results in negative norms are also true, and can be obtained in a classical way, by further refering to the Aubin-Nitsche duality argument. Hence, Theorem 5 is also valid for negative values of s.

These results allow us to prove that the approximation of most elliptic variational problems by spectral methods is optimal. As an example, let us consider the (non-constant) Laplace problem on a cube $\Omega = (-1,1)^3$: given a 3×3 matrix, symmetric and uniformly positive definite, we consider the problem of finding $u \in H_0^1(\Omega)$ such that

$$-\mathrm{div}[A\mathbf{grad}]u = f. \tag{12}$$

The approximation then consists in finding an element u_N in $X_N \equiv \mathbb{P}_N(\Omega) \cap H_0^1(\Omega)$ such that

$$\int_\Omega A\nabla u_N \nabla v_N = \int_\Omega f v_N, \quad \forall v_N \in X_N. \tag{13}$$

Assuming that $u \in H^r(\Omega)$, we deduce from (3) and (11) that

$$\|u - u_N\|_{H^1(\Omega)} \le cN^{1-r}\|u\|_{H^r(\Omega)}.$$

As hinted in the introduction, this problem is numerically intractable; indeed the implementation of (13) requires the computation of the two integrals appearing on the left- and the right-hand sides of this equation. The exact computation is most often impossible, and certainly numerically not fast enough. The use of numerical integration rules is the cure to this problem, but in order to combine efficiency and precision, following Gottlieb [9] and Mercier [17], we refer to the use of Gauss type quadrature rule. Indeed, they are well known to be well suited for the integration of polynomials.

§3. Interpolation Operators

Between the different numerical quadrature rules over $(-1,1)$, well suited for polynomial integration, we shall quote here the Legendre-Gauss and Legendre-Gauss-Lobatto ones. We refer to [2] for more details. For the sake of completeness, we recall the definition of these formulae:

Theorem 6. *(Gauss formula) For any real number n, there exists a unique set of points $-1 < \zeta_1^n < \zeta_2^n < \cdots < \zeta_n^n < 1$, and a unique set of positive weights ω_i^n such that for any polynomial $\phi \in \mathbb{P}_{2n-1}(-1,1)$, the following equality holds:*

$$\int_{-1}^{1} \phi(\zeta)d\zeta = \sum_{i=1}^{n} \phi(\zeta_i^n)\omega_i^n.$$

Theorem 7. *(Gauss-Lobatto formula) For any real number n, there exists a unique set of points $-1 = \xi_0^n < \xi_1^n < \cdots < \xi_n^n = 1$, and a unique set of positive weights ρ_i^n such that for any polynomial $\phi \in \mathbb{P}_{2n-1}(-1,1)$, the following equality holds:*

$$\int_{-1}^{1} \phi(\zeta) d\zeta = \sum_{i=0}^{n} \phi(\xi_i^n) \rho_i^n.$$

From now on, we shall assume that the degree of the polynomials for the approximation is fixed to be N, and we shall use $N+1$ points either of Gauss or Gauss-Lobatto type. For the sake of simplicity, these points will be denoted with no superscript, i.e. in all of what follows, we set $\zeta_i \equiv \zeta_i^{N+1}$ and $\xi_i \equiv \xi_i^N$. We recall that these points are the roots (resp. the extrema) of the Legendre polynomials; more precisely, we have

$$\forall i, \quad L_{N+1}(\zeta_i) = 0, \quad (\text{resp. } (1 - \xi_i^2) L'_{N+1}(\xi_i) = 0).$$

After tensorization, these one dimensional quadrature rules easily provide quadrature rules on the square and on the cube defined as follows (e.g. in 2D for the Gauss Lobatto formula):

$$\sum_{\mathrm{GL}} \phi \equiv \sum_{i=0}^{N} \sum_{j=0}^{N} \phi(\xi_i, \xi_j) \omega_i \omega_j.$$

The problem that is actually implemented is then the following: find an element u_N in X_N such that

$$\sum_{\mathrm{GL}} A \nabla u_N \nabla v_N = \sum_{\mathrm{GL}} f v_N, \quad \forall v_N \in X_N. \tag{14}$$

Even in the case where A is constant, at least in more than one dimension, the left-hand side is not exactly computed. The problem is no longer of the form (1), and the abstract theory has to be generalized in order to handle this problem as well.

Here is not the place to detail this generalization (see [3], where the complete analysis is performed) but it is natural that the α-ellipticity of the bilinear form on the left-hand side of (14) is again one of the key ingredients and has to be satisfied. This follows from the property, proven in [7]

$$\forall \phi_N \in \mathbb{P}_N(-1,1), \quad \sum_{\mathrm{GL}} \phi_N^2 \geq \int_{-1}^{1} \phi_N^2(\zeta) d\zeta.$$

From this property it can be easily derived that the solution u_N to (14) exists and is unique.

The approximation properties of the polynomial interpolation operator over the Gauss-Lobatto nodes is of great importance in the error bounds. Let i_N denote this operator in one dimension:

$$\forall \phi \in \mathcal{C}^0([-1,1]), \quad i_N(\phi) \in \mathbb{P}_N(-1,1) \text{ and } \forall i, 0 \leq i \leq N, \quad i_N(\phi)(\xi_i) = \phi(\xi_i)$$

and let us tensorize it in order to get a two (resp. a three) dimensional operator $\mathcal{I}_N \equiv i_n \otimes i_n$ (resp. $\mathcal{I}_N \equiv i_n \otimes i_n \otimes i_n$). The properties of this operator have been established in [12] and [2], and read as follows:

Theorem 8. *For any real numbers s and r satisfying $r > (d+s)/2$ and $0 \le s \le 1$, there exists a positive constant c depending only on r such that for any ϕ in $H^r((-1,1)^d)$ the following estimate holds*

$$\|\phi - \mathcal{I}_N(\phi)\|_{H^s((-1,1)^d)} \le cN^{s-r}\|\phi\|_{H^r((-1,1)^d)}. \tag{15}$$

It has to be noticed that this operator requires more regularity than the L^2 projection operator, but it is optimal both in the L^2 and the H^1 norms. It has also to be recalled that in the classical approximation properties in the L^∞ norm, the Lebesgue constant appears as a pollution of the approximation properties of the interpolation operator as regards the optimality provided by the corresponding best fit. This is not the case in the L^2–norm. In this direction what we have more precisely is that, for any function ϕ in $H_0^1(-1,1)$,

$$\|i_N\phi\|_{L^2(-1,1)} \le c(\|\phi\|_{L^2(-1,1)} + \frac{1}{N}\|\frac{d\phi}{d\zeta}\|_{L^2(-1,1)}),$$

and for any function ϕ in $H^1(-1,1)$,

$$\|i_N\phi\|_{H^1(-1,1)} \le c\|\phi\|_{H^1(-1,1)}.$$

Another nice property of this operator, that has some importance for nonlinear PDE's, is the following result: for any polynomial $\phi_M \in \mathbb{P}_M(-1,1)$,

$$\|i_N\phi_M\|_{L^2(-1,1)} \le c(1+\frac{M}{N})\|\phi_M\|_{L^2(-1,1)}.$$

Here no duality argument allows us to derive from the previous theorem improved approximation properties in negative norms. It is an open problem to derive such results.

The numerical analysis of problem (13) then continues by noticing that

$$\sum_{\mathrm{GL}} fv_N = \sum_{\mathrm{GL}} \mathcal{I}_N(f)v_N,$$

which is one of the ingredients that allows to prove (see [2]):

Theorem 9. *Assume that the solution u of (12) belongs to $H^r(\Omega)$, that the coefficients in A are very regular, and that the data f belongs to $H^\rho(\Omega)$. Then the solution u_N to (13) satisfies*

$$\|u - u_N\|_{H^1(\Omega)} \le c(N^{1-r}\|u\|_{H^r(\Omega)} + N^{-\rho}\|f\|_{H^\rho(\Omega)}).$$

The case where A is not so regular can be handled with the same type of arguments, but more technical tools are involved; we refer to [16] for more details. It is interesting also to note at this level that, taking into account non-homogeneous Dirichlet boundary condition is very simple thanks to the nice properties of the interpolation operator. Indeed, assume that the solution to our problem (12) has to satisfy (instead of zero Dirichlet boundary conditions) the following condition: $u_{|\partial\Omega} = g$ where g is a given function on the boundary of Ω. Then, naturally, for the approximation, we look for a polynomial u_N in $\mathbb{P}_N(\Omega)$ such that (12) holds, and in addition

$$u_{N|\partial\Omega} = \widetilde{\mathcal{I}}_N g,$$

where $\widetilde{\mathcal{I}}_N$ is the operator of interpolation defined edge by edge (respectively face by face) from i_N (resp. from $\mathcal{I}_N$). Since the interpolation operator is optimal both in L^2 and in H^1, it results by an argument of interpolation between Sobolev spaces that it is also optimal with respect to the $H^{1/2}(\partial\Omega)$–norm. This fractional order norm is the natural one for the treatment of the boundary terms. It has also to be stressed that neither the L^2–projection operator nor the H^1–projection operator allow such an optimality nor such ease in the implementation.

Next, associated with the Gauss quadrature formula, we can also define an interpolation operator, denoted as j_N and defined as follows: $\forall\phi \in \mathcal{C}^0([-1,1])$,

$$j_N(\phi) \in \mathbb{P}_{N+1}(-1,1) \text{ and } \forall i, 1 \leq i \leq N+1, \quad j_N(\phi)(\zeta_i) = \phi(\zeta_i).$$

The $L^2(-1,1)$–approximation properties of this second interpolation operator are also optimal. Unfortunately, in the H^1–norm it is not optimal; for instance it is readily checked that $j_N(L_{N+1} - L_{N-1}) = L_{N-1}$. Recalling that

$$\forall n, \quad L_{n+1} - L_{n-1} = -\frac{2n+1}{n(n+1)}(1-\zeta^2)L'_n,$$

it is then easily proven that $\|L_{N+1} - L_{N-1}\|_{H^1(-1,1)}$ scales like $\mathcal{O}(N^{1/2})$ while $\|L_{N-1}\|_{H^1(-1,1)}$ scales like $\mathcal{O}(N)$; j_N is thus not stable in the H^1 norm.

For similar reasons, the interpolation operator i_N on the Gauss-Lobatto nodes does not have optimal approximation properties in the $H^2(-1,1)$–norm. In order to achieve such a property, we have to refer to generalized Gauss-Lobatto rules as is done e.g.in [1].

§4. An "Ideal" Operator

At this stage there is no operator from $L^2(-1,1)$ onto the set of polynomials that has optimal approximation properties and is stable both in the L^2 and the H^1 norm. Such an operator is useful, as will be explained below, in the analysis of the Stokes problem. In order to define this "ideal" operator, we fix a positive real number λ and a cut-off function χ of class $\mathcal{C}^1$ on $\mathbb{R}^+$ such that χ is equal to 1 on $[0, 1-\lambda]$, decreases from 1 to 0 on $[1-\lambda, 1]$ and vanishes on $[1,\infty]$. Next, with each positive integer N, we associate as in [18] an operator π_N^χ with values in $\mathbb{P}_N(-1,1) \cap H_0^1(-1,1)$ as follows: since each function ϕ in $H_0^1(-1,1)$ can be written as $\phi = \sum_{n=1}^{\infty} \widehat{\phi}^n(L_{n+1} - L_{n-1})$, we set $\pi_N^\chi\phi = \sum_{n=1}^{\infty} \chi(\frac{n}{N})\widehat{\phi}^n(L_{n+1} - L_{n-1})$. Note that the sum above is finite since χ has a bounded support. It is proven in [4] that this operator is stable both in the H_0^1 and the L^2 norms:

Theorem 10. *There exists a constant c, independent of N such that for any function* $\phi \in H_0^1(-1,1)$,

$$\|\pi_N^\chi \phi\|_{L^2(-1,1)} \le c\|\phi\|_{L^2(-1,1)}, \quad \|\pi_N^\chi \phi\|_{H^1(-1,1)} \le \|\phi\|_{H^1(-1,1)}. \tag{16}$$

It is an easy matter to verify that the operator π_N^χ leaves invariant all polynomials of $\mathbb{P}_{\lambda N}(-1,1) \cap H_0^1(-1,1)$. The previous stability and the best fit estimates (9), (10) imply

Theorem 11. *For any positive real number r and any real number* $0 \le s \le r$, *there exists a constant* $c > 0$, *depending only on r and s such that, for any function* $\phi \in H_0^r(-1,1)$ *if* $r \le 1$ *and any function* $\phi \in H_0^1(-1,1) \cap H^r(-1,1)$ *if* $r \ge 1$,

$$\|\phi - \pi_N^\chi \phi\|_{H^s(-1,1)} \le cN^{s-r}\|\phi\|_{H^r(-1,1)}.$$

As an application of the previous result, we can consider the problem of finding compatible spaces for the approximation of the Stokes equation. Under variational formulation, this problem consists in finding a pair $(\boldsymbol{u}, p)$ in $(H_0^1(\Omega))^d \times L_0^2(\Omega)$ of velocity and pressure such that

$$\int_\Omega \nabla \boldsymbol{u} \nabla \boldsymbol{v} - \int_\Omega p \mathrm{div} \boldsymbol{v} = \int_\Omega f \boldsymbol{v}, \quad \forall \boldsymbol{v} \in (H_0^1(\Omega))^d, \tag{17}$$

$$\int_\Omega q \mathrm{div} \boldsymbol{u} = 0, \quad \forall q \in L_0^2(\Omega), \tag{18}$$

where $L_0^2(\Omega)$ is the set of L^2 functions with zero average. It is well understood now that the spectral approximation of the Stokes problem based on polynomials of the *same* degree leads to instabilities. This is due to the fact that the pressure space is too rich in comparison to the velocity space. Indeed, there exist polynomials q_N in $\mathbb{P}_N((-1,1)^d)$ such that $\int_\Omega q_N \mathrm{div} \boldsymbol{v}_N = 0$ for all $\boldsymbol{v}_N$ in $(\mathbb{P}_N((-1,1)^d) \cap H_0^1((-1,1)^d))^d$ (e.g. $q_N(x,y,z) = L_N(x)L_N(y)L_N(z)$). Of course such polynomials (called spurious modes) prevent the discrete problem from being well-posed since it prevents the definition of a unique pressure. The cure is well known, and consists in depleting the pressure space for a given velocity space. In [14] the pair $(\mathbb{P}_N((-1,1)^d) \cap H_0^1((-1,1)^d))^d \times \mathbb{P}_{N-2}((-1,1)^d) \cap L_0^2((-1,1)^d)$ has been proposed, and gets rid of the spurious modes. It is known as the $\mathbb{P}_N \times \mathbb{P}_{N-2}$–method. Actually, what is looked for is a pair $X_N \times M_N$ approximating $(H_0^1((-1,1)^d))^d \times L_0^2((-1,1)^d)$ well and such that not only $\forall q_N \in M_N, \exists \boldsymbol{v}_N, \ \int_\Omega p \mathrm{div} \boldsymbol{v}_N \ne 0$, but more precisely, in order to get a stable method, we require that

$$\forall q_N \in M_N, \exists \boldsymbol{v}_N, \quad \int_\Omega p \mathrm{div} \boldsymbol{v}_N \ge \beta \|\boldsymbol{v}_N\|_{H^1((-1,1)^d)} \|q_N\|_{L^2((-1,1)^d)},$$

where β is known as the constant of the inf-sup condition. The behaviour of β for the $\mathbb{P}_N \times \mathbb{P}_{N-2}$–method scales as $\mathcal{O}(N^{-\frac{d-1}{2}})$ (see [2]), and it has been a long standing question whether there is a uniformly stable spectral approximation of the Stokes problem. It has to be said that the nonuniform behaviour of the inf-sup constant pollutes the accuracy of the pressure, but also pollutes the convergence properties of some classical solvers for the Stokes problem (see [13]). The "ideal" operator introduced above allows us to prove that a compatible choice is the $\mathbb{P}_N \times \mathbb{P}_{\lambda N}$–method that proposes, for the same choice of velocity space, $\mathbb{P}_{\lambda N}((-1,1)^d) \cap L^2_0((-1,1)^d)$ to be the pressure space. The following result is due to Bernardi and Maday [4]:

Theorem 12. *For any real number λ, $0 < \lambda < 1$, there exists a positive constant β independent of N such that, for any integer $N \geq 2/(1-\lambda)$ and any $q_N \in \mathbb{P}_{\lambda N}((-1,1)^d) \cap L^2_0((-1,1)^d)$,*

$$\sup_{\boldsymbol{v}_N \in (\mathbb{P}_N((-1,1)^d) \cap H^1_0((-1,1)^d))^d} \frac{\int_\Omega p \,\mathrm{div} \boldsymbol{v}_N}{\|\boldsymbol{v}_N\|_{H^1((-1,1)^d)}} \geq \beta \|q_N\|_{L^2((-1,1)^d)}.$$

Proof: Let q_N be any polynomial in $\mathbb{P}_{\lambda N}((-1,1)^d) \cap L^2_0((-1,1)^d)$. It is a standard matter (see e.g. Corollary 2.4 in [8]) that, to q_N, can be associated a (continuous) element $\boldsymbol{v}$ in $[H^1_0((-1,1)^d)]^d$ such that

$$\mathrm{div} \boldsymbol{v} = q_N \text{ and } \|\boldsymbol{v}\|_{H^1((-1,1)^d)} \leq c \|q_N\|_{L^2((-1,1)^d)}.$$

The problem is that $\boldsymbol{v}$ is not a polynomial. We define $\boldsymbol{v}_N = \pi^\chi_N \otimes \pi^\chi_N \boldsymbol{v}$ in 2D and $\boldsymbol{v}_N = \pi^\chi_N \otimes \pi^\chi_N \otimes \pi^\chi_N \boldsymbol{v}$ in 3D for which we derive thanks to (13) that

$$\|\boldsymbol{v}_N\|_{H^1((-1,1)^d)} \leq c \|q_N\|_{L^2((-1,1)^d)}.$$

Due to the fact that π^χ_N leaves invariant all polynomials of $\mathbb{P}_{\lambda N}(-1,1) \cap H^1_0(-1,1)$, we deduce that $\int_\Omega q_N \mathrm{div}(\boldsymbol{v}_N - \boldsymbol{v}) = 0$, and thus

$$\int_\Omega q_N \mathrm{div} \boldsymbol{v}_N = \int_\Omega q_N \mathrm{div} \boldsymbol{v} = \int_\Omega q_N^2,$$

which concludes the proof with $\beta = \frac{1}{c}$. □

§5. Extension to Domain Decompositions

In the spectral method history, the need to tackle more general domains was recognized early. In this direction, Patera has proposed in [19] the spectral element method that combines the accuracy of the spectral method with the flexibility of the domain decomposition methods. The idea is to introduce a partition of the domain Ω as a union of nonoverlapping subdomains:

$$\overline{\Omega} = \cup_{k=1}^K \overline{\Omega}^k, \quad \Omega^k \cap \Omega^\ell = \emptyset.$$

In addition, we assume that each subdomain Ω^k is associated with a regular one-to-one mapping $\mathcal{F}^k$ that maps the brick $(-1,1)^d$ onto Ω^k and, for the time being at least, we make the following assumptions:

Assumption 1.

$$\overline{\Omega}^k \cap \overline{\Omega}^\ell = \begin{cases} \text{an entire common face (in 3D),} & \text{or} \\ \text{an entire common edge,} & \text{or} \\ \text{a common vertex,} & \text{or} \\ \emptyset \end{cases}$$

Assumption 2. The two parametrizations of the previous intersection $\overline{\Omega}^k \cap \overline{\Omega}^\ell$, resulting from $\mathcal{F}^k$ and $\mathcal{F}^\ell$, coincide.

This allows us to define the discrete space

$$X_N = \{v_N \in H_0^1(\Omega), v_{N|\Omega^k} \circ \mathcal{F}^k \in \mathbb{P}_N((-1,1)^d)\}$$

and the discrete associated problem (2) (or its implementable version involving the Gauss-Lobatto quadrature rule over each Ω^k as in (13)).

The main ingredient that allows us to prove that the previous scheme is again optimal lies in the definition of a element in X_N that approximates well a given regular function u. This is done easily by considering the element v_N, defined locally over each subdomain as $v_{N|\Omega^k} \circ \mathcal{F}^k = \mathcal{I}_N[u_{|\Omega^k} \circ \mathcal{F}^k]$. It results from Assumptions 1 and 2 that v_N is actually continuous and vanishes over $\partial\Omega$. From (15) it is an optimal approximation of u in the sense that

$$\|u - v_N\|_{H^1(\Omega)} \leq cN^{1-r}\|u\|_{H^r(\Omega)}. \tag{19}$$

The best fit in $H^1(\Omega)$ is certainly as good as the proposed v_N, and the spectral element method can be proven to be an optimal approximation. We have only sketched the numerical analysis of this approximation, since the main purpose of this paper is to discuss projection operators. It is fundamental to have used here the interpolation operator to construct v_N, since it provides a globally continuous function. As an example, the use of the H^1–projection operator would not have given rise to a continuous function since, for a given function ϕ over the brick $(-1,1)^d$, the value of $\Pi_N^{1,0}(\phi)$ over any face depends not only on the value of ϕ on the given face, but depends on ϕ inside the whole domain.

We want to end this section by giving some hints on the "mortar spectral element method" due to Bernardi, Maday and Patera, that allows to relax assumptions 1 and 2 (and even, more generally, allows to combine spectral methods on some subdomains with different finite element methods on others see [5]). Due to lack of space, but also in order to better understand the main feature of the projection operators that is at the basis of the method, we shall consider a simple two dimensional domain $\Omega = (-1,2) \times (-1,1)$ decomposed into 3 subdomains $\Omega^1 = (-1,1) \times (-1,1)$, $\Omega^2 = (1,2) \times (-1,0)$ and $\Omega^3 = (1,2) \times (0,1)$. This decomposition violates assumption 1 since the intersection $\overline{\Omega}^1 \cap \overline{\Omega}^2$ is not a common *whole* edge. We want nevertheless to propose a discrete method that will allow to provide an optimal approximation of the solution u of (12) (with A =Id for the sake of simplicity). The discrete space X_N^* that we propose is imbedded in

$$Y_N = \{v_N \in L^2(\Omega), v_{N|\Omega^k} \in \mathbb{P}_N, v_{N|\partial\Omega} = 0, v_{N|\Omega^2} = v_{N|\Omega^3} \text{ over } \overline{\Omega}^2 \cap \overline{\Omega}^3\}$$

but it is readily checked that imposing continuity at the level of the interface $x = 1$ will rigidify the approximation and, in the general case, will spoil the accuracy of the method. In order to relax this continuity condition (remind that it is inherited from the requirement that $X_N^* \subset X$), we resort to nonconforming approximations. We shall replace the continuity condition by requiring that, over the interface $x = 1$, we impose for each element in Y_N

$$\int_{-1}^{1} [v^-(1,y) - v^+(1,y)]\psi_N(y)dy = 0, \quad \forall \psi_N \in \mathbb{P}_{N-2}(-1,1), \tag{20}$$

where $v^- = v_{|\Omega^1}$ and

$$v^+ = \begin{cases} v_{|\Omega^2} & \text{for } (x,y) \in \Omega^2, \\ v_{|\Omega^3} & \text{for } (x,y) \in \Omega^3. \end{cases}$$

Since $v^-(1,y)$ has to vanish for $y = \pm 1$ (due to the homogeneous boundary conditions), it is entirely defined by the $N-1$ conditions in (20); in particular choosing ψ in $\mathbb{P}_N(-1,1)$ would be much too stringent. The elements of Y_N that satisfy (20) constitute the space X_N^* of approximation. The method is then: find $u_N^* \in X_N^*$ such that

$$a_N(u_N^*, v_N) \equiv \sum_{k=1}^{K} \int_{\Omega^k} \nabla u_N^* \nabla v_N = \sum_{k=1}^{K} \int_{\Omega^k} f v_N, \quad \forall v_N \in X_N^*. \tag{21}$$

Since X_N^* is no longer a subspace of X, the ellipticity of the bilinear form of this problem is not straightforward. Nevertheless, it is true (and here it is particularly obvious since $\partial\Omega^k \cap \partial\Omega \neq \emptyset$). This argument allows us to check that there exists a unique solution u_N^* to (21). In order to derive the error bound we proceed as follows: for any $w_N \in X_N^*$,

$$\begin{aligned}
\alpha \|u_N^* - w_N\|_*^2 &\leq a_N(u_N^* - w_N, u_N^* - w_N) \\
&= \sum_{k=1}^{K} \int_{\Omega^k} f(u_N^* - w_N) - \sum_{k=1}^{K} \int_{\Omega^k} \nabla w_N \nabla (u_N^* - w_N) \\
&= \sum_{k=1}^{K} \int_{\Omega^k} -\Delta u(u_N^* - w_N) - \sum_{k=1}^{K} \int_{\Omega^k} \nabla w_N \nabla (u_N^* - w_N) \\
&= \sum_{k=1}^{K} \int_{\Omega^k} \nabla u \nabla (u_N^* - w_N) - \sum_{k=1}^{K} \int_{\Omega^k} \nabla w_N \nabla (u_N^* - w_N) \\
&\quad - \int_{x=1} \frac{\partial u}{\partial x}[(u_N^* - w_N)^- - (u_N^* - w_N)^+],
\end{aligned}$$

so that, from (20) we derive that for any $\psi \in \mathbb{P}_{N-2}(-1,1)$

$$\begin{aligned}
\alpha \|u_N^* - w_N\|_*^2 \leq& \sum_{k=1}^{K} \int_{\Omega^k} \nabla (u - w_N) \nabla (u_N^* - w_N) \\
&- \int_{x=1} [\frac{\partial u}{\partial x} - \psi][(u_N^* - w_N)^- - (u_N^* - w_N)^+].
\end{aligned}$$

It follows from the previous inequality that

$$\alpha\|u_N^* - w_N\|_* \le c\|u - w_N\|_* + \sup_{v_N \in X_N^*} \frac{\int_{x=1}[\frac{\partial u}{\partial x} - \psi][v_N^- - v_N^+]dy}{\|v_N\|_*}. \quad (22)$$

By choosing ψ equal to $\pi_{N-2}\frac{\partial u}{\partial x}$, it results that

$$\sup_{v_N \in X_N^*} \frac{\int_{x=1}[\frac{\partial u}{\partial x} - \psi][v_N^- - v_N^+]dy}{\|v_N\|_*} \le cN^{-r}\|u\|_{H^r(\Omega)}.$$

It remains to choose a good approximation w_N of u in X_N^* to take into account the first term on the right hand side of (22). This is done by noticing that, for any $\phi \in H_0^1(-1,1)$, the element ϕ_N of $\mathbb{P}_N(-1,1) \cap H_0^1(-1,1)$ that satisfies

$$\int_{-1}^{1} [\phi_N - \phi]\psi_N(y)dy = 0, \quad \forall \psi_N \in \mathbb{P}_{N-2}(-1,1),$$

is nothing other than $\phi_N = \pi_N^{1,0}(\phi)$. Indeed, we remark that, for any $\chi_N \in \mathbb{P}_N(-1,1) \cap H_0^1(-1,1)$, then $\chi_N'' \in \mathbb{P}_{N-2}(-1,1)$, thus

$$\int_{-1}^{1} [\phi_N - \phi]\chi_N''(y)dy = -\int_{-1}^{1} [\phi_N - \phi]'\chi_N'(y)dy.$$

The choice of a good element w_N is done as follows. We first set $\tilde{w}_{N|\Omega^k} = I_N(u_{N|\Omega^k})$ that is an element of Y_N. We then set $w_{N|\Omega^k} = \tilde{w}_{N|\Omega^k}$ for $k = 2, 3$, and build $w_{N|\Omega^1}$ by adding to $\tilde{w}_N^-$ the correction $\pi_N^{1,0}(\tilde{w}_N^+ - \tilde{w}_N^-)(y)\frac{(1+x)L_N'(x)}{2L_N'(1)}$ so that it satisfies (20). Due to the optimal approximation properties of the operator $\pi_N^{1,0}$ both in the L^2 and in the H^1–norms, we deduce that the mortar spectral approximation (21) is optimal in the sense that (19) still holds.

References

1. Bernardi, C. and Y. Maday , Some spectral approximations of one-dimensional fourth-order problems, in *Progress in Approximation Theory*, P. Nevai and A. Pinkus (eds.), Academic Press, San Diego, 1991, 43–116.
2. Bernardi, C. and Y. Maday, Polynomial interpolation results in Sobolev spaces, J. Comput. and Applied Math. **43** (1992), 53–80.
3. Bernardi, C. and Y. Maday, Spectral Methods, in *Handbook of Numerical Analysis*, Vol. **V**, P.G. Ciarlet & J.-L. Lions eds., North-Holland, 1997, 209–485.
4. Bernardi, C. and Y. Maday, Uniform inf-sup conditions for the spectral discretization of the Stokes problem, Math. Models and Methods in Applied Sciences **3(9)** (1999), 395–414.

5. Bernardi, C., Y. Maday, and A. T. Patera, A New Non Conforming Approach to Domain Decomposition: The Mortar Element Method. *Collège de France Seminar*, (1990), Pitman, H. Brézis, J.-L. Lions.
6. Canuto, C., M. Y. Hussaini, A. Quarteroni, and T. Zang, *Spectral Methods in Fluid Dynamics*, Springer-Verlag, Berlin, Heidelberg, 1987.
7. Canuto, C. and A. Quarteroni, Approximation results for orthogonal polynomials in Sobolev spaces, Math. Comput. **38** (1982), 67–86.
8. Girault, V. and P.-A. Raviart, *Finite Element Methods for the Navier–Stokes Equations, Theory and Algorithms*, Springer–Verlag, 1986.
9. Gottlieb, D., The stability of pseudospectral Chebyshev methods, Math. Comput. **36** (1981), 107–118.
10. Gottlieb, D. and S. A. Orszag, *Numerical Analysis of Spectral Methods, Theory and Applications*, SIAM Publications, 1977.
11. Maday, Y., Analysis of spectral projectors in one-dimensional domains, Math. Comput. **55** (1990), 537–562.
12. Maday, Y., Résultats d'approximation optimaux pour les opérateurs d'interpolation polynomiale, C.R. Acad. Sc. Paris **312** série I (1991), 705–710.
13. Maday, Y., D. Meiron, A. T. Patera, and E. M. Rønquist, Analysis of iterative methods for the steady and unsteady Stokes problem: application to spectral element discretizations, SIAM J. Numer. Anal. **14** (1993), 310–337.
14. Maday, Y., A. T. Patera, and E. M. Rønquist, The $\mathbb{P}_N - \mathbb{P}_{N-2}$ method for the approximation of the Stokes problem, Internal Report of the Laboratoire d'Analyse Numérique, Université Paris 6 (1992).
15. Maday, Y. and A. Quarteroni, Legendre and Chebyshev spectral approximations of Bürgers' equation, Numer. Math. **37** (1981), 321–332.
16. Maday, Y. and E. M. Rønquist, Optimal error analysis of spectral methods with emphasis on non-constant coefficients and deformed geometries, Comp. Methods in Applied Mech. and Eng. **80** (1990), 91–115.
17. Mercier, B., *An Introduction to the Numerical Analysis of Spectral Methods*, Springer-Verlag, Berlin, Heidelberg, 1989.
18. Nessel, R. J. and G. Wilmes, on Nikolkii-type inequalities for orthogonal expansions, in *Approximation Theory II*, G. G. Lorentz, C. K. Chui, and L. L. Schumaker (eds.), Academic Press, New York, (1976), 479–484.
19. Patera, A.T., A spectral element method for fluid dynamics: laminar flow in a channel expansion, J. Comput. Physics **54** (1984), 468–488.

Yvon Maday
Laboratoire d'Analyse Numérique
Université Pierre et Marie Curie
4, Place Jussieu, 75252, Paris Cedex 05, France
maday@ann.jussieu.fr

H–Bases I: The Foundation

H. Michael Möller and Thomas Sauer

Abstract. The H–basis concept allows an investigation of multivariate polynomial spaces degree by degree. In this paper, we mention its connection to the Gröbner basis concept, characterize H–bases, show how to construct them, and present a procedure for simplifying polynomials to their normal forms. Applications will be given in [8].

§1. Introduction

We consider Π, the ring of polynomials in $x_1, \ldots, x_n$ with coefficients from an infinite field $\mathbb{K}$, i.e. $\Pi = \mathbb{K}[x_1, \ldots, x_n]$, and the subsets Π_d of all polynomials of degree at most d. In many applications, one is interested in getting a basis or a generating set for the linear vector space $I \cap \Pi_d$, where $I \subseteq \Pi$ is an ideal. Having an H–basis $\{f_1, \ldots, f_s\}$ for I, then the set of all $p_i \cdot f_i$ with $p_i \in \Pi_{d-deg(f_i)}$, $i = 1, \ldots, s$, generates $I \cap \Pi_d$ as a linear vector space. Thus the H–basis concept is a tool for transforming a non-linear problem in Π into a problem in one (or in a series of) finite dimensional linear space(s) Π_d.

H–bases were introduced first by Macaulay [4]. His original motivation was the transformation of systems of polynomial equations into simpler ones. The power of this concept was not really understood, presumably because of the lack of facilities for symbolic computations. When Computer Algebra Systems came up, Gröbner bases (G–bases for short) were used instead of H–bases. These bases, originally invented by Buchberger [2] for computing multiplication tables for factor rings, are now also applied for simplifying some problems in Numerical Analysis, see [5].

The G–bases give generating systems not to $I \cap \Pi_d$ but to $I \cap \mathcal{F}_i$, where $\mathcal{F}_i \subset \Pi$ is a linear vector space of dimension i, and $\mathcal{F}_i \subset \mathcal{F}_{i+1}$ for all i and $\Pi = \cup_{i \geq 0} \mathcal{F}_i$. This finer decomposition has some drawbacks. For instance if an ideal is invariant under an affine symmetry group, its G–bases are typically not invariant. Since the spaces Π_d are invariant under affine symmetry groups, H–bases do not destroy such symmetries.

Many of the problems in applications which can be solved by Gröbner techniques can also be treated successfully with H–bases. In [7] we gave an

Curve and Surface Fitting: Saint-Malo 1999
Albert Cohen, Christophe Rabut, and Larry L. Schumaker (eds.), pp. 325–332.

ISBN 0-8265-1357-3.

overview of such problems. In the present paper, we describe briefly the underlying concept of grading rings, which leads to G– and H–bases, and present some properties characterizing H–bases. In contrast to [7], where we only gave a class of examples of H–bases, we present here the construction of an H–basis for zero-dimensional ideals I. A useful tool for our procedure is the so called normal form mapping NF, presented in Section 4, which projects Π orthogonally to the ideal I provided an H–basis of I is given. In [8] we show how these normal forms can be applied in numerical applications.

§2. H–bases and G–bases

In ring theory, rings can be graded by an ordered monoid, i.e. by an abelian semigroup Γ with addition $+$ and total ordering $\prec$ satisfying

$$\gamma_1 \prec \gamma_2 \Rightarrow \gamma_0 + \gamma_1 \prec \gamma_0 + \gamma_2, \quad \forall \gamma_0, \gamma_1, \gamma_2 \in \Gamma.$$

There are two major examples for grading Π by an ordered monoid Γ:

$$\Pi = \bigoplus_{\gamma \in \Gamma} \Pi_\gamma^{(\Gamma)}, \quad \Pi_{\gamma_1}^{(\Gamma)} \Pi_{\gamma_2}^{(\Gamma)} \subseteq \Pi_{\gamma_1+\gamma_2}^{(\Gamma)} \ \forall \gamma_1, \gamma_2 \in \Gamma.$$

The first one is the H–grading with $\Gamma := \mathbb{N}_o$,

$$\Pi_\gamma^{(\Gamma)} := \{p \in \Pi \mid p \ \textit{homogeneous of order} \ \gamma\}.$$

The ordering of $\Gamma = \mathbb{N}_o$ is the natural one. The second example for gradings is the G–grading, where $\Gamma := \mathbb{N}_o^n$ and

$$\Pi_{(\gamma_1,\dots,\gamma_n)}^{(\Gamma)} := \{c x_1^{\gamma_1} \cdots x_n^{\gamma_n} \mid c \in \mathbb{K}\}.$$

$\Gamma = \mathbb{N}_o^n$ is ordered by an admissible term ordering,

$$0 \preceq i, \ i \prec j \Rightarrow i + k \prec j + k.$$

Since the decomposition of Π into the sets $\Pi_\gamma^{(\Gamma)}$ is a direct sum, every $f \in \Pi$ has a unique representation $f = \sum f_\gamma$. The maximal γ with $f_\gamma \neq 0$ is called the maximal part of $f \neq 0$, $M^{(\Gamma)}(f)$ for short. It is also called the maximal form in the H–case, or leading monomial in the G–case. In the G–case, $M^{(\Gamma)}(f) = lc(f)lt(f)$, where $lc(f) \in \mathbb{K}$ is the leading coefficient and $lt(f) = x_1^{i_1} \cdots x_n^{i_n}$ the leading term. The maximal form of $f \neq 0$ is also denoted by $M_H(f)$.

Definition 1. *$\{p_1, \dots, p_m\} \subset I$ is called* a basis *of an ideal $I \subseteq \Pi$, briefly $I = \langle p_1, \dots, p_m \rangle$, if $\forall \ p \in I$*

$$\exists g_1, \dots, g_m \in \Pi \ : \ p = \sum_{k=1}^{m} g_k p_k.$$

It is also a G-basis *or* H-basis *resp. if* $g_1, \ldots, g_m$ *satisfy in addition*

$$\max_{k=1}^{m} lt(g_k)lt(p_k) = lt(p) \quad (G - basis),$$

$$or \ \max_{k=1}^{m} \deg(g_k p_k) = \deg(p) \quad (H - basis).$$

Theorem 1. *Let* $I = \langle p_1, \ldots, p_m \rangle$. *Then* $\{p_1, \ldots, p_m\}$ *is an H–basis (G–basis resp.) if and only if the least ideal containing all* $M_H(f)$, $0 \neq f \in I$ *(or all* $lt(f)$, $0 \neq f \in I$ *resp.) is generated by* $M_H(p_1), \ldots, M_H(p_m)$ *(or by* $lt(p_1), \ldots, lt(p_m)$ *resp.).*

This theorem, which holds *mutatis mutandis* for arbitrary graded rings, is proved for instance in [6]. An immediate consequence of it is that every ideal $I \neq (0)$ has an H– and a G–basis.

G–bases are now a standard tool in Computer Algebra. They are covered by nearly all textbooks, and are contained in almost all Computer Algebra Systems. The grading by one-dimensional linear spaces $\Pi_\gamma^{(\Gamma)}$ often reduces the computation to solving a series of one-dimensional problems. On the other hand, the construction of G–bases is often difficult or even impossible because of the high complexity of Buchberger's algorithm for computing G–bases. In addition, in many applications the G–bases allows only little insight into the structure of a solution by the artificial ordering term by term.

§3. Characterization of H–bases and Normal Forms

Macaulay introduced H–bases using homogenizations and dehomogenizations of polynomials. The name *H–basis* originates from the first letter of homogenization.

Definition 2. *Let* $f \in \mathbb{K}[x_1, \ldots, x_n]$ *have degree* d,

$$f = \sum_{i=0}^{d} f_i, \quad f_i \ homogeneous \ of \ degree \ i, \quad f_d \neq 0.$$

Then introducing a new variable x_0, *the* homogenization *of* f *is a homogeneous degree* d *polynomial in* $\mathbb{K}[x_0, x_1, \ldots, x_n]$,

$$\Phi(f) := \sum_{i=0}^{d} x_0^{d-i} f_i.$$

A homogeneous $F \in \mathbb{K}[x_0, x_1, \ldots, x_n]$ *can be dehomogenized to an* $f \in \Pi$ *by* $x_0 = 1$.

For more details on homogenizations and their connection to projective coordinates, we refer to [3].

Theorem 2. (Macaulay [4]). *Let $I = \langle h_1, \ldots, h_s \rangle$. Then the following statements are equivalent*

1) *The least ideal containing all $\Phi(h)$, $0 \neq h \in I$, is $\langle \Phi(h_1), \ldots, \Phi(h_s) \rangle$.*
2) $x_0 F \in \langle \Phi(h_1), \ldots, \Phi(h_s) \rangle \Rightarrow F \in \langle \Phi(h_1), \ldots, \Phi(h_s) \rangle$.
3) $\{h_1, \ldots, h_s\}$ *is an H–basis of I.*

The power of the G–basis concept is mainly based on the possibility of reducing a polynomial to a simpler one by subtracting suitable multiples of elements of the G–basis. A consequent application of this reduction strategy gives the so called normal form, in a sense the simplest polynomial obtainable by the reductions. We translated this technique to H–bases in [7], and give here for consistency a short résumé of the main results.

Definition 3. *We denote by $\Pi_d^{(H)}$ the space of all homogeneous degree d polynomials for $d \in \mathbb{N}_o$ and $\Pi_d^{(H)} := \{0\}$ for $d < 0$. Let $h_1, \ldots, h_s \in \Pi$. Then we define a finite dimensional linear subspace of $\Pi_d^{(H)}$ by*

$$V_d(h_1, \ldots, h_s) := \{\sum_{i=1}^{s} g_i M_H(h_i) \mid g_i \in \Pi_{d-deg(h_i)}^{(H)}\}.$$

Analogously for an ideal $I \subset \Pi$,

$$V_d(I) := \{M_H(p) \mid p \in I,\ deg(p) = d\} \cup \{0\}.$$

We introduce an inner product $\langle\ .\ ,\ .\ \rangle$ in Π, for instance, by the inner product of the (weighted) coefficient vectors, or by a strictly positive linear functional J and $\langle f, g \rangle := J(f \cdot g)$ if $\mathbb{K} \subseteq \mathbb{R}$ or $:= J(f\bar{g})$ if $\mathbb{K} = \mathbb{C}$. Then we can define orthogonal complements $W_d(h_1, \ldots, h_s)$ and $W_d(I)$ in $\Pi_d^{(H)}$. Hence

$$V_d(h_1, \ldots, h_s) \oplus W_d(h_1, \ldots, h_s) = \Pi_d^{(H)} \text{ and } V_d(I) \oplus W_d(I) = \Pi_d^{(H)}.$$

Let us consider a polynomial f of degree d. Then

$$M_H(f) \in V_d(h_1, \ldots, h_s) \oplus W_d(h_1, \ldots, h_s).$$

Let w_d denote its natural projection on $W_d(h_1, \ldots, h_s)$. This homogeneous polynomial can be computed by solving a finite linear system of equations because $\Pi_d^{(H)}$ has a finite dimension. Hence there are homogeneous polynomials $g_1, \ldots, g_s$ such that

$$f = w_d + \sum_{i=1}^{s} g_i h_i + f_1, \quad g_i \in \Pi_{d-deg(h_i)}^{(H)},\ f_1 \in \Pi_{d-1}.$$

We say f reduces to $w_d + f_1$ modulo $\{h_1, \ldots, h_s\}$ and call f_1 then the remainder of f.

In the reduction modulo $\{h_1, \ldots, h_s\}$ the degree of the remainder f_1 is less than $deg(f)$. Hence this reduction can be applied recursively reducing f_{i-1} constructively to $w_{d+1-i} + f_i$ modulo $\{h_1, \ldots, h_s\}$ for $i = 1, \ldots, d+1$ starting with $f = f_0 \in \Pi_d$ and terminating with $f_{d+1} = 0$, because the constant f_d is either in $V_0(h_1, \ldots, h_s)$ or in $W_0(h_1, \ldots, h_s)$. Combining these reductions modulo $\{h_1, \ldots, h_s\}$, one obtains for f

$$f = \sum_{i=0}^{d} w_i + \sum_{i=0}^{d} \sum_{j=1}^{s} g_{ij} h_j, \quad g_{ij} \in \Pi^{(H)}_{i-deg(h_j)}.$$

Then $\sum_{i=0}^{d} w_i$ is uniquely determined by f, by $\{h_1, \ldots, h_s\}$, and by the underlying inner product.

Definition 4. *Let $h_1, \ldots, h_s \in \Pi$. We say $f \in \Pi_d$* reduces fully modulo *$\{h_1, \ldots, h_s\}$ to $\sum_{i=0}^{d} w_i$ if every $w_i \in W_i(h_1, \ldots, h_s)$ is constructed as described above. $\sum_{j=0}^{d} w_j$ is called the* normal form of f modulo $\{h_1, \ldots, h_s\}$, *for short*

$$\mathtt{NF}(f, \{h_1, \ldots, h_s\}) := \sum_{j=0}^{d} w_j.$$

If $\{h_1, \ldots, h_s\}$ is not an H–basis of $I := \langle h_1, \ldots, h_s \rangle$, then $M_H(f)$ is not necessarily contained in $V_{deg(f)}(h_1, \ldots, h_s)$, although $f \in I$. This means, that eventually the first homogeneous polynomial w_d is not 0 if $f \in I$. Hence at most if $\{h_1, \ldots, h_s\}$ is an H–basis, then $\mathtt{NF}(f, \{h_1, \ldots, h_s\}) = 0$. In fact, as quoted in [7] but shown already in [9], $\{h_1, \ldots, h_s\}$ is an H–basis if and only if $\mathtt{NF}(f, \{h_1, \ldots, h_s\}) = 0$ for every $f \in \langle h_1, \ldots, h_s \rangle$.

Another characterization of H–bases given in [7] is as follows.

Theorem 3. *Let I be an ideal and $h_1, \ldots, h_s \in I$. $\{h_1, \ldots, h_s\}$ is an H–basis of I if, and only if, for all $d \in \mathbb{N}$,*

$$V_d(I) = V_d(h_1, \ldots, h_s).$$

§4. On the Construction of H–bases

Macaulay proposed in [4] a procedure for computing H–bases of ideals given by a basis. However, his description was only by an example. He claimed "*This procedure is a general one*". But he needs in his example the computation of certain modules of syzygies. These can be constructed only in special cases or by computing first a G–basis and then applying techniques as in [1].

On the other hand, if an admissible term ordering $\prec$ is compatible with degrees,

$$deg(x_1^{\gamma_1} \cdots x_n^{\gamma_n}) < deg(x_1^{\beta_1} \cdots x_n^{\beta_n}) \Rightarrow x_1^{\gamma_1} \cdots x_n^{\gamma_n} \prec x_1^{\beta_1} \cdots x_n^{\beta_n},$$

then a G–basis with respect to $\prec$ is also an H–basis. Hence Buchberger's algorithm for computing G–bases also serves for computing H–bases. This seems a more direct access than via syzygies. However, if one wants to use H–bases instead of G–bases, this way is still a detour.

In case the number n of variables coincides with the number of given polynomials, then there is an easy test for H–bases as proved in [7].

Theorem 4. *Let $h_1, \ldots, h_n$ be n polynomials such that their maximal forms $M_H(h_1), \ldots, M_H(h_n)$ have only the point $(0, \ldots, 0)$ as common zero. Then $\{h_1, \ldots, h_n\}$ is an H–basis.*

For an arbitrary zero-dimensional ideal, i.e. for an ideal I such that the polynomials in I (equivalently: the polynomials in an arbitrary basis of I) have only a finite number of common zeros in $\overline{\mathbb{K}}^n$, $\overline{\mathbb{K}}$ the algebraic closure of $\mathbb{K}$, see [3], we present here a procedure which computes an H–basis from a given basis.

Procedure for computing H–bases.
In : $\mathcal{H}_o$, a finite polynomial set generating a zero-dimensional ideal I.
Out : $\mathcal{H}$, an H-basis of I.
Start: $\mathcal{H} := \mathcal{H}_o$, $d = 0$.
Loop: Check the finite dimensional linear vector space

$$V_d(\mathcal{H}) := \{\sum_{h \in \mathcal{H}} g_h M_H(h) \mid g_h M_H(h) \in \Pi_d^{(H)}\}$$

for linear dependencies. If $\sum_{h \in \mathcal{H}} g_h M_H(h) = 0$, then compute $p := \mathtt{NF}(\sum_{h \in \mathcal{H}} g_h h, \mathcal{H}))$. If $p \neq 0$, then enlarge $\mathcal{H}$ by p, and modify consequently $V_0(\mathcal{H}), \ldots, V_{d-1}(\mathcal{H})$. Lower d to the least k where $V_k(\mathcal{H})$ is changed and go to Loop. If for no linear dependency such p is nonzero, then then enlarge d by 1. If now $V_d(\mathcal{H}) = \Pi_d^{(H)}$ holds true, then return $\mathcal{H}$ otherwise go to Loop.

This informal description can be extended easily to a correct algorithm. One has to observe that the checking of $V_d(\mathcal{H})$ for linear dependencies needs a basis of the nullspace

$$\{(g_1, \ldots, g_s) \in \Pi^{(H)}_{d-deg(h_1)} \times \ldots \times \Pi^{(H)}_{d-deg(h_s)} \mid \sum_{i=1}^{s} g_i M_H(h_i) = 0\},$$

where $\mathcal{H} = \{h_1, \ldots, h_s\}$. If for every basis element $(g_1, \ldots, g_s)$ the normal form of $\sum_{i=1}^{s} g_i h_i$ is 0, then it holds for every element of the nullspace, i.e. for every dependency. As a byproduct of the basis computation one obtains $dim V_d(\mathcal{H})$. Then the test $V_d(\mathcal{H}) = \Pi_d^{(H)}$ reduces to a comparison of the dimensions because of $V_d(\mathcal{H}) \subseteq \Pi_d^{(H)}$.

For proving correctness and termination, we consider first $f := \sum_{h \in \mathcal{H}} g_h h$ with $g_h M_H(h) \in \Pi_d^{(H)}$ for all $h \in \mathcal{H}$. If $\mathtt{NF}(f, \mathcal{H}) = 0$, then especially $M_H(f) \in V_k(\mathcal{H})$ for a $k \leq d$, and hence

$$M_H(f) \in \langle M_H(h_1), \ldots, M_H(h_s) \rangle, \textit{ where } \mathcal{H} = \{h_1, \ldots, h_s\}.$$

In case $p := \mathtt{NF}(f, \mathcal{H}) \neq 0$ either $M_H(f) \neq M_H(p)$ holds, i.e. again

$$M_H(f) \in \langle M_H(h_1), \ldots, M_H(h_s) \rangle,$$

or $M_H(f) = M_H(p)$ holds, i.e.

$$M_H(f) \in \langle M_H(h_1), \ldots, M_H(h_s), M_H(p)\rangle.$$

Therefore, if in the procedure d is increased by 1 (and $\mathcal{H}$ is updated), then for every $0 \neq \sum_{h\in\mathcal{H}} g_h h \in I$, $deg(g_h) + deg(h) \leq d$ the relation

$$M_H(\sum_{h\in\mathcal{H}} g_h h) \in \langle M_H(h_1), \ldots, M_H(h_s)\rangle$$

holds where again $\mathcal{H} =: \{h_1, \ldots, h_s\}$. This is our inductive hypothesis.

The ideal I has an H–basis, say $\{\varphi_1, \ldots, \varphi_m\}$. $\mathcal{H}$ is a basis of I. Hence every φ_i has a representation $\varphi_i = \sum_{j=1}^{s} g_{ij}h_j$, $g_{ij} \in \Pi$. If the inductive hypothesis holds for d, then one obtains at least for $d \geq M :=\max_{ij} deg(g_{ij}h_j)$ that

$$M_H(\varphi_i) \in \langle M_H(h_1), \ldots, M_H(h_s)\rangle,\ i = 1, \ldots, m.$$

Hence for those d $V_d(\varphi_1, \ldots, \varphi_m) \subseteq V_d(\mathcal{H})$. But $V_d(\varphi_1, \ldots, \varphi_m) = V_d(I)$, since $\{\varphi_1, \ldots, \varphi_m\}$ is an H–basis of I. Therefore $V_d(\mathcal{H}) = V_d(I)$ for $d \geq M$. By the inductive hypothesis, also $V_k(\mathcal{H}) = V_k(I)$ holds for $k < M$. Hence $\mathcal{H}$ is an H–basis of I if we arrived at a $d \geq M$ in the procedure.

The ideal I has dimension 0. Then there is a D such that $V_d(I) = \Pi_d^{(H)}$ for all $d \geq D$, see for instance [3, Ch 9.4,Prop.6] and [3, Ch 5.3,Thm.6]. Hence for $d \geq max\{D, M\}$ one has $V_d(\mathcal{H}) = \Pi_d^{(H)}$. Thus in the course of the procedure, one arrives once, not knowing M, at a d_0 with $V_{d_0}(\mathcal{H}) = \Pi_{d_0}^{(H)}$. Then also $V_k(\mathcal{H}) = \Pi_k^{(H)}$ for all $k > d_0$. Therefore, for every polynomial $f \in I$ with

$$f = \sum_{i=1}^{s} g_i h_i,\ g_i \in \Pi_{k-deg(h_i)}^{(H)}$$

the assumption $M_H(f) \notin \langle M_H(h_1), \ldots, M_H(h_s)\rangle$ leads to $deg(M_H(f)) < d_0$. But then the inductive hypothesis gives a contradiction. Therefore the procedure gives no new $p \neq 0$ enlarging the set $\mathcal{H}$. This ensures termination.

An implementation of an algorithm based on this procedure and a complexity analysis is still a work under progress.

Acknowledgments. The second author was supported by the Deutsche Forschungsgemeinschaft with a Heisenberg fellowship, Grant Sa–627/6.

References

1. Adams, W. W. and P. Loustaunau, *An Introduction to Gröbner Bases*, Graduate Studies in Mathematics 3, AMS 1994.
2. Buchberger, B., Ein Algorithmus zum Auffinden der Basiselemente des Restklassenringes nach einem nulldimensionalen Polynomideal, (Doctoral thesis), Univ. Innsbruck, 1965.

3. Cox, D., J. Little, and D. O'Shea, *Ideals, Varieties, and Algorithms*, Springer Verlag, New York, 1992.
4. Macaulay, F. S., *The Algebraic Theory of Modular Systems*, Cambridge Tracts in Math. and Math. Phys. 19, Cambridge Univ. Press, 1916.
5. Möller, H. M., Gröbner bases and Numerical Analysis, in: *Groebner Bases and Applications, (Proc. of Conf. 33 Years of Groebner Bases)*, B. Buchberger and F. Winkler (eds.), Cambridge University Press 1998, 159–178.
6. Möller, H. M., and F. Mora: New constructive methods in classical ideal theory, J. of Algebra 100 (1986), 138–178.
7. Möller, H. M. and T. Sauer, H–bases for polynomial interpolation and system solving, Advances Computat. Math., to appear.
8. Möller, H. M. and T. Sauer, H–Bases II: Applications to numerical problems, *Curve and Surface Fitting: Saint-Malo 1999*, Albert Cohen, Christophe Rabut, and Larry L. Schumaker (eds.), Vanderbilt University Press, Nashville, 2000, 333–342.
9. Sauer, T., Gröbner bases, H-bases and interpolation, Proc. Amer. Math. Soc., to appear.

H. Michael Möller
Fachbereich Mathematik
der Universität
44221 Dortmund, Germany
`hmm@mathematik.uni-dortmund.de`

Thomas Sauer
Mathematisches Institut
der Universität Erlangen–Nürnberg
Bismarckstr. 1 $\frac{1}{2}$
91054 Erlangen, Germany
`sauer@mi.uni-erlangen.de`

H–Bases II: Applications to Numerical Problems

H. Michael Möller and Thomas Sauer

Abstract. We show how H–bases can be applied to polynomial interpolation and for the solution of systems of nonlinear equations. We will give an example of a system of polynomial equations where the H–basis leads to more stable computations than with the Gröbner basis.

§1. Introduction

In the preceding paper [12], we introduced the notion of H–bases for polynomial ideals, and showed how to construct H–bases in the numerically most interesting case of a zero dimensional ideal. In this paper we consider two problems from Numerical Analysis, namely polynomial interpolation and solving systems of polynomial equations, and point out how H–bases can be applied to both. More precisely, in both cases the computation of normal forms with respect to an ideal plays a crucial role, and with the basic results from [12] available, H–bases yield a perfect replacement for the Gröbner bases which are normally and frequently used to do this job [8]. Finally, we will consider an example where a properly chosen H–basis leads to a significant stabilization of the computations in comparison with the use of Gröbner bases.

§2. Interpolation

A finite set $\Theta \subset \Pi'$ of *linearly independent* functionals on Π is said to define an ideal interpolation scheme if its kernel, $\ker \Theta \subset \Pi$, is an ideal in Π. Given an ideal interpolation scheme Θ and a polynomial $f \in \Pi$, the interpolation problem consists of finding $p \in \Pi$ such that

$$\Theta(p) = \Theta(f), \qquad \text{i.e.,} \qquad \vartheta(p) = \vartheta(f), \quad \vartheta \in \Theta. \tag{1}$$

Curve and Surface Fitting: Saint-Malo 1999
Albert Cohen, Christophe Rabut, and Larry L. Schumaker (eds.), pp. 333–342.

ISBN 0-8265-1357-3.

So far, we have put no restrictions on p; hence, there are infinitely many solutions to (1). More precisely, if p is any solution of (1), hence $f - p \in \ker\Theta$, then the set of all solutions is the equivalence class

$$[p] = p + \ker\Theta = f + \ker\Theta\,[f].$$

We denote the linear space of all equivalence classes by $\Pi/\ker\Theta$, and remark that $(\dim\Pi/\ker\Theta) = \#\Theta$. Of course, in order to compute interpolation polynomials, we must find a way to choose a specific element from the equivalence class $[f]$. A "natural" choice is to take the normal form $\mathtt{NF}\ (f,\mathcal{H})$, where $\mathcal{H}$ is an H–basis for $\ker\Theta$. Since $[f] = [g]$ implies that $f - g \in \langle\mathcal{H}\rangle$, and since $\mathtt{NF}\ (\cdot,\mathcal{H})$ is a linear operator, we have that

$$[f] = [g] \qquad \Longrightarrow \qquad \mathtt{NF}\ (f,\mathcal{H}) = \mathtt{NF}\ (g,\mathcal{H}) + \underbrace{\mathtt{NF}\ (f-g,\mathcal{H})}_{=0} = \mathtt{NF}\ (g,\mathcal{H}).$$

Hence, $\mathtt{NF}\ ([f],\mathcal{H}) = \mathtt{NF}\ (f,\mathcal{H})$, that is, the normal form is the same for any element of the same equivalence class. This algebraic approach also allows for interpolation of functionals which are only given *implicitly*, that is, by an ideal $\mathcal{I} \subset \Pi$: compute an H–basis $\mathcal{H}$ for $\mathcal{I}$ and the interpolation operator is the "remainder of division" $\mathtt{NF}\ (\cdot,\mathcal{H})$. It is worthwhile to remark that one of the oldest papers on multivariate interpolation, namely [6], starts with implicitly given interpolation nodes.

Another approach is to look for a polynomial space $\mathcal{P} \subset \Pi$ which allows for unique interpolation with respect to Θ; to restrict the number of solutions to this problem, one usually demands the interpolation operator $L_{\mathcal{P}} : \Pi \to \mathcal{P}$ to be degree reducing [3], that is,

$$\deg L_{\mathcal{P}} f \le \deg f, \qquad f \in \Pi.$$

Such an interpolation space with a degree reducing interpolation operator is called a minimal degree interpolation space. The most prominent minimal degree interpolation spaces is the least interpolation space introduced by de Boor et al in [2], and is the unique degree reducing interpolation space which satisfies the additional condition

$$\mathcal{P} = \bigcap_{q\in\ker\Theta} \ker q(D), \qquad q(D) := q\left(\frac{\partial}{\partial x_1},\dots,\frac{\partial}{\partial x_n}\right).$$

On the other hand, it is obvious that the operator $\mathtt{NF}\ (\cdot,\mathcal{H})$ is *degree reducing*, linear and interpolating, hence all the spaces $\mathcal{P} = \mathtt{NF}\ (\Pi,\mathcal{H})$, for any H–basis $\mathcal{H}$, are minimal degree interpolation spaces with interpolation operator $L_{\mathcal{P}} = \mathtt{NF}\ (\cdot,\mathcal{H})$. Moreover, it is even possible to recover known minimal degree interpolation spaces by this algebraic process.

Theorem 1. [15] *The least interpolation space is given as* $\mathtt{NF}\ (\Pi,\mathcal{H})$, *where* $\mathcal{H}$ *is an orthogonal H–basis with respect to the inner-product*

$$(p,q) = (p(D)\,q)\,(0), \qquad p,q \in \Pi.$$

§3. Polynomial System Solving

Probably the best–known and most frequent use of Gröbner bases is for solving polynomial systems of equations, where they form a core part of literally all available computer algebra systems. These systems of equations arise naturally in a geometric context, such as finding solutions of geometric constraints (for example, any Euclidean distance constraint yields a quadratic equation) or "simply" computing the intersection of algebraic curves/surfaces given in implicit form. So, given any finite set $\mathcal{F} \in \Pi$ one wants to find the associated algebraic variety $X \in \overline{\mathbb{K}}$ (some algebraic closure of our underlying field $\mathbb{K}$) such that

$$\mathcal{F}(X) = 0, \tag{2}$$

that is,

$$f(x) = 0, \qquad x \in X,\ f \in \mathcal{F}.$$

Note that the emphasis here is not on finding *one* solution (which could, at least in the case that $\#\mathcal{F} = n$, be done by a Newton method), but on finding *all* solutions and obtaining *structural* information about the variety. It is easy to see that the variety is not a property of the specific set $\mathcal{F}$, but of the ideal $\langle \mathcal{F} \rangle$:

$$\mathcal{F}(X) = 0 \qquad \Longleftrightarrow \qquad \langle \mathcal{F} \rangle (X) = 0.$$

Therefore, it may be helpful to find particular bases for $\langle \mathcal{F} \rangle$ which allow for an efficient solution of (2). The "classical" implementation in most Computer Algebra systems relies on the computation of elimination ideals, which means the computation of a basis for the subideals

$$\langle \mathcal{F} \rangle_k = \langle \mathcal{F} \rangle \cap \mathbb{K}[x_1, \ldots, x_k], \qquad k = 1, \ldots, n,$$

where $\langle \mathcal{F} \rangle_n = \langle \mathcal{F} \rangle$. In fact, this corresponds to transforming the original problem $\mathcal{F}(X) = 0$ into a *triangular* system

$$\begin{array}{llllll} g_1(& x_1 & & & &) = 0, \\ g_2(& x_1, & x_2 & & &) = 0, \\ & & & \vdots & & \\ g_m(& x_1, & x_2, & \ldots, & x_n &) = 0. \end{array} \tag{3}$$

Once such a triangular system is available, the solution strategy is obvious: determine the zeros of the univariate polynomial $g_1(x_1)$ and substitute them into $g_2(\cdot, x_2)$ which is now, for for any such zero, again a univariate polynomial in x_2, and go on with this procedure. Moreover, such a triangular basis can indeed be computed: $\mathcal{G}_{lex}$, the reduced Gröbner basis for $\langle \mathcal{F} \rangle$ with respect to the lexicographical term order where $x_1 \prec x_2 \prec \cdots \prec x_n$ has the property that

$$\mathcal{G}_k = \mathcal{G} \cap \mathbb{K}[x_1, \ldots, x_k] \subset \langle \mathcal{F} \rangle_k$$

is a Gröbner basis for $\langle \mathcal{F} \rangle_k$ (cf. [4, p. 114]).

However, as nice as this idea of successive elimination of variables sounds, there are numerous drawbacks:

(i) The complexity of computing a lexicographical Gröbner basis is tremendous, and even relatively "simple" problems still exceed the limitations of existing computing facilities.

(ii) There are often several polynomials in a certain number of variables, that is, the system is not as triangular as one would want it to be.

(iii) The degree of the polynomial g_1 is usually very high. This makes it impossible to compute its zeros exactly.

(iv) The tempting idea to find g_1's zeros *approximately* and substitute these values will not lead very far since it is well–known that the zeros of a polynomial are usually quite ill–conditioned with respect to its coefficients (cf. [5,17]).

So, the summary is fairly disappointing: elimination methods do not provide a good tool to tackle polynomial systems of equations. In particular, they rely too much on *symbolic* methods (with exact computations) to become a useful tool in *numerical* applications.

A different approach has been proposed quite recently by Stetter [16] (see also [10]; in [7] this method is partly attributed to Stickelberger) which is based on transforming the nonlinear system of equations into an eigenvalue problem for which a huge library of powerful routines is available. For that purpose, let us assume that the set of solutions X is *finite* (that is, the associated ideal $\langle \mathcal{F} \rangle$ is zero dimensional) and that all the common zeros are simple. The latter restriction is made to keep the presentation simple; details on how to handle multiplicities can be found in [10]. We first note that for any $f \in \Pi$, the mapping

$$\Phi_f : \begin{cases} \Pi / \langle \mathcal{F} \rangle & \rightarrow & \Pi / \langle \mathcal{F} \rangle \\ [\,p\,] & \mapsto & [\,f \cdot p\,] \end{cases}$$

is a homomorphism on the $\#X$–dimensional linear space $\Pi / \langle \mathcal{F} \rangle$. Now, suppose for a moment that we know X. Then there are polynomials $p_x \in \Pi$, $x \in X$, defined by

$$p_x\left(x'\right) = \delta_{x,x'}, \qquad x, x' \in X,$$

which form a basis for $\Pi / \langle \mathcal{F} \rangle$, i.e.,

$$\Pi / \langle \mathcal{F} \rangle = \operatorname{span} \left\{ \, [p_x] : x \in X \, \right\}.$$

Obviously, for any $x \in X$, the polynomial $g_x = (f - f(x))\, p_x$ satisfies $g_x(X) = 0$, and therefore

$$[0] = [g_x] = [\,(f - f(x))\, p_x\,] = \Phi_f\, [p_x] - f(x)\, [p_x]\,.$$

What we have proved with this simple argument is the following crucial theorem.

Theorem 2. *The polynomials p_x, $x \in X$, are joint eigenvectors of* all *homomorphisms Φ_f, $f \in \Pi$, with respect to the eigenvalue $f(x)$.*

This result again suggests a strategy to solve polynomial systems of equations: compute a set of representers for $\Pi / \langle \mathcal{F} \rangle$, that is, a finite set $\mathcal{P} \subset \Pi$ of linearly independent polynomials such that

$$\Pi / \langle \mathcal{F} \rangle = \operatorname{span} \{ [p] : p \in \mathcal{P} \},$$

and compute the matrix M_f which describes the action of Φ_f with respect to the basis $\mathcal{P}$. The eigenvalues of such matrices yield, when combined appropriately, the solutions X. We remark that the (transpose of the) matrix M_f is called the multiplication table for f with respect to $\mathcal{P}$, and that the original goal for Buchberger's doctoral thesis (supervised by Gröbner) was not the invention of Gröbner bases but the computation of multiplication tables. Of course, the most natural approach would be to compute the multiplication tables M_{x_j}, $j = 1, \ldots, n$, for the coordinate functions and thus compute the respective coordinates of the elements of X as the eigenvalues of the multiplication table. Note that the different components are finally "glued together" by the requirement that they must correspond to the same eigenvector.

What we now have is the possibility of reducing the search for the solutions of a polynomial system of equations to an eigenvalue problem, provided that we are able to perform two operations:

(i) Given a basis $\mathcal{F}$ for an ideal $\langle \mathcal{F} \rangle$ compute a basis $\mathcal{P}$ of representers for $\Pi / \langle \mathcal{F} \rangle$.

(ii) Having this basis available and given any $f \in \Pi$, compute the multiplication table M_f with respect to $\mathcal{P}$.

Fortunately, this is where [12] enters – the answer are *normal forms*: if $\mathcal{H}$ is an H–basis for $\langle \mathcal{F} \rangle$, then any basis for $\mathtt{NF}\,(\Pi, \mathcal{H})$ is exactly the desired $\mathcal{P}$, and the action of Φ_f can be computed by expanding $\mathtt{NF}\,(f \cdot p, \mathcal{H})$ for all $p \in \mathcal{P}$, which yields the multiplication table M_f. The remaining question is "why H–bases?", and this question is justified since the computation of normal forms and thus of multiplication tables is perfectly possible with the help of Gröbner bases as well. To give a partial answer to this question, we look at an example.

§4. When Two Ellipses Meet

In this section we consider a simple example which will show that also the eigenvalue method can encounter serious obstacles, in particular when Gröbner bases are involved. The important thing here is *simplicity*, as it will not be too surprising if extremely complicated and difficult examples cause problems.

We consider the two ellipses

$$\begin{aligned} f(x,y) &= \frac{1}{3}x^2 + \frac{2}{3}y^2 - 1, \\ g(x,y) &= \frac{2}{3}x^2 + \frac{1}{3}y^2 - 1. \end{aligned}$$

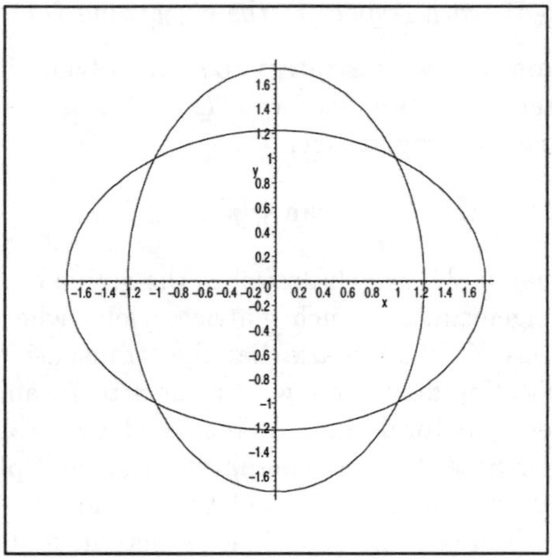

Fig. 1. The two ellipses.

Clearly, these two ellipses intersect in the four well–separated points $(\pm 1, \pm 1)$ as can be seen in Fig. 1.

Now, we are going to perturb g a little bit and replace it by $g_\phi = g(A_\phi(x,y))$, where A_ϕ denotes the rotation

$$A_\phi(x,y) = \begin{bmatrix} \cos\phi & \sin\phi \\ -\sin\phi & \cos\phi \end{bmatrix} \begin{bmatrix} x \\ y \end{bmatrix}.$$

Note that we have in mind *small* values of ϕ, so the intersections should still be close to $(\pm 1, \pm 1)$ and the problem should still be well–conditioned.

Recalling that lexicographic Gröbner bases are known as troublemakers, we first try some "better" Gröbner basis, namely the one which is based on the graded lexicographic term order with $x \prec y$. Note that this ideal basis is not only a Gröbner basis, but also an H–basis. In this case the Gröbner bases $\mathcal{G}_\phi$ consists, for $\phi \neq 0$, of the three polynomials

$$\begin{aligned} &4\sin\phi\, xy + 3\cos\phi\, x^2 - 3\cos\phi, \\ &x^2 + 2y^2 - 3, \\ &\cos\phi\left(\cos^2\phi + 8\right)\, x^3 - 3\cos\phi\left(\cos^2\phi + 2\right)\, x + 12\,\sin\phi\left(\sin^2\phi - 1\right)\, y, \end{aligned}$$

while

$$\mathcal{G}_0 = \left\{x^2 - 1, y^2 - 1\right\}.$$

Here we already observe that some singularity must appear for $\phi = 0$, since $\mathcal{G}_0$ is not just a limit $\phi \to 0$ of $\mathcal{G}_\phi$, although the basis changes continuously with respect to ϕ. The singularity becomes more apparent if we look at the normal forms, which are

$$\mathcal{P}_\phi = \begin{cases} \left\{1, x, y, x^2\right\} & \text{if } \phi \neq 0, \\ \left\{1, x, y, xy\right\} & \text{if } \phi = 0. \end{cases}$$

Finally, the multiplication tables $M_{x,\phi}$ for the multiplication by x take the form

$$M_{x,0} = \begin{bmatrix} 0 & 1 & 0 & 0 \\ 1 & 0 & 0 & 0 \\ 0 & 0 & 0 & 1 \\ 0 & 0 & 1 & 0 \end{bmatrix},$$

while the multiplication table

$$M_{x,\phi} = \begin{bmatrix} 0 & 0 & \frac{3\cos\phi}{4\sin\phi} & 0 \\ 1 & 0 & 0 & -3\frac{5\cos^2\phi-8}{\cos^2\phi+8} \\ 0 & 0 & 0 & -12\frac{\sin\phi\,\cos\phi}{\cos^2\phi+8} \\ 0 & 1 & -\frac{3\cos\phi}{4\sin\phi} & 0 \end{bmatrix}, \qquad \phi \neq 0,$$

provides us with difficulties. Not only does this matrix *not* converge to $M_{x,0}$ for $\phi \to 0$, but some entries in this matrix even *diverge* to $\pm\infty$, respectively. Indeed, if one tries to compute the eigenvalues and eigenvectors of this matrix for small values of ϕ, things become disastrous: A `Maple` computation with 10 digits worked until about $\phi \sim 10^{-5}$, where an error message reported that the QR algorithm did not work. For smaller values, like $\phi \sim 10^{-6}$, `Maple` invented *complex* zeros with an imaginary part of the magnitude 0.5×10^{-5} which by far exceeds any negligible machine number. On the other hand, `Octave`, a free `Matlab` clone whose Linear Algebra facilities are based on `LAPACK` [1], reproduced the eigenvalues correctly, but gave eigenvectors which were practically 0.

Hence, we end up with some kind of paradox which is due to a singularity at $\phi = 0$: though the original problem of solving the polynomial system of equations is very well–conditioned, the graded lexicographical Gröbner basis is extremely sensitive to *very small* perturbations ($|\phi| \le 10^{-5}$), but by far not so sensitive to relatively "large" ($|\phi| > 10^{-5}$) perturbations.

Similar problems appear when we replace the graded lexicographical Gröbner basis by a purely lexicographical one with $x \prec y$ which yields the normal forms

$$\mathcal{P}_\phi = \begin{cases} \{1, x, x^2, x^3\} & \text{if } \phi \neq 0, \\ \{1, x, y, xy\} & \text{if } \phi = 0. \end{cases}$$

Though the components of the multiplication table $M_{x,\phi}$ at least are continuous functions in ϕ and remain bounded in this case, the limit $\phi \to 0$ again is not $M_{x,0}$. But the multiplication tables $M_{y,\phi}$ with respect to the purely lexicographical Gröbner basis is even worse: its entries are either zero or diverge for $\phi \to 0$.

The behavior of the Gröbner bases at $\phi = 0$ raises the question of whether this singularity is *systematic*, that is, intrinsic to the problem, or if it is a *representation singularity* generated by the Gröbner bases. Systematic singularities appear, for example, if several zeros "collapse" into one multiple zero which leads to extremely intricate problems in the multivariate case [9]. Here, however, the good separation of the zeros suggests the conjecture that we only face a representation singularity.

Indeed, since H–bases leave more degrees of freedom, we can try another one which is now based on orthogonalization. For this purpose, we use the inner-product

$$(p,q) = (p(D)\, q)\,(0)$$

and recall from [11, Theorem 5.3] that the set $\{f, g_\phi\}$ is already an H–basis. Moreover, the normal form space, which is, according to Theorem 1, the least interpolation space, is spanned by

$$\mathcal{P}_\phi^* = \left\{1, x, y, 2\sin\phi\, x^2 - 3\cos\phi\, xy - \sin\phi\, y^2\right\}$$

and depends *continuously* on ϕ with

$$\lim_{\phi\to 0} \mathcal{P}_\phi^* = \mathcal{P}_0 = \{1, x, y, xy\}\,.$$

Then one can compute the respective multiplication table as

$$M_{x,\phi}^* = \begin{bmatrix} 0 & 1+\varepsilon_1(\phi) & \varepsilon_2(\phi) & \varepsilon_3(\phi) \\ 1 & 0 & 0 & \varepsilon_4(\phi) \\ 1 & 0 & 0 & 1+\varepsilon_5(\phi) \\ 0 & \varepsilon_6(\phi) & 1+\varepsilon_7(\phi) & \varepsilon_8(\phi) \end{bmatrix},$$

where $\varepsilon_j\,(\cdot)$, $j = 1,\ldots,8$, are continuous functions which vanish at the origin. In particular, $M_{x,\phi}^* \to M_{x,0}$ as $x \to 0$ and the computation of eigenvalues and eigenvectors of $M_{x,\phi}^*$ can now be done with sufficient accuracy. However, we remark that the fact that the matrices $M_{x,\phi}^*$ and $M_{y,\phi}^*$ have two approximately double eigenvalues ± 1, requires some extra care when connecting these individual values in the final determination of the intersections.

§5. Summary

We have given examples of numerical applications which can be reduced to the computation of normal forms with respect to a certain polynomial ideal, an operation which is usually performed using a Gröbner basis. On the other hand, H–bases could as well be used for normal form computations, and their greater flexibility may yield stabilizing effects which are highly desired in numerical computations.

Acknowledgments. The second author was supported by the Deutsche Forschungsgemeinschaft with a Heisenberg fellowship, Grant Sa–627/6.

References

1. Anderson, E., Z. Bai, C. Bischof, J. Demmel, J. Dongarra, J. Du Croz, A. Greenbaum, S. Hammarling, A. McKenney, S. Ostrouchov and D. Sorensen, *LAPACK User's Guide, Second Edition*, SIAM, 1995.
2. de Boor, C. and A. Ron, On multivariate polynomial interpolation, Constr. Approx. **6** (1990), 287–302.
3. de Boor, C. and A. Ron, The least solution for the polynomial interpolation problem, Math. Z. **210** (1992), 347–378.
4. Cox, D., J. Little and D. O'Shea, *Ideals, Varieties and Algorithms*, Springer, 1992.
5. Gautschi, W., Questions of numerical conditions related to polynomials, in *Studies in Numerical Analysis*, G. H. Golub (ed.), The Mathematical Association of America, 1984, 140–177.
6. Kronecker, L.,Über einige Interpolationsformeln für ganze Funktionen mehrerer Variabeln (Lecture at the academy of sciences, December 21, 1865), in *L. Kroneckers Werke*, volume I, H. Hensel (ed.), 133–141. Teubner 1895, reprinted by Chelsea Publishing Company, 1968.
7. Gonzales–Vega, L., F. Rouillier and M.–F. Roy, Symbolic recipes for polynomial system solving, in *Some Tapas in Computer Algebra*, A. M. Cohen, H. Cuypers and M. Sterk (eds.), Springer, 1999, 34–65.
8. Möller, H. M., Gröbner bases and numerical analysis, in *Gröbner Bases and Applications (Proc. of the Conf. 33 Years of Gröbner Bases)*, B. Buchberger and F. Winkler (eds.), Cambridge University Press 1998, 159–178.
9. Marinari, M. G., H. M. Möller and T. Mora, On multiplicities in polynomial system solving, Trans. Amer. Math. Soc. **348** (1996), 3283–3321.
10. Möller, H. M. and H. J. Stetter, Multivariate polynomial equations with multiple zeros solved by matrix eigenproblems, Numer. Math. **70** (1995), 311–329.
11. Möller, H. M. and T. Sauer, H–bases for polynomial interpolation and system solving, Advances Comput. Math., to appear.
12. Möller, H. M. and T. Sauer, H–Bases I: The foundation, *Curve and Surface Fitting: Saint-Malo 1999*, Albert Cohen, Christophe Rabut, and Larry L. Schumaker (eds.), Vanderbilt University Press, Nashville, 2000, 325–332.
13. Sauer, T., Polynomial interpolation of minimal degree, Numer. Math. **78** (1997), 59–85.
14. Sauer, T., Polynomial interpolation of minimal degree and Gröbner bases, in *Gröbner Bases and Applications (Proc. of the Conf. 33 Years of Gröbner Bases)*, B. Buchberger and F. Winkler (eds.), Cambridge University Press 1998, 483–494.

15. Sauer, T., Gröbner bases, H–bases and interpolation, Trans. Amer. Math. Soc., to appear.
16. Stetter, H. J., Matrix Eigenproblems at the heart of polynomial system solving, SIGSAM Bull. **30** (1995), 22–25.
17. Wilkinson, J. H., The perfidious polynomial, in *Studies in Numerical Analysis*, G. H. Golub (ed.), The Mathematical Association of America, 1984, 1–28.

H. Michael Möller
Fachbereich Mathematik
Universität Dortmund
Vogelpothsweg 87
D–44221 Dortmund, Germany
moeller@math.uni-dortmund.de

Thomas Sauer
Mathematisches Institut
Universität Erlangen–Nürnberg
Bismarckstr. 1 $\frac{1}{2}$
D–91054 Erlangen, Germany
sauer@mi.uni-erlangen.de

Dependence Structure of Random Wavelet Coefficients in Terms of Cumulants

Philippe Naveau, Peter Brockwell, and Doug Nychka

Abstract. When the Gaussian assumption for a times series no longer holds, second order moment properties such as the covariance and the spectrum are not necessarily sufficient to describe the dependence structure. Although wavelet models have been proposed to de-correlate the signal, this strategy must be reexamined when applied to non-Gaussian processes. The process of interest is a continuous parameter, mean-squared continuous real-valued process that is not necessarily Gaussian or linear. To study the departures from linearity and Gaussianity, we consider joint cumulants, which are linear combinations of higher order moments, and their associated spectra. A specific objective is to obtain new expressions for cumulants of the random discrete wavelet coefficients instead of the second order moments, and to study their higher order polyspectra. Conditions on the polyspectrum to give null wavelet cumulants within and across wavelet coefficient levels are derived. Expressions of the original cumulants as a function of the wavelets cumulants are also given.

§1. Introduction

The covariance and spectral properties of the discrete wavelet coefficients for random continuous real-valued processes have been extensively studied in the past. Among others, Donoho et al. [4], Flandrin [6], Mallat et al. [13], Masry [14], and Walter [17] have investigated the correlation within and across wavelet coefficients. Focusing exclusively on the second order properties of wavelet coefficients for a Gaussian process is a reasonable task since the dependence structure of Gaussian processes is entirely characterized by the covariance. When the normality assumption no longer holds, higher order cumulants are necessary.

Exploring some of the links that exist between wavelets and cumulants is fairly new. Brillinger [1] studied a non-parametric regression problem with cumulants and wavelets. In geophysics and astrophysics, Lazear [11] and Ferreira et al. [5] applied wavelets and cumulants to seismic data sets and to the Cosmic Microwave Background problem. The dependence structure

Curve and Surface Fitting: Saint-Malo 1999
Albert Cohen, Christophe Rabut, and Larry L. Schumaker (eds.), pp. 343–350.

ISBN 0-8265-1357-3.

between wavelet packets and in particular wavelet coefficients has also been studied by D. Leporini and J.C. Pesquet [12]. In this paper, we derive new and general results about the dependence structure of random wavelet coefficients via its cumulants.

The process of interest $X(t)$, indexed by the real parameter t, is supposed to be a continuous real-valued process that is mean-square continuous, and such that moments of some order $l \geq 2$ exist, i.e.,

$$\sup_t E|X(t)|^l < \infty \qquad \text{and} \qquad \lim_{h \to 0} E|X(t+h) - X(t)|^2 = 0. \tag{1}$$

§2. Cumulant Definition and Properties

If some useful information of the signal is not contained in the second-order covariances (and the second order spectra), then one can still calculate some meaningful linear combination of higher order moments, called cumulants. Some early work on higher order cumulants and their Fourier transform was proposed by Hasselman et al. [8] for investigating nonlinear interaction of ocean waves, and Godfrey [7] used it for the analysis of economic time series. Rosenblatt with Lii and Van Atta in a series of papers have described how higher cumulants could be used to study nonlinear transfer of energy in turbulence.

The mth joint cumulant of the set of random variables $\{X(t_1), ..., X(t_m)\}$, denoted by $CUM(X(t_1), ..., X(t_m))$, with $m \leq l$, is given by

$$CUM(X(t_1), ..., X(t_m)) = \sum (-1)^p (p-1)!(E \prod_{r \in v_1} X(t_r))...(E \prod_{r \in v_p} X(t_r)),$$

where the summation extends over all partitions $\{v_1, ..., v_p\}$ of $\{1, ..., m\}$ with $p = 1, ..., m$. From this definition, we can notice that the information contained in the first m cumulants is exactly the same as that contained in the first m moments. However, cumulants have some advantages over moments. For example, cumulants have useful linear properties,

$$CUM(Z + X(t_1), ..., X(t_m)) = CUM(Z, ..., X(t_m)) + CUM(X(t_1), ..., X(t_m)),$$
$$CUM(aX(t_1), ..., X(t_m)) = aCUM(X(t_1), ..., X(t_m)),$$

for any real a. Another important property of cumulants concerns the dependence structure of the process: if some subset of $\{X(t_1), ..., X(t_m)\}$ is independent of the remainder, then $CUM(X(t_1), ..., X(t_m))$ is identically equal to zero. Hence, the cumulant, $CUM(X(t_1), ..., X(t_m))$ can be interpreted as a measure of dependence of $\{X(t_1), ..., X(t_m)\}$. For the special case of Gaussian processes, cumulants of order higher than two are zero.

In the remainder of this section, we suppose that the process $\{X(t)\}$ is stationary up to order l, i.e.,

$$E(X(t_0)X(t_1)...X(t_l)) = E(X(t_0 + h)X(t_1 + h)...X(t_l + h)), \qquad \forall h.$$

Stationarity as just defined is frequently referred to in the literature as weak stationarity. For us however the term stationarity, without further qualification, will always refer to the above equality.

Under stationarity, $CUM(X(t), X(t+s_1), ..., X(t+s_{m-1}))$ does not depend on t, and can be denoted by $\gamma_m(s_1, ..., s_{m-1})$. With these notations, the second order cumulant $\gamma_1(u)$ is just the covariance function. The third order cumulant $\gamma_2(u,v)$ is the same as the third-order central moment,

$$E((X(t)-\mu)(X(t+u)-\mu)(X(t+v)-\mu)),$$

where μ is the mean value of the process.

From the covariance function, one may define the power spectrum, i.e., the Fourier transform of $\gamma_1(t)$, $f_1(\omega) = \int \gamma_1(t)\exp(-i\omega t)dt$. A natural extension is the mth-order polyspectrum defined by

$$f_m(\omega_1, ..., \omega_{m-1}) = \int ... \int \gamma_m(t_1, ..., t_{m-1})\exp(-i\sum_{j=1}^{m-1}\omega_j t_j)dt_1...dt_{m-1},$$

assuming that the above Fourier transforms exist. An important property of the polyspectra is that all polyspectra of higher order than second order vanish when $\{X(t)\}$ is a Gaussian process. Another characteristic of the polyspectra is that the ratio

$$\frac{|f_2(\omega_1,\omega_2)|^2}{f_1(\omega_1)f_1(\omega_2)f_1(\omega_1+\omega_2)}$$

is constant whenever the process $\{X(t)\}$ is linear. Hence, the simplest higher order spectrum, called bispectrum, can be regarded as deviation measures from Gaussianity and linearity. Different statistical tests have been derived from it (see Subbua Rao and Gabr [9] and Hinich [16]).

§3. Random Wavelet Coefficients

Consider a discrete orthonormal wavelet decomposition of a stochastic process $\{X(t)\}$ that satisfies condition (1). The corresponding wavelet coefficients

$$W_{j,k} = \int X(t)\psi_{j,k}(t)dt \tag{2}$$

are random variables. Here the equality sign is to be understood in the mean-square sense, and $\psi_{j,k}(t) = 2^{j/2}\psi(2^j t - k)$ is an orthonormal wavelet basis function with the mother wavelet ψ. A rigorous framework concerning the construction for wavelet orthonormal basis can be found in Meyer [15] and Daubechies [3]. There exist many candidates for the mother wavelet. The simplest example of an orthonormal wavelet basis is provided by the Haar system for which

$$\psi(x) = \begin{cases} 1, & \text{if } 0 \le x < 0.5, \\ -1, & \text{if } 0.5 \le x < 1, \\ 0, & \text{otherwise.} \end{cases}$$

Our main interest is to understand the dependence structure between the different wavelet coefficients defined by (2).

In this article, we will either use the Haar system as a simple example of compactly support wavelets, or a particular type of band-limited wavelets called Meyer-type wavelets (see Walter [17], Zayed and Walter [18]). Having a compact support in the frequency domain can facilitate the computation of wavelet cumulants. These Meyer-type wavelets have some additional attractive features, such as being highly smooth (they can be made C^∞) and having fast decay in the time domain. They are introduced as follows.

Let F be any probability measure supported on $[-\epsilon, \epsilon]$ for some $\epsilon \leq \pi/3$. Then the mother wavelet $\psi(\cdot)$ is defined by its Fourier transform

$$\tilde{\psi}(\omega) = \exp(-i\omega/2)[\int_{|\omega/2|-\pi}^{|\omega|-\pi} dF]^{1/2}. \tag{3}$$

From this definition, it is possible to check that the orthogonality and dilation conditions are satisfied for the wavelet basis generated from this mother wavelet. There is a large class of distributions that can be chosen in equality (3), and the edges of the support, $[-\epsilon, \epsilon]$, can be made highly smooth.

§4. Dependence Structure

The dependence structure between wavelet coefficients is closely related to the dependence inside the original signal. Hence, our first problem is to explain how to obtain the joint cumulants of the wavelet coefficients from the joint cumulants of the process. The first proposition takes care of this problem.

Because of space limitations, the proof of our propositions will not be included in this paper. However, complete details of the proofs can be requested from the authors.

Proposition 1. *Let $\{X(t)\}$ be a stochastic process that satisfies condition (1). Suppose that the joint cumulants of order $m \leq l$ of $\{X(t)\}$ exist. Then*

$$CUM(W_{j_1k_1}, ..., W_{j_mk_m}) = \int ... \int CUM(X(t_1), ..., X(t_m)) \prod_{n=1}^{m} \psi_{j_n,k_n}(t_n) dt_n.$$

Proposition 1 is directly applicable to compactly supported wavelets, since the product

$$\prod_{n=1}^{m} \psi_{j_n,k_n}(t_n)$$

is null except at the intersection of the translated and dilated supports. For example, suppose that the wavelet basis corresponds to the Haar system. The expression of cumulants of the wavelet coefficients becomes

$$CUM(W_{j_1k_1}, ..., W_{j_mk_m}) = \sum_{l=1}^{2^m} (-1)^l 2^{j_l/2} \int_{A_l} CUM(X(t_1), ..., X(t_m)) dt_n. \tag{4}$$

The A_l are rectangular boxes defined by the tensor product

$$A_l = \bigotimes_{n=1,..,m} I_{n,c_{nl}},$$

where the vectors $\tilde{c}_l = (c_{1l}, , ..., c_{ml})$ represent the set of all 2^m possible configurations of $\{0, 1/2\}^m$, and the interval $I_{n,c_{nl}}$ is defined by

$$I_{n,c_{nl}} = \{t_n : 2^{-j_n}(k_n + c_{nl}) \leq t_m < 2^{-j_n}(k_n + c_{nl} + 1/2), \text{ for } n = 1, .., m\}.$$

In order to apply (4) to a more specific example, we suppose that the process $\{X(t)\}$ is a zero-mean stationary process with standard deviation σ and covariance $\gamma_1(h) = b\exp(-a|h|)$, and $\gamma_1(h_1, h_2) = c\exp(-a|h_1 + h_2|)$, where a, b, c are constants that depends on the second and third moment and other parameters describing the original process. Processes with such a cumulant function correspond to Continuous Auto-Regressive processes (CAR) (see Brockwell [2]) or equivalently solutions of particular stochastic differential equations with non-necessarly Gaussian noise. After some algebra, wavelet cumulants simplify to

$$CUM(W_{j_1k_1}, W_{j_2k_2}) = -\frac{bK_a(j_1, k_1)}{a^2K_a(j_2, k_2)}[\sum_{(u_1,u_2)\in\{-1,1\}^2} H_{j_1}(u_1a)H_{j_2}(u_2a)],$$

for $2^{-j_2}(k_2 + 1) \leq 2^{-j_1}k_1$ and

$$CUM(W_{j_1k_1}, W_{j_2k_2}, W_{j_3k_3}) = -\frac{cK_{2a}(j_1, k_1)}{2a^3K_a(j_2, k_2)K_a(j_3, k_3)} \times [\sum_{(u_1,u_2,u_3)\in\{-1,1\}^3} H_{j_1}(u_1a)H_{j_2}(u_2a)H_{j_3}(u_3a)]$$

for $2^{-j_3}(k_3 + 1) \leq \min(2^{-j_1}k_1, 2^{-j_2}k_2)$ with $K_a(j, k) = \exp(a(2^{-j}(k + 0.5)))$ and $H_j(a) = 1 - \exp(a2^{-j-1})$. The previous formulas can be easily extended to higher dimensions, and can be used to derive asymptotic behavior, e.g $|j_1 - j_2| \uparrow \infty$ and so on.

Another possible application of Proposition 1 is to non-stationary processes. A large variety of models, such as the bilinear model, autoregressive models with random coefficients, and the threshold model, have been proposed to take into account of the non-stationarity. To illustrate the use of cumulants, we restrict attention to piecewise stationary processes, i.e. a sum of independent stationary processes:

$$X(t) = \sum_{l=1}^{r} \mathcal{I}(u_l \leq t < u_{l+1})X^{(l)}(t), \text{ where } \mathcal{I}(A) = \begin{cases} 1, & \text{if } t \in A \\ 0, & \text{otherwise,} \end{cases}$$

and $X^{(l)}(t)$ are independent stationary processes and the change-points are equal to $-\infty = u_0 < u_1 < \cdots < u_r < u_{r+1} = \infty$. Because of linear properties of the cumulants, we have immediately that

$$CUM(X(t_1), ..., X(t_m)) = \sum_{l=1}^{r}\prod_{n=1}^{m} \mathcal{I}(u_l \leq t_n < u_{l+1})CUM(\underline{\mathbf{X}}^{(l)}(t)),$$

where $CUM(\underline{\mathbf{X}}^{(l)}(t)) = CUM(X^{(l)}(t_1), ..., X^{(l)}(t_m))$. Using Proposition 1, it follows that

$$CUM(\underline{\mathbf{W}}_{jk}) = \sum_{l=1}^{r} \int CUM(\underline{\mathbf{X}}^{(l)}(t)) \prod_{n=1}^{m} \mathcal{I}(u_l \le t_n < u_{l+1}) \psi_{j_n,k_n}(t_n) dt_n$$

with $CUM(\underline{\mathbf{W}}_{jk}) = CUM(W_{j_1k_1}, ..., W_{j_mk_m})$. The above expression shows that wavelet cumulants for piecewise stationary processes can be easily computed for compactly supported wavelets such as the Haar system, and when each process $X^{(l)}$ has simple cumulant functions (e.g. the CARMA process).

From the CAR example, we saw that wavelet cumulants are computable for the Haar system, but the resulting formula are not so easy to manipulate. Another approach is to use band-limited wavelets. Simpler expression of the wavelet cumulants can be derived. To illustrate this point, we look at the Meyer-type wavelet in the next proposition. In this case, the Meyer-type wavelet gives null or small wavelet cumulants within and across wavelet coefficient levels under simple conditions.

Proposition 2. *Let $\{X(t)\}$ be a stationary process that satisfies condition (1). Suppose that its mth-order polyspectrum f_m is well defined, and the orthonormal basis $\{\psi_{jk}\}$ is generated by a mother Meyer-type wavelet. If there exists some integer j^* in $\{j_1, ..., j_m\}$ such that $\sum_{j \neq j^*} 2^j \le 2^{j^*-2}$, then $CUM(W_{j_1k_1}, ..., W_{j_mk_m}) = 0$. In addition, if $f_m(\omega)$ and $\tilde{\psi}(\omega)$ are both in C^p, then the cumulant at a fixed resolution level satisfies*

$$CUM(W_{j_1k_1}, ..., W_{j_1k_m}) = O(\max_r \prod_{s \neq r} |k_r - k_s|^{-p}).$$

Proposition 2 shows that the Meyer-type wavelet transform not only can remove the correlation inside the original signal, but in addition the higher-order cumulants are either null at distant scales or very small at a fixed scale. Walter's result [17] for the covariance is a special case of Proposition 2:

$$CUM(W_{j_1k_1}, W_{j_2k_2}) = \begin{cases} 0, & \text{for } |j_1 - j_2| > 1, \\ O(|k_1 - k_2|^{-p}), & \text{for } j_1 = j_2. \end{cases}$$

It is interesting to note that the results stated in Proposition 2 hold for any choice of F in the definition of the Meyer-type wavelet. An open problem is to determine if there exist some distributions F which will significantly reduce cumulants between wavelet coefficients. In this direction, Zayed and Walter [18] minimized the covariance between wavelet coefficients by using a bi-orthonormal wavelet basis that is a function of the original covariance.

In Propositions 1 and 2, different expressions of the wavelet cumulants were derived. A natural question is whether or not the cumulants of the original process can be expressed in terms of $CUM(W_{j_1k_1}, ..., W_{j_mk_m})$. Thus, our next result is the converse of Proposition 1.

Proposition 3. *Suppose that $\{W_{jk}\}$ is a sequence of variables with all finite moments. If the process*

$$X(t) = \sum_{j=-\infty}^{\infty} \sum_{k=-\infty}^{\infty} W_{jk}\psi_{jk}(t)$$

is well defined (in the mean-square sense), then we have

$$CUM(X(t_1), ..., X(t_m)) = \sum_{j_1,k_1=-\infty}^{\infty} \cdots \sum_{j_m,k_m=-\infty}^{\infty} \prod_{n=1}^{m} \psi_{j_n k_n}(t_n) CUM(\underline{\mathbf{W}}_{jk}).$$

§5. Conclusion and Future Work

In this paper, different relationships between wavelets and cumulants have been presented. Results show that the wavelet transform is not only a good tool to de-correlate a Gaussian process, but it also gives small higher-order cumulants of a non-Gaussian signal. The Meyer-type wavelet is particularly well-adapted for stationary processes since they give null wavelet cumulants at distant scales.

The combination of wavelets and cumulants has not yet been fully exploited. The statistical study of estimators of the bispectrum based on wavelets is of particular interest for application with real data sets. Also investigating the properties of non-linear and non-stationary times series models using bi-spectral methods and a wavelet decomposition approach needs further research.

Acknowledgments. This work was supported by NSF grants DMS-9815344 and DMS-9312686.

References

1. Brillinger, D. R., Some uses of cumulants in wavelet analysis, Nonparametric Statistics **6** (1996), 93–114.
2. Brockwell, P., Continuous-time ARMA processes, preprint.
3. Daubechies, I., *Ten Lectures on Wavelets*, SIAM, Philadelphia, 1992.
4. Donoho, D., S. Mallat and R. Von Sachs, Estimating covariances of locally stationary processes: rates of convergence of best basis methods, submitted to Annals of Statistics.
5. Ferreira, P. G., J. Magueijo and J. Silk, Cumulants as non gaussian Qualifiers, preprint.
6. Flandrin, P., On the spectrum of fractional Brownian motion, IEEE Trans. Info. Th., **35** (1989), 197–199.

7. Godfrey, M. D., An explorationary study of the bispectrum of economic time series, Appl. Statist. **14** (1965), 48–69.
8. Hasselman, K., W. Munk and G. MacDonald, Bispectra of Ocean Waves, in *Time Series Analysis*, M. Rosenblatt (ed.), John Wiley, New-York, 1963, 125–139.
9. Hinich, M. J., Testing for gaussianity and linearity of a stationary time series, J. Time Ser. Anal. **3** (1982), 169–178.
10. Johnstone, I. M. and B. W. Siverman, Wavelet threshold estimators for data with correlated noise, J. Roy. Statist. Soc. Ser. B **59** (1997), 319–351.
11. Lazear, G. D., Mixed-Phase wavelet estimation using fourth-order cumulants, Geophysics **58** (1993), 1042–1051.
12. Leporini, D., and J.C. Pesquet, High Order Properties of M-band Wavelet Packet Decompositions, IEEE Workshop on High-order Statistics, Banff, Canada, 1997.
13. Mallat, S., G. Papanicolaou and Z. Zhang, Adaptive covariance estimation of locally stationary processes, Annals of Statistics **26** (1998), 1–47.
14. Masry, E., The wavelet transform of stochastic processes with stationary increments and its application to fractional Brownian motion, IEEE Trans. Info. Th. **39** (1993), 260–264.
15. Meyer, Y., *Wavelets: Algorithms and Applications*, SIAM, Philadelphia, 1993.
16. Subba Rao, T. and M. Gabr, *An Introduction to Bispectral Analysis and Biliniear Time Series Models*, Lecture Notes in Statistics Vol. 24, Springer-Verlag, Berlin, 1984.
17. Walter, G. G., *Wavelets and Other Orthogonal Systems with Applications*, CRC Press, Boca Raton, 1994.
18. Zayed, A. I. and G. G. Walter, Characterization of analytic functions in terms of their wavelet coefficients, Complex Variables **29** (1996), 265–276.

Philippe Naveau and Doug Nychka
Geophysical Statistics Project
National Center for Atmospheric Research
B.P. Box 3000, Boulder CO 80307, USA
pnaveau@ucar.edu
nychka@ucar.edu

Peter Brockwell
Department of Statistics
Colorado State University
Ft. Collins, CO 80523-1877, USA
pjbrock@stat.colostate.edu

Interpolating Functions on Lines in 3-Space

Martin Peternell and Helmut Pottmann

Abstract. Given straight lines L_i, $i = 1, \ldots, N$, in Euclidean 3–space with associated function values f_i, we study the interpolation problem of constructing a smooth real valued function F which interpolates values f_i at given data lines L_i. The function F shall be defined on the entire set of lines or at least on lines contained in a domain of interest in 3–space.

§1. Introduction

The problem of constructing an interpolating function F for data lines L_i and corresponding function values f_i is a scattered data interpolation problem in the set of lines $\mathcal{L}$ in Euclidean 3-space E^3.

A variety of solutions of scattered data interpolation problems for data points $X_i \in U$ with $U = \mathbb{R}^n$ or $U \subset \mathbb{R}^n$ are known, see [3]. Extensions to spheres and other surfaces in $\mathbb{R}^3$ are described in [2] and references therein.

Scattered data interpolation on lines is quite different, since the set of lines $\mathcal{L}$ is not a Euclidean space. It is a result of classical geometry that the set of lines $\bar{\mathcal{L}}$ of projective extension P^3 of Euclidean 3–space E^3 is a 4–dimensional quadratic variety M_2^4 in projective P^5. Thus, the general formulation of the problem is as follows: Construct a function $F : M_2^4 \to \mathbb{R}$ interpolating values f_i to corresponding data lines L_i. For practical purposes it is sufficient to construct (or represent) functions on subsets of M_2^4 which correspond to domains of interest in E^3, containing all data lines.

The solution presented here will be the following. We restrict to specific four-dimensional subsets $\mathcal{L}_0$ of M_2^4. These subsets possess parametrizations $\mathbb{R}^4 \to \mathcal{L}_0$ with the property that distances between lines in $\mathcal{L}_0$ are induced by special positive quadratic forms in $\mathbb{R}^4$. This fact allows us to apply well–known methods in $\mathbb{R}^4$ to solve interpolation (or also approximation) problems.

Applications include light field rendering in computer graphics [4]. Considering motion planning in robotics, the method applies to represent a distance function of robot arms (lines) to obstacles. The first motivation for studying functions on lines came from five axis milling. There, the question occurs of how to represent axis positions (lines) of the cutting tool.

Curve and Surface Fitting: Saint-Malo 1999
Albert Cohen, Christophe Rabut, and Larry L. Schumaker (eds.), pp. 351–358.

ISBN 0-8265-1357-3.

§2. Lines in Space

An oriented line L in Euclidean 3-space E^3 is determined by a point $\mathbf{p}$ and a unit direction vector $\mathbf{l}$ ($\|\mathbf{l}\| = 1$). Together with the moment vector

$$\bar{\mathbf{l}} = \mathbf{p} \times \mathbf{l}, \tag{1}$$

we obtain a representation of L by a sixtuple

$$\mathbf{L} = (\mathbf{l}; \bar{\mathbf{l}}) = (l_1, l_2, l_3; l_4, l_5, l_6). \tag{2}$$

These l_i's are called normalized Plücker coordinates of L. By (1), these coordinates are not independent, but satisfy the Plücker relation

$$\mathbf{l} \cdot \bar{\mathbf{l}} = l_1 l_4 + l_2 l_5 + l_3 l_6 = 0. \tag{3}$$

Substituting $\mathbf{l}$ by $-\mathbf{l}$ leads to coordinate vector $-\mathbf{L}$ which defines the same line but with opposite orientation. To get more information about the structure of lines in space, it is necessary to study the set of lines $\bar{\mathcal{L}}$ in the projective extension P^3 of E^3.

E^3 is extended to P^3 by adding points and lines at infinity. Using the analytical model $\mathbb{R}^4$, points in P^3 are one dimensional subspaces of $\mathbb{R}^4$. Thus, we will use the following notation for points in P^3,

$$(x_0, x_1, x_2, x_3)\mathbb{R} := (\lambda x_0, \ldots, \lambda x_3), \lambda \in \mathbb{R}.$$

Let $\omega : x_0 = 0$ be the plane at infinity. We write briefly $(x_0, \mathbf{x})\mathbb{R}$, with $\mathbf{x} \in \mathbb{R}^3$ for points in P^3. The transition from homogeneous to Cartesian coordinates is given by

$$(x_0, x_1, x_2, x_3) \mapsto (\frac{x_1}{x_0}, \frac{x_2}{x_0}, \frac{x_3}{x_0}),$$

which is obviously only possible for points not at infinity.

A line L in P^3 usually is spanned by two points $(p_0, \mathbf{p})\mathbb{R}$ and $(q_0, \mathbf{q})\mathbb{R}$. Homogeneous Plücker coordinates are obtained by

$$\mathbf{L} = (l_1, \ldots, l_6) = (p_0 \mathbf{q} - q_0 \mathbf{p}, \mathbf{p} \times \mathbf{q}). \tag{4}$$

If we substitute $(p_0, \mathbf{p})$ by $\lambda(p_0, \mathbf{p})$, we get $\lambda \mathbf{L}$ such that the l_i's are only determined up to a scalar multiple. This proves homogeneity of $\mathbf{L}$.

If L is not in ω, the relation to definition (2) is obtained as follows. Let $(p_0, \mathbf{p})\mathbb{R}$ be a proper point on L such that we can switch to Cartesian coordinates $\mathbf{p}$ by letting $p_0 = 1$. Further, let $(q_0, \mathbf{q})\mathbb{R}$ be the intersection point $\omega \cap L$ which implies $q_0 = 0$. Inserting this into (4) gives (2) up to a normalization of the direction vector $\mathbf{q} = \mathbf{l}$ of the line L.

If L is in ω, its Plücker coordinates are $(\mathbf{o}, \mathbf{a})\mathbb{R}$ with $\mathbf{o} = (0, 0, 0)$ and some not vanishing vector $\mathbf{a}$. We can interprete L as the line at infinity of a pencil of parallel planes $\mathbf{a} \cdot \mathbf{x} = c$, with $c \in \mathbb{R}$. All these planes possess $\mathbf{a}$ as normal vector.

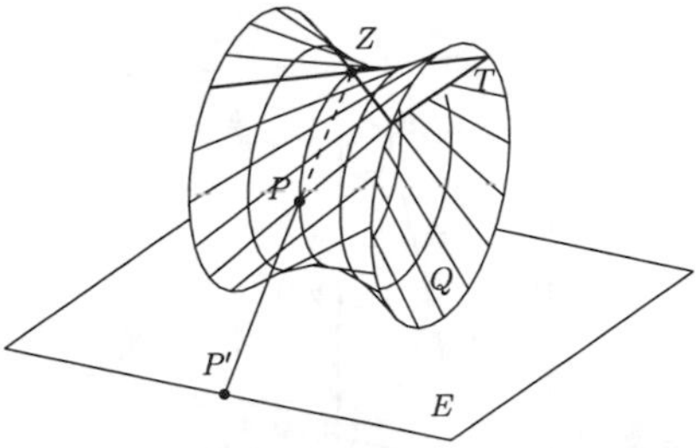

Fig. 1. Stereographic projection of a hyperboloid Q.

Since $\mathbf{L}$ and $\lambda\mathbf{L}$ define the same line in P^3, homogeneous Plücker coordinates (4) define points $\mathbf{L}\mathbb{R}$ in P^5. But only those 6-tuples $(x_1, \ldots, x_6)\mathbb{R}$ are Plücker coordinates of a line X in P^3, which satisfy

$$x_1x_4 + x_2x_5 + x_3x_6 = 0.$$

This quadratic variety is called the Klein quadric M_2^4, where upper and lower indices denote dimension and degree of this variety. The maximal dimension of its subspaces is 2. It is a point model of the set of lines $\bar{\mathcal{L}}$ of P^3. The bijection

$$\gamma : \bar{\mathcal{L}} \to M_2^4$$

from lines $L \subset P^3$ to points $\mathbf{L}\mathbb{R}$ of M_2^4 is called Klein mapping.

The image points $(\mathbf{o}, \mathbf{a})\mathbb{R}$ of lines at infinity lie in the plane $E_\omega : x_1 = x_2 = x_3 = 0$ which is entirely contained in M_2^4. All lines passing through the origin $O = (1, 0, 0, 0)\mathbb{R}$ have Plücker coordinates $\mathbf{L} = (\mathbf{l}, \mathbf{o})$. This can be checked by letting $\mathbf{p} = (0, 0, 0)$ in formula (1). The corresponding image points in P^5 lie in the plane $E_o : x_4 = x_5 = x_6 = 0$. In general, all lines through an arbitrary point in P^3 possess γ-images which lie in a 2-dimensional subspace of M_2^4. The same holds for lines contained in an arbitrary plane in P^3. Thus, M_2^4 contains two 3-parametric families of 2-dimensional subspaces.

We emphasize that $\mathcal{L}$ and $\bar{\mathcal{L}}$ are *not* Euclidean, affine or projective spaces.

Local coordinates of lines

We have seen that $\mathcal{L}$ is isomorphic to $M_2^4 - E_\omega$, where E_ω consists of image points of all lines at infinity. Let T be the tangent hyperplane of M_2^4 at a point Z and let $\tau = M_2^4 \cap T$. It is known that τ is the γ-image of all lines intersecting the line $L = Z\gamma^{-1}$.

Lemma 1. *$M_2^4 - \tau = A^4$ is an affine space.*

Proof: This lemma is a result of classical geometry, and is proved by stereographic projection. Let Q be a regular quadric in P^n. Let Z be a point in Q and T its tangent hyperplane, see Figure 1. Further, consider E to be a hyperplane in P^n, not incident with Z. The intersection $\tau = Q \cap T$ is a quadratic cone with vertex Z. The intersection $e = E \cap T$ is a hyperplane

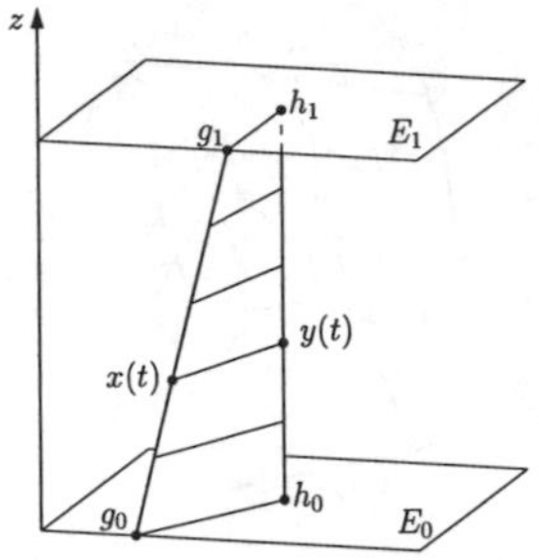

Fig. 2. Local coordinates, distance function.

in E. This says that $E - e$ is an affine space. The stereographic projection $\sigma : Q - \tau \to E - e$ with center Z is bijective and maps points $P \in Q - \tau$ to points P' in affine space $E - e$. □

Figure 1 shows a low dimensional example. Q is a hyperboloid, and τ is a pair of lines. Planes E and T are parallel such that e is at infinity.

We come back to line geometry and the Klein quadric M_2^4. Let $Z = (0,0,0,0,0,1)\mathbb{R}$ be the center of a stereographic projection. It is the γ-image of the line at infinity which is determined by horizontal planes z =const. with normal vector $(0,0,1)$. The tangent hyperplane T at Z with respect to M_2^4 is given by the equation $x_3 = 0$. The exceptional set $\tau = M_2^4 \cap T$ consists of γ-images $(l_1, l_2, 0; \ldots)\mathbb{R}$ of all horizontal lines. Lemma 1 says that all non-horizontal lines form an affine space A^4.

Consider two horizontal planes $E_0 : z = 0$ and $E_1 : z = 1$. The intersection points $\mathbf{g}_0 = (g_1, g_2, 0)$ and $\mathbf{g}_1 = (g_3, g_4, 1)$ of a line G and planes E_i (Figure 2) define a parametrization of all non-horizontal lines by

$$\begin{aligned} \mathbb{R}^4 = \mathbb{R}^2 \times \mathbb{R}^2 &\to \mathcal{L} \\ (g_1, g_2, g_3, g_4) &\mapsto G. \end{aligned} \tag{5}$$

Plücker coodinates of G are $\mathbf{G} = (g_3 - g_1, g_4 - g_2, 1; g_2, -g_1, g_1 g_4 - g_2 g_3)$. The stereographic projection with center Z onto $x_6 = 0$ gives

$$\mathbf{G}' = (g_3 - g_1, g_4 - g_2, 1; g_2, -g_1, 0).$$

This equals (5) up to a linear mapping. Hence, the mapping (5) from non-horizontal lines to points in $\mathbb{R}^4$ is geometrically equivalent to a stereographic projection of $M_2^4 - \tau$.

Distance function of lines

For practical purposes, it is sufficient to consider distances of lines within a *domain of interest.* To specify this domain, we will consider only lines which enclose an angle $\leq \phi_0$ with a fixed unit vector $\mathbf{z}$. The unit direction vector $\mathbf{g}$ of such a line G satisfies

$$\mathbf{g} \cdot \mathbf{z} \geq \cos \phi_0.$$

We have chosen a Cartesian coordinate system with $\mathbf{z}$ as third axis. Further, we will consider only segments of lines between two planes E_0, E_1, bounding the domain of interest. This is motivated by the fact that we are interested in particular in distances between points lying between those planes. Let $\mathbf{g}_i, \mathbf{h}_i$ be intersection points of lines G, H with E_i, and consider points $\mathbf{x}, \mathbf{y}$ on G and H, respectively,

$$\begin{aligned} \mathbf{x}(t) &= (1-t)\mathbf{g}_0 + t\mathbf{g}_1, \\ \mathbf{y}(t) &= (1-t)\mathbf{h}_0 + t\mathbf{h}_1. \end{aligned} \tag{6}$$

The square of a useful distance between lines G, H within the above domain of interest is defined by

$$\begin{aligned} d(G,H)^2 &= \int_0^1 \|\mathbf{x}(t) - \mathbf{y}(t)\|^2 dt \\ &= (\mathbf{g}_0 - \mathbf{h}_0)^2 + (\mathbf{g}_1 - \mathbf{h}_1)^2 + (\mathbf{g}_0 - \mathbf{h}_0) \cdot (\mathbf{g}_1 - \mathbf{h}_1). \end{aligned} \tag{7}$$

It measures horizontal distances between corresponding points $\mathbf{x}, \mathbf{y}$ of G, H. We will not distinguish between a line X and its coordinate vector $X = (x_1, x_2, x_3, x_4)$ in $\mathbb{R}^4$ according to parametrization (5). Formula (7) is a positive definite quadratic form in $\mathbb{R}^4$ with the following coordinate representation

$$\langle X, X \rangle = x_1^2 + x_2^2 + x_3^2 + x_4^2 + x_1 x_3 + x_2 x_4.$$

Remark 2. *These distances differ from orthogonal distances (from a point to a line G) only by a factor $\leq \cos\phi_0$. So, taking ϕ_0 relatively small will control the difference between these distances and the Euclidean distances in E^3.*

Summary 3. *The restriction to specific subsets $\mathcal{L}_0$ of line space allows parametrizations $\mathbb{R}^4 \to \mathcal{L}_0$. A positive definite quadratic form in $\mathbb{R}^4$ serves to define distances between lines in a useful manner.*

Choice of local coordinates

Distance d is not invariant under motions in E^3, but depends on the choice of $\mathbf{z}$ and planes E_0, E_1. Consider oriented lines $L_i, i = 1, \ldots, N$ with unit direction vectors $\mathbf{l}_i$. Assume that $\mathbf{l}_j \cdot \mathbf{l}_k < C$. This expresses that the angle between any two lines is bounded by $\arccos(C)$. A good choice for the vector $\mathbf{z}$ can be computed as solution of a regression problem. Assuming $\|\mathbf{l}_i\| = 1$, we want to maximize

$$\sum_{i=1}^{N} (\mathbf{l}_i \cdot \mathbf{z})^2 \tag{8}$$

over all unit vectors $\mathbf{z}$. Maximizing the quadratic form (8) under the quadratic side condition $\mathbf{z} \cdot \mathbf{z} = 1$ leads to an eigenvalue problem in $\mathbb{R}^3$. Thus, we found a possibility to construct $\mathbf{z}$ with respect to a set of lines L_i. Planes E_0, E_1 perpendicular to $\mathbf{z}$ bounding the domain of interest have to be chosen depending on the problem. In this sense we can say that the coordinate system is connected with the problem in an invariant way.

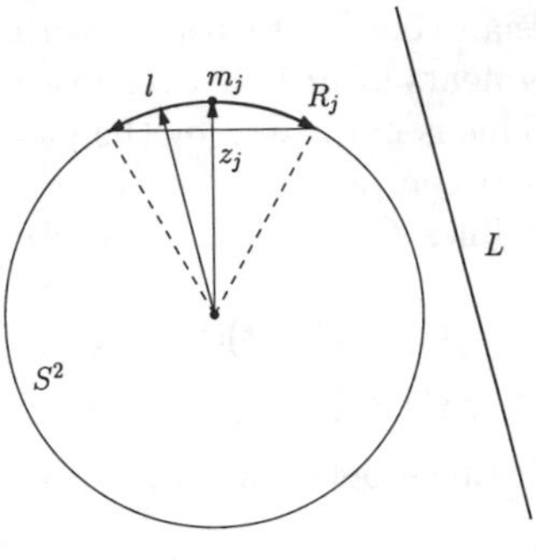

Fig. 3. Definition of domains.

If direction vectors $\mathbf{l}_i$ of lines L_i are distributed over a whole hemisphere or more, we have to split the set of lines into subsets and perform the construction of coordinate systems for the subsets. Remark 2 gives information about the deviation of distances compared to usual distances in E^3.

§3. Representation of Functions on $\mathcal{L}$

Given N lines L_i with corresponding function values f_i, we would like to compute a function $F : \mathcal{L} \to \mathbb{R}$ with $F(L_i) = f_i$. This is a scattered data interpolation problem on $\mathcal{L}$ (or M_2^4). With help of local parametrizations we obtain scattered data interpolation problems on $\mathbb{R}^4$. The given algorithm consists of three steps.

1) Find a covering $\{U_j, j = 1, \dots, M\}$ of $\mathcal{L}$ with domains U_j which are parametrized over $\mathbb{R}^4$. Decide the membership of lines and domains.
2) Compute partial solutions F_j of the interpolation problem for all domains U_j.
3) Merge all partial solutions F_j in a global solution F with required continuity.

First of all we want to find a covering of lines L_i by domains U_j with $1 < j < M$. We choose M unit vectors $\mathbf{z}_j$ and real numbers R_j which serve as centers and spherical radii of caps of the unit sphere S^2. These caps determine domains U_j in the following way. A line L belongs to U_j if and only if

$$\mathbf{l} \cdot \mathbf{z}_j \geq \cos R_j$$

holds for its direction vector, see Figure 3. Clearly, L can be contained in more than one domain. We determine the membership of all lines L_i for domains U_j.

In a second step we compute partial solutions F_j of the interpolation problem for each domain U_j. This is done by letting

$$F_j(X) = \sum_{k=1}^{N_j} a_{jk} B_k(X),$$

where N_j shall be the number of lines L_i belonging to domain U_j. $X = (x_1, x_2, x_3, x_4) \in \mathbb{R}^4$ is coordinate vector of a line X according to parametrization (5). $B_k(X)$ are (for instance) radial basis functions and depend only on the distance $d(X, L_k)$. The coefficients a_{jk} are solutions of linear systems. The problem of regularity of such systems dependent on the type of basis function is solved in [5]. So we get partial solutions F_j valid in domains U_j.

In the last step we have to merge all partial solutions to a unique one. This can be done by forming a weighted sum

$$F(X) = \sum_{j=1}^{M} w_j(X) F_j(X).$$

The weights can be chosen as

$$w_j(X) = \frac{(1 - \arccos(\mathbf{x} \cdot \mathbf{m}_j)/R_j)_+^r}{\sum_{l=1}^{M}(1 - \arccos(\mathbf{x} \cdot \mathbf{m}_l)/R_l)_+^r},$$

where $\mathbf{m}_j$ and R_j are center and radius of the spherical cap which defines U_j and $\mathbf{x}$ denotes the normalized direction vector of the line X. The notation $(q)_+^r$ expresses that $w_j(X)$ is positive in the interior of U_j and is zero outside. This says that $(q)_+^r = q^r$ for positive q, and $(q)_+^r = 0$ otherwise.

Weights $w_j(X)$ are in the differentiability class C^{r-1}. If partial solutions F_j possess the same smoothness, then also F is in C^{r-1}.

§4. Visualization of Functions on Lines

Since the dimension of $\mathcal{L}$ is four, visualization of function values is an advanced topic. In general, displaying functions on low dimensional subsets seems to be promising. We decided to choose several bundles of lines for evaluation and want to describe two methods of visualization.

We choose an appropriate number of points $\mathbf{v}_i$ within the domain of interest, and evaluate F at sufficiently many lines passing through vertices $\mathbf{v}_i$. Let F_{max} be an (existing!) upper bound of the absolute function values. Consider lines L_{ij} with function values $F(L_{ij}) = f_{ij}$ passing through vertex $\mathbf{v}_i$. Assume that L_{ij} are oriented lines. Displaying the star-shaped surfaces

$$\mathbf{p}_i = \mathbf{v}_i + (1 + \frac{f_{ij}}{F_{max}})\mathbf{l}_{ij}$$

for all chosen vertices $\mathbf{v}_i$ is one possibility to visualize function values. If function values for L and $-L$ are equal, the $\mathbf{p}_i$ will be centrally symmetric surfaces. For functions on nonoriented lines, we will use both direction vectors $\mathbf{l}_{ij}$ and $-\mathbf{l}_{ij}$ for the definition of $\mathbf{p}_i$, and assign the same function value f_{ij} to them. Thus we always get centrally symmetric surfaces. Figure 4 shows an interpolant. The test function is a function of the distances between lines L_i and points (not displayed).

For the second method we use spheres S_i, centered at vertices $\mathbf{v}_i$. All lines L_{ij} of the bundle $\mathbf{v}_i$ with constant function values form a cone C with vertex $\mathbf{v}_i$. Intersecting these cones $C(c_i)$ for several constants c_i gives level curves on spheres S_i (not displayed).

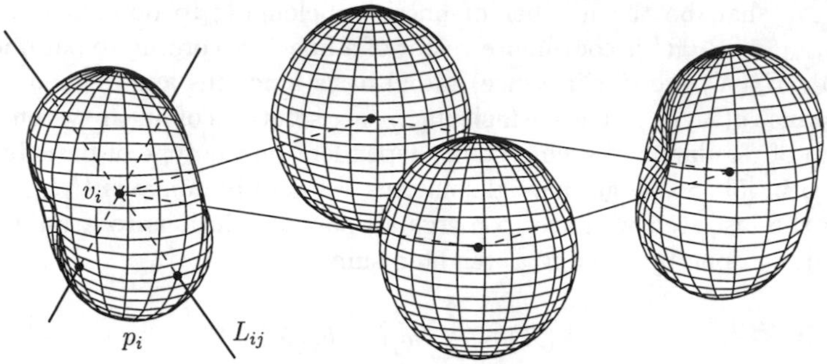

Fig. 4. Visualization of functions on lines.

Acknowledgments. This work has been supported by project P13648-MAT of the Austrian Science Foundation.

References

1. Chen, H-Y. and H. Pottmann, Approximation by ruled surfaces, J. Comput. Appl. Math. **102** (1999), 143–156.
2. Fasshauer, G. E. and L. L. Schumaker, Scattered data fitting on the sphere, *Mathematical Methods for Curves and Surfaces II*, M. Dæhlen, T. Lyche, and L. L. Schumaker (eds.), Vanderbilt University Press, Nashville, 1998, 117–166.
3. Hoschek, J. and D. Lasser, *Fundamentals of Computer Aided Geometric Design*, AK Peters, Wellesley, MA, 1993.
4. Levoy, M. and P. Hanrahan, Light field rendering, *SIGGRAPH 96, Annual Conference Series*, 1996, 31–41.
5. Micchelli, C. A., Interpolation of scattered data: distance matrices and conditionally positive definite functions, Constructive Approximation **2** (1986), 11–22.
6. Powell, M. J. D., The theory of radial basis function approximation in 1990, in *Advances in Numerical Analysis Vol.2*, W. Light (ed), Claredon Press, Oxford, 1992, 105–210.
7. Wendland, H., Konstruktion und Untersuchung radialer Basisfunktionen mit kompaktem Träger, Dissertation, Universität Göttingen, 1996.

Martin Peternell and Helmut Pottmann
Institut für Geometrie, TU Wien
Wiedner Hauptstr. 8-10
A-1040 Wien, Austria
`peternell@geometrie.tuwien.ac.at`
`pottmann@geometrie.tuwien.ac.at`

Numerical Techniques Based on Radial Basis Functions

Robert Schaback and Holger Wendland

Abstract. Radial basis functions are tools for reconstruction of multivariate functions from scattered data. This includes, for instance, reconstruction of surfaces from large sets of measurements, and solving partial differential equations by collocation. The resulting very large linear $N \times N$ systems require efficient techniques for their solution, preferably of $\mathcal{O}(N)$ or $\mathcal{O}(N \log N)$ computational complexity. This contribution describes some special lines of research towards this future goal. Theoretical results are accompanied by numerical examples, and various open problems are pointed out.

§1. Introduction

Many problems of numerical analysis take the form of a generalized interpolation in spaces of multivariate functions [21]. Due to the Mairhuber-Curtis theorem [12], such spaces cannot be fixed beforehand, but must necessarily depend on the given data. For a plain multivariate interpolation problem on a finite set $X = \{x_1, \ldots, x_N\}$ of pairwise different points in a domain $\Omega \subseteq \mathbb{R}^d$, there is an easy possibility to generate a data-dependent space via linear combinations of something that depends on a free variable $x \in \Omega \subseteq \mathbb{R}^d$ and the data locations x_j, namely

$$S_{X,\Phi} := \text{span} \{\Phi(x, x_j) \ : \ 1 \le j \le N\} \tag{1}$$

with a fixed function $\Phi \ : \ \Omega \times \Omega \to \mathbb{R}$. The numerical generation of the space can be simplified considerably in the special situations

1) $\Phi(x, y) = \phi(x - y)$ with $\phi \ : \ \mathbb{R}^d \to \mathbb{R}$ (translation invariance),
2) $\Phi(x, y) = \phi(\|x - y\|_2)$ with $\phi \ : \ [0, \infty) \to \mathbb{R}$ (radiality),

Curve and Surface Fitting: Saint-Malo 1999
Albert Cohen, Christophe Rabut, and Larry L. Schumaker (eds.), pp. 359–374.

ISBN 0-8265-1357-3.

and this is how the notion of a radial basis function came up. To assure that the interpolation in the points of $X = \{x_1, \dots, x_N\}$ is uniquely defined, the matrix

$$A_{\Phi,X} := (\Phi(x_j, x_k))_{1 \le j,k \le N} \tag{2}$$

must be nonsingular. By definition, positive definite functions Φ even make this matrix symmetric and positive definite, and the positive definite radial functions

$$\phi(r) = \exp(-r^2) \text{ on } \mathbb{R}^d \text{ for all } d \text{ (Gaussian)},$$
$$\phi(r) = (1-r)_+^4(1+4r) \text{ on } \mathbb{R}^d, d \le 3, \text{ see [25]},$$

are typical examples. See the review articles [4,15,18,16] for details. Though the above functions are scalar, the positive definiteness property of the second depends on the dimension d of the space containing x and y when forming the scalar argument $r = \|x - y\|_2$.

These two examples already show that the matrix $A_{\Phi,X}$ in (2) can be sparse or have a strong off-diagonal decay. It will be very large if we consider real-world problems with many data, arising e.g. from inverse engineering or terrain modelling. If the data points are rather densely scattered over the domain Ω, the approximation power of the space (1) will be very good, but the matrix in (2) will have lots of similar rows and columns, yielding a bad condition number. The connection between these phenomena is described in some detail in [17].

This paper concentrates on the numerical solution of large symmetric positive definite systems with matrices of the form (2). There are some additional goals:

1) $\mathcal{O}(N)$ complexity of solving the $N \times N$ system,
2) $\mathcal{O}(1)$ complexity of evaluating an element of (1) with N terms and
3) getting away with $n << N$ terms at a tolerable loss of accuracy, when interpolating N data.

We shall describe greedy algorithms as recently studied by deVore [5] and Temlyakov [24], but we shall omit multiple scales as proposed by Floater and Iske [9] and continued later in [13] and [6]. The techniques will be partially based on Krylov subspaces as recently and independently studied by Faul and Powell [8].

§2. Splitting the Native Space Energy

Dropping Φ in the notation, we can write functions from (1) in the form

$$s_{c,X} := \sum_{j=1}^{N} c_j \Phi(\cdot, x_j) \tag{3}$$

with $c \in \mathbb{R}^N$, $X = \{x_1, \dots, x_N\} \subseteq \Omega \subseteq \mathbb{R}^d$ and arbitrary N. Our main tool will be the natural inner-product

$$(s_{c,X}, s_{d,Y})_\Phi = \left(\sum_j c_j \Phi(\cdot, x_j), \sum_k d_k \Phi(\cdot, y_k) \right)_\Phi := \sum_j \sum_k c_j d_k \Phi(x_j, y_k) \tag{4}$$

introduced by a positive definite function Φ on the union of spaces (1) for arbitrary data sets X. We note in passing that one can form the corresponding Hilbert space closure to get the native Hilbert space of Φ, but we refer the reader to [11] and [19] for details. If we fix the set $X = \{x_1, \dots, x_N\}$ and the corresponding positive definite matrix $A_{\Phi,X}$, we get an inner-product

$$(c, d)_{A_{\Phi,X}} := c^T A_{\Phi,X} d \tag{5}$$

on $\mathbb{R}^N$ which is familiar from the theory of the conjugate gradient method based on Krylov subspaces.

Note that the inner-product (4) is zero if the function $s_{c,X}$ vanishes on Y or if $s_{d,Y}$ vanishes on X. If we assume $Y \subseteq X$ and make sure that $s_{d,Y}$ agrees with $s_{c,X}$ on Y (e.g. by interpolation on Y), then

$$\begin{aligned} (s_{d,Y}, s_{c,X} - s_{d,Y})_\Phi &= 0 \\ \|s_{d,Y}\|_\Phi^2 + \|s_{c,X} - s_{d,Y}\|_\Phi^2 &= \|s_{c,X}\|_\Phi^2. \end{aligned} \tag{6}$$

The second identity can be viewed as a split of the energy of the function $s_{c,X}$ into the energy of its interpolant $s_{d,Y}$ on a subset Y of X and the residual $s_{c,X} - s_{d,Y}$. We shall use this energy split over and over again.

§3. Interpolation in Native Spaces

At this point, we digress a little and study the interpolation of an *arbitrary* real-valued function f on a domain $\Omega \subseteq \mathbb{R}^d$. On each fixed finite subset $X \subset \Omega$ we can interpolate the values of f by a function $s_{f,X}$ from (1). Due to (6), the energy $\|s_{f,X}\|_\Phi^2$ is a monotonic function of X with respect to addition of points, and it can be easily evaluated using (4). The energy is bounded independent of f if and only if [22] the function f lies in the native space of Φ, and in this case we have $\|f\|_\Phi = \sup_X \|s_{f,X}\|_\Phi$.

This observation has some consequences for applications. If the user does not have any a-priori information on f, the proper choice of Φ is a problem. But if the behaviour of $\|s_{f,X}\|_\Phi^2$ with respect to X is monitored for larger and larger sets X, the user can switch to a less smooth Φ if the energy values grow dramatically with X. By (6) the behaviour of $\|s_{f,X}\|_\Phi^2$ is related to the energy $\|f - s_{f,X}\|_\Phi^2$ of the residual, and further study of this as a function of X is needed, especially for f with additional smoothness properties. Current starting points are in [20] and [10], and readers are encouraged to proceed from there.

§4. Iteration on Residuals

We now fix a positive definite function Φ and a function f_0 from the native space of Φ or, at least, from some space of the form (1) for a rather large finite set X. Our goal is to reconstruct f_0 by an iterative process. Note that this solves a large system with the matrix from (2), if we start from $f_0 := s_{f,X}$ interpolating some given function f on X. In this case, we do not reconstruct f but rather its interpolant. In both cases we need not worry about the existence of $\|f_0\|_\Phi$.

If we pick some numerically manageable finite set Y_0 and interpolate f_0 on Y_0, we can define $f_1 := f_0 - s_{f_0,Y_0}$ and proceed iteratively by

$$f_{j+1} := f_j - s_{f_j,Y_j}, \quad Y_j \subset \Omega, \quad j = 0, 1, \ldots \tag{7}$$

using finite sets Y_j that we shall have to deal with later. Anyway, the energy splitting (6) yields

$$\begin{aligned} \|s_{f_j,Y_j}\|_\Phi^2 + \|f_j - s_{f_j,Y_j}\|_\Phi^2 &= \|f_j\|_\Phi^2 \\ \|f_j - f_{j+1}\|_\Phi^2 + \|f_{j+1}\|_\Phi^2 &= \|f_j\|_\Phi^2, \end{aligned} \tag{8}$$

and by summation we get the telescoping sum

$$\sum_{j=0}^{k} \|s_{f_j,Y_j}\|_\Phi^2 = \sum_{j=0}^{k} \|f_j - f_{j+1}\|_\Phi^2 = \|f_0\|_\Phi^2 - \|f_{k+1}\|_\Phi^2 \le \|f_0\|_\Phi^2 \tag{9}$$

which must necessarily converge for $k \to \infty$ even if the choice of Y_j is bad, e.g. $Y_j = Y_0$ for all j. By standard Hilbert space argumentation via Cauchy sequences, the functions

$$g_{k+1} := \sum_{j=0}^{k} s_{f_j,Y_j} = \sum_{j=0}^{k} (f_j - f_{j+1}) = f_0 - f_{k+1}$$

must converge in the norm $\|.\|_\Phi$ to some element g in the native space, but we do not want to use this fact. Our goal is to prove that the residuals f_k converge to zero, and this would imply that the functions g_k converge to f_0, yielding the desired reconstruction.

Of course we shall need some additional assumptions on the sets Y_j to be successful. Equations (8) and (9) suggest that we should let s_{f_j,Y_j} take up as much energy from f_j as possible, and this will be our guideline for the convergence analysis in the following sections.

§5. Conditions for Linear Convergence

For simplicity, let us first assume that s_{f_j,Y_j} picks up at least a fixed percentage of the energy of f_j, i.e.

$$\|s_{f_j,Y_j}\|_\Phi^2 \ge \gamma \|f_j\|_\Phi^2 \tag{10}$$

with some fixed $\gamma \in (0, 1]$. This is a disguised hypothesis on the proper choice of Y_j, and we have to prove later how to satisfy this assumption. From (8) and (10) we conclude linear convergence of f_j to zero via $\|f_{j+1}\|_\Phi^2 \le (1-\gamma)\|f_j\|_\Phi^2$. This proves

Theorem 1. *If the choice of sets Y_j satisfies (10), the residual iteration (7) converges linearly in the native space norm, and there is an error bound*

$$\|f_0 - g_k\|_\Phi^2 = \|f_k\|_\Phi^2 \le (1-\gamma)^k \|f_0\|_\Phi^2.$$

Assumption (10) is not easy to handle, because the norm involves Φ and the value of the right-hand side is not explicitly known. But in case of $f_0 := s_{f,X}$ for some large finite set $X \subset \Omega \subseteq \mathbb{R}^d$ we can restrict ourselves to sets $Y_j \subseteq X$, and all functions f_j will stay in the finite-dimensional space (1). On this space, we can pick any norm $\|.\|_X$, for instance any discrete L_p norm of functions on X, and make use of the norm equivalence

$$c\|s\|_X \le \|s\|_\Phi \le C\|s\|_X \tag{11}$$

for all functions s from the space (1), where the constants satisfy $0 < c \le C$. Then we can try to get away with

$$\|s_{f_j,Y_j}\|_X^2 \ge \delta \|f_j\|_X^2 \tag{12}$$

with some $\delta \in (0,1]$ instead of (10). But since this equation implies (10) with $\gamma \ge \delta c^2/C^2 > 0$, we get

Theorem 2. *If the choice of sets Y_j satisfies (12) for some norm $\|.\|_X$ of functions on X, the residual iteration (7) for reconstruction of $f_0 := s_{f,X}$ converges linearly in $\|.\|_X$ and there is an error bound*

$$\|f - g_k\|_X \le \frac{C}{c}\left(1 - \delta\frac{c^2}{C^2}\right)^{k/2} \|f\|_X.$$

§6. Maximizing Energy of Interpolants

Our argument at the end of Section 4 leads to the problem of finding a finite set Y such that the energy $\|s_{f,Y}\|_\Phi^2$ of the interpolant of some function (or residual) f is large. If $f_Y \in \mathbb{R}^{|Y|}$ is the vector of values of f on Y, the interpolant $s_{f,Y}$ solves a system with a matrix $A_{\Phi,Y}$ defined as in (2), and the energy is given by the quadratic form

$$\|s_{f,Y}\|_\Phi^2 = f_Y^T {A_{\Phi,Y}}^{-1} f_Y \ge \|f_Y\|_2^2 \lambda_{min}\left({A_{\Phi,Y}}^{-1}\right) = \|f_Y\|_2^2 / \lambda_{max}\left(A_{\Phi,Y}\right)$$

as a function of f and Y. The maximal eigenvalue $\lambda_{max}\left(A_{\Phi,Y}\right)$ is hard to discuss in general (see Narcowich and Ward [14] for results), and we simply view this quantity as a factor that depends on the geometry of Y and the number $|Y|$ of points in Y. It is an interesting open problem to design some Remes-type algorithm based on exchanges of points to arrive at the best choice of a set Y with a prescribed number of points.

In the special case $|Y| = 1$, $Y = \{y\}$ things are easy. We get

$$\|s_{f,Y}\|_\Phi^2 = f(y)^2\Phi(y,y) \geq f(y)^2 \min_{z\in\Omega} \Phi(z,z)$$

and the maximum of f^2 will be the best choice, especially if Φ is translation-invariant or radial.

If we have $Y \subseteq X$ for a large finite set X, we can invoke Courant's minimum-maximum principle to get $\lambda_{max}(A_{\Phi,Y}) \leq \lambda_{max}(A_{\Phi,X})$ as an upper bound that does not depend on the choice of Y. A reasonable strategy for maximizing $\|s_{f,Y}\|_\Phi^2$ then is to pick the $|Y|$ points of X where f takes its largest absolute values.

In the general situation, we have to face the fact that coalescing points are not allowed. A reasonable strategy is to mimic the previous situation, i.e. to take some large set X of well-distributed points and pick the points of X where f is largest in absolute value.

Another possible strategy is the iterative greedy collection of more and more points, forming a recursive Cholesky factorization. Since this possibility does not seem to be familiar to researchers in this area, we outline the process here. Assume that an interpolant to f on Y is available together with the inverse B of $A_{\Phi,Y}$. We now want to add another point $z \in \Omega \setminus Y$ to Y, thus enlarging the energy of the interpolant. A naive choice of z is via the maximum of the absolute value of $f - s_{f,Y}$, but since we have

$$\|s_{f,Y\cup\{z\}}\|_\Phi^2 = \|s_{f,Y}\|_\Phi^2 + 2f(z)\sum_{y\in Y} f(y)\Phi(z,y) + f^2(z)\Phi(z,z), \tag{13}$$

the best choice of z for fixed Y is obtained by maximizing the right-hand side of this equation. Having found z, one has to update B in a suitable way. First, calculate the vector $v \in \mathbb{R}^{|Y|}$ with components $\Phi(z,y)$, $y \in Y$ and form $w := Bv$. The number

$$1/\alpha := \Phi(z,z) - v^T w = \Phi(z,z) - v^T Bv$$

can be shown to be positive, because Φ is positive definite and z does not belong to Y. Then form $u := -\alpha w$ and $C := B + u^T u/\alpha$. The matrix

$$\begin{pmatrix} C & u \\ u^T & \alpha \end{pmatrix}$$

then is the inverse of $A_{\Phi,Y\cup\{z\}}$ needed for the interpolation on $Y \cup \{z\}$. Unfortunately, there is no numerical experience in this direction so far, especially for the maximization of (13). A more careful calculation of the numerical complexity reveals that we have nothing else here than a special formulation of the partial Cholesky algorithm with pivoting. The choice of pivots, however, is adapted to the setup of the problem as an interpolation.

Altogether, this section was intended to motivate readers to look at the problem of finding good finite sets Y for improvement of interpolants.

§7. A Simple Greedy Algorithm

Among other things, the previous section showed how to work on subsets Y consisting of a single point y each. The best possible choice is to take the point where f takes its maximum absolute value, and the interpolant is $s_{f,\{y\}} = f(y)\Phi(y,\cdot)/\Phi(y,y)$. We now do this iteratively in the sense of (7) by picking $Y_j := \{y_j\}$ with $|f_j(y_j)| = \|f_j\|_\infty$. In the "discrete" case $f_0 := s_{f,X}$ we take the Chebyshev norm on X, while in the "continuous" case $f_0 := f \in C(\Omega)$ with Ω being a compact subset of $\mathbb{R}^d$, we take the Chebyshev norm on Ω.

Due to Theorem 2, the discrete case leads to linear convergence towards $f_0 := s_{f,X}$, because (12) is satisfied with $\delta = 1$. From a theoretical viewpoint, this is much better than the non-quantitative convergence result of Faul and Powell in [7]. On the other hand, there always is the conjugate gradient method as a competitor, and it has linear convergence, too. But it needs to form matrix-vector products, while our greedy algorithm does not even store the matrix. It simply needs two arrays of length $|X|$ for the residuals and the coefficients, and in each cycle it updates one cofficient and runs once over the residuals to update them and find the maximum for the next iteration. This extreme numerical simplicity must come at a price, and the price is very slow convergence after some good progress in the first few iterations. We report on numerical experiments and adaptive multiscale improvements in [23], but at this point we want to direct the reader's attention to extend the above strategy, e.g. via some suitable preconditioning.

Before we look more closely at the greedy algorithm in the discrete case, let us digress a little into the continuous case.

Theorem 3. *If Φ is a continuous translation-invariant positive definite function on a compact domain $\Omega \subset \mathbb{R}^d$, the greedy algorithm for interpolation of a function f from the native space of Φ converges uniformly.*

Proof: We have

$$\|s_{f_j,Y_j}\|_\Phi^2 = f_j^2(y_j)\phi(0) = \|f_j\|_\infty^2\phi(0)$$

and (9) shows that the quantities $\|f_j\|_\infty^2$ are summable. Consequently, the residuals f_j converge uniformly to zero on the compact set Ω. □

Corollary 4. *Under the assumptions on Φ as in Theorem 3, the native space norm is expressible via a series*

$$\|f\|_\Phi^2 = \phi(0)\sum_{j=0}^{\infty} f_j^2(y_j) = \phi(0)\sum_{j=0}^{\infty}\|f_j\|_\infty^2,$$

where

$$f_0 := f,\ |f_j(y_j)| = \|f_j\|_\infty,\ f_{j+1} := f_j - f_j(y_j)\phi(\cdot - y_j)/\phi(0).$$

This result may look complicated at first sight, but it should be compared to other definitions of the native space norm, e.g. via Fourier transforms, by abstract completion of a pre-Hilbert space, or by the supremum of the action of certain functionals.

We do not want to go into details here (see [11], for instance), but prefer to give an illustrative example. If specialized to Sobolev space $W_2^k(\Omega)$ with $k > d/2$, one has to take $\phi(x) = \|x\|_2^{k-d/2} K_{k-d/2}(\|x\|_2)$ in order to recover Sobolev space as a native space for $\Phi(x,y) = \phi(\|x-y\|_2)$. Now by Corollary 4 one gets the Sobolev norm of a function as a series containing just function values on the domain in a numerically accessible way, using neither derivatives nor integration (but, of course, maximization). It should be pointed out that this technique provides some means to assess the Sobolev smoothness of a given function numerically. Readers are encouraged to proceed from here.

§8. Dual Techniques

Another possible approach to solving a large $N \times N$ system with a large symmetric and positive definite coefficient matrix $A_{\Phi,X}$ via smaller finite subproblems is to define certain finite-dimensional subspaces S_j of the native space and to approximate the exact solution on $X = \{x_1, \ldots, x_N\}$ by approximation in the native space norm. More precisely, the iteration starts like in Section 4 with some function f_0 and $j := 0$, and iterates like (7) according to

$$\begin{aligned} \|f_j - s_j\|_\Phi &:= \inf_{s \in S_j} \|f_j - s\|_\Phi \\ f_{j+1} &:= f_j - s_j. \end{aligned} \tag{14}$$

By standard arguments, this iteration also satisfies (8) and the rest of Section 4, including the summability condition (9). Note that if a space $S_j = S_{Y_j,\Phi}$ has the form (1) for some finite set Y_j, then the best approximation solution s_j in (14) coincides with s_{f_j,Y_j} and we are back to the method in Section 4. This observation follows from Theorem 7 in Section 9.

But there are other possible choices for the spaces S_j. In particular, Faul and Powell [7] pick certain one-dimensional spaces $S_j = \text{span}\{u_j\}$ for all $j \geq 0$. Then $s_j := \alpha_j u_j$ with $\alpha_j := (u_j, f_j)_\Phi / (u_j, u_j)_\Phi$ solves the approximation problem, and we have the summability condition

$$\sum_{j=0}^{k} \|s_j\|_\Phi^2 = \sum_{j=0}^{k} \|u_j\|_\Phi^2 \alpha_j^2 = \sum_{j=0}^{k} \left(\frac{u_j}{\|u_j\|_\Phi}, f_j\right)_\Phi^2 = \|f_0\|_\Phi^2 - \|f_{k+1}\|_\Phi^2 \leq \|f_0\|_\Phi^2. \tag{15}$$

§9. Cyclic and Greedy Dual Strategies

In [7], Faul and Powell fix N such functions u_j by a certain precalculation that we shall discuss later. These functions are used periodically, i.e. u_j is used in step $j + kN$ for all $k \geq 0$. The periodic reuse has the advantage that one can precalculate and store the u_j, if their construction is somewhat involved. We start with a generalized and simplified version of the convergence result in [7]:

Theorem 5. *If f_0 is in the span of the functions u_j for $1 \le j \le N$, then the cyclic dual method of Faul and Powell converges to f_0.*

Proof: Since everything takes place in a finite-dimensional space, and since the technique involves an energy split, the functions $g_j = f_0 - f_j$ converge to some function g in the span of the u_j, and the f_j converge to $f_0 - g$. But as (15) implies

$$\lim_{k\to\infty} \left(\frac{u_j}{\|u_j\|_\Phi}, f_{j+kN}\right)_\Phi = 0 = \left(\frac{u_j}{\|u_j\|_\Phi}, f_0 - g\right)_\Phi$$

for all j, the functions f_0 and g must coincide. □

There are lots of choices of u_j that satisfy the hypothesis of Theorem 5. Conjugate directions and $u_j := \Phi(\cdot, x_j)$ would do the job. The latter strategy coincides with the greedy method, if the cyclic choice is given up in favour of picking the point where the residual is maximal in absolute value. A linear convergence result is possible, if such a modification is made in general:

Theorem 6. *If f_0 is in the span of the functions u_j for $1 \le j \le N$, then the iteration (14) with*

$$\begin{aligned} (f_j, u_{k_j})_\Phi^2 &:= \max_k (f_j, u_k)_\Phi^2 \\ S_j &:= \text{span}\,\{u_{k_j}\} \end{aligned} \tag{16}$$

converges linearly to f_0.

Proof: We can proceed as in Section 5, using

$$\|u\|_X^2 := \max_j (u, u_j)_\Phi^2$$

for all functions u in the span U of all u_j. The assumption (12) is satisfied for s_j instead of s_{f_j,Y_j} due to

$$\begin{aligned} s_j &= \frac{(f_j, u_{k_j})_\Phi}{(u_{k_j}, u_{k_j})_\Phi} u_{k_j} \\ \|s_j\|_X^2 &\ge (s_j, u_{k_j})_\Phi^2 = (f_j, u_{k_j})_\Phi^2 = \|f_j\|_X^2, \end{aligned}$$

and the rest follows easily. □

The inner-products in (16) can be evaluated explicitly, if we work in the space (1) and use (4) and (5) in the form

$$(s_{c,X}, s_{d,X})_\Phi = \sum_k c_k s_{d,X}(x_k). \tag{17}$$

This is particularly efficient if the functions u_j have only a small number of nonzero coefficients in their representation of the form $u_j = s_{c^j,X}$. Another possibility, exploiting the dual nature of the algorithm, is to store and update the inner-products $(f_j, u_k)_\Phi$ instead of the values $f_j(x_k)$. So far, there is no numerical experience with dual greedy algorithms, unfortunately.

One has a lot of leeway for picking suitable functions u_j, especially when preconditioning arguments come into play. Faul and Powell use local Lagrange functions u_j based on relatively small subsets Y_j of X that contain x_j. In particular, $u_j \in S_{Y_j,\Phi}$ is defined by the interpolation conditions

$$\begin{aligned} u_j(x_j) &= 1, \\ u_j(x_k) &= 0 \text{ for all } x_k \in Y_j,\ k \neq j, \end{aligned} \tag{18}$$

and is expressible in the form $u_j = s_{c^j,Y_j} = s_{d^j,X}$ with at most $|Y_j|$ nonzero coefficients. The precalculation involves the solution of N systems with $|Y_j| \times |Y_j|$ matrices A_{Φ,Y_j},and it can be kept at $\mathcal{O}(N)$, if the values $|Y_j|$ are bounded independent of N. Our arguments in Section 10 will show how this technique can be interpreted as preconditioning the matrix $A_{\Phi,X}$. For a fixed accuracy to be obtained, and for their special choice of the sets Y_j, Faul and Powell then observe that they need only a small fixed number of cycles of the dual algorithm. Each cycle has N one-dimensional subproblems, but there are techniques to keep each subproblem at a reasonable complexity, provided that techniques like multipole expansions [1] or compactly supported radial basis functions [26, 25] are used.

The selection of functions u_j is particularly good if there are orthogonality or conjugacy relations among them. Let us look at an inner-product $(u_j, u_k)_\Phi$ in case of (18), using (17) and $u_j = s_{c^j,X}$. We get

$$(u_j, u_k)_\Phi = \sum_{m: x_m \in Y_j} c^j_m u_k(x_m),$$

and this quantity vanishes if $Y_j \subseteq Y_k \setminus \{x_k\}$.

This can be seen as a motivation for choosing

$$x_j \in Y_j \subseteq \{x_j, x_{j+1}, \ldots, x_N\} \tag{19}$$

as done by Faul and Powell. Even if the functions u_j are in general not mutually orthogonal they are at least linear independent as needed for Theorems 5 and 6. To see this note that the matrix $C = (c^j_i)$ which describes the transition from the basis $(\Phi(\cdot, x_j), 1 \le j \le N)$ to $(u_j, 1 \le j \le N)$ is an upper triangle matrix and thus invertible if $c^i_i \neq 0$ for $1 \le i \le N$. This is indeed the case because of

$$0 \neq \|u^i_i\|^2_\Phi = \sum_{m: x_m \in Y_j} c^i_m u_i(x_m) = c^i_i.$$

We finish this section by pointing out how to make optimal use of solving N systems with $|Y_j| \times |Y_j|$ matrices A_{Φ,Y_j} for subsets Y_j in a precalculation. If the full inverses of the A_{Φ,Y_j} are stored instead of the coefficients of u_j, one can use the cyclic dual algorithm with $S_j := S_{Y_j,\Phi}$. The energy split at each step of the algorithm will then be better or equal to the split obtained by the dual cyclic algorithm using a single $u_j \in S_{Y_j,\Phi}$ like the one used by Faul and Powell. This is clear from (14), and the following theorem, which is well known since the advent of splines, shows that we end up with a cyclic interpolatory method of the form (7).

Theorem 7. *If Y is a finite subset of Ω, the approximation problem*

$$\inf_{s \in S_{Y,\Phi}} \|f - s\|_\Phi$$

for any f in the native space of Φ is solved by the interpolant $s_{f,Y}$.

Proof: Equations (6) generalize via continuous transition to the Hilbert space completion to

$$(s_{f,Y}, f - s_{f,Y})_\Phi = 0$$
$$\|s_{f,Y}\|_\Phi^2 + \|f - s_{f,Y}\|_\Phi^2 = \|f\|_\Phi^2$$

for all f in the native space, and the assertion follows. □

Consequently, algorithms using interpolants on finite subsets make optimal use of the information contained in the space $S_{Y,\Phi}$. This links the dual techniques back to the interpolatory methods in Section 4. Numerical results concerning the above cyclic interpolatory method, e.g. using the sets Y_j of Faul and Powell, are still missing. The progress must be better due to Theorem 7, but at the expense of much more storage. And, an incorporation of greedy selections using the good preconditioning power of the Faul-Powell approach seems worth investigating.

§10. Quasi-Interpolation

There is a hidden link between the Faul-Powell technique, preconditioning of $A_{\Phi,X}$, and certain quasi-interpolation methods using local Lagrange functions, as investigated by Beatson, Powell and their coworkers (see for example [2]). If we write the interpolant $s_{f,X}$ to some function f in Lagrange representation

$$s_{f,X} = \sum_{j=1}^{N} f(x_j) v_j \tag{20}$$

with N Lagrange basis functions $v_k \in S_{X,\Phi}$ satisfying $v_j(x_k) = \delta_{jk}$, we can relax (20) to a quasi-interpolation formula

$$s_{f,u,X} := \sum_{j=1}^{N} f(x_j) u_j \tag{21}$$

for any other choice of functions u_j that approximate the global Lagrange basis functions v_j. The choice (18) for certain subsets Y_j is quite natural, because one can often [3] observe that local Lagrange functions based on a set Y_j of neighbouring points to $x_j \in Y_j$ decay quickly away from x_j. Assuming (18) (but not (19)) from now on, the representation (21) can be rewritten in terms of $u_j = s_{c^j,X}$ and (3) as

$$s_{f,u,X} = \sum_{j=1}^{N} f(x_j) \sum_{k:x_k \in Y_j} c_k^j \Phi(\cdot, x_k) = \sum_{k=1}^{N} \Phi(\cdot, x_k) \sum_{j:x_k \in Y_j} f(x_j) c_k^j.$$

The coefficients of the second representation can be evaluated locally, and the computational advantage is particularly evident in case of compactly supported radial basis functions.

We now want to look at the quality of such quasi-interpolants on the discrete set X itself. The operator that maps the vector

$$f_{|X} := (f(x_1), \ldots, f(x_N))^T \in \mathbb{R}^N$$

to $s_{f,u,X}{}_{|X} \in \mathbb{R}^N$ can be written as the matrix product $A_{\Phi,X} \cdot C$, where $C = (c_k^j)$ is the nonsymmetric $N \times N$ matrix with row index k and column index j containing the coefficients c_k^j of the u_j columnwise. The operator that generates the residuals on X then is $E_N - A_{\Phi,X} \cdot C$ with the $N \times N$ identity matrix E_N. In case of $Y_j = X$ for all j we have $C = A_{\Phi,X}{}^{-1}$, and there are good reasons to expect that there are numerically interesting cases where some matrix norm of $E_N - A_{\Phi,X} \cdot C$ is smaller than one. In such cases one can solve the problem on X by successive quasi-interpolation via a Neumann series. In terms of vectors f^j and s^j containing the values of residuals f_j and quasi-interpolants to f_j on X, we have the linearly convergent iteration

$$\begin{aligned} f^0 &:= f_{0|X} \\ s^j &:= A_{\Phi,X} \cdot C f^j \\ f^{j+1} &:= f^j - s^j = (E_N - A_{\Phi,X} \cdot C)^{j+1} f^0 \end{aligned}$$

calculating the interpolant to the data of f_0 on X as the sum over the s^j. Note that we cannot use the energy split here, because we have left the context of interpolation and approximation. Note further that C acts as a (nonsymmetric!) preconditioner or an approximate inverse to $A_{\Phi,X}$.

§11. Experiments Concerning Quasi-Interpolation

To calculate the norm of $E_N - A_{\Phi,X} \cdot C$ numerically, we observe that the matrix $A_{\Phi,X} \cdot C$ has the entries $u^j(x_i)$, where i is the row index. Thus the entry at (i,j) of $E_N - A_{\Phi,X} \cdot C$ vanishes for $x_i \in Y_j$, and the column-sum norm of $E_N - A_{\Phi,X} \cdot C$ can be written as

$$\max_j \sum_{i:x_i \notin Y_j} |u^j(x_i)|. \tag{22}$$

Again it turns out that the decay of local Lagrange functions is essential.

In case of data on the uniform grid $(h\mathbb{Z})^2$, a radial basis function ϕ_c with support in $[0,c]$, and sets $Y_j := \{y \in (h\mathbb{Z})^2 \ : \ \|x_j - y\|_2 \le R\}$ of neighbours to x_j within a radius R, the norm in (22) can be evaluated by looking at the local Lagrange function u_0 with respect to the origin and the set $Y_0 := \{y \in (h\mathbb{Z})^2 \ : \ \|y\|_2 \le R\}$ of local interpolation points. Since both Y_0 and the support of ϕ are bounded, the function u_0 is zero on integer grid

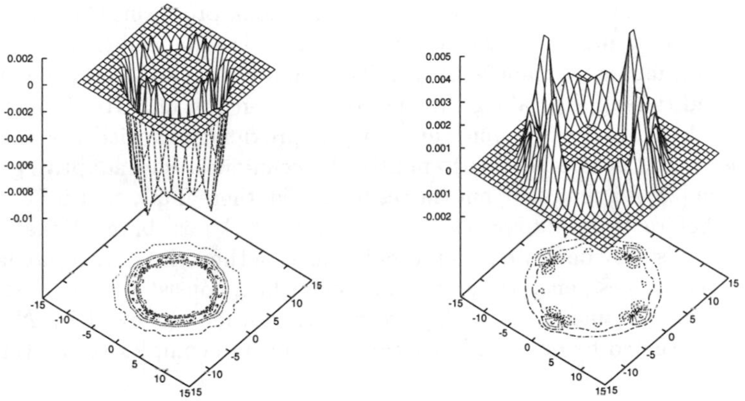

Fig. 1. $C = 4, R = 6$, norm $= 0.48$, and $C = 5, R = 8$, norm $=0.29$.

N	5	9	13	21	25	29	37	45	49	57	61
$M_{0.9}$	5	5	5	9	29	49	69	81	97	145	145
$M_{0.1}$	5	5	21	29	109	137	149				

Tab. 1. Point numbers M_p required for norm $\leq p$ and N points in support of ϕ.

points outside the disk around zero with radius $R + c$. Omitting the value 1 at the origin for scaling reasons, Figure 1 shows the behaviour of u_0 on integer gridpoints around the origin. We picked two cases for the C^2 function $\phi_c(r) := (1 - r/c)_+^4(1 + 4r/c)$ from [25] where the norm of $E_N - A_{\Phi,X} \cdot c$ is smaller than one, and the corresponding numbers of local interpolation points in Y_0 are 113 and 197, respectively.

For applications, it is necessary to know how large R must be for fixed c and h in order to make the norm of $E_N - A_{\Phi,X} \cdot c$ smaller than 0.9 or 0.1, say. Since R and c scale with h, the numbers M and N of points in Y_0 and the support of ϕ depend on R/h and C/h, respectively. Given a support radius c and a maximal meshwidth h such that the support of ϕ_c contains $N = 1, 5, 9, 13, \ldots$ points, we provide in Table 1 the minimal number M_p of points in Y_0 that are necessary to keep the norm of $E_N - A_{\Phi,X} \cdot c$ below p. Another way of reading Table 1 is that if the matrix $A_{\Phi,X}$ for interpolation by $\Phi(x, y) := \phi_c(\|x - y\|_2)$ on a regular grid has bandwidth N, then it has an approximate inverse with bandwidth M_p that leads to a residual matrix of norm p. The quasi-interpolant is to be calculated via subproblems with $M_p \times M_p$ matrices. It is an interesting challenge to provide sparse approximate inverses for sparse symmetric positive definite matrices, because normally the exact inverses will not be sparse.

§12. Conclusions

At first sight, our results on linear convergence look promising, but they still are too weak to provide a convergence rate that is independent of N, since no preconditioning techniques are involved. Improvements should thus focus on preconditioning, e.g. along the lines of Faul and Powell. Greedy methods for fixed Φ are limited to quick-and-dirty approximations with few nonzero coefficients and need extension to multiscale techniques. The adaptive greedy method in [23] is a first step, but the results shown there imply that it has to be stopped before it runs into scales that are too small. A possible continuation at small scales is provided by quasi-interpolation as outlined here. A combination of both techniques generates approximations which consist first of $K << N$ global terms obtained by an adaptive greedy method, followed by N local terms constructed by quasi-interpolation. The overall complexity can thus be kept at $\mathcal{O}(N)$.

Acknowledgments. The authors are grateful to Fabien Hinault for detecting an error in the first version of the paper.

References

1. Beatson, R. K. and G. N. Newsam, Fast evaluation of radial basis functions. I, Comput. Math. Appl. **24** (1992), 7–19.
2. Beatson, R. K., G. Goodsell, and M. J. D. Powell, On multigrid techniques for thin plate spline interpolation in two dimensions, in *Lectures in Applied Mathematics*, Vol. 32, 77–97.
3. Buhmann, M. D., Multivariate cardinal interpolation with radial-basis functions, Constr. Approx. **6** (1990), 225–255.
4. Buhmann, M. D., New developments in the theory of radial basis function interpolation, in *Multivariate Approximation: From CAGD to Wavelets*, K. Jetter and F. I. Utreras (eds.), World Scientific, Singapore, 1993, 35–75.
5. DeVore, R. A. and V. N. Temlyakov, Some remarks on greedy algorithms, Advances in Comp. Math. **5** (1996), 173–187.
6. Fasshauer, G. and J. Jerome, Multistep approximation algorithms: improved convergence rates through postconditioning with smoothing kernels, Advances in Comp. Math. **10** (1999), 1–27.
7. Faul, A. C. and M. J. D. Powell, Proof of convergence of an iterative technique for thin plate spline interpolation in two dimensions, Advances in Comp. Math. **11** (1999), 183–192.
8. Faul, A. C. and M. J. D. Powell, Krylov subspace methods for radial basis function interpolation, preprint.
9. Floater, M.S. and A. Iske: Multistep scattered data interpolation using compactly supported radial basis functions, J. Comput. Appl. Math. **73** (1996), 65–78.

10. Johnson, M., Overcoming the boundary effects in suface spline interpolation, preprint.

11. Luh, L.-T., Characterizations of native spaces, PhD. Dissertation, Göttingen, 1998.

12. Mairhuber, J. C., On Haar's theorem concerning Chebychev approximation problems having unique solutions, Proc. Amer. Math. Soc. **7** (1956), 609–615.

13. Narcowich F. J., R. Schaback, and J. D. Ward, Multilevel interpolation and approximation, Applied and Computational Harmonic Analysis, to appear.

14. Narcowich, F. J., N. Sivakumar, and J. D. Ward, On condition numbers associated with radial-function interpolation, J. Math. Anal. Appl. **186** (1994), 457–485.

15. Powell, M. J. D., The theory of radial basis function approximation in 1990, in *Advances in Numerical Analysis Vol. 2,* W. Light (ed.), Clarendon Press, Oxford, 1992, 105–210.

16. Schaback, R., Multivariate interpolation and approximation by translates of a basis function, in *Approximation Theory VIII, Vol. 1: Approximation and Interpolation*, Charles K. Chui and Larry L. Schumaker (eds.), World Scientific Publishing Co., Inc., Singapore, 1995, 491–514.

17. Schaback, R., Error estimates and condition numbers for radial basis function interpolation, Advances in Comp. Math. **3** (1995), 251–264.

18. Schaback, R., Creating surfaces from scattered data using radial basis functions, *Surface Fitting and Multiresolution Methods*, A. Le Méhauté, C. Rabut, and L. L. Schumaker (eds.), Vanderbilt University Press, Nashville, 1997, 477–496.

19. Schaback, R., Native spaces for radial basis functions I, in *New Developments in Approximation Theory*, M. W. Müller, M. D. Buhmann, D. H. Mache, and M. Felten (eds.), Birkhäuser, Basel, 1999, 255–282.

20. Schaback, R., Improved error bounds for scattered data interpolation by radial basis functions, Math. Comp. **68** (1999), 201–216.

21. Schaback, R. and H. Wendland, Using compactly supported radial basis functions to solve partial differential equations, in *Boundary Element Technology XIII*, C. Chen, C. A. Brebbia, and D. W. Pepper (eds.), Wit-Press, Southampton, Boston, 1999, 311–324

22. Schaback, R. and H. Wendland, Inverse and saturation theorems for radial basis function interpolation, preprint.

23. Schaback, R. and H. Wendland, Adaptive greedy techniques for approximate solution of large RBF systems, Numerical Algorithms, to appear.

24. Temlyakov, V. N. The best m-term approximation and greedy algorithms, Advances in Comp. Math. **8** (1998), 249–265.

25. Wendland, H., Piecewise polynomial, positive definite and compactly supported radial functions of minimal degree, Advances in Comp. Math. **4** (1995), 389–396.

26. Wu, Z., Multivariate compactly supported positive definite radial functions, Advances in Comp. Math. **4**, (1995), 283–292.

Robert Schaback & Holger Wendland
Institut für Numerische und Angewandte Mathematik
Universität Göttingen, Lotzestraße 16-18
D-37083 Göttingen, Germany
`schaback@math.uni-goettingen.de`
`wendland@math.uni-goettingen.de`

Parametric Polynomial Curves of Local Approximation of Order 8

K. Scherer

Abstract. Parametric approximation of curves offers the possibility of increasing the order of approximation by using the additional parameters in the parametrization of the curve. This has been studied in several papers, see e.g.[1-8]. The resulting problems are highly nonlinear. Here the cases of approximation order $\mathcal{O}(h^8)$ are studied which need piecewise quartic curves in the plane and piecewise quintic curves in space.

§1. Introduction

A general conjecture concerning the local approximation order by polynomial curves can be formulated as follows (see e.g. Mørken-Scherer [6]):

- For sufficiently small $h > 0$ and a sufficiently smooth curve $\underline{f}(t) : t \in [0, h] \to \underline{f}(t) \in \mathbf{R}^d$, there exists a polynomial curve $\underline{p}(t)$ of degree n and a reparametrization φ of $\underline{f}$ on $[a, b]$ such that

$$\sup_{0 \le t \le h} \|(\underline{f} \circ \varphi)(t) - \underline{p}(t)\| \le C(\underline{f}) h^m ,$$

where $m := n + 1 + [\frac{n-1}{d-1}]$.

The increased order m is explained in [6] by the principle of degree reducing. It comes from the idea of approximating with an interpolating polynomial curve $\underline{p}(t)$ of degree $m - 1$ such that

$$\underline{p}(t_i) = (\underline{f} \circ \phi)(t_i), \qquad 1 \le i \le m, \tag{1}$$

for points t_i in $[0, h]$ (multiplicities allowed). The additional parameters occuring via the reparametrization ϕ are used to reduce the degree $m - 1$ of $\underline{p}(t)$ by requiring

$$[t_1, \ldots, t_{m-i}](\underline{p}) = [t_1, \ldots, t_{m-i}](\underline{f} \circ \phi) = 0, \quad i = 0, \ldots, k - 1, \tag{2}$$

Curve and Surface Fitting: Saint-Malo 1999
Albert Cohen, Christophe Rabut, and Larry L. Schumaker (eds.), pp. 375–384.

ISBN 0-8265-1357-3.

with k as large as possible. Since we have to normalize ϕ such that $\phi(0) = 0, \phi(h) = h$, there are $m-2$ parameters left at our disposal for this goal. Thus $m-1$ can be reduced to $n = m-1-k$, where $k \cdot d \leq m-2$ or $k \leq (n-1)/(d-1)$.

From classical approximation theory it is then clear that a solution of (1),(2) yields a polynomial curve $\underline{p}(t)$ of degree $m-1-k$ satisfying

$$\sup_{0 \leq t \leq h} \|(\underline{f} \circ \phi)(t) - \underline{p}(t)\| \leq h^m \sup_{0 \leq t \leq h} \|D^m(\underline{f} \circ \phi)(t)\|. \quad (3)$$

Thus conjecture (•) is true if one can guarantee in addition that the parameters $\phi^{(2)}(0), \dots, \phi^{(m)}(0)$ of a solution of (2) remain bounded for $h \to 0$. This question of stability is also discussed in [6]. Note also that in this case equations (1) can be written as

$$\underline{p} \circ \phi^{-1}(s_i) = \underline{f}(s_i), \quad 1 \leq i \leq m,$$

where the nodes s_i are defined by $s_i := \phi(t_i)$.

The most interesting case of the conjecture is when $(n-1)/(d-1)$ is an integer k, i.e. $m = n + 1 + k = kd + 2$. In this case there are kd equations in (2), and the degree $m - 1$ is reduced by k. Then ϕ is determined by

$$[t_1, \dots, t_{n+1+i}](\underline{f} \circ \phi) = 0, \quad i = 1, \dots, k, \quad (4)$$

and $\underline{p}(t)$ by the first $n+1$ equations in (1).

So far the conjecture seems to be proved only in the case $n = d$ or $k = 1$ (see [5–6]) and for $k = 2 = d$ (see [1,3]). Here we treat the next most difficult cases $k = 3, d = 2$ and $k = 2, d = 3$, which amount to six equations in (4), and will lead to quartic curves in the plane and quintic curves in space with approximation rates of order 8, respectively.

§2. Reduction to 2 × 2 Systems

In Morken -Scherer [6] equations (4) were studied in particular for the Taylor case

$$D^{n+i}(\underline{f} \circ \phi) = 0, \quad i = 1, \dots, k.$$

The crucial point is then the formula of Faa di Bruno in the form given by T. Goodman. It reads (cf. [4])

$$D^l(\underline{f} \circ \phi)(0) = \sum_{j=1}^{l} a_{l,j} \underline{f}^{(j)}(0), \qquad \beta_l := \phi^{(l)}(0), \quad (5)$$

where

$$a_{l,j} = \sum_{l_1 + \dots l_j = l, l \geq 1} \begin{bmatrix} l \\ l_1 \cdots l_j \end{bmatrix} \beta_{l_1} \cdots \beta_{l_j}$$

and

$$\begin{bmatrix} l \\ l_1 \cdots l_j \end{bmatrix} = \frac{l!}{l_1! \cdots l_j! m_1! \cdots m_r!}.$$

Here the integer r denotes the number of distinct integers among the $l_1, \ldots, l_j$, where $m_1, \ldots, m_r$ are the multiplicities of them. Specific examples are

$$a_{l,1} = \beta_l, \quad a_{l,2} = \sum_{i=1}^{[l/2]} \binom{l}{i} \beta_i \beta_{l-i}, \qquad a_{l,l} = \beta_1^l.$$

At first we consider the planar case $d = 2$ for $k = 3$. Then the equations (4) specialize to

$$D^j(\underline{f} \circ \phi)(0) = 0, \quad j = 5, 6, 7. \tag{6}$$

Under the normalization $\phi(0) = 0$, $\beta_1 := \phi'(0) = 1$, this yields six equations for the unknowns $\beta_2, \ldots, \beta_7$. In [8] this system has been reduced to a 2×2 system for β_2, β_3. The basic idea was to simplify system (6) by determining a preliminary reparametrization ψ with $\psi(0) = 0, \psi'(0) = 1$ such that

$$D^{2i+1}(\underline{f} \circ \psi)(0) = 0, \quad i = 1, 2, 3.$$

It can easily be shown that this system is uniquely solvable in the unknowns $\gamma_2, \ldots, \gamma_7$, where $\gamma_j := \psi^{(j)}(0)$ provided

$$\operatorname{span}(\underline{f}'(0), \underline{f}''(0)) = \mathbb{R}^2. \tag{7}$$

Thus one can assume that $D^3\underline{f}(0) = D^5\underline{f}(0) = D^7\underline{f}(0)$ in (6). Denoting $d_{i,j}$ as the cross product of $f^{(j)}$ and $f^{(i)}$ in $\mathbb{R}^2$, i.e.

$$d_{i,j} := \underline{f}^{(i)} \times \underline{f}^{(j)}.$$

Further straightforward computation (cf. [9]) leads to

Lemma 1. *Under the assumption (7) and the normalization $\phi(0) = 0$, $\beta_1 := \phi'(0) = 1$, the vector $\beta_2, \ldots, \beta_7$ is a solution of the equations (7) iff β_2, β_3 is a solution of the 2×2 system*

$$\begin{aligned} 0 = d_{6,1} &+ 20 d_{4,1}\beta_3 + 10 d_{2,1}\beta_3^2 \\ &+ 60\beta_2 d_{4,2} + 15(d_{4,1} - 2d_{2,1}\beta_3)\beta_2^2, \end{aligned} \tag{8}$$

and

$$\begin{aligned} 0 = 15\beta_2^3 d_{4,1} &+ 75\beta_2^2 d_{4,2} + (3d_{6,1} - 10d_{4,1}^2/d_{2,1})\beta_2 + d_{6,2} \\ &+ 10(2d_{4,2} + \beta_2 d_{4,1})\beta_3 - 10 d_{2,1}\beta_2\beta_3^2. \end{aligned} \tag{9}$$

The elimination of the variable β_3 in this system along the lines of the resultant method yields an equation of degree 9 in β_2. However, it was overlooked in [8] that the constant term in this equation vanishes, so that it is in essence of degree 8 . Thus, existence of a (real) solution of (8)–(9) cannot be derived in this way. We will close this gap in the next section.

In the **space case** $d = 3$ we have to find a reparametrization ϕ such that

$$D^j(\underline{f} \circ \phi)(0) = 0, \quad j = 6, 7. \tag{10}$$

We simplify this again by determining a reparametrization ψ with $\psi(0) = 0$, $\psi'(0) = 1$, $\psi^{(j)}(0) := \gamma_j$ and

$$D^j(\underline{f} \circ \psi)(0) = 0, \quad j = 4, 7.$$

This is possible since by (5), in the three equations forming $D^4(\underline{f} \circ \psi)(0) = 0$ the coefficients of $\underline{f}', \underline{f}''$ and $\underline{f}'''$ are linear in the unknowns $\gamma_2, \gamma_3, \gamma_4$, and in the latter three are linear with respect to the γ_5, γ_6, γ_7. Therefore, these equations are uniquely solvable under the assumption

$$\operatorname{span}(\underline{f}'(0), \underline{f}''(0), \underline{f}'''(0)) = \mathbb{R}^3, \tag{11}$$

and we can consider (10) without loss under the assumption $D^4(\underline{f})(0) = D^7(\underline{f})(0) = 0$. This yields (with $\beta_1 = 1$) the equations

$$\begin{aligned} 0 = \beta_6 \underline{f}' &+ (6\beta_5 + 15\beta_2\beta_4 + 10\beta_3^2)\underline{f}'' \\ &+ (15\beta_4 + 60\beta_2\beta_3 + 15\beta_2^3)\underline{f}''' + 15\beta_2 \underline{f}^{(5)} + \underline{f}^{(6)}, \end{aligned} \tag{12}$$

$$\begin{aligned} 0 = \beta_7 \underline{f}' &+ (7\beta_6 + 21\beta_2\beta_5 + 35\beta_3\beta_4)\underline{f}'' + (105\beta_2^2 + 35\beta_3)\underline{f}^{(5)} \\ &+ (21\beta_5 + 105\beta_2\beta_4 + 70\beta_3^2 + 105\beta_2^2\beta_3)\underline{f}''' + 21\beta_2 \underline{f}^{(6)}. \end{aligned} \tag{13}$$

The next step is to take in (12) the scalar product with cross products $\underline{f}' \times \underline{f}''$, $\underline{f}' \times \underline{f}'''$ and $\underline{f}'' \times \underline{f}'''$, respectively. We obtain the equivalent equations

$$0 = (15\beta_4 + 60\beta_2\beta_3 + 15\beta_2^3)d_{1,2,3} + 15\beta_2 d_{5,1,2} + d_{6,1,2} \tag{14}$$

$$0 = (6\beta_5 + 15\beta_2\beta_4 + 10\beta_3^2)d_{2,1,3} + 15\beta_2 d_{5,1,3} + d_{6,1,3} \tag{15}$$

$$0 = \beta_6 d_{1,2,3} + 15\beta_2 d_{5,2,3} + d_{6,2,3}, \tag{16}$$

where

$$(\underline{f}^{(i_1)} \times \underline{f}^{(i_2)}, \underline{f}^{(j)}) := \det(\underline{f}^{(i_1)}, \underline{f}^{(i_2)}, \underline{f}^{(j)}) := d_{i_1,i_2,j}.$$

These equations serve for eliminating the variables β_4, β_5 and β_6 since they appear linearly. Before doing this, we transform (13) into three equivalent scalar equations analogously to (12). We obtain the three equations

$$0 = \beta_7 d_{1,2,3} + (105\beta_2^2 + 35\beta_3)d_{5,2,3} + 21\beta_2 d_{6,2,3}, \tag{17}$$

$$0 = (\beta_6 + 3\beta_2\beta_5 + 5\beta_3\beta_4)d_{2,1,3} + (15\beta_2^2 + 5\beta_3)d_{5,1,3} + 3\beta_2 d_{6,1,3}, \tag{18}$$

$$0 = (3\beta_5 + 15\beta_2\beta_4 + 10\beta_3^2 + 15\beta_2^2\beta_3)d_{3,1,2} + (15\beta_2^2 + 5\beta_3)d_{5,1,2} + 3\beta_2 d_{6,1,2}. \tag{19}$$

Equation (17) determines β_7 directly in terms of β_2, β_3. Now we eliminate β_5, β_6 in (18) via (15),(16). This gives at first

$$0 = -(3\beta_2\beta_5 + 5\beta_3\beta_4)d_{2,1,3} + (15\beta_2^2 + 5\beta_3)d_{5,1,3} + 3\beta_2(d_{6,1,3} + 5d_{5,2,3}) + d_{6,2,3}$$

and then

$$\begin{aligned} 0 = &\frac{15}{2}\beta_2^2 d_{1,2,3} + 5\beta_3 d_{5,1,3} + \frac{5}{2}\beta_2(d_{6,1,3} + 6d_{5,2,3}) + d_{6,2,3} \\ &+ (\frac{15}{2}\beta_2^2\beta_4 + 5\beta_3^2\beta_2)d_{1,2,3} - 5\beta_3\beta_4 d_{1,2,3}. \end{aligned}$$

Now we eliminate the variable β_4 by (14). The result is the equation

$$0 = \tilde{q}_0(\beta_2) + \tilde{q}_1(\beta_2)\beta_3 + \tilde{q}_2(\beta_2)\beta_3^2, \tag{20}$$

where

$$\begin{aligned} \tilde{q}_0(\beta_2) :=\ &(15/2)d_{1,2,3}\beta_2^5 - (15/2)d_{1,2,5}\beta_2^3 + [(15/2)d_{1,3,5} - (1/2)d_{1,2,6}]\beta_2^2 \\ &+ (15d_{2,3,5} - (5/2)d_{1,3,6})\beta_2 + d_{2,3,6}, \\ \tilde{q}_1(\beta_2) :=\ &(1/3)d_{1,2,6} + 5d_{1,3,5} + 5d_{1,2,5}\beta_2 - 25d_{1,2,3}\beta_2^3, \\ \tilde{q}_2(\beta_2) :=\ &-25d_{1,2,3}\beta_2. \end{aligned}$$

Analogously we reduce equation (19) to an equation in β_2, β_3 by eliminating β_5 and then β_4. Using (15), we obtain

$$\begin{aligned} 0 = &(5\beta_3^2 + 15\beta_2^2\beta_3)d_{1,2,3} + (15\beta_2^2 + 5\beta_3)d_{5,1,2} \\ &+ 3\beta_2(d_{6,1,2} + \frac{5}{2}d_{5,1,3}) + \frac{d_{6,1,3}}{2} + \frac{15}{2}\beta_2\beta_4 d_{1,2,3}, \end{aligned}$$

and then by (14)

$$\begin{aligned} 0 = &d_{1,2,3}(-15\beta_2^4 + 30\beta_3\beta_2^2 + 10\beta_3^2) + d_{5,1,2}(15\beta_2^2 + 10\beta_3) \\ &+ \beta_2(15d_{5,1,3} + 5d_{6,1,2}) + d_{6,1,3}. \end{aligned} \tag{21}$$

In order to get rid of the term with β_2^4 in (21), we make a final substitution

$$\beta_3 := \bar{\beta}_3 + \alpha\beta_2^2, \qquad \alpha := (3/2) + \sqrt{15}/2.$$

Then

$$-15\beta_2^4 - 30\beta_3\beta_2^2 + 10\beta_3^2 = [-30\beta_2^2 + 20\alpha\beta_2^2]\bar{\beta}_3 + 10\bar{\beta}_3{}^2,$$

and (21) simplifies to

$$0 = p_0(\bar{\beta}_3) + p_1(\bar{\beta}_3)\beta_2 + p_2(\bar{\beta}_3)\beta_2^2, \tag{22}$$

where

$$p_0(\bar{\beta_3}) := 10d_{1,2,3}\bar{\beta_3}^2 + 10d_{1,2,5}\bar{\beta_3} + d_{1,3,6},$$
$$p_1(\bar{\beta_3}) := 5d_{1,2,6} + 15d_{1,3,5}, \tag{23}$$
$$p_2(\bar{\beta_3}) := 10\sqrt{15}d_{1,2,3}\bar{\beta_3} + (15+10\alpha)d_{1,2,5}.$$

With the new variable $\bar{\beta_3}$, (20) transforms into

$$0 = q_0(\beta_2) + q_1(\beta_2)\bar{\beta_3} + q_2(\beta_2)\bar{\beta_3}^2, \tag{24}$$

with

$$q_0(\beta_2) := \tilde{q}_0(\beta_2) - 25\alpha^2 d_{1,2,3}\beta_2^5 + \alpha\beta_2^2\tilde{q}_1(\beta_2),$$
$$q_1(\beta_2) := \tilde{q}_1(\beta_2) - 50d_{1,2,3}\alpha\beta_2^3,$$
$$q_2(\beta_2) := \tilde{q}_2(\beta_2) = -25d_{1,2,3}\beta_2,$$

and the $\tilde{q}_i(\beta_2)$ defined as above. We summarize all this in

Lemma 2. *Under the assumption $d_{1,2,3} \neq 0$, i.e. assumption (11), and the normalization $\phi(0) = 0, \beta_1 := \phi'(0) = 1$, the vector $\beta_2, ..., \beta_7$ is a solution of the equations (22), (24) iff β_2, β_3 is a solution of the 2×2 system (22), (24).*

Remark: The systems (8)–(9) in the planar case, and (22)–(24) in the space case possess a similar structure. In (8) and (22) the coefficients of β_2 are polynomials of the same degree in β_3 and β_2, respectively. The same is true for (9) and (24), except that the corresponding polymials have different degrees.

§3. Existence Theorems

In view of the last remark, we treat in detail only the planar case.

Theorem 1. *The system (8)–(9) has at least one and at most 5 (real) solution pairs β_2, β_3 outside the line $\beta_3 = d_{4,1}/2d_{2,1}$.*

Proof: Let us write for shortness $x := \beta_2$ and $y := \beta_3$ as well as

$$A(y) := p_0(y), \quad 2B := p_1(y) = 60d_{4,2}, \quad C(y) := p_2(y) = 15d_{4,1} - 30d_{2,1}y.$$

Then (8) reads $0 = A(y) + 2Bx + C(y)x^2$. Formal solution for x gives

$$x = \varphi_\pm(y) := \frac{-B \pm \sqrt{B^2 - A(y)C(y)}}{C(y)} := \frac{-B \pm \sqrt{R(y)}}{C(y)}, \tag{25}$$

with the cubic polynomial

$$R(y) = 15[20d_{2,1}^2 y^3 + 30d_{4,1}d_{2,1}y^2 + (2d_{6,1}d_{4,1} - 20d_{4,1}^2)y + 240d_{4,2}^2 - d_{4,1}d_{6,1}].$$

Then write (9) as $0 = \sum_{i=0}^3 a_i x^i + b_0 y + b_1 xy + b_2 xy^2$, and insert (25), since by assumption $C(y) \neq 0$. After multiplication with $C(y)^3$, we obtain

$$0 = \sum_{i=0}^{3} a_i(-B \pm \sqrt{R(y)})^i C(y)^{3-i} + b_0 y C(y)^3$$
$$+ b_1(-B \pm \sqrt{R(y)})C(y)^2 + b_2 y^2(-B \pm \sqrt{R(y)})C(y)^2. \tag{26}$$

Now observe that

$$(-B \pm \sqrt{R(y)})^2 = B^2 + R(y) \pm 2B\sqrt{R(y)},$$
$$(-B \pm \sqrt{R(y)})^3 = -B^3 \pm 3B^2\sqrt{R(y)} - 3BR(y) \pm R(y)\sqrt{R(y)},$$

and sort all terms with and without $\sqrt{R(y)}$, respectively. Then (26) can be written as

$$U(y) = \pm V(y)\sqrt{R(y)}, \tag{27}$$

where $U(y)$ is a polynomial with leading term $-11 \cdot 15 \cdot \cdot 9000\, d_{4,2}\, d_{2,1}^3\, y^4$ and $V(y)$ also of degree 4 with leading term $9000\, d_{2,1}^3\, y^4$.

Hence under the above assumption, β_2, β_3 is a solution of the system (8)–(9) iff $y = \beta_3$ is a solution of (27) with sign either + or - on the right hand side. Suppose that n_1 and n_2 are the numbers of solutions of these two equations (including multiplicities). Then the squared equation

$$U^2(y) = V^2(y) \cdot R(y)$$

is of degree 11, and has $2(n_1 + n_2)$ solutions. Hence we conclude $n_1 + n_2 \leq 5$.

To prove existence, write (9) as

$$H(\beta_2, \beta_3) = 0, \tag{28}$$

where H is of degree 3 in β_2 and with $-10\beta_2\beta_3^2 d_{2,1}$ as leading term in β_3. Then introduce y^* as the largest zero of $R(y)$, so that in view of $R(+\infty) = +\infty$

$$R(y^*) = 0, \qquad R(y) > 0 \quad \text{for} y^* < y < \infty.$$

Now insert <u>both</u> functions $\varphi_{\pm}(y)$ in (28), and obtain the functions

$$H_{\pm}(y) := H(\varphi_{\pm}(y), y).$$

In order to guarantee existence of a solution of (8)–(9), it suffices therefore to show that the ranges of $H_{\pm}$ satisfy

$$H_+[y^*, \infty) \cup H_-[y^*, \infty) = \mathbb{R}. \tag{29}$$

For this, observe at first the properties

$$\varphi_{\pm}(y^*) = -B/C(y^*) = 2d_{4,2}/(d_{4,1} - 2y^*\ d_{2,1}) := \varphi^*$$

and

$$\varphi_{\pm}(y) \approx \pm|\frac{y}{3}|^{1/2}\text{sign }(d_{2,1}), \quad y \to \infty.$$

Then we distinguish the cases (assume without loss $d_{2,1} > 0$):

$$\text{i)} \quad y^* > \tilde{y} := d_{4,1}/2d_{2,1}, \qquad \text{ii)} \quad y^* < \tilde{y}.$$

In case i), we consider the ranges of $\varphi_{\pm}$ for (y^*, ∞), and have

$$\varphi_+[y^*, \infty) = (-\infty, \varphi^*], \qquad \varphi_-[y^*, \infty) = [\varphi^*, \infty). \tag{30}$$

Since both $\varphi_{\pm}$ are well defined and continuous on (y^*, ∞), so are the functions $H_{\pm}(y)$, and furthermore

$$H_{\pm}(y^*) = H(\varphi^*, y^*).$$

In combination with

$$H_{\pm}(y) \approx 10|d_{2,1}||y|^2 \varphi_{\pm}(y), \quad y \to \infty, \tag{31}$$

it follows that

$$H_+[y^*, \infty) = (-\infty, H(\varphi^*, y^*)], \qquad H_-[y^*, \infty) = [H(\varphi^*, y^*), \infty),$$

and hence the desired assertion (29) in case i).

In case ii), the function $\varphi_-(y)$ has a singularity at $\tilde{y}$ which lies in (y^*, ∞). However, we can restrict its domain to $[y^*, \tilde{y})$ and still have $(d_{2,1} > 0)$

$$\varphi_-[y^*, \tilde{y}) = [\varphi^*, \infty). \tag{32}$$

On the other hand the function $\varphi_+(y)$ remains continuous on the whole interval (y^*, ∞) since

$$\varphi_+(y) = A(y)/(B + \sqrt{B^2 - A(y)C(y)}).$$

Hence we have

$$\varphi_+[y^*, \infty) = (-\infty, \varphi^*],$$

so that together with (32) we have the same situation as in (30) and can proceed further exactly as before in order to prove (29). □

It remains to discuss whether there exist solutions of (8)–(9) if $\beta_3 = d_{4,1}/2d_{2,1} := \tilde{y}$. In this case, (8) gives $\beta_2 = A(\tilde{y})/B$ if $B \neq 0$, and (9) can have a solution only under some additional constraint on the parameters $d_{2,1}, d_{4,1}, d_{4,2}, d_{6,1}, d_{6,2}$. We omit it here, as well as the one which results from (8) if in addition $B = 60d_{4,2} = 0$.

Further, we remark that there can indeed exist 5 solutions of (8)–(9). To this end, one can consider the case $d_{4,1} = 0$, where these equations simplify in such a way that solving (8) for $z := \beta_3 - (3/2)\beta_2^2$ gives

$$\pm z = \frac{3}{2}\beta_2^2 - \frac{2d_{4,2}\beta_2^{-1}}{d_{2,1}} - \frac{d_{6,1}\beta_2^{-2}}{30d_{2,1}} + O(\beta_2^{-4}).$$

Inserting this into (9) with sign $+$, it follows that

$$0 = -90d_{2,1}\beta_2^5 + 255d_{4,2}\beta_2^2 + 5d_{6,1}\beta_2 + d_{6,2} + O(1/\beta_2).$$

Here the polynomial of degree 5 dominates for large β_2, and it is clear that the 4 parameters $d_{2,1}, d_{4,2}, d_{6,2}, d_{6,1}$ can be chosen such that there exist 5 zeros outside some bounded interval containing 0. The situation for the system (22)–(24) in the space case is similar.

Theorem 2. *The system (22)–(24) has at least one and at most 7 (real) solution pairs $\beta_2, \bar{\beta}_3$ if $\bar{\beta}_3 \neq -(6+\sqrt{15})d_{1,2,5}/5\sqrt{15}d_{1,2,3}$.*

Proof: Concerning existence, the argument is the same as in Theorem 1. We define $x := \beta_2, y := \bar{\beta}_3$ and $A(y) := p_0(y), B := p_1(y), C(y) := p_2(y)$. Then $R(y)$ in (25) has the form

$$R(y) = 50\sqrt{15}d_{1,2,3}^2 y^3 + 50\sqrt{15}d_{1,2,5}y^2 \ + \text{ linear term.}$$

Again define $\tilde{y} := -(6+\sqrt{15})d_{1,2,5}/5\sqrt{15}d_{1,2,3}$ as the zero of $C(y)$ and y^* as the largest zero of $R(y)$, and let $\varphi_\pm(y) := (B \pm \sqrt{R(y)})/C(y)$. Its asymptotic behaviour is described by

$$\varphi_\pm(y) \approx \text{sgn}\ d_{1,2,3}|\frac{2y}{\sqrt{15}}|^{1/2}, \quad y \to \infty. \tag{33}$$

Now we distinguish as in Theorem 1 the cases i) and ii) and conclude that either (30) holds or (31), respectively.

Next write (24) similarly as in (28) as

$$\tilde{H}(\beta_2, \bar{\beta}_3) = 0$$

and define $\tilde{H}_\pm(y) := \tilde{H}(\varphi_\pm(y), y)$. Its asymptotic behaviour is somewhat more complicated to determine than in (31), since by the definition of the $q_i(y)$, $i = 0, 1, 2$, in (24), the leading term of $\tilde{H}(\beta_2, \bar{\beta}_3)$ now has the form

$$(\frac{15}{2} - 25\alpha^2)d_{1,2,3}\beta_2^5 + 25\alpha d_{1,2,3}\beta_2^5 + (25d_{1,2,3} - 50\alpha d_{1,2,3})\bar{\beta}_3\beta_2^3 - 25d_{1,2,3}\beta_2\bar{\beta}_3^{\,2}.$$

Inserting (33) with $\bar{\beta}_3 = y$, one derives from this

$$\tilde{H}_\pm(y) \approx \text{const.}\ |y|^{5/2}, \quad y \to \infty.$$

Hence we obtain the same property (29) for $\tilde{H}_\pm$ as for $H_\pm$ in Theorem 1, and existence of a (real) solution pair of (22), (24) follows.

In order to show the bound on the number of solutions, we proceed again as in Theorem 1. We solve (22) for $x = \beta_2$ as in (25) and insert it into (24) written as

$$0 = \sum_{i=0}^{5} a_i x^i + y \sum_{i=0}^{3} b_i x^i + cxy^2,$$

with constants a_i, b_i and c. This gives after multiplication with $C(y)^5$

$$0 = \sum_{i=0}^{5} a_i(-B \pm \sqrt{R(y)})^i C(y)^{5-i}$$
$$+ y\sum_{i=0}^{3} b_i(-B \pm \sqrt{R(y)})^i C(y)^{5-i} + cy^2(-B \pm \sqrt{R(y)})C(y)^4.$$

Then sort by terms with and without $\pm\sqrt{R(y)}$. The result is an equation of type (27), this time with polynomials $U(y)$ of degree 7 and $V(y)$ of degree 6. Therefore we conclude by the same argument as in Theorem 1 that the system (22), (24) has at most 7 solutions . □

The discussion of the degenerate case $\bar{\beta}_3 \neq -(6+\sqrt{15})d_{1,2,5}/5\sqrt{15}d_{1,2,3}$ is omitted. It can be done similarly as for Theorem 1.

§4. Final remarks

We have shown that local approximation order 8 can be achieved with parametric polynomial curves of degree 4 in the planar case, and degree 5 in the space case. The method for this consists in determining a suitable reparametrization ϕ, and then the Taylor-polynomial with respect to $\underline{f} \circ \phi$. For practical purposes, however, it is important to consider also Lagrange or Hermite interpolation in the sense of the equations (1). This has been done in case $k = 2$ for $d = 2$ in [1,3,6] and for $d = 3$, in [5], but results for higher k do not seem to be available so far. In this respect another interesting open question is which order of geometric continuity can be preserved when a piecewise polynomial curve is constructed by pieces of such local aproximations.

References

1. de Boor, C., K. Höllig, and M. Sabin, High accuracy geometric Hermite interpolation, Comput. Aided Geom. Design **4** (1987), 269–278.
2. Degen, W. L. F., High accuracy approximation of parametric curves, in *Mathematical Methods for Curves and Surfaces*, M. Dæhlen, T. Lyche, and L. L. Schumaker (eds.), Vanderbilt University Press, Nashville, 1995, 83–98.
3. Feng, Y. Y. and J. Kozak, On G^2 continuous cubic spline interpolation, BIT **37** (1997), 312–332.
4. Gregory, J. A., Geometric Continuity, in *Mathematical Methods in Computer Aided Geometric Design*, T. Lyche and L. L. Schumaker (eds.), Academic Press, New York, 1989, 353–371.
5. Höllig, K. and J. Koch, Geometric Hermite interpolation, Comput. Aided Geom. Design **12** (1995), 567–580.
6. Mørken, K. and K. Scherer, A general framework for high-accuracy parametric interpolation, Mathematics of Computation **66** (1997), 237–260.
7. Rababah, A., High order approximation method for curves, Comput. Aided Geom. Design **12** (1995), 89–102.
8. Schaback, R., Geometrical differentiation and high-accuracy curve interpolation, in *Approximation Theory, Spline Functions and Applications*, S. P. Singh (eds.), Kluwer Academic Publishers 1992, 581–584.
9. Scherer, K., On local parametric approximation by polynomial curves, in *Approximation Theory*, M. W. Mueller, M. Felten, and D. H. Mache (eds.), Akademie-Verlag, 1995, 285–292.

Karl Scherer
Institut für Angewandte Mathematik
Universität Bonn
Bonn, 53115, Germany
scherer@iam.uni-bonn.de

A B–Spline Approach to Hermite Subdivision

U. Schwanecke and B. Jüttler

Abstract. We present a new approach to Hermite subdivision schemes. It is based on the observation that a sequence of second order Hermite data define a unique interpolating cubic C^1 spline. The B-Spline form of this interpolating spline leads to a stationary binary subdivision scheme with 4 different subdivision rules for the control points. We construct a generalized 4–point scheme which leads to a new family of C^2 Hermite subdivision schemes.

§1. Introduction

Starting from an initial sequence $\{h_i^{(0)}\}_{i\in\mathbb{Z}}$ of second order Hermite elements (i.e. vectors containing function values and associated first derivatives), a Hermite subdivision scheme (cf. [4,5,6,7]) of order two recursively generates finer sequences $\{h_i^{(k)}\}_{i\in\mathbb{Z}}$ of Hermite elements associated with the dyadic points $\{t_i^{(k)} = i\,2^{-k}\}_{i\in\mathbb{Z}}$. The refinement is based on two rules,

$$h_{2i}^{(k+1)} = \sum_{j=0}^{m} A_j^{(k)} h_{i+j}^{(k)}, \quad h_{2i+1}^{(k+1)} = \sum_{j=0}^{m} B_j^{(k)} h_{i+j}^{(k)}, \quad k = 0, 1, 2, \ldots, \tag{1}$$

where the matrix masks $\mathbf{A}^{(k)} = \{A_0^{(k)}, \ldots, A_m^{(k)}\}$, $\mathbf{B}^{(k)} = \{B_0^{(k)}, \ldots, B_m^{(k)}\}$ of the scheme consist of real 2×2 matrices $A_j^{(k)}, B_j^{(k)}$ depending on the subdivision level k. Merrien [7] considered Hermite–type 2–point–schemes (i.e. with $m = 1$), generating C^1 functions. By introducing an auxiliary point subdivision scheme, Dyn and Levin [4,5] analyzed stationary Hermite–interpolatory subdivision schemes of arbitrary order. Using this approach, Kuijt [6] constructed several C^2 Hermite interpolatory subdivision schemes of order two. Kuijt derived the refinement rules by considering the polynomials interpolating neighboring Hermite elements, and sampling Hermite data from them.

By considering the interpolating splines associated with the Hermite elements, this paper introduces a new approach to Hermite subdivision. We analyze the smoothness of the limit function, and present a family of C^2 Hermite subdivision schemes generalizing the 4-point scheme [3]. This spline–based approach can be generalized to Hermite elements of arbitrary order.

Curve and Surface Fitting: Saint-Malo 1999
Albert Cohen, Christophe Rabut, and Larry L. Schumaker (eds.), pp. 385–392.

ISBN 0-8265-1357-3.

§2. Spline Subdivision Schemes

At each subdivision level k, the Hermite data $\{h_i^{(k)}\}_{i\in\mathbb{Z}}$ define a unique interpolating cubic C^1 spline, having the B-spline representation

$$X^{(k)}(t) = \sum_i p_i^{(k)} N_{i,4}(t) \text{ with knots } T^{(k)}{=}(\ldots,\underbrace{t_i^{(k)},t_i^{(k)}}_{2\times},\underbrace{t_{i+1}^{(k)},t_{i+1}^{(k)}}_{2\times},\ldots). \qquad (2)$$

The control points $p_i^{(k)} \in \mathbb{R}$ are associated with the Greville–abscissas (see e.g. [8]) $\xi_{2i}^{(k)} = t_i^{(k)} - \frac{1}{3\,2^k}$ and $\xi_{2i+1}^{(k)} = t_i^{(k)} + \frac{1}{3\,2^k}$, forming a nonuniform sequence. Control points and Hermite elements are related by the transformations $(p_{2i}^{(k)}, p_{2i+1}^{(k)})^\top = H^{(k)} h_i^{(k)}$ and $h_i^{(k)} = (H^{(k)})^{-1}(p_{2i}^{(k)}, p_{2i+1}^{(k)})^\top$, where

$$H^{(k)} = \begin{pmatrix} 1 & -\frac{1}{3\,2^k} \\ 1 & \frac{1}{3\,2^k} \end{pmatrix}, \qquad (H^{(k)})^{-1} = \begin{pmatrix} \frac{1}{2} & \frac{1}{2} \\ -3\,2^{k-1} & 3\,2^{k-1} \end{pmatrix}. \qquad (3)$$

Clearly, the spline function $X^{(k)}$ can be represented with respect to the refined knot vector $T^{(k+1)}$. Knot insertion leads to the following 4 refinement rules for the B-Spline control points:

$$\begin{aligned} \tilde{p}_{4i}^{(k+1)} &= \tfrac{3}{4} p_{2i}^{(k)} + \tfrac{1}{4} p_{2i+1}^{(k)}, & \tilde{p}_{4i+2}^{(k+1)} &= \tfrac{1}{8} p_{2i}^{(k)} + \tfrac{5}{8} p_{2i+1}^{(k)} + \tfrac{2}{8} p_{2i+2}^{(k)}, \\ \tilde{p}_{4i+1}^{(k+1)} &= \tfrac{1}{4} p_{2i}^{(k)} + \tfrac{3}{4} p_{2i+1}^{(k)}, & \tilde{p}_{4i+3}^{(k+1)} &= \tfrac{2}{8} p_{2i+1}^{(k)} + \tfrac{5}{8} p_{2i+2}^{(k)} + \tfrac{1}{8} p_{2i+3}^{(k)}. \end{aligned} \qquad (4)$$

The affine combinations (4) describe a 4-rule stationary binary subdivision scheme for the (nonuniformly parametrized) B-spline control points. This scheme generalizes the splitting step of a binary uniform subdivision scheme. The sequence of control polygons converges to the C^1 limit function $X^{(0)}$. Generalizing (4) leads to the notion of a spline subdivision scheme:

Definition 1. *A* spline subdivision scheme $S(\mathbf{a}^0, \mathbf{a}^1, \mathbf{a}^2, \mathbf{a}^3)$ *with the coefficient masks* $\mathbf{a}^h = (a_0^h, \ldots, a_{2m+1}^h)$, *generating a sequence of cubic* C^1 *spline functions* $X^{(k)}(t)$, *is given by the four subdivision rules*

$$p_{4i+h}^{(k+1)} = \sum_{j=0}^{2m+1} a_j^h p_{2i+j}^{(k)}, \quad h = 0,1,2,3, \qquad k = 0,1,2,\ldots. \qquad (5)$$

With the help of the transformations (3), the matrix masks of Hermite subdivision schemes (1) can be transformed into the coefficient masks of spline subdivision scheme (5), thus motivating the following definition.

Definition 2. *A Hermite scheme is said to be* stationary *if the matrices* $A_j := H^{(k+1)} A_j^{(k)} (H^{(k)})^{-1}$, $B_j := H^{(k+1)} B_j^{(k)} (H^{(k)})^{-1}$, *are constant for all* k $(j = 0, \ldots, m)$. *The coefficients* $\mathbf{a}^h$ *of the associated spline subdivision scheme are obtained from*

$$A_j = \begin{pmatrix} a_{2j}^0 & a_{2j+1}^0 \\ a_{2j}^1 & a_{2j+1}^1 \end{pmatrix}, \quad B_j = \begin{pmatrix} a_{2j}^2 & a_{2j+1}^2 \\ a_{2j}^3 & a_{2j+1}^3 \end{pmatrix}.$$

Consequently, every stationary Hermite subdivision scheme $S(\mathbf{A}^{(k)}, \mathbf{B}^{(k)})$ is equivalent to a spline subdivision scheme $S(\mathbf{a}^0, \mathbf{a}^1, \mathbf{a}^2, \mathbf{a}^3)$. Note that a spline subdivision scheme can also be seen as a special matrix subdivision scheme (see [1]) acting on vectors of 2 consecutive control points.

§3. Convergence Analysis

In the sequel we generalize the approach introduced by Dyn, Gregory and Levin [2] to the spline subdivision case. Consider a spline subdivision scheme $S(\mathbf{a}^0, \mathbf{a}^1, \mathbf{a}^2, \mathbf{a}^3)$ on the finite domain $[0, n] \in \mathbb{R}$. The scheme is well defined on this domain, for all $k \geq 0$, if the associated Hermite elements at stage k are defined on the set $\{t_i^{(k)} | i \in Z_k\}$, where

$$Z_k = \{0, 1, \ldots, 2^k n + n_1\}, \quad n_1 = \begin{cases} 2m - 1 & \text{if} \quad A_m \neq 0 \\ 2m - 2 & \text{if} \quad A_m = 0. \end{cases}$$

The spline function (2) has the knots

$$T^{(k)} = (t_{-1}^{(k)}, t_0^{(k)}, t_0^{(k)}, t_1^{(k)}, t_1^{(k)}, \ldots, t_{2^k n + n_1}^{(k)}, t_{2^k n + n_1}^{(k)}, t_{2^k n + n_1 + 1}^{(k)})$$

and the control points $(p_i^{(k)})_{i=0,\ldots,2(2^k n + n_1)+1}$.

Consider an interval $I_i^{(k)} = [t_i^{(k)}, t_{i+1}^{(k)}]$ at the k-th subdivision step. The control points which govern the future behavior of the process in this interval are gathered in the vector $\mathbf{p}_{i,k} = (p_{2i}^{(k)}, \ldots, p_{2(i+n_1+1)+1}^{(k)})^\top$.

The control point vectors $\mathbf{p}_{2i,k+1}, \mathbf{p}_{2i+1,k+1}$ at the subdivision level $k+1$, associated with the subintervals $I_{2i}^{(k+1)}$ and $I_{2i+1}^{(k+1)}$, are obtained from $\mathbf{p}_{i,k}$ by two linear transformations,

$$\mathbf{p}_{2i,k+1} = G_0 \, \mathbf{p}_{i,k} \quad \text{and} \quad \mathbf{p}_{2i+1,k+1} = G_1 \, \mathbf{p}_{i,k} \tag{6}$$

with $G_0 = G\binom{1 \ldots M-2}{1 \ldots M-2}$ and $G_1 = G\binom{3 \cdots M}{1 \ldots M-2}$, where $G\binom{i_1 \ldots i_p}{j_1 \cdots j_p}$ is the matrix comprised of the elements of G, at rows $i_1 < \cdots < i_p$ and columns $j_1 < \cdots < j_p$. These linear transformations are expressed as submatrices of the $M \times M$ generator matrix G, where $M = 2(n_1 + 3)$. If $A_m \neq 0$, then $M = 4(m+1)$, and we get the generator matrix

$$G = \begin{pmatrix} A_0 & \cdots & \cdots & A_m & 0 & \cdots & \cdots & 0 \\ B_0 & \cdots & \cdots & B_m & 0 & \cdots & \cdots & 0 \\ 0 & A_0 & \cdots & \cdots & A_m & 0 & \cdots & 0 \\ 0 & B_0 & \cdots & \cdots & B_m & 0 & \cdots & 0 \\ \vdots & \vdots & \vdots & \vdots & \vdots & \vdots & \vdots & \vdots \\ 0 & \cdots & 0 & A_0 & \cdots & \cdots & A_m & 0 \\ 0 & \cdots & 0 & B_0 & \cdots & \cdots & B_m & 0 \end{pmatrix}. \tag{7}$$

Otherwise, if $A_m = 0$, $M = 4m + 2$ and the generator matrix is as above but with the last two rows and columns deleted.

3.1. Continuity

The following necessary condition is analogous to [2, Prop. 2.3]. Alternatively it can be formulated using the eigenstructure of the masks of the associated matrix subdivision scheme, cf. [1].

Proposition 3 (Affine invariance). *A necessary condition for the uniform convergence of the spline subdivision process to a continuous nonzero limit function on* $[0, n]$, *for arbitrary initial data, is that* $\sum_{j=0}^{2m+1} a_j^h = 1$ *for all* $h = 0, 1, 2, 3$.

In order to analyze the convergence to a continuous limit function, we examine the difference scheme $\Delta S(\mathbf{a}^0, \mathbf{a}^1, \mathbf{a}^2, \mathbf{a}^3)$ generating the differences $\Delta_i^{(k)} = p_{i+1}^{(k)} - p_i^{(k)}$. If the necessary condition of Proposition 3 is satisfied, then this process can again be described with the help of another generator matrix

$$C = E_M \, G \, (E_M)^{-1} \binom{1 \dots M-1}{1 \dots M-1}, \tag{8}$$

which is obtained using the upper triangular matrices $E_M = (-\delta_{i,j} + \delta_{i+1,j})$ and $(E_M)^{-1} = (-\sum_{h=0}^{M-1} \delta_{i+h,j})$, cf. [2, Prop. 3.2]. The $M-3$ differences governing the future behavior of the process in the interval $I_i^{(k)}$ are again collected in a vector $\mathbf{\Delta}_{i,k} = [\Delta_{2i}^{(k)}, \dots, \Delta_{2(i+n_1+1)}^{(k)}]^\top$. The analogues of the transformation (6) are

$$\mathbf{\Delta}_{2i,k+1} = C_0 \mathbf{\Delta}_{i,k} \quad \text{and} \quad \mathbf{\Delta}_{2i+1,k+1} = C_1 \mathbf{\Delta}_{i,k}, \tag{9}$$

where $C_0 = C\binom{1 \dots M-3}{1 \dots M-3}$ and $C_1 = C\binom{3 \dots M-1}{1 \dots M-3}$. Note that the row and column ranges of the sub–matrices C_0, C_1 are different from those in [2], as we analyze a difference process with 4 (rather than 2) rules. We get (cf. [2, Theorem 3.1])

Theorem 4. *Let the spline subdivision process satisfy the necessary convergence condition of Proposition 3. Then the following are equivalent:*

(*i*) *The spline subdivision process* $S(\mathbf{a}^0, \mathbf{a}^1, \mathbf{a}^2, \mathbf{a}^3)$ *converges uniformly to a continuous limit function on* $[0, n]$ *for arbitrary initial data.*

(*ii*) *The difference process* $\Delta S(\mathbf{a}^0, \mathbf{a}^1, \mathbf{a}^2, \mathbf{a}^3)$ *is contracting, i.e. it converges uniformly to zero on* $[0, n]$ *for arbitrary initial data.*

(*iii*) *There exists an integer* $L > 0$ *and a real number* $0 \le \alpha < 1$ *such that* $\|C_{i_1} \cdots C_{i_L}\|_\infty \le \alpha$, $\forall\, i_j \in \{0, 1\}$ *and* $j = 1, \dots, L$.

In the sequel we have to analyze other point processes with four different refinement rules. The continuity of the limit function can then be analyzed in an analogous way, where the generator matrix is obtained as in (8).

3.2. Derivative process

In order to investigate the differentiability of the limit function f, we analyze the first derivative of the cubic C^1 splines (2). Clearly, we obtain a sequence of quadratic C^0 splines with knots $T^{(k)}$, see Figure 1. If the necessary condition of Prop. 3 is satisfied, then the quadratic splines are generated by another spline subdivision scheme, again with four different rules for the control points. This scheme will be called the derivative scheme $\partial S(\mathbf{a}^0, \mathbf{a}^1, \mathbf{a}^2, \mathbf{a}^3)$.

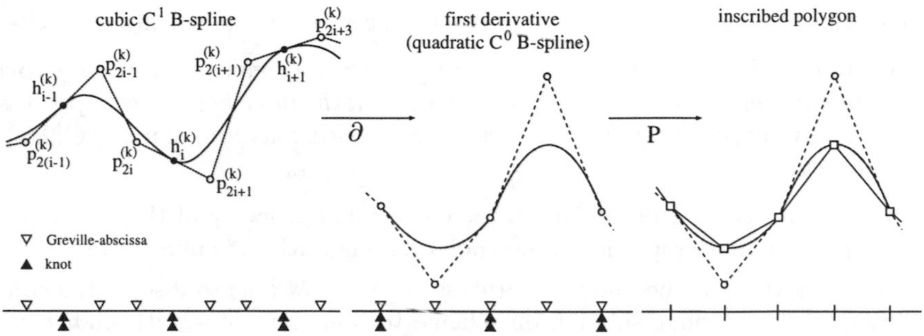

Fig. 1. Derivative scheme and inscribed polygon process.

Proposition 5. *If the derivative process* $\partial S(\mathbf{a}^0, \mathbf{a}^1, \mathbf{a}^2, \mathbf{a}^3)$ *converges uniformly to* $d \in C[0,n]$*, then the spline subdivision scheme* $S(\mathbf{a}^0, \mathbf{a}^1, \mathbf{a}^2, \mathbf{a}^3)$ *converges uniformly to* $f \in C^1[0,n]$*, and* $f' = d$.

Using similar techniques as in Section 3.1, we define control point vectors and a generator matrix $\mathcal{D}$. We omit the details, giving only the main result:

Proposition 6. *The derivative scheme* $\partial S(\mathbf{a}^0, \mathbf{a}^1, \mathbf{a}^2, \mathbf{a}^3)$ *has the* $(M-1) \times (M-1)$ *generator matrix*

$$\mathcal{D} = 2E_M^1 \; G \; (E_M^1)^{-1} \binom{1 \ldots M-1}{1 \ldots M-1},$$

with $E_M^1 = \operatorname{diag}(1, 2, 1, \ldots, 2)\, E_M$ *and* $(E_M^1)^{-1} = (E_M)^{-1} \operatorname{diag}(1, \frac{1}{2}, 1, \ldots, \frac{1}{2})$.

The continuity of the limit function generated by the derivative scheme can now easily be analyzed as in Section 3.1 by discussing the associated difference scheme $\Delta\partial S(\mathbf{a}^0, \mathbf{a}^1, \mathbf{a}^2, \mathbf{a}^3)$. This leads to criteria for C^1 continuity of Hermite subdivision schemes.

3.3. C^k convergence analysis via inscribed polygons

In order to examine higher order continuity, we inscribe a polygon into the quadratic C^0 spline and analyze the resulting subdivision scheme, called the inscribed polygon process $P\partial S(\mathbf{a}^0, \mathbf{a}^1, \mathbf{a}^2, \mathbf{a}^3)$. More precisely, at the subdivision level k, we consider the piecewise linear function with the vertices $(\, t_i^{(k+1)}, \dot{X}^{(k)}(t_i^{(k+1)})\,)$, see Figure 1.

Proposition 7. *The inscribed polygon process* $P\partial S(\mathbf{a}^0, \mathbf{a}^1, \mathbf{a}^2, \mathbf{a}^3)$ *of the derivative scheme has the generator matrix* $P = L_{M-1}\, \mathcal{D}\, (L_{M-1})^{-1}$ *which is obtained using the* $(M-1) \times (M-1)$ *auxiliary matrices*

$$L_{M-1} = \begin{pmatrix} 1 & & & & & & \\ \frac{1}{4} & \frac{1}{2} & \frac{1}{4} & & & & \\ & & 1 & & & & \\ & & & \ddots & & & \\ & & & & \frac{1}{4} & \frac{1}{2} & \frac{1}{4} \\ & & & & & & 1 \end{pmatrix}, \quad (L_{M-1})^{-1} = \begin{pmatrix} 1 & & & & & & \\ -\frac{1}{2} & 2 & -\frac{1}{2} & & & & \\ & & 1 & & & & \\ & & & \ddots & & & \\ & & & & -\frac{1}{2} & 2 & -\frac{1}{2} \\ & & & & & & 1 \end{pmatrix}.$$

The derivative and the inscribed polygon processes are equivalent:

Lemma 8. *The derivative process $\partial S(\mathbf{a}^0, \mathbf{a}^1, \mathbf{a}^2, \mathbf{a}^3)$ converges uniformly to a continuous limit function f on $[0, n]$ if and only if the inscribed polygon process $P\partial S(\mathbf{a}^0, \mathbf{a}^1, \mathbf{a}^2, \mathbf{a}^3)$ converges uniformly to a continuous function $g \in C[0, n]$. Moreover, $f = g$.*

Proof: This equivalence is due to the convex hull property of B-splines, and to approximation properties of interpolating quadratic C^0 splines. □

Using the inscribed polygon process, we are now able to discuss the convergence of the spline subdivision scheme to limit functions with higher order differentiability, by the same analysis as in the non-Hermite case (point schemes). We simply have to analyze the divided difference processes $D^\nu P\partial S(\mathbf{a}^0, \mathbf{a}^1, \mathbf{a}^2, \mathbf{a}^3)$ (see [2, Theorem 4.2]) of the inscribed polygons, as follows.

Theorem 9. *If the l-th order divided difference scheme of the inscribed polygon process $P\partial S(\mathbf{a}^0, \mathbf{a}^1, \mathbf{a}^2, \mathbf{a}^3)$ exists and converges uniformly to $f_l \in C[0, n]$, then also the divided difference processes $D^\nu P\partial S(\mathbf{a}^0, \mathbf{a}^1, \mathbf{a}^2, \mathbf{a}^3)$ exist and converge uniformly to $f_\nu \in C^{l-\nu}[0, n]$ for $\nu = 0, 1, \ldots, l$, and $f_0^{(\nu)} = f_\nu$. Hence, the spline subdivision scheme $S(\mathbf{a}^0, \mathbf{a}^1, \mathbf{a}^2, \mathbf{a}^3)$ converges uniformly to $g \in C^{l+1}[0, n]$ with $g^{(\nu+1)} = f_\nu$.*

For instance, in order to prove that the limit function generated by the spline subdivision scheme is C^2, the difference process $\Delta DP\partial S(\mathbf{a}^0, \mathbf{a}^1, \mathbf{a}^2, \mathbf{a}^3)$ of the divided difference scheme has to be shown to be contractive, analogously to Section 3.1. The divided difference scheme $DP\partial S(\mathbf{a}^0, \mathbf{a}^1, \mathbf{a}^2, \mathbf{a}^3)$ has the generator matrix $D = 2E_{M-1}\, P\, (E_{M-1})^{-1}\binom{1\ldots M-2}{1\ldots M-2}$. From this matrix we get the generator matrix $C^* = E_{M-2}\, D\, (E_{M-2})^{-1}\binom{1\ldots M-3}{1\ldots M-3}$ of the associated difference scheme $\Delta DP\partial S(\mathbf{a}^0, \mathbf{a}^1, \mathbf{a}^2, \mathbf{a}^3)$. In order to guarantee a C^2 limit function, the matrix norms $\|C^*_{i_1} \cdots C^*_{i_L}\|_\infty$, $\forall\, i_j \in \{0, 1\}$ and $j = 1, \ldots, L$ have to be less than 1 for some L, where $C^*_0 = C^*\binom{1\ldots M-5}{1\ldots M-5}$, and $C^*_1 = C^*\binom{3\ldots M-3}{1\ldots M-5}$.

§4. A Generalized 4–Point Scheme

Based on a geometric construction, Dyn, Gregory, and Levin [3] derived a family of interpolating 4-point schemes. This family can also be obtained from an optimization–based approach, as follows. Let the subdivision scheme generate a sequence of piecewise linear functions $Y^{(k)}$ with knots $t_i^{(k)}$ and control points $q_i^{(k)}$. In order to derive the refinement rules, we replace one segment of $Y^{(k)}$ with two new ones (shown as dashed lines in Figure 2, left), subject to C^0 boundary conditions. The new vertex $q_{2i+1}^{(k+1)}$ is placed by minimizing the jumps of the first derivatives between new and old polygons. In fact, minimizing the weighted linear combination

$$F(q_{2i+1}^{(k+1)}) = \tfrac{1-4\omega}{4}\,[\,\dot{Y}_-^{(k+1)}(t_{2i+1}^{(k+1)}) - \dot{Y}_+^{(k+1)}(t_{2i+1}^{(k+1)})\,]^2$$
$$+\, 2\omega\,[\,\dot{Y}_-^{(k)}(t_i^{(k)}) - \dot{Y}_+^{(k+1)}(t_i^{(k)})\,]^2 + 2\omega\,[\,\dot{Y}_+^{(k+1)}(t_{i+1}^{(k)}) - \dot{Y}_-^{(k)}(t_{i+1}^{(k)})\,]^2$$

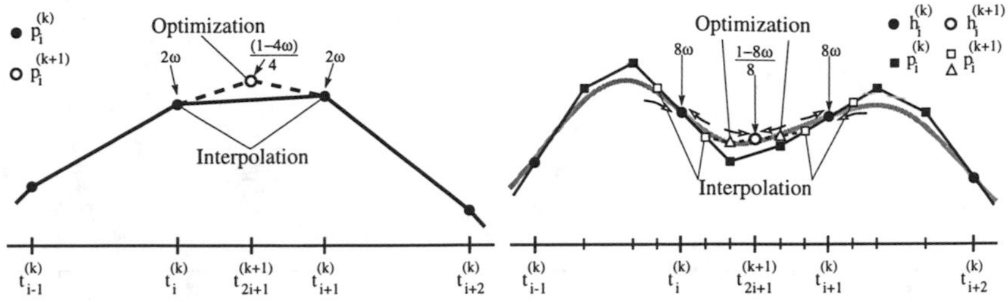

Fig. 2. Weights of objective functions in the point (left) and spline (right) case.

of squared differences of derivatives produces exactly the refinement rules of the interpolating 4-point scheme.

This approach can be generalized to the spline case. In order to derive the refinement rules, we replace one segment of $X^{(k)}$ with two new ones (shown with dashed control polygons in Figure 2, right), subject to C^1 boundary conditions. The inner new control points $p_{4i+2}^{(k+1)}$, $p_{4i+3}^{(k+1)}$ are placed by minimizing the jumps of the second derivatives between new and old splines. Minimizing the weighted linear combination

$$F(p_{4i+2}^{(k+1)}, p_{4i+3}^{(k+1)}) = \tfrac{1-8\omega}{8}\,[\ddot{X}_-^{(k+1)}(t_{2i+1}^{(k+1)}) - \ddot{X}_+^{(k+1)}(t_{2i+1}^{(k+1)})]^2$$
$$+\, 8\omega\,[\ddot{X}_-^{(k)}(t_i^{(k)}) - \ddot{X}_+^{(k+1)}(t_i^{(k)})]^2 + 8\omega\,[\ddot{X}_i^{(k+1)}(t_{i+1}^{(k)}) - \ddot{X}_+^{(k)}(t_{i+1}^{(k)})]^2$$

of the squared differences of the second derivatives gives the refinement masks

$$\mathbf{a}^0 = (0, 0, \tfrac{3}{4}, \tfrac{1}{4}, 0, 0, 0, 0), \qquad \mathbf{a}^1 = (0, 0, \tfrac{1}{4}, \tfrac{3}{4}, 0, 0, 0, 0),$$
$$\mathbf{a}^2 = (0, \tfrac{1}{8} + \omega, -\tfrac{1}{8} - 2\omega, \tfrac{3}{4} + 3\omega, \tfrac{3}{8} - 3\omega, -\tfrac{1}{4} + 2\omega, \tfrac{1}{8} - \omega, 0),$$
$$\mathbf{a}^3 = (0, \tfrac{1}{8} - \omega, -\tfrac{1}{4} + 2\omega, \tfrac{3}{8} - 3\omega, \tfrac{3}{4} + 3\omega, -\tfrac{1}{8} - 2\omega, \tfrac{1}{8} + \omega, 0),$$

see (5). In order to analyze C^2 continuity of the limit function, we compute the generator matrix of the difference process $\Delta DP\partial S(\mathbf{a}^0, \mathbf{a}^1, \mathbf{a}^2, \mathbf{a}^3)$, which has to be contractive. Using the techniques of Section 3.1, we estimate the C^2 convergence range of the parameter ω, see Figure 3.

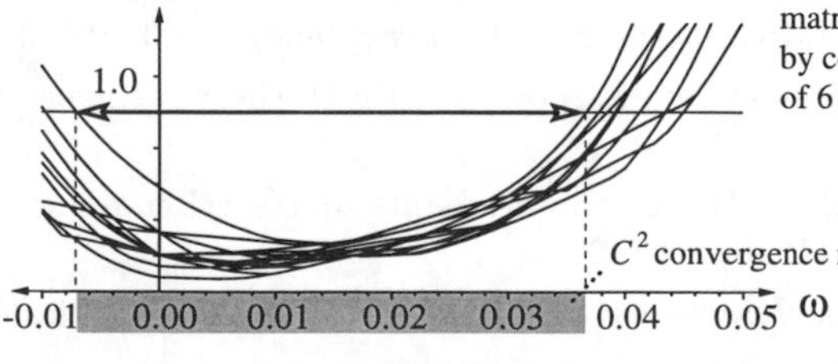

Fig. 3. Estimating the C^2 convergence range of the generalized 4 point scheme.

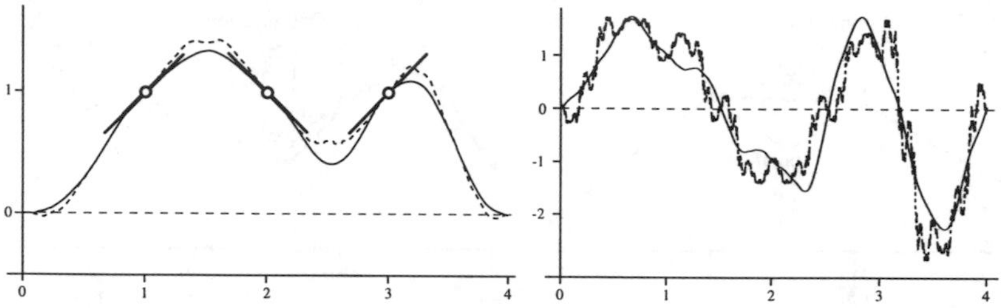

Fig. 4. Interpolatory limit functions (left) and derivatives (right).

Two limit functions interpolating three given Hermite elements have been drawn in Figure 4 (left). The functions have been generated with the help of Merrien's C^1 scheme ($\alpha = 0.2$, dashed curve, cf. [4,7]), and using the generalized 4-point scheme ($\omega = 0.015$, solid curve). As can clearly be seen from the associated first derivatives (right), the generalized 4-point scheme produces a C^2 limit function.

References

1. Cohen, A., N. Dyn, and D. Levin, Stability and inter–dependence of matrix subdivision, *Advanced Topics in Multivariate Approximation*, F. Fontanella, K. Jetter and L. L. Schumaker (eds.), World Scientific Publishing Co., Inc., Singapore, 1993, 33–46.
2. Dyn, N., J. Gregory, and D. Levin, Analysis of Uniform Binary Subdivision Schemes for Curve Design, Constr. Approx. **7** (1991), 127–147.
3. Dyn, N., J. Gregory, and D. Levin, A 4–point interpolatory subdivision scheme for curve design, Comput. Aided Geom. Design **4** (1987), 257–268.
4. Dyn, N., and D. Levin, Analysis of Hermite–type subdivision schemes, in *Approximation Theory VIII, Vol. 2: Wavelets*, Charles K. Chui and Larry L. Schumaker (eds.), World Scientific Publishing Co., Inc., Singapore, 1995, 117–124.
5. Dyn, N., and D. Levin. Analysis of Hermite–interpolatory subdivision schemes, *Proceedings of the CRM workshop on splines and wavelets*, Serge Dubuc and Gilles Deslauriers (eds.), CRM Proceedings and Lecture Notes **18**, American Mathematical Society, Providence, 1999, 105–113.
6. Kuijt, F., *Convexity preserving interpolation*, Ph.D. thesis, University of Twente, 1998.
7. Merrien, J.-L., A family of Hermite interpolants by bisection algorithms, Numer. Algorithms **2** (1990), 187–200.
8. Schumaker, L. L., *Spline Functions: Basic Theory*, Wiley, New York, 1981.

An Interpolation Method with Weights and Relaxation Parameters

Chiew-Lan Tai, Brian A. Barsky, and Kia-Fock Loe

Abstract. This paper presents a new interpolation method that is based on blending a nonuniform rational B-spline (NURB) curve with a singularly reparametrized (SR) linear spline. The resulting curve is called the α-spline. It has weights and relaxation parameters. Given the data points to be interpolated, a NURB curve is obtained by using these data points as its control points. The SR linear spline is then determined by imposing constraints on the α-spline to interpolate the data points. The α-spline is parametrically continuous. It involves only simple computation, and does not require solving linear systems. This approach is extended to produce interpolatory surfaces, and can be used as a modeling tool for deforming polygonal shapes into smooth spline surface models.

§1. Introduction

Cubic interpolating splines are known to exhibit undesirable oscillations due to "extraneous" inflection points. This undesirable property motivated the introduction of tension applied to interpolating splines. Barsky analyzed two well-known approaches for applying tension to interpolating curves [1]. The first approach is the exponential-based spline in tension [12]; the second is the more efficient polynomial alternative, the ν-spline [11]. The tension parameters are associated with data points in these splines.

Tension has also been introduced as a shape parameter to non-interpolating splines. The Beta-spline [2] incorporates bias and tension parameters, based on the theory of geometric continuity. Another spline technique that has bias and tension parameters (and also a "continuity" parameter) was proposed by Kochanek and Bartels [8]. Unlike those in the Beta-spline, these parameters control only the first derivatives, and the resulting curve is C^1 and interpolatory. The weighted splines by Foley [6] provide tension control on curve segments between interpolation points, rather than at the interpolation points.

Curve and Surface Fitting: Saint-Malo 1999
Albert Cohen, Christophe Rabut, and Larry L. Schumaker (eds.), pp. 393–402.

ISBN 0-8265-1357-3.

In addition to spline methods that allow the user to modify the shape parameters, a related research area is the automatic determination of shape parameters to produce curves and surfaces that satisfy certain shape-preserving criteria [5,7]. Another more subjective criterion used is fairness, or "visual appeal".

NURBS are also often used to solve interpolation problems. However, using the NURB as a surface interpolant requires performing $O(mn)$ operations to solve the $O(m+n)$ linear systems for a dataset of $m \times n$ points. Although these systems are tridiagonal, they still incur substantial computations. Moreover, the interpolation method is global; thus, changes to any data point will require solving again all the linear systems.

The idea of using singular blending to solve the interpolation problem, with tension control, was proposed by Loe and Tai [10]. In this paper, we use this idea to propose another interpolation method that has weights and relaxation parameters. A high relaxation value has the effect of reducing tautness. The method is based on blending a NURB curve with a sequence of line segments, or a linear spline, that is reparametrized such that each of the line segments has zero derivatives at both ends.

This work is closely related to some earlier work of Coons [4]. He used this blending technique to modify any given piecewise curve by letting the line segments be those connecting the joints of the given curve, and proved that if the given curve is C^k, then the blending function only has to be C^{k-1} for the resulting curve to retain the continuity of the original curve. More precisely, Coons' aim was not to solve the interpolation problem: for the new curve to be interpolatory, the original curve has to be interpolatory.

§2. Singular Blending

Given a smooth piecewise curve $\boldsymbol{C}(u)$, a tension control can be introduced by blending the smooth curve with a linear spline that approximates the curve [4,9]. This linear spline can simply be obtained by connecting the joints of the smooth curve, but in general it need not interpolate the smooth curve. Let $\boldsymbol{L}(u)$ denotes a linear spline. The blending then gives

$$\boldsymbol{Q}(u) = (1-\alpha)\boldsymbol{C}(u) + \alpha \boldsymbol{L}(u), \qquad u \in [u_0, u_n], \qquad 0 \le \alpha \le 1. \tag{1}$$

It is easy to see that as α increases, the contribution of $\boldsymbol{L}(u)$ to $\boldsymbol{Q}(u)$ increases; thus, the resulting curve is more taut, simulating a higher tension.

Assuming that the smooth curve $\boldsymbol{C}(u)$ is C^2 (generalization to higher order curves is straightforward), then for $\boldsymbol{Q}(u)$ to retain this continuity, in general $\boldsymbol{L}(u)$ must be at least C^2. We define $\boldsymbol{L}(u)$ as

$$\boldsymbol{L}(u) = (1-s(t))\boldsymbol{V}_j + s(t)\boldsymbol{V}_{j+1}, \qquad u \in [u_j, u_{j+1}), \tag{2}$$

where $t = \frac{u-u_j}{u_{j+1}-u_j}$, $\boldsymbol{V}_j$, $j = 0, ..., n$, are the vertices of the linear spline, and $s(t)$ is a monotonically increasing function yet to be defined. Since $\boldsymbol{L}(u)$ must be C^2, it must satisfy the following conditions:

$$\boldsymbol{L}'(u_j) = \boldsymbol{L}'(u_{j+1}) = 0 \qquad \text{and} \qquad \boldsymbol{L}''(u_j) = \boldsymbol{L}''(u_{j+1}) = 0.$$

These conditions can be satisfied by letting $s(t)$ be the quintic Hermite polynomial $10t^3 - 15t^4 + 6t^5$ [3]. We call a linear spline that satisfies these conditions a singularly reparametrized (SR) linear spline.

For the technique to be useful, it should have local tension control; that is, the α in (1) must be a function of u,

$$\boldsymbol{Q}(u) = (1 - \alpha(u))\boldsymbol{C}(u) + \alpha(u)\boldsymbol{L}(u), \qquad u \in [u_0, u_n]. \tag{3}$$

The $\alpha(u)$ function must satisfy three criteria: (1) interpolate the local parameters α_j^*, $j = 0, ..., n$; (2) employ a local interpolation method so that modifying a particular α_j^* will affect only the neighboring curve segments; (3) be at least C^2 so that the blended curve is also C^2 (Coons [4] showed that if the SR line segments are restricted to those connecting the joints of the given curve, then $\alpha(u)$ only need to be C^1). We observe that $s(t)$ can be used to define $\alpha(u)$, with all three criteria satisfied; hence, we define

$$\alpha(u) = (1 - s(t))\alpha_j^* + s(t)\alpha_{j+1}^*, \qquad u \in [u_j, u_{j+1}), \tag{4}$$

where $t = \frac{u-u_j}{u_{j+1}-u_j}$. We then have $\alpha'(u_j)=\alpha'(u_{j+1})=\alpha''(u_j)=\alpha''(u_{j+1})=0$. A drawback of this definition of $\alpha(u)$ is that drastic changes in adjacent α_j^* values can cause the curve to undulate. An alternate method is to estimate the first and second derivatives at the joints by some approximation method, and obtain $\alpha(u)$ by Hermite interpolation.

This idea of singular blending can be applied to many combinations of curves and SR linear splines. Since NURBS are prevalent in industry, we let the smooth curve $\boldsymbol{C}(u)$ be a NURB curve and call the resulting curve the α-spline. If the SR linear spline is the control polygon of the NURBS curve, then the resulting α-spline is non-interpolatory. The non-interpolatory α-spline includes the NURB as a special case; when all tension parameters are zero, the α-spline reduces to a NURB. Some other geometric properties inherited from NURB include convex hull, affine and projective invariance, and local control.

§3. The Interpolating α-spline Curve

The α-spline is non-interpolatory when the SR linear spline is the control polygon of the NURB curve. However, there is no reason to restrict the SR linear spline to be the control polygon. In this paper, given a NURB curve, we will determine a new SR linear spline such that when they are blended, the resulting α-spline interpolates a given set of data points.

Let the data points be $\boldsymbol{P}_j, j = 0, ..., n$. The cubic NURB curve must somehow approximate the data points. A simple way to achieve this is by letting the given data points serve as the control points. Next, we must introduce two new control points so that the number of NURB curve segments is n, where each curve segment corresponds to an interval between two data points. Since we want the NURB curve to interpolate the endpoints $\boldsymbol{P}_0$ and

$\boldsymbol{P}_n$, we let the new points be coincident with the two endpoints, and set the first four and last four end knot values to be equal; that is,

$$\boldsymbol{P}_{-1} = \boldsymbol{P}_0 \qquad \text{and} \qquad \boldsymbol{P}_{n+1} = \boldsymbol{P}_n,$$

and $u_{-3}=u_{-2}=u_{-1}=u_0$, and $u_n=u_{n+1}=u_{n+2}=u_{n+3}$ in $\boldsymbol{U}=\{u_{-3},...,u_{n+3}\}$. The interior knot values $u_0,...,u_n$ are computed using methods such as the chord-length or centripetal method. All the examples in this paper use the centripetal parametrization. The cubic NURB curve sequence is given by

$$\boldsymbol{C}(u) = \sum_{j=-1}^{n+1} R_{j,4}(u)\boldsymbol{P}_j, \qquad u \in [u_0, u_n],$$

where $R_{j,4}(u)$ are the cubic rational B-spline basis functions defined over the knot vector $\boldsymbol{U}$.

To determine the unknown vertices $\boldsymbol{V}_j$, $j = 0, ..., n$, of the SR linear spline, we impose constraints on the resulting α-spline $\boldsymbol{Q}(u)$ to interpolate all the data points: $\boldsymbol{P}_j = \boldsymbol{Q}(u_j), j = 0, ..., n$. Substituting $\boldsymbol{Q}(u_j)$ from (3),

$$\boldsymbol{P}_j = (1-\alpha(u_j))\boldsymbol{C}(u_j) + \alpha(u_j)\boldsymbol{L}(u_j) = (1-\alpha_j^*)\boldsymbol{C}(u_j) + \alpha_j^*\boldsymbol{V}_j.$$

Solving for $\boldsymbol{V}_j$, then adding $\boldsymbol{P}_j$ and subtracting $\frac{\alpha_j^*\boldsymbol{P}_j}{\alpha_j^*}$ yields equations that require only simple computations:

$$\boldsymbol{V}_j = \frac{\boldsymbol{P}_j - (1-\alpha_j^*)\boldsymbol{C}(u_j)}{\alpha_j^*} = \boldsymbol{P}_j + \frac{1-\alpha_j^*}{\alpha_j^*}(\boldsymbol{P}_j - \boldsymbol{C}(u_j)).$$

Defining $\rho_j = \frac{1-\alpha_j^*}{\alpha_j^*}$ yields

$$\boldsymbol{V}_j = \boldsymbol{P}_j + \rho_j\ (\boldsymbol{P}_j - \boldsymbol{C}(u_j)), \qquad j = 0, ..., n. \tag{5}$$

We call ρ_j the relaxation parameters. The geometric interpretation of (5) is that $\boldsymbol{C}(u_j)$, $\boldsymbol{P}_j$, and $\boldsymbol{V}_j$ are collinear, and the distance between $\boldsymbol{C}(u_j)$ and $\boldsymbol{P}_j$, and between $\boldsymbol{P}_j$ and $\boldsymbol{V}_j$ are in the ratio $1 : \rho_j$. We know that $\rho_j \geq 0$, because $\alpha_j^* \leq 1$. From empirical study, we have found $0 \leq \rho_j \leq 4$ to be a useful range. The midpoint value $\rho_j = 2$ yields visually appealing shapes for most datasets; thus, we use that value as the default relaxation value. The effect of ρ_j can be interpreted from noting that $\alpha_j^* = \frac{1}{\rho_j+1}$ and observing the role of α in (3). The value $\rho_j = 0$ corresponds to the maximum tension $\alpha_j^* = 1$; when all $\rho_j = 0$, only the SR linear spline contributes to the α-spline and we have a linear interpolant. When the relaxation value ρ_j increases, α_j^* decreases, and the contribution of the SR linear spline to the α-spline decreases; thus, the resulting curve is being relaxed locally. By using different relaxation values, we can easily obtain rounder or sharper corners without specifying multiple knots or multiple control points. Fig. 1 shows the effect of

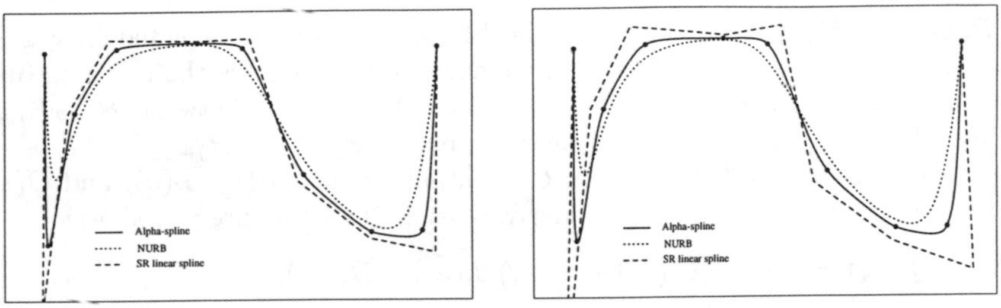

Fig. 1. Effect of global relaxation: (left) 1, (right) 2.

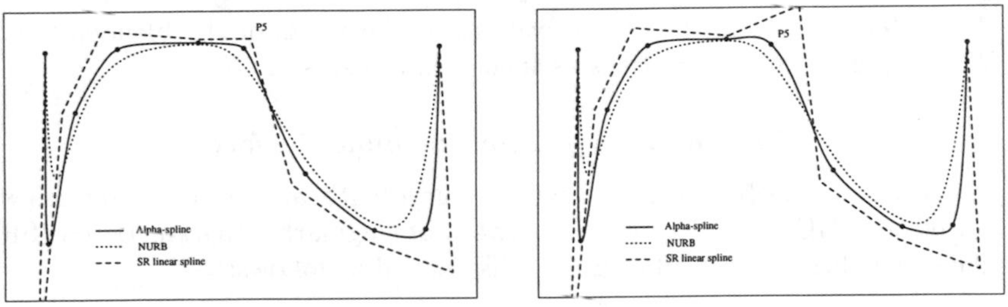

Fig. 2. Effect of varying ρ_5: 1, 4; the other ρ_j are 2.

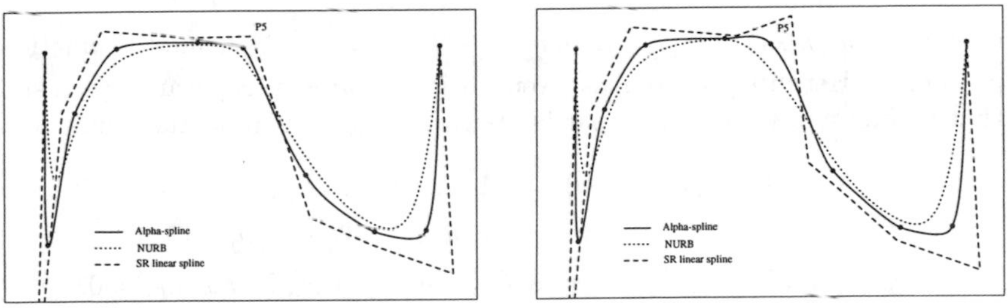

Fig. 3. Effect of varying w_5: 2, 0.5; the other w_j are unity. All ρ_j are 2.

applying the relaxation globally. It can be observed that while the SR linear spline found is dependent on the global relaxation value specified, the cubic NURB curve remains fixed. Fig. 2 shows the effect of varying the relaxation locally (while the other part of the curve has the default relaxation value). Note that the effect of each ρ_j is very localized; only the nearest two curve segments are affected.

The effect of the weight is less obvious. Each w_j affects four neighboring segments, i.e., $[u_{j-2}, u_{j+2}]$. When w_j decreases, $\boldsymbol{C}(u_j)$ moves further from $\boldsymbol{P}_j$ (assuming ρ_j is fixed); thus, from (5), $\boldsymbol{V}_j$ found is further from $\boldsymbol{P}_j$. Since the weights are relative in nature, decreasing w_j also causes $\boldsymbol{C}(u_{j-1})$ and $\boldsymbol{C}(u_{j+1})$ to move closer to the edge $\boldsymbol{P}_{j-2}\boldsymbol{P}_{j-1}$ and $\boldsymbol{P}_{j+1}\boldsymbol{P}_{j+2}$, respectively, and to be on that edge when w_j=0. Hence, when $w_j = 0$, from (5) again, $\boldsymbol{P}_{j-2}$, $\boldsymbol{C}(u_{j-1})$, $\boldsymbol{P}_{j-1}$, and $\boldsymbol{V}_{j-1}$ are all collinear, and so are $\boldsymbol{P}_{j+2}$, $\boldsymbol{C}(u_{j+1})$,

$\boldsymbol{P}_{j+1}$, and $\boldsymbol{V}_{j+1}$. This effect on the SR linear spline is depicted in Fig. 3; only w_5 is varied here. To summarize, decreasing w_j causes the interpolating α-spline to be rounder near $\boldsymbol{P}_j$, but causes its segments between $\boldsymbol{P}_{j-2}\boldsymbol{P}_{j-1}$ and $\boldsymbol{P}_{j+1}\boldsymbol{P}_{j+2}$ to be more taut near the points $\boldsymbol{P}_{j-1}$ and $\boldsymbol{P}_{j+1}$.

The α-spline $\boldsymbol{Q}(u)$ is clearly C^2 continuous since $\alpha(u)$, $\boldsymbol{L}(u)$, and $\boldsymbol{C}(u)$ are all C^2. The first and second derivatives at the knots are as follows:

$$\boldsymbol{Q}'(u_j) = (1-\alpha_j^*)\boldsymbol{C}'(u_j) + \alpha'(u_j)(\boldsymbol{L}(u_j) - \boldsymbol{C}(u_j)),$$
$$\boldsymbol{Q}''(u_j) = (1-\alpha_j^*)\boldsymbol{C}''(u_j) - 2\alpha'(u_j)\boldsymbol{C}'(u_j) + \alpha''(u_j)(\boldsymbol{L}(u_j) - \boldsymbol{C}(u_j)).$$

If $\alpha(u)$ is defined as in (4), then $\boldsymbol{Q}'(u_j) = (1-\alpha_j^*)\boldsymbol{C}'(u_j)$ and $\boldsymbol{Q}''(u_j) = (1-\alpha_j^*)\boldsymbol{C}''(u_j)$. That is, the derivatives at the joints of the blended curve $\boldsymbol{Q}(u)$ are in the same directions as their counterparts of $\boldsymbol{C}(u)$.

§4. The Interpolating α-spline Surface

Analogous to the blending of a NURB curve with an SR linear spline, we can blend a NURB surface with a network of singularly reparametrized (SR) bilinear patches. An SR bilinear patch is defined as follows:

$$\begin{aligned}\boldsymbol{L}(u,v) =& (1-s(r))(1-s(t))\boldsymbol{V}_{i,j} + (1-s(r))s(t)\boldsymbol{V}_{i,j+1} \\ &+ s(r)(1-s(t))\boldsymbol{V}_{i+1,j} + s(r)s(t)\boldsymbol{V}_{i+1,j+1},\end{aligned} \tag{6}$$

where $u \in [u_i, u_{i+1})$, $v \in [v_j, v_{j+1})$, $r = \frac{u-u_i}{u_{i+1}-u_i}$, $t = \frac{v-v_j}{v_{j+1}-v_j}$, and the functions $s(\cdot)$ are the Hermite polynomials given earlier. That is, it is parametrized such that its first and second order partial derivatives go to zero at the boundaries:

$$\partial_u \boldsymbol{L}(u_i,v) = \partial_u \boldsymbol{L}(u_{i+1},v) = \partial_v \boldsymbol{L}(u,v_j) = \partial_v \boldsymbol{L}(u,v_{j+1}) = 0,$$
$$\partial_u^2 \boldsymbol{L}(u_i,v) = \partial_u^2 \boldsymbol{L}(u_{i+1},v) = \partial_v^2 \boldsymbol{L}(u,v_j) = \partial_v^2 \boldsymbol{L}(u,v_{j+1}) = 0,$$
$$\partial_{uv}^2 \boldsymbol{L}(u_i,v) = \partial_{uv}^2 \boldsymbol{L}(u_{i+1},v) = \partial_{uv}^2 \boldsymbol{L}(u,v_j) = \partial_{uv}^2 \boldsymbol{L}(u,v_{j+1}) = 0.$$

The α-spline surface is then given by

$$\boldsymbol{Q}(u,v) = (1-\alpha(u,v))\boldsymbol{S}(u,v) + \alpha(u,v)\boldsymbol{L}(u,v), \qquad u \in [u_0,u_m];\ v \in [v_0,v_n],$$

where $\alpha(u,v)$ is the blending function that interpolates the local tension parameters $\alpha_{i,j}^*$, defined by an equation similar to (6).

To find an α-spline surface interpolating a given network of data points $\boldsymbol{P}_{i,j}, i = 0,...,m,\ j = 0,...,n$, we must first find a NURB surface, then determine a network of SR bilinear patches to be blended with the NURB surface. The NURB surface is defined by simply letting the data points be the control points. As in the case of curves, we repeat the boundary vertices and let the first and last four knot values be equal. Repeating the boundary vertices along the j-index, then repeating those along the i-index, we obtain

$$\boldsymbol{P}_{-1,j} = \boldsymbol{P}_{0,j} \quad \text{and} \quad \boldsymbol{P}_{m+1,j} = \boldsymbol{P}_{m,j}, \qquad j = 0,...,n,$$
$$\boldsymbol{P}_{i,-1} = \boldsymbol{P}_{i,0} \quad \text{and} \quad \boldsymbol{P}_{i,n+1} = \boldsymbol{P}_{i,n}, \qquad i = -1,...,m+1.$$

Setting the end knot values to be equal in the knot vectors $\boldsymbol{U}=\{u_{-3},...,u_{m+3}\}$ and $\boldsymbol{V}=\{v_{-3},...,v_{n+3}\}$,

$$u_{-3}=u_{-2}=u_{-1}=u_0, \qquad u_m=u_{m+1}=u_{m+2}=u_{m+3},$$
$$v_{-3}=v_{-2}=v_{-1}=v_0, \qquad v_n=v_{n+1}=v_{n+2}=v_{n+3}.$$

The interior knot values can be determined using any good parametrization method. The NURB surface is then defined by

$$\boldsymbol{S}(u,v)=\sum_{i=-1}^{m+1}\sum_{j=-1}^{n+1} R_{i,4}(u)R_{j,4}(v)\boldsymbol{P}_{i,j}, \qquad u\in[u_0,u_m];\ v\in[v_0,v_n],$$

where $R_{i,4}(u)$ and $R_{j,4}(v)$ are cubic rational B-spline basis functions defined over $\boldsymbol{U}$ and $\boldsymbol{V}$.

To determine the vertices $\boldsymbol{V}_{i,j}$'s that define the SR bilinear patches, we impose constraints on the α-spline surface to interpolate the data points:

$$\begin{aligned}\boldsymbol{P}_{i,j} &= \boldsymbol{Q}(u_i,v_j)\\ &= (1-\alpha^*_{i,j})\boldsymbol{S}(u_i,v_j)+\alpha^*_{i,j}\boldsymbol{V}_{i,j}, \qquad i=0,...,m;\ j=0,...,n.\end{aligned}$$

Solving for $\boldsymbol{V}_{i,j}$, we obtain

$$\boldsymbol{V}_{i,j}=\boldsymbol{P}_{i,j}+\frac{1-\alpha^*_{i,j}}{\alpha^*_{i,j}}(\boldsymbol{P}_{i,j}-\boldsymbol{S}(u_i,v_j)).$$

Defining $\rho_{i,j}=\frac{1-\alpha^*_{i,j}}{\alpha^*_{i,j}}$ yields

$$\boldsymbol{V}_{i,j}=\boldsymbol{P}_{i,j}+\rho_{i,j}(\boldsymbol{P}_{i,j}-\boldsymbol{S}(u_i,v_j)), \qquad i=0,...,m;\ j=0,...,n.$$

This enables the direct evaluation of $\boldsymbol{V}_{i,j}$, and avoids the necessity of solving linear systems required by interpolation with NURB surface.

§5. Smoothing Polygonal Shapes

In addition to fitting a smooth surface over a dataset, the proposed interpolation method can also be viewed as an interactive modeling tool for deforming polygonal shapes (with an underlying rectangular topology) into smooth objects. The vertices of the polygonal shape are the data points to be interpolated by a smooth surface. With this modeling tool, the user only has to specify the polygonal vertices, which are fewer than the number of control points of most spline schemes. Manipulating polygonal shapes is also simple and easy for novice designers.

Figures 4 and 5 show some modeling examples, all of which are obtained from the same input polygonal shape, shown in the top left corner of Fig. 4. In Fig. 4, the relaxation parameters are varied: globally in the top row, and locally in the bottom row (the third row of vertices have their relaxation parameters varied while the other vertices have fixed $\rho_{i,j}=1$). In Fig. 5, the weights of the third row of vertices are varied, while the other weights are fixed at unity, and the global relaxation is set at $\rho_{i,j}=1$ for all i,j.

Fig. 4. Effect of global (top) and local (bottom) relaxation: 0, 0.25, 0.67, 1.5, 2.3.

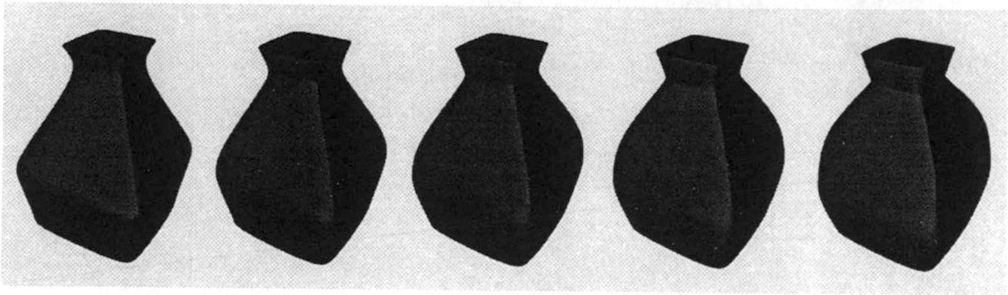

Fig. 5. Effect of local weight (third-row vertices): 2.5, 1, 0.5, 0.25, 0.

§6. Homogeneous Representation of α-spline

It is well known that the rational B-spline can be viewed as the projection of a polynomial B-spline in homogeneous space. This property is important because it implies that all the algorithms for the polynomial B-spline can also be applied to the rational B-spline. The rational B-spline is given by

$$\boldsymbol{C}(u) = \sum R_{j,4}(u)\boldsymbol{P}_j, \tag{7}$$

where $\boldsymbol{P}_j = (x_j, y_j, z_j)$ are the control points and $R_{j,4}(u) = \frac{N_{j,4}(u)w_j}{\sum N_{i,4}(u)w_i}$, $w_j \geq 0$, are the weights, and $N_{j,4}(u)$ denotes the cubic B-spline basis functions. The polynomial B-spline curve in the homogeneous space, $\boldsymbol{C}^h(u)$, whose projection yields the rational B-spline $\boldsymbol{C}(u)$ is given by

$$\boldsymbol{C}^h(u) = \sum N_{j,4}(u)\boldsymbol{P}^h_j, \tag{8}$$

where $\boldsymbol{P}^h_j = (w_jx_j, w_jy_j, w_jz_j, w_j)$, since $\boldsymbol{C}(u) = (\frac{C^h_x(u)}{C^h_w(u)}, \frac{C^h_y(u)}{C^h_w(u)}, \frac{C^h_z(u)}{C^h_w(u)})$, and the x, y, z and w subscripts denote the respective components of $\boldsymbol{C}^h(u)$.

We can show that this property is also true for the α-spline. Substituting $\boldsymbol{C}(u)$ from (7) into the alpha-spline equation in (3), and rewriting it as a rational function, yields

$$\boldsymbol{Q}(u) = \frac{(1-\alpha(u))\sum N_{j,4}(u)w_j\boldsymbol{P}_j + \alpha(u)C_w^h(u)\boldsymbol{L}(u)}{C_w^h(u)}.$$

The denominator can be expressed as $(1-\alpha(u))C_w^h(u) + \alpha(u)C_w^h(u)$; hence, the polynomial form of the α-spline in homogeneous space is

$$\boldsymbol{Q}^h(u) = (1-\alpha(u))\boldsymbol{C}^h(u) + \alpha(u)\boldsymbol{L}^h(u), \qquad u \in [u_0, u_n],$$

where $\boldsymbol{C}^h(u)$ is given in (8), and $\boldsymbol{L}^h(u)$ is given by

$$\boldsymbol{L}^h(u) = C_w^h(u)((1-s(t))\boldsymbol{V}_j^h + s(t)\boldsymbol{V}_{j+1}^h), \qquad u \in [u_j, u_{j+1}),$$

where $\boldsymbol{V}_j^h = (x_j', y_j', z_j', 1)$ is obtained from $\boldsymbol{V}_j = (x_j', y_j', z_j')$.

§7. Conclusion

We have proposed a new interpolation scheme based on blending a non-interpolatory NURB curve (surface) with an SR linear spline (SR bilinear patches). The resulting interpolating α-spline inherits the continuity, and the affine and projective invariant properties of the NURB. The method provides weight and relaxation control, involves only simple computations, and supports the modeling paradigm of deforming polygonal shapes into smooth spline surfaces.

Several issues have yet to be investigated for the α-spline. One example is knot insertion which is useful for shape refinement and rendering. Another issue is the automatic determination of the parameters to satisfy certain shape-preserving conditions.

Acknowledgments. Gratitude is extended to Yim-Hung Chan for the implementation. This work is partially supported by the grant HKUST6215/99 awarded by the Hong Kong Research Grant Council.

References

1. Barsky, B. A., Exponential and polynomial methods for applying tension to an interpolating spline curve, Computer Vision, Graphics, and Image Processing **27** (1984), 1–18.
2. Barsky, B. A., *Computer Graphics and Geometric Modeling Using Beta-splines,* Springer-Verlag, Heidelberg, 1988.
3. Bartels, R. H., J. C. Beatty, and B. A. Barsky, *An Introduction to Splines for Use in Computer Graphics & Computer Modeling,* Morgan Kaufmann Publ., San Francisco, CA, 1987.

4. Coons, S. A., Modification of the shape of piecewise curves, Computer-Aided Design **9** (1977), 178–180.
5. Costantini, P., T. N. T. Goodman, and C. Manni, Constructing C^3 shape preserving interpolating space curves, manuscript.
6. Foley, T. F., Interpolation with interval and point tension controls using cubic weighted ν-splines, ACM Trans. Math. Software **13** (1987), 68–96.
7. Kaklis, P. D. and N. S. Sapidis, Convexity-preserving interpolatory parametric splines of non-uniform polynomial degree, Comput. Aided Geom. Design **11** (1995), 1–26.
8. Kochanek, D. H. U. and R. H. Bartels, Interpolating splines with local tension, continuity, and bias control, ACM SIGGRAPH '84 **18** (1984), 33–41.
9. Loe, K. F., αB-spline: a linear singular blending B-spline, Visual Comp. **12** (1996), 18–25.
10. Loe, K. F. and C. L. Tai, Interpolation with tension control using singular blending, submitted for publication.
11. Nielson, G. M., Some piecewise polynomial alternatives to splines in tension, in *Computer Aided Geometric Design*, R. E. Barnhill and R. F. Riesenfeld (eds), Academic Press, New York, 1974, 209–235.
12. Schweikert, D. G., An interpolation curve using a spline in tension, J. Math. Phys. **45** (1966), 312–317.

Chiew-Lan Tai
Department of Computer Science
The Hong Kong University of Science & Technology
Clear Water Bay, Kowloon, Hong Kong
taicl@cs.ust.hk

Brian A. Barsky
Computer Science Division
University of California
Berkeley, CA 94720-1776, U.S.A.
barsky@cs.berkeley.edu

Kia-Fock Loe
Department of Computer Science
School of Computing
National University of Singapore
Lower Kent Ridge, Singapore 119260
loekf@comp.nus.edu.sg

On Properties of Contours of Trilinear Scalar Fields

Holger Theisel

Abstract. We study properties of contour surfaces of trilinear scalar fields, and give a classification based on how many unconnected surface parts they consist of. Furthermore, we introduce the concept of the segment number of a voxel. The segment number is a threshold-independent measure which estimates how complicated the contours inside the voxel are expected to be. Finally, we give necessary and sufficient conditions for a voxel to have a segment number of 1. These conditions are applied to analyze a computer tomography data set.

§1. Introduction

Contours (isosurfaces) of trilinear scalar fields are treated in a variety of applications. For instance, the data used in volume visualization usually consists of a number of scalars defined at certain grid points; between the grid points a piecewise trilinear interpolation of the scalar field is applied.

Given a voxel $\boldsymbol{V} = [0,1]^3$, the trilinear scalar field is defined by setting the values $c_{ijk}(i,j,k \in \{0,1\})$ of the field at the corners of $\boldsymbol{V}$. Then the scalar field is defined as

$$\begin{aligned} s(u,v,w) = &(1-u)\cdot(1-v)\cdot(1-w)\cdot c_{000} + (1-u)\cdot(1-v)\cdot w\cdot c_{001} \\ &+ (1-u)\cdot v\cdot(1-w)\cdot c_{010} + (1-u)\cdot v\cdot w\cdot c_{011} \\ &+ u\cdot(1-v)\cdot(1-w)\cdot c_{100} + u\cdot(1-v)\cdot w\cdot c_{101} \\ &+ u\cdot v\cdot(1-w)\cdot c_{110} + u\cdot v\cdot w\cdot c_{111}. \end{aligned} \tag{1}$$

Figure 1a illustrates this. A contour of $\boldsymbol{V}$ is defined by $s(u,v,w) = r$ =const for a certain threshold r. Figure 1b shows an example of a contour of (1).

There are a number of algorithms to produce a triangular approximation of a contour of (1). Of these, the Marching Cubes (MC) method ([3] and [4])

Curve and Surface Fitting: Saint-Malo 1999
Albert Cohen, Christophe Rabut, and Larry L. Schumaker (eds.), pp. 403–410.

ISBN 0-8265-1357-3.

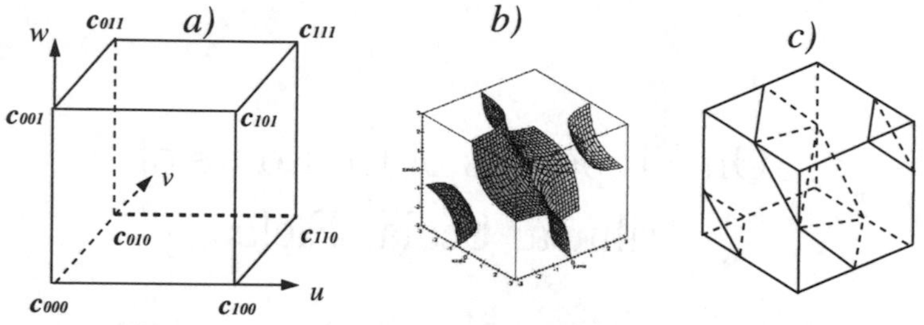

Fig. 1. a) Voxel $\boldsymbol{V}$; b) a contour in $\boldsymbol{V}$; c) result of MC.

is the most popular. Figure 1c shows the resulting triangular approximation of the contour shown in Figure 1b using the Marching Cubes method.

The Marching Cubes algorithm distinguishes several cases where some of them are harder to treat than others. In this paper we introduce a measure of how costly in terms of computing time the MC algorithm inside a certain voxel is expected to be. This characterization of a voxel – called segment number – is independent of a particular threshold. It estimates the costs of a Marching Cubes algorithm for varying thresholds.

As already stated in [2], the contour of (1) is a rational cubic surface. In [2] this surface is approximated by a collection of rational quadratic triangular patches.

Section 2 of this paper studies the contours of (1) in the domain $\mathbb{R}^3$. We give a classification based on how many unconnected surface parts the contours consist of. Sections 3 and 4 focus on contours of (1) inside a certain voxel. Section 3 introduces the concept of segment number as a measure of how simply a voxel can be treated by an MC algorithm. In Section 4, necessary and sufficient geometric conditions for a voxel to have a segment number of 1 are shown. In Section 5, the number of voxels with a segment number of 1 are computed for a real volume data set.

§2. Classification of the Contour in $\mathbb{R}^3$

In this section we consider the contour of (1) not in a particular voxel but in the domain $\mathbb{R}^3$. In general, the contour consists of a number of surface parts which are not connected to each other. Before we classify the contours of (1) by the number of unconnected surface parts, we apply a translation of the coordinate system as shown in Figure 2. Choosing

$$p = c_{001} + c_{010} + c_{100} + c_{111} - c_{000} - c_{011} - c_{101} - c_{110}$$

$$\boldsymbol{p}_0 = \frac{1}{p} \cdot \begin{pmatrix} c_{000} + c_{011} - c_{001} - c_{010} \\ c_{000} + c_{101} - c_{001} - c_{100} \\ c_{000} + c_{110} - c_{010} - c_{100} \end{pmatrix}$$

we obtain for (1)

$$s = a \cdot u + b \cdot v + c \cdot w + d \cdot u \cdot v \cdot w + e \tag{2}$$

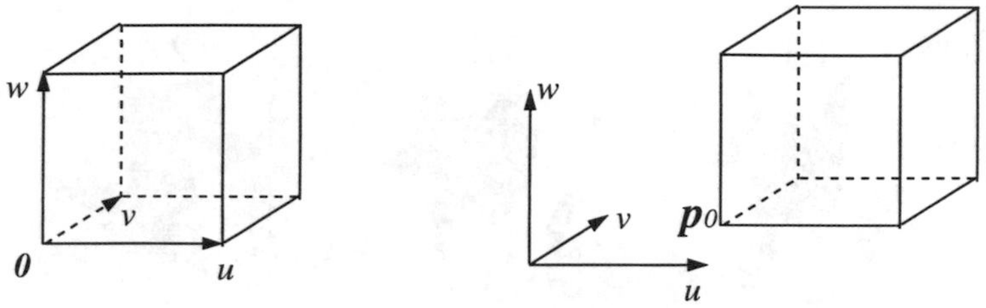

Fig. 2. Translating the coordinate system of a voxel.

with

$$a = \frac{(c_{111} - c_{011}) \cdot (c_{100} - c_{000}) - (c_{110} - c_{010}) \cdot (c_{101} - c_{001})}{p}$$

$$b = \frac{(c_{111} - c_{101}) \cdot (c_{010} - c_{000}) - (c_{110} - c_{100}) \cdot (c_{011} - c_{001})}{p}$$

$$c = \frac{(c_{111} - c_{110}) \cdot (c_{001} - c_{000}) - (c_{101} - c_{100}) \cdot (c_{011} - c_{010})}{p}$$

$$d = p,$$

where e is a certain constant. Thus, we only have to analyze

$$s(u, v, w) = a \cdot u + b \cdot v + c \cdot w + d \cdot u \cdot v \cdot w = r = const \tag{3}$$

in $\mathbb{R}^3$. A classification of (3) can be achieved by rewriting (3) as $w = \frac{r - a \cdot u - b \cdot v}{c + d \cdot u \cdot v}$ and comparing the zeros of the numerator and denominator function. The zeros of the numerator function form a line in the $u - v$–plane, whereas the zeros of the denominator function give a hyperbola. Studying their interplay gives the following classification:

- case 1: $abcd \leq 0, d \neq 0$:
  - case 1.1: $r^2 \geq -\frac{4abc}{d}$: (3) gives 3 unconnected surface parts
  - case 1.2: $r^2 < -\frac{4abc}{d}$: (3) gives 2 unconnected surface parts
- case 2: $abcd < 0$: (3) consists of 1 connected part
- case 3: $abcd = 0, d \neq 0$:
  - case 3.1: $r \neq 0$:
    - case 3.1.1: $ab \neq 0, c = 0$: (3) gives 2 unconnected surface parts
    - case 3.1.2: $a \neq 0, b = c = 0$: (3) gives 3 unconnected surface parts
    - case 3.1.3: $a = b = c = 0$: (3) gives 4 unconnected surface parts
  - case 3.2: $r = 0$:
    - case 3.2.1: $ab \neq 0, c = 0$: (3) gives 3 unconnected surface parts
    - case 3.2.2: $a \neq 0, b = c = 0$: (3) gives 3 parts intersecting each other
    - case 3.2.3: $a = b = c = 0$: (3) gives 3 perpendicular planes.

Figure 3 illustrates these cases.

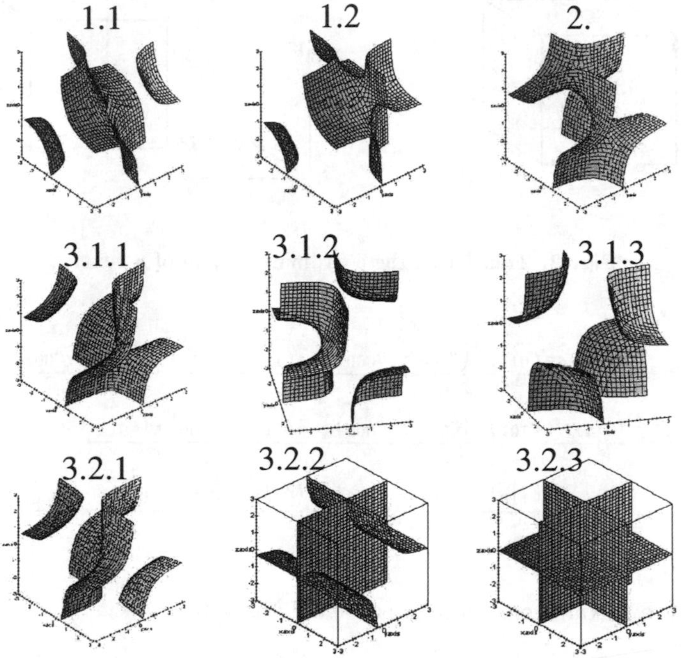

Fig. 3. Classification of the contours of (3) in $\mathbb{R}^3$.

§3. Segment Number of a Voxel

We now study the contour of (3) in a particular voxel $\boldsymbol{V} = [u_0, u_0 + 1] \times [v_0, v_0 + 1] \times [w_0, w_0 + 1]$. Unfortunately, the results of Section 2 are not directly applicable here because one connected surface part may intersect $\boldsymbol{V}$ more than once.

Varying the threshold r in (3), the contours change. So does the number of unconnected surface parts of the contour.

Definition 1. *Given the trilinear scalar field $s(u, v, w) = a \cdot u + b \cdot v + c \cdot w + d \cdot u \cdot v \cdot w$ in the domain of the voxel $\boldsymbol{V} = [u_0, u_0 + 1] \times [v_0, v_0 + 1] \times [w_0, w_0 + 1]$, the* segment number *$S(\boldsymbol{V})$ of $\boldsymbol{V}$ is the maximal number of unconnected surface parts of the contour $s(u, v, w) = r$ =const in $\boldsymbol{V}$ for any threshold r.*

Figure 4 gives an example of a voxel $\boldsymbol{V}$ with $S(\boldsymbol{V}) = 1$. Increasing the value of r, the isosurface "moves" through the voxel. It consists of at most one connected part for any r. Figure 5 shows a voxel with $S(\boldsymbol{V}) = 4$. Here the contours consist of up to 4 unconnected parts.

The segment number is a threshold-independent characterization of a voxel $\boldsymbol{V}$. For any $\boldsymbol{V}$ we get $S(\boldsymbol{V}) \in \{1, 2, 3, 4\}$. For visualization purposes, voxels with a segment number 1 are of particular interest. As shown in the example of Figure 4, they have a nice behavior while varying r. In fact, for

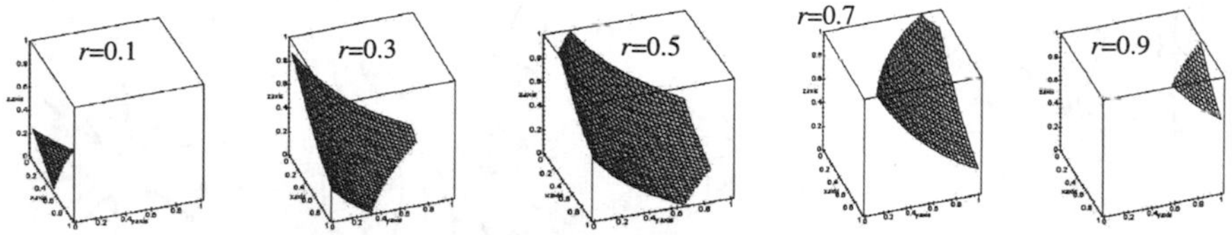

Fig. 4. Contours of a voxel with $S(\boldsymbol{V}) = 1$.

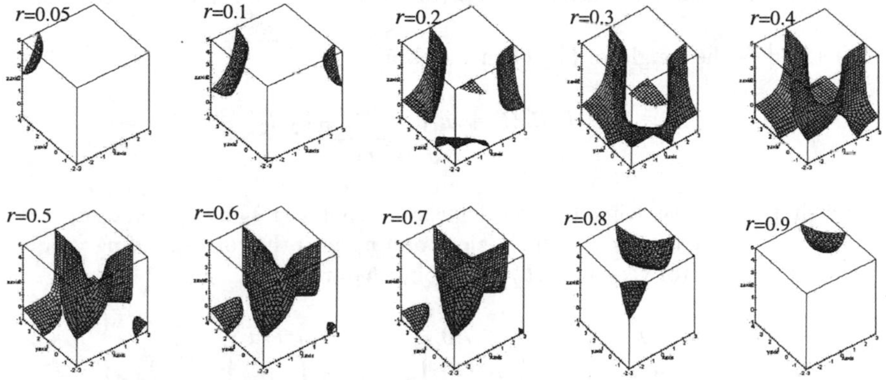

Fig. 5. Contours of a voxel with $S(\boldsymbol{V}) = 4$.

any r the contour consists of only one connected surface part inside $\boldsymbol{V}$. Thus, accelerated Marching Cubes methods may apply to them. Moreover, adjacent voxels with $S(\boldsymbol{V}) = 1$ may be merged to form one bigger voxel before applying Marching Cubes methods. So it makes sense to search for geometric conditions for a voxel $\boldsymbol{V}$ to have $S(\boldsymbol{V}) = 1$.

§4. Geometric Conditions for $S(\boldsymbol{V}) = 1$

In this section we give necessary and sufficient geometric conditions for a voxel to have $S(\boldsymbol{V}) = 1$. Again, we consider the contour of (3) in the voxel $\boldsymbol{V} = [u_0, u_0 + 1] \times [v_0, v_0 + 1] \times [w_0, w_0 + 1]$.

To formulate the conditions for $S(\boldsymbol{V}) = 1$, we need to introduce the concept of characteristic hyperbolas. The first characteristic hyperbola $\boldsymbol{h}_1$ in $\mathbb{R}^3$ is defined by the condition $s_{vw}(u, v, w) = 0$ in (3). $\boldsymbol{h}_1$ can be written as a rational quadratic Bézier curve described by two control vectors $\boldsymbol{b}_0^1$, $\boldsymbol{b}_2^1$ and a control point $\boldsymbol{b}_1^1$ (see [1]). For $\boldsymbol{h}_1$ we obtain

$$\boldsymbol{b}_0^1 = \begin{pmatrix} (-4bc)/d \\ 0 \\ 0 \end{pmatrix}, \quad \boldsymbol{b}_1^1 = \begin{pmatrix} 0 \\ 0 \\ 0 \end{pmatrix}, \quad \boldsymbol{b}_2^1 = \begin{pmatrix} 0 \\ 1/b \\ 1/c \end{pmatrix}, \quad w_1^1 = 1,$$

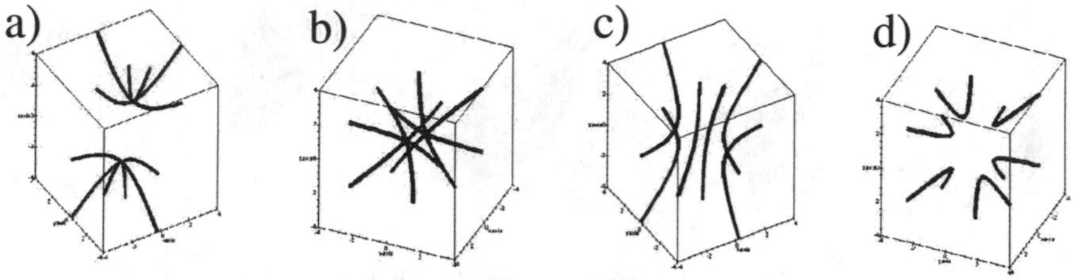

Fig. 6. Location of characteristic hyperbolas; **a)**,**b)**: $abcd < 0$; **c)**,**d)**: $abcd > 0$.

where w_1^1 is the weight of $\boldsymbol{b}_1^1$. Then we obtain

$$\boldsymbol{h}_1(t) = \frac{\boldsymbol{b}_0^1 B_0^2(t) + w_1^1 \boldsymbol{b}_1^1 B_1^2(t) + \boldsymbol{b}_2^1 B_2^2(t)}{w_1^1 B_1^2(t)}.$$

In a similar way we define the characteristic hyperbola $\boldsymbol{h}_2$ by $s_{uw}(u,v,w) = 0$, and $\boldsymbol{h}_3$ by $s_{uv}(u,v,w) = 0$. The Bézier point $\boldsymbol{b}_1^2$ with the corresponding weight w_1^2 and the control vectors $\boldsymbol{b}_0^2, \boldsymbol{b}_2^2$ describing $\boldsymbol{h}_2$ are

$$\boldsymbol{b}_0^2 = \begin{pmatrix} 0 \\ (-4ac)/d \\ 0 \end{pmatrix}, \quad \boldsymbol{b}_1^2 = \begin{pmatrix} 0 \\ 0 \\ 0 \end{pmatrix}, \quad \boldsymbol{b}_2^2 = \begin{pmatrix} 1/a \\ 0 \\ 1/c \end{pmatrix}, \quad w_1^2 = 1.$$

$\boldsymbol{h}_3$ is described by

$$\boldsymbol{b}_0^3 = \begin{pmatrix} 0 \\ 0 \\ (-4ab)/d \end{pmatrix}, \quad \boldsymbol{b}_1^3 = \begin{pmatrix} 0 \\ 0 \\ 0 \end{pmatrix}, \quad \boldsymbol{b}_2^3 = \begin{pmatrix} 1/a \\ 1/b \\ 0 \end{pmatrix}, \quad w_1^3 = 1.$$

If $a \cdot b \cdot c \cdot d < 0$ then $\boldsymbol{h}_1, \boldsymbol{h}_2, \boldsymbol{h}_3$ intersect in two common points. Figures 6a and 6b illustrate this situation from two different viewpoints. If $a \cdot b \cdot c \cdot d > 0$, then $\boldsymbol{h}_1, \boldsymbol{h}_2, \boldsymbol{h}_3$ do not have any intersections. Figures 6c and 6d show this from different viewpoints. The degenerate case $a \cdot b \cdot c \cdot d = 0$ is omitted here.

To formulate conditions for $S(\boldsymbol{V}) = 1$, we have to classify the faces of $\boldsymbol{V}$. Given the voxel $\boldsymbol{V} = [u_0, u_0+1] \times [v_0, v_0+1] \times [w_0, w_0+1]$, let $\boldsymbol{f}_1 = \{(u,v,w) \in \boldsymbol{V} : u = u_0 \vee u = u_0 + 1\}$, $\boldsymbol{f}_2 = \{(u,v,w) \in \boldsymbol{V} : v = v_0 \vee v = v_0 + 1\}$, and $\boldsymbol{f}_3 = \{(u,v,w) \in \boldsymbol{V} : w = w_0 \vee w = w_0 + 1\}$. See Figure 7 for an illustration of the faces.

Theorem 1. *Let $\boldsymbol{V} = [u_0, u_0+1] \times [v_0, v_0+1] \times [w_0, w_0+1]$ be a voxel in the scalar field defined by (3). Then the condition $S(\boldsymbol{V}) = 1$ is equivalent to the three conditions $\boldsymbol{h}_1 \cap \boldsymbol{f}_1 = \emptyset$ and $\boldsymbol{h}_2 \cap \boldsymbol{f}_2 = \emptyset$ and $\boldsymbol{h}_3 \cap \boldsymbol{f}_3 = \emptyset$.*

Figure 8 illustrates the idea of the proof. Suppose $\boldsymbol{h}_3$ intersects $\boldsymbol{f}_3$ as shown in Figure 8a. Figure 8b is a magnification of the voxel and $\boldsymbol{h}_3$ in

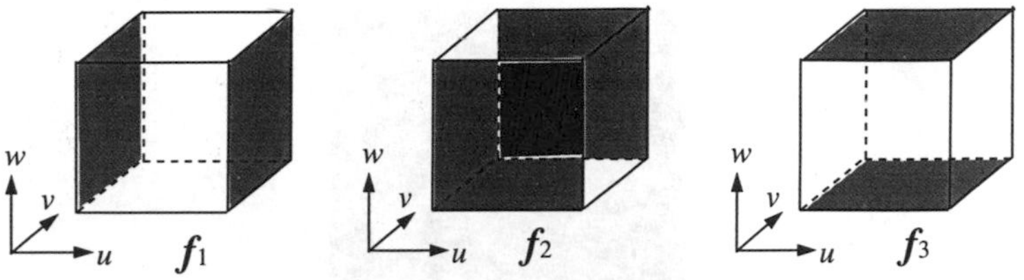

Fig. 7. The faces $\boldsymbol{f}_1, \boldsymbol{f}_2, \boldsymbol{f}_3$ of a voxel.

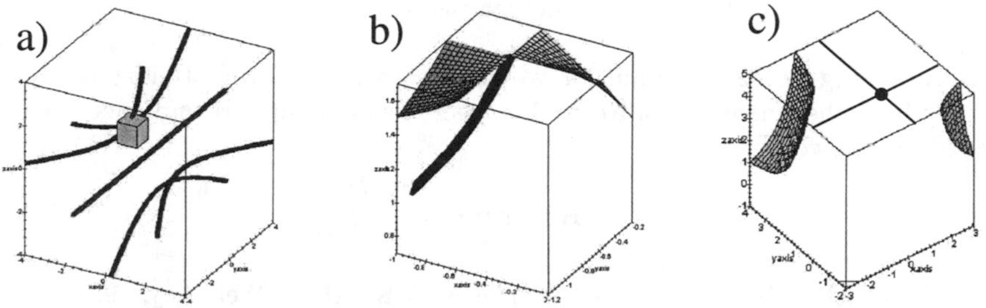

Fig. 8. Proof idea of Theorem 1.

Figure 8a. We compute the intersection point of $\boldsymbol{h}_3$ and $\boldsymbol{f}_3$, and consider the contour passing through this point. As shown in Figure 8a, this contour consists of at least two surface parts.

For the proof of the converse statement of Theorem 1, we assume that for a certain threshold r the contour consists of at least two unconnected surface parts. Then we can find a face of $\boldsymbol{V}$ which has two intersection curves with the contour. (In the worst case we have to vary r to find such a face). (Figure 8c shows two surface parts of the contour which produce two intersection curves in the upper face of $\boldsymbol{f}_3$). Then we can find a point on this face which is the intersection point with the corresponding characteristic hyperbola. (In Figure 8c, the marked point on the upper part of $\boldsymbol{f}_3$ is the intersection with $\boldsymbol{h}_3$).

§5. Results and Future Work

We have tested the voxels of a CT test data set for the property $S(\boldsymbol{V}) = 1$. The data set consists of $255 \times 255 \times 108 = 7,022,700$ voxels. Figure 9 shows a slice through the data set.

In the raw data we found 1,978,711 voxels with $S(\boldsymbol{V}) = 1$ (28 %). After some noise reducing filter operations on the data, we detected 4,833,063 voxels with $S(\boldsymbol{V}) = 1$ (69 %). This shows that there is a reasonable number of voxels with $S(\boldsymbol{V}) = 1$ to pay special attention to them.

In the future we plan to develop algorithms to merge voxels with $S(\boldsymbol{V}) = 1$ to form bigger voxels before starting the Marching Cubes algorithm.

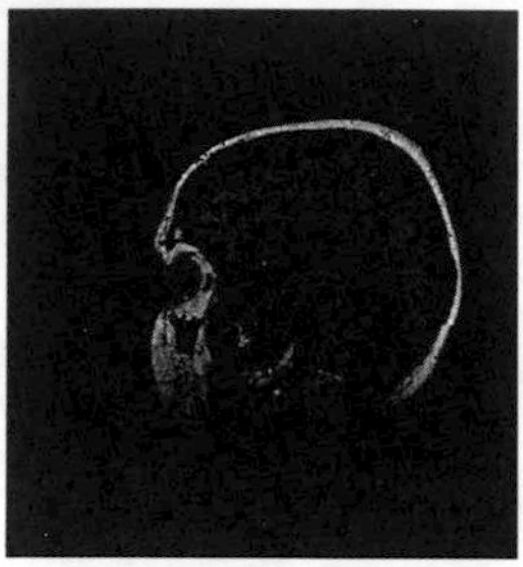

Fig. 9. Slice through the test data set.

Acknowledgments. The author would like to thank Prof. Heidrun Schumann from the University of Rostock for her constant support and encouragement of this work.

References

1. Farin, G., *NURB Curves and Surfaces*, A K Peters, Wellesley, 1995.
2. Hamann, B., I. J. Trotts, and G. Farin, On approximating contours of the piecewise trilinear interpolant using triangular rational quadratic Bézier patches, IEEE Transactions on Visualization and Computer Graphics, **3(3)** (1997), 215–227.
3. Lorensen, G. M. and H. E. Cline, Marching Cubes: a high resolution 3D surface reconstruction algorithm, Computer Graphics **21** (1987), 163–169.
4. Nielson, G. M. and B. Hamann, The asymptotic decider: resolving the ambiguity in marching cubes, proceedings IEEE Visualization 91, 1991, 83–91.

Holger Theisel
University of Rostock, Computer Science Department
PostBox 999, 18051 Rostock, Germany
`theisel@informatik.uni-rostock.de`
`http://wwwicg.informatik.uni-rostock.de/~theisel/`

Non-Stationary Subdivision for Inhomogeneous Order Differential Equations

Joe Warren and Henrik Weimer

Abstract. This paper provides a methodology for the systematic derivation of subdivision schemes that model solutions to inhomogeneous order linear differential equations. In previous work, we showed that subdivision can be used to capture very efficiently the solutions of homogeneous order, linear differential equations. The resulting subdivision masks are stationary and can be precomputed, allowing for very simple and fast application of these schemes. In this paper, we show that this method can be extended to express solutions of systems of inhomogeneous order, linear differential equations. Even though the resulting subdivision masks may be non-stationary, the masks can again be precomputed. Thus, the resulting subdivision schemes capture very efficiently solutions of inhomogeneous order, linear partial differential equations.

§1. Subdivision for the Modeling of Shapes

Subdivision is a popular and efficient method for modeling shapes. In particular, subdivision describes a continuous shape $\boldsymbol{p}$ as the limit of a sequence $p_k,\ k \geq 0$ of discrete shapes,

$$\lim_{k\to\infty} p_k = \boldsymbol{p}.$$

The beauty of subdivision lies in the fact that these discrete shapes p_k are linked by a simple linear transformation S which is based on splitting and averaging,

$$p_k = S_{k-1}p_{k-1}.$$

Figure 1 shows an example of a subdivision scheme. Starting from the coarse shape p_0 on the left, application of the subdivision matrix S_0 yields the denser shape p_1. As we continue the process, the sequence of discrete shapes

Curve and Surface Fitting: Saint-Malo 1999
Albert Cohen, Christophe Rabut, and Larry L. Schumaker (eds.), pp. 411–420.

ISBN 0-8265-1357-3.

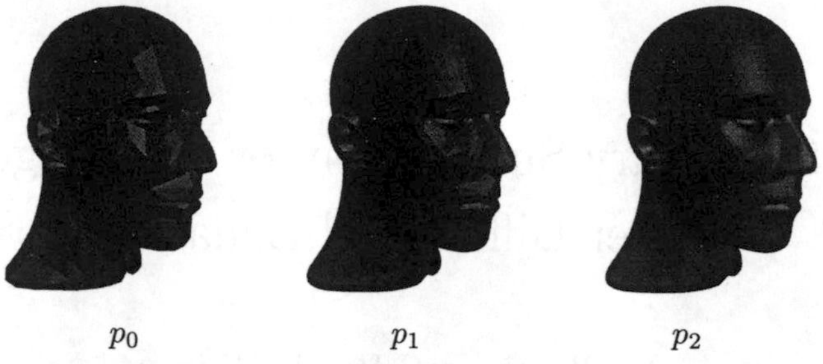

Fig. 1. Subdivision models a shape as the limit of a sequence of discrete shapes.

converges rapidly to a continuous shape $\boldsymbol{p}$ that follows the original coarsest shape p_0 and whose properties are determined by the subdivision matrix.

Subdivision's popularity for the modeling of curves is due to the algorithms by Chaikin [9], and Lane and Riesenfeld [6]. The breakthrough for the modeling of surfaces via subdivision was marked by the papers by Catmull and Clark [2] and by Doo and Sabin [3]. A popular subdivision scheme for modeling with triangular meshes has been proposed by Loop [7], which was also used for creating Figure 1.

§2. Shape Modeling through Differential Equations

Alternatively, shapes can be characterized as solutions to partial differential equations. For example, any polynomial spline $\boldsymbol{p}[\boldsymbol{x}]$ of degree m satisfies the differential equation $\boldsymbol{p}^{(m+1)}[\boldsymbol{x}] = 0$, requiring the $(m+1)$st derivative of the spline to be zero everywhere except at a fixed number of knots [1]. Other examples of shapes based on partial differential equations are the polyharmonic surfaces, including Thin Plate Splines, as well as many different classes of fluid flows.

When modeling with differential equations, we determine a continuous shape $\boldsymbol{p}$ that is a solution to a set of partial differential equations

$$\boldsymbol{D}\,\boldsymbol{p} = \boldsymbol{b}, \tag{1}$$

where $\boldsymbol{D}$ denotes a continuous differential operator and $\boldsymbol{b}$ encodes the boundary conditions for the problem. For the example of natural cubic splines, we have $\boldsymbol{D} = \frac{\partial^4}{\partial \boldsymbol{x}^4}$ and $\boldsymbol{b} = 0$ almost everywhere. If all differential operators in $\boldsymbol{D}$ are of the same, fixed order, we call the differential equation homogeneous order. Otherwise, the equation is called inhomogeneous order.

To handle such problems in a computational environment, one commonly discretizes the continuous problem. To this end, a domain grid T_k is chosen and all entities of the continuous partial differential equation (1) are discretized over this domain grid. The result is a system of linear equations

$$D_k p_k = b_k, \tag{2}$$

where p_k denotes an approximation of the continuous solution $\boldsymbol{p}$ over the grid T_k, b_k denotes a discretization of the boundary conditions, and D_k is a discrete approximation of the continuous differential operators $\boldsymbol{D}$ on the domain grid T_k.

Relying on the theory for finite elements or finite differences [11], the discrete solutions p_k can be formally guaranteed to converge to the continuous solution $\boldsymbol{p}$ of the original continuous problem (1) if the discretizations T_k are chosen carefully and the discrete representations D_k and b_k are well chosen.

At this point, the problem of finding the continuous solution $\boldsymbol{p}$ of the system of continuous partial differential equations (1) has been reduced to the problem of solving denser an denser systems of linear equations (2).

The links between mesh modeling and differential equations were previously investigated by Mallet [8], Taubin [12], and Kobbelt [5]. The method presented here is new because subdivision schemes that model solutions of inhomogeneous order differential equations are precomputed entirely, enabling very efficient modeling of shapes guided by inhomogeneous order differential equations. In particular, the actual application of the subdivision schemes does not require any computational solving whatsoever.

§3. Subdivision for Homogeneous Order Differential Equations

In our previous work [13,14] we characterized subdivision schemes for the solutions of homogeneous order linear partial differential equations. In this framework, the subdivision matrix S_{k-1} is determined as the solution to the system of linear equations

$$D_k S_{k-1} = 2^d U_{k-1} D_{k-1}, \tag{3}$$

where d is the dimension of the domain. Recall that the differencing operator D_k is the discrete approximation of the continuous differential operator $\boldsymbol{D}$ of the original, continuous problem (1) on the level k grid T_k. Further, U_{k-1} denotes a very simple linear transformation, called replication or upsampling, that carries coefficients over the grid T_{k-1} into coefficients over the next denser grid T_k. The action of U_{k-1} is very simple: Coefficients centered over knots in T_{k-1} are replicated over the same knots in the denser grid T_k while coefficients centered over the remaining knots $T_k - T_{k-1}$ are set to zero. Thus, U_{k-1} is a matrix whose rows are either zero or a standard unit vector, and U_{k-1} can be constructed easily and efficiently.

We visualize the meaning of equation (3) in Figure 2: The subdivision matrix is determined so that a certain commutativity relationship holds between subdivision, upsampling and differencing. Differencing coefficients on the coarse grid and upsampling those differences to the finer grid by inserting zero for all new grid points ($U_{k-1}D_{k-1}$, the right hand side of equation (3)) should yield the same result as subdividing the coefficients using the subdivision scheme and then differencing on the finer grid ($D_k S_{k-1}$, the left hand side of equation (3)).

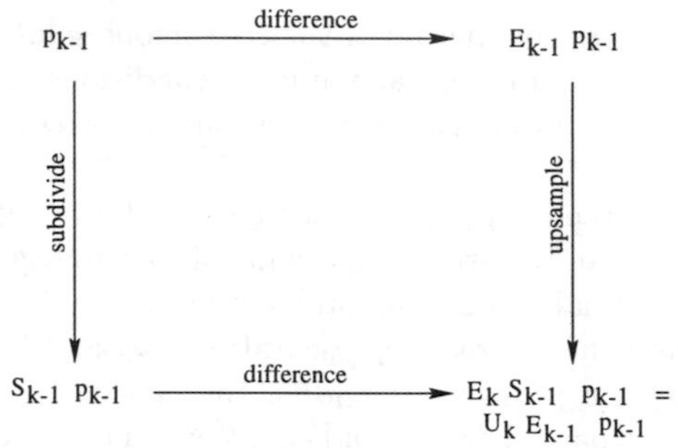

Fig. 2. The subdivision scheme is determined such that this commutativity relationship holds.

The subdivision matrices S_k are the only unknowns of equation (3), and we can use linear algebra to systematically solve for these subdivision matrices. In our previous work [13] we showed that solutions produced by these subdivision matrices are related to solutions produced by an interpolating finite element solver using a simple, fixed change of basis. Therefore, if the finite element solver converges, then the subdivision solution is also well defined.

As a side note, in our previous work [13] we establish that the right-hand side of the system, as solved by the subdivision scheme from relation (3), is $D_k p_k = U_{k-1} U_{k-2} \ldots U_0 D_0 p_0$ where $p_k = S_{k-1} p_{k-1}$ and p_0 is a user-given set of initial control coefficients. In other words, the subdivision scheme leads to a specific combination of the integer shifts of the Green function of the differential operator.

Further, computation of the subdivision matrix S_{k-1} based on relation (3) requires the inversion of the differencing operator D_k. Consequently, the computational work required for finding the subdivision matrix is at least the same as inverting the finite difference system. However, later we will see that the subdivision matrices can be precomputed. Thus, in contrast to a conventional finite difference solver, new shapes can be generated extremely efficiently.

As an example, we briefly derive subdivision schemes for piecewise polynomial splines. Recall from deBoor [1] that the piecewise polynomial spline $\boldsymbol{p}[\boldsymbol{x}]$ of degree m satisfies the differential equation $\boldsymbol{D}[\boldsymbol{x}]^{m+1}\boldsymbol{p}[\boldsymbol{x}] = 0$ where $\boldsymbol{D}[\boldsymbol{x}]$ denotes the first derivative in the variable $\boldsymbol{x}$.

We employ generating functions [4] for concise and convenient encoding of discrete coefficient sequences. To this end, we choose the domain grids for our analysis as the dilates $\frac{1}{2^k}\mathbb{Z}$ of the integer grid $\mathbb{Z}$. A generating function $p_k[x]$ is a power series that associates the ith coefficient of the discrete shape p_k as the coefficient of x^i. For example, the coefficient sequence $\{1, 2, 3, 4, 5\}$ is represented by $1 + 2x + 3x^2 + 4x^3 + 5x^4$.

Recall the definition of the first derivative operator,

$$\boldsymbol{D}[\boldsymbol{x}]\boldsymbol{p}[\boldsymbol{x}] = \lim_{\boldsymbol{t}\to 0} \frac{\boldsymbol{p}[\boldsymbol{x}] - \boldsymbol{p}[\boldsymbol{x}+\boldsymbol{t}]}{\boldsymbol{t}}.$$

Substituting $\boldsymbol{t} = \frac{1}{2^k}$ yields

$$\boldsymbol{D}[\boldsymbol{x}]\boldsymbol{p}[\boldsymbol{x}] = \lim_{k \to \infty} \frac{\boldsymbol{p}[\boldsymbol{x}] - \boldsymbol{p}\left[\boldsymbol{x} + \frac{1}{2^k}\right]}{\frac{1}{2^k}}.$$

Thus, for $\boldsymbol{x} \in \frac{1}{2^k}\mathbb{Z}$, the approximation of the first derivative is given by the difference between two adjacent discretizations, normalized by the grid spacing. In terms of generating functions, this differencing operation is represented by the Laurent polynomial

$$D_k[x] = 2^k \frac{1-x}{x^{1/2}}.$$

Higher order derivatives and differences are obtained by repeated application of the respective continuous or discrete operator.

In terms of generating functions, the action of the upsampling operator U_k can be captured very concisely: The expression $p\left[x^2\right]$ represents the upsampled coefficient sequence of $p[x]$ as a generating function. Thus, in our example of polynomial splines, the generating function $s_k[x]$ for the subdivision scheme satisfies

$$D_k[x]^{m+1} s_k[x] = 2 D_{k-1}\left[x^2\right]^{m+1},$$

which can be simplified to

$$s_k[x] = 2\left(\frac{D_{k-1}[x^2]}{D_k[x]}\right)^{m+1}.$$

Fortunately,

$$\frac{D_{k-1}[x^2]}{D_k[x]} = \frac{1}{2}\,\frac{1+x}{x^{1/2}},$$

i.e. the generating functions for the differencing operations on the level $k-1$ and level k grids divide out yielding a simple expression independent of k. As a result, the subdivision mask for the degree m polynomial splines are exactly the coefficients of

$$s[x] = 2\left(\frac{1+x}{2x^{1/2}}\right)^{m+1}.$$

Remarkably, these are precisely the known subdivision schemes for piecewise linear functions ($m = 1$), Chaikin's algorithm [9] ($m = 2$) and the Lane/Riesenfeld algorithm [6] ($m = 3$).

Previously we applied this strategy to derive subdivision schemes modeling solutions of homogeneous order linear differential equations yielding local, stationary subdivision masks [13,14]. In this paper, we show that largely the same strategy can be used to determine subdivision schemes for inhomogeneous order linear partial differential equations. As we will see, in this case the actual subdivision masks may depend on the particular level of subdivision, i.e. are non-stationary. However, the masks can still be precomputed as a closed form algebraic expression in the level of subdivision, which can then be evaluated very efficiently during the actual application of the scheme.

§4. Subdivision for Inhomogeneous Order Differential Equations

In this section we extend our systematic construction of subdivision schemes to handle inhomogeneous order linear partial differential equations. We consider the simple yet interesting problem of splines in tension [10]. The continuous spline in tension $\boldsymbol{p}[\boldsymbol{x}]$ for tension parameter α satisfies the differential equation

$$\left(\boldsymbol{D}[\boldsymbol{x}]^4 - \alpha^2\ \boldsymbol{D}[\boldsymbol{x}]^2\right)\boldsymbol{p}[\boldsymbol{x}] = 0, \tag{4}$$

where $\boldsymbol{D}[\boldsymbol{x}]$ again represents the continuous first derivative operator with respect to the variable $\boldsymbol{x}$. Note that equation (4) incorporates both second and fourth derivatives of $\boldsymbol{p}[\boldsymbol{x}]$, i.e. is inhomogeneous order.

Following the same strategy as in the derivations for polynomial splines, we use generating functions to encode the discrete approximation p_k of the spline in tension on grid T_k as well as for the representation of the differencing operation $D_k[x] = 2^k \frac{1-x}{x^{1/2}}$. Next, we apply equation (3) to characterize the subdivision scheme $s_k[x]$ as the solution to

$$\left(D_k[x]^4 - \alpha^2 D_k[x]^2\right) s_{k-1}[x] = 2\left(D_{k-1}\left[x^2\right]^4 - \alpha^2 D_{k-1}\left[x^2\right]^2\right), \tag{5}$$

which can be simplified to

$$s_{k-1}[x] = \frac{2\left(D_{k-1}[x^2]^4 - \alpha^2 D_{k-1}[x^2]^2\right)}{D_k[x]^4 - \alpha^2 D_k[x]^2}. \tag{6}$$

However, at this point we note that there is no simple closed-form expression for the quotient $\frac{D_{k-1}[x^2]^4 - \alpha^2 D_{k-1}[x^2]^2}{D_k[x]^4 - \alpha^2 D_k[x]^2}$ (unless $\alpha = 0$). In other words, there is no finitely-supported subdivision scheme $s_{k-1}[x]$ for splines in tension. Moreover, the coefficients of the Laurent series expansion of the quotient $s_{k-1}[x]$ depend on the level of subdivision k, i.e. the subdivision scheme has to be non-stationary.

Fortunately, due to the structure of equation (3) the coefficients of this expansion decrease very rapidly away from the origin. Thus, we can approximate the infinite Laurent expansion of the subdivision mask well by a locally supported scheme. To this end, we construct the generating function $s_{k-1}[x]$ of desired support symbolically with the actual coefficients s_{k-1}^i as unknowns,

$$s_{k-1}[x] = \sum_{i=-n}^{n} s_{k-1}^i x^i$$

for a user-defined support n. We then construct a generating function for the residual of equation (5),

$$\begin{aligned} r_k[x] &= \left(D_k[x]^4 - \alpha^2 D_k[x]^2\right) s_{k-1}[x] - 2\left(D_{k-1}\left[x^2\right]^4 - \alpha^2 D_{k-1}\left[x^2\right]^2\right) \\ &= \sum_i r_k^i x^i. \end{aligned} \tag{7}$$

Using linear algebra, we can now solve for the unknowns s^i_{k-1} of (7) symbolically by minimizing the least squares residual of the coefficients r^i_k. The motivation behind our strategy is to construct a best solution for the characteristic equation (3) of given support. The results of this process are actual, symbolic coefficients for the local subdivision scheme $s_{k-1}[x]$, depending on the tension parameter α as well as on the level k. As an example, the approximation to (6) with the same support as the Lane-Riesenfeld algorithm ($n = 2$) has

$$\frac{2^{2\,k}\left(693\,2^{1+10\,k}+891\,4^{1+4\,k}\,\alpha^2+3525\,64^k\,\alpha^4+399\,4^{1+2\,k}\,\alpha^6+333\,4^k\,\alpha^8+26\,\alpha^{10}\right)}{8\,(693\,2^{1+12\,k}+891\,4^{1+5\,k}\,\alpha^2+3861\,256^k\,\alpha^4+273\,2^{3+6\,k}\,\alpha^6+675\,16^k\,\alpha^8+27\,4^{1+k}\,\alpha^{10}+7\,\alpha^{12})}$$

as the coefficient for $x^{\pm 2}$,

$$\frac{\left(4^{1+k}+\alpha^2\right)\left(693\,1024^k+171\,4^k\,\alpha^8+2\,\alpha^2\left(891\,256^k+219\,4^{1+3\,k}\,\alpha^2+399\,16^k\,\alpha^4+7\,\alpha^8\right)\right)}{4\,(693\,2^{1+12\,k}+891\,4^{1+5\,k}\,\alpha^2+3861\,256^k\,\alpha^4+273\,2^{3+6\,k}\,\alpha^6+675\,16^k\,\alpha^8+27\,4^{1+k}\,\alpha^{10}+7\,\alpha^{12})}$$

as the coefficient for $x^{\pm 1}$, and finally

$$\frac{2079\,4^{1+6\,k}+6039\,4^{1+5\,k}\,\alpha^2+1755\,16^{1+2\,k}\,\alpha^4+8265\,2^{1+6\,k}\,\alpha^6+5235\,16^k\,\alpha^8+213\,4^{1+k}\,\alpha^{10}+56\,\alpha^{12}}{8\,(693\,2^{1+12\,k}+891\,4^{1+5\,k}\,\alpha^2+3861\,256^k\,\alpha^4+273\,2^{3+6\,k}\,\alpha^6+675\,16^k\,\alpha^8+27\,4^{1+k}\,\alpha^{10}+7\,\alpha^{12})}$$

as the coefficient associated with x^0. Note that for $\alpha = 0$ these coefficients exactly reduce to the subdivision scheme for natural cubic splines based on the Lane-Riesenfeld algorithm.

During an actual application of the subdivision scheme, the user-defined tension parameter α and the current level of subdivision k are substituted into the symbolic solution $s_{k-1}[x]$, yielding a simple generating function in only the variable x. The coefficients of this generating function encode the subdivision masks for the spline in tension for the given tension parameter α at the current level k. Again, application of this subdivision scheme is very efficient. For example, given $\alpha = 0$, the above expression simplifies to the generating function for natural cubic spline subdivision, independent of k.

$k = 1$:	0.11063	0.55261	0.88274	0.55261	0.11063
$k = 2$:	0.12345	0.52441	0.80178	0.52441	0.12345
$k = 3$:	0.12489	0.50733	0.76487	0.50733	0.12489
$k = 4$:	0.12499	0.50192	0.75386	0.50192	0.12499
$k = 5$:	0.125	0.50049	0.75097	0.50049	0.125
$k = 6$:	0.125	0.50012	0.75024	0.50012	0.125
$k = 7$:	0.125	0.50003	0.75006	0.50003	0.125
$k = 8$:	0.125	0.50001	0.75002	0.50001	0.125
$k = 9$:	0.125	0.5	0.75	0.5	0.125
$k = 10$:	0.125	0.5	0.75	0.5	0.125
$k = 11$:	0.125	0.5	0.75	0.5	0.125

Fig. 3. Subdivision masks for $\alpha = 1$, $k = 0$.

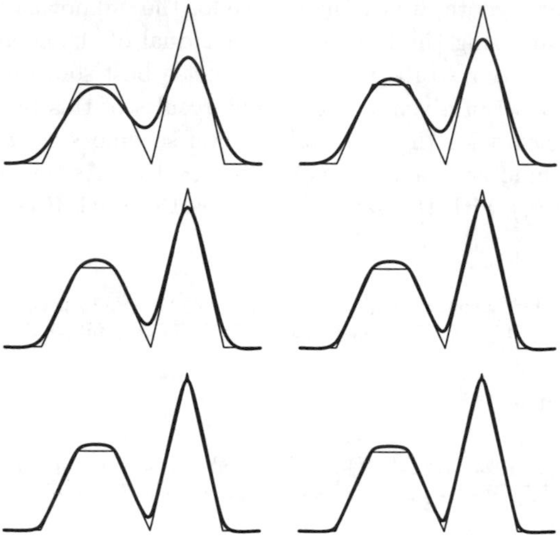

Fig. 4. Splines in tension for varying α.

Figure 3 shows the actual coefficients of a locally supported generating function ($n = 2$) for $\alpha = 1$ and $k = 1, \ldots, 11$. Coefficients were rounded to five significant digits. Note that the coefficient sequence rapidly converges to the subdivision scheme for natural cubic splines. Indeed, after a few rounds of subdivision, a spline in tension behaves like a natural cubic spline over a denser initial grid with its initial control coefficients determined by the first few rounds of subdivision.

Figure 4 depicts application of four rounds of the local subdivision scheme (support $n = 4$) for α ranging from 0 to 5. The initial control polygon is shown as a thin line, while the subdivided curve is shown in solid. Note that as α is increased, the curve follows the control polygon more closely. In the limit, $\alpha \to \infty$, the curve is actually the piecewise linear interpolant of the initial control points.

Figure 5 shows the least squares residuals $\sum_i (r_k^i)^2$ of approximations of different sizes for $\alpha = 1$ and $k = 0$ (the residual is largest for $k = 0$) on a logarithmic scale. Note that for the approximation of size $n = 4$ the residual is already very small.

At a higher level, we follow these steps in the derivation of non-stationary subdivision schemes for inhomogeneous order linear partial differential equations:

Starting from the continuous, inhomogeneous order, linear partial differential equations we discretize the continuous differential operators to yield appropriate differencing operators over the respective subdivision grids T_k. We then characterize the subdivision scheme s_{k-1} as the only unknown of equation (3) using these differencing operators as well as simple replication/upsampling. Next, we construct a representation of the subdivision scheme s_{k-1} in terms

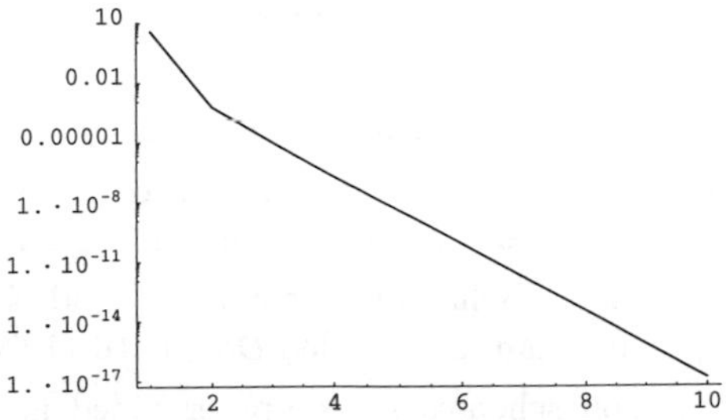

Fig. 5. Residuals of local approximations with varying support.

of unknowns and symbolically build a residual expression representing the difference between left-hand side and right-hand side of equation (3). Finally, we use linear algebra to solve symbolically for the unknowns of the subdivision scheme s_{k-1}, which may depend on the level of subdivision and possible parameters to the original partial differential equations. As a result, application of the subdivision scheme only involves instantiation of these constants, yielding a locally supported, approximating subdivision scheme for solutions of the original inhomogeneous order partial differential equations.

§5. Summary and Conclusion

In this paper, we showed that subdivision can be used to model solutions of inhomogeneous order differential equations. Using the characterization of the subdivision scheme based on the commutativity relationship (3), we can systematically solve for these schemes. Even though the exact subdivision schemes may be globally supported, locally supported schemes approximate the solution well enough for practical purposes. Non-stationary schemes can be handled using the same methodology by allowing the locally supported subdivision masks to change between levels. Because these subdivision schemes can be precomputed, the modeling of solutions of inhomogeneous order linear partial differential equations can be handled very efficiently.

The proposed method for modeling solutions to inhomogeneous order linear differential equations is quite general and promises to be useful in a variety of applications. First of all, approximations based on local subdivision schemes are often sufficient for modeling applications. Indeed, the approximate solutions are qualitatively indistinguishable from the exact solution. Second, if the accuracy of the subdivision solution is not satisfactory, the subdivision scheme can be used to produce very good initial estimates for more traditional solution methods. Third, the results of traditional solution methods often need to be refined locally for visualization and analysis. A local subdivision scheme can be used to refine solutions to any desired accuracy and provide better accuracy than traditional polynomial fits.

Acknowledgments. This work was supported in part under NSF grant number CCR-9732344.

References

1. de Boor, C., *A Practical Guide to Splines.* Springer, 1978.
2. Catmull, E. and J. Clark, Recursively generated B-spline surfaces on arbitrary topological meshes, Computer Aided Design **10** (1978), 350–355.
3. Doo, D. and M. Sabin: Behavior of recursive subdivision surfaces near extraordinary points, Computer Aided Design **10** (1978), 356–360.
4. Dyn, N., Subdivision schemes in computer aided geometric design, in *Advances in Numerical Analysis II,* W. Light (ed.), Oxford University Press, 1992, 36–104.
5. Kobbelt, L., Fairing by finite difference methods, in *Mathematical Methods for Curves and Surfaces II,* M. Daehlen, T. Lyche, and L. L. Schumaker (eds.), Vanderbilt University Press, 1998.
6. Lane, J. and R. Riesenfeld, A theoretical development for the computer generation and display of piecewise polynomial surfaces, IEEE Transactions on Pattern Analysis and Machine Intelligence **2** (1980), 35–46.
7. Loop, C. T., *Smooth Subdivision Surfaces Based on Triangles,* Master's Thesis, Department of Mathematics, University of Utah, 1987.
8. Mallet, J.-L., Discrete smooth interpolation, Transactions on Graphics **4** No. 2 (1985), 74–123.
9. Riesenfeld, R., On Chaikin's algorithm, Computer Graphics and Image Processing **4** (1975), 304–310.
10. Schweikert, D. G., An interpolating curve using a spline in tension, J. Math. and Physics **45** (1966), 312–317.
11. Shaidurov, V. V., *Multigrid Methods for Finite Elements,* Kluwer Academic Publishers, 1995.
12. Taubin, G., A signal processing approach to fair surface design, Proceedings of SIGGRAPH 95, Computer Graphics Proceedings, Annual Conference Series, Addison Wesley (1995), 351–358.
13. Warren, J. and H. Weimer, Subdivision schemes for variational problems, preprint available online under `http://www.cs.rice.edu/~henrik/`.
14. Weimer, H. and J. Warren, Subdivision schemes for fluid flow, Proceedings of SIGGRAPH 99, Computer Graphics Proceedings, Annual Conference Series, Addison Wesley (1995), 111–120.

Joe Warren and Henrik Weimer
Rice University, Department of Computer Science
P.O. Box 1892
Houston, TX 77251-1892, USA
`{jwarren,henrik}@rice.edu`